Plants, Genes and Crop Biotechnology

Plants, Genes and Crop Biotechnology

Editor: Isabel Nelson

www.callistoreference.com

Callisto Reference,
118-35 Queens Blvd., Suite 400,
Forest Hills, NY 11375, USA

Visit us on the World Wide Web at:
www.callistoreference.com

ISBN: 978-1-64116-115-2 (Hardback)

Cataloging-in-Publication Data

Plants, genes and crop biotechnology / edited by Isabel Nelson.
    p. cm.
Includes bibliographical references and index.
ISBN 978-1-64116-115-2
1. Agricultural biotechnology. 2. Plants. 3. Genes. 4. Plant biotechnology.
5. Plant genetic engineering. I. Nelson, Isabel.
S494.5.B563 P53 2019
630--dc23

Table of Contents

Preface

The study of plant genetics helps in understanding the structure and functions of genes in plants. These studies are used in crop biotechnology to modify plants and crops. Crop biotechnology uses the techniques of tissue culture, molecular markers and genetic engineering to produce desired traits in crops. The modification of crops aims to improve characteristics like disease resistance, flavor, size, color, etc. This book explores all the important aspects of plant genetics and crop biotechnology. It attempts to understand the multiple branches that fall under these disciplines and how such concepts have practical applications. Researchers, experts and students in these fields will be assisted by this book.

The information shared in this book is based on empirical researches made by veterans in this field of study. The elaborative information provided in this book will help the readers further their scope of knowledge leading to advancements in this field.

Finally, I would like to thank my fellow researchers who gave constructive feedback and my family members who supported me at every step of my research.

Editor

Mouse fat storage-inducing transmembrane protein 2 (FIT2) promotes lipid droplet accumulation in plants

Yingqi Cai[1], Elizabeth McClinchie[1], Ann Price[1], Thuy N. Nguyen[2,†], Satinder K. Gidda[2], Samantha C. Watt[2], Olga Yurchenko[3], Sunjung Park[3,‡], Drew Sturtevant[1], Robert T. Mullen[2], John M. Dyer[3] and Kent D. Chapman[1,*]

[1]Center for Plant Lipid Research, University of North Texas, Denton, TX, USA
[2]Department of Molecular and Cellular Biology, University of Guelph, Guelph, ON, Canada
[3]US Arid-Land Agricultural Research Center, USDA-ARS, Maricopa, AZ, USA

*Correspondence
email chapman@unt.edu
†Present address: Department of Molecular Genetics, University of Toronto, Toronto, ON, Canada.
‡Present address: Biology Department, Central Arizona College, Maricopa, AZ 85138, USA

Keywords: endoplasmic reticulum, fat storage-inducing transmembrane protein 2, lipid droplets, lipid storage, triacylglycerol, lipid partitioning.

Summary

Fat storage-inducing transmembrane protein 2 (FIT2) is an endoplasmic reticulum (ER)-localized protein that plays an important role in lipid droplet (LD) formation in animal cells. However, no obvious homologue of FIT2 is found in plants. Here, we tested the function of FIT2 in plant cells by ectopically expressing mouse (Mus musculus) FIT2 in Nicotiana tabacum suspension-cultured cells, Nicotiana benthamiana leaves and Arabidopsis thaliana plants. Confocal microscopy indicated that the expression of FIT2 dramatically increased the number and size of LDs in leaves of N. benthamiana and Arabidopsis, and lipidomics analysis and mass spectrometry imaging confirmed the accumulation of neutral lipids in leaves. FIT2 also increased seed oil content by ~13% in some stable, overexpressing lines of Arabidopsis. When expressed transiently in leaves of N. benthamiana or suspension cells of N. tabacum, FIT2 localized specifically to the ER and was often concentrated at certain regions of the ER that resembled ER-LD junction sites. FIT2 also colocalized at the ER with other proteins known to be involved in triacylglycerol biosynthesis or LD formation in plants, but not with ER resident proteins involved in electron transfer or ER-vesicle exit sites. Collectively, these results demonstrate that mouse FIT2 promotes LD accumulation in plants, a surprising functional conservation in the context of a plant cell given the apparent lack of FIT2 homologues in higher plants. These results suggest also that FIT2 expression represents an effective synthetic biology strategy for elaborating neutral lipid compartments in plant tissues for potential biofuel or bioproduct purposes.

Introduction

Lipid droplets (LDs) are organelles found in all types of plant cells where they function to compartmentalize neutral lipids such as triacylglycerols (TAGs), steryl esters (SEs) and even isoprenoids (e.g. carotenoids, rubber) (Chapman and Ohlrogge, 2012; Murphy, 2001). Once considered as simply a repository for lipid storage molecules, these organelles are becoming more appreciated for their roles in other dynamic cellular processes, including signalling, trafficking and membrane remodelling (Brasaemle and Wolins, 2012; Chapman et al., 2012). LDs have a 'half-unit' membrane of phospholipids covering a hydrophobic core of nonpolar lipids, and in different cell types, different proteins localize to the LD surface (Murphy, 2012). Perhaps best studied in plants are the oleosins associated with LDs in seed tissues (Chapman et al., 2012; Frandsen et al., 2001; Huang, 1996), but a number of other proteins are located on the surface of LDs as well, including the recently identified lipid droplet-associated proteins (LDAPs) that appear to stabilize LDs in many nonseed (i.e. vegetative) cell types (Chapman et al., 2012; Gidda et al., 2016; Horn et al., 2013; Kim et al., 2016).

Cytosolic LDs, in both seeds and nonseed tissues, have also been referred to in the literature as lipid bodies, oil bodies, oleosomes and spherosomes, but more recently, the term lipid droplet has been used to emphasize a more cohesive and evolutionary relationship with neutral-lipid-containing organelles

from various cell types and organisms (Murphy, 2012). Cytosolic LDs in plant cells are distinctly different in location and composition from plastoglobuli, which are LDs that originate from the thylakoid membrane and accumulate in the plastid stroma (Austin et al., 2006; Nacir and Bréhélin, 2013). By contrast, cytosolic LDs are believed to originate from the endoplasmic reticulum (ER), where the neutral lipids are synthesized and partitioned somehow between the membrane bilayer leaflets before being packaged for release from the ER surface (Wanner et al., 1981). In plants, storage lipid accumulation into LDs has mostly been studied in seed tissues (Chapman et al., 2012), and transcriptional regulation of lipid synthesis and compartmentation is modulated through several key transcription factors, including WRINKLED1 (WRI1) and LEAFY COTYLEDON2 (LEC2) (Sreenivasulu and Wobus, 2013). Although a number of specific proteins in yeast, invertebrates and mammals have been identified that participate in LD ontogeny (Beller et al., 2010; Gao and Goodman, 2015; Pol et al., 2014), relatively little is known mechanistically about how plant cells form LDs at the ER and facilitate their release into the cytosol in seeds or nonseed tissues.

There are several models for cytosolic LD biogenesis in eukaryotic cells, and these all involve the ER in some manner. A widely accepted 'lens' model indicates that TAGs collect within the ER membrane bilayer and then bud from the ER surface into the cytosol to form an individual LD (Choudhary et al., 2015). In another model, pre-existing vesicles are filled with neutral lipids from the ER to form LDs (Walther and Farese, 2009). Yet another model suggests that LDs that are formed from ER-derived TAGs remain attached to the ER, perhaps even being interconnected with the ER lumen, and enlarging or reducing in size and prevalence based on cellular needs (Farese and Walther, 2009; Jacquier et al., 2011; Mishra et al., 2016). Others have suggested the possibility that LDs grow by small LDs fusing together to form large LDs (Wilfling et al., 2014). And finally, an 'egg-in-cup' model has been proposed where LDs originate at a particular location on the cytosolic surface of the ER membrane, whereby the ER partially surrounds the developing LD in such a way that it resembles an LD-'egg' inside an ER-membrane 'cup' (Robenek et al., 2006). Although these various models represent different mechanistic views of how the LD organelle is produced, one shared feature is that LDs are formed in association with the ER. Indeed, the ways in which LDs form in cells likely include several mechanisms that are cell type- or stage-specific and may not be mutually exclusive (Chapman et al., 2012; Pol et al., 2014; Wilfling et al., 2014).

The ER is the subcellular location for the majority of glycerolipid assembly within cells and includes enzymes for the synthesis of the neutral lipid LD core, as well as the phospholipid monolayer of the LD surface (Chapman and Ohlrogge, 2012; Chapman et al., 2012). In addition, a number of membrane-bound proteins, mostly identified in animals and yeast, have been shown to be involved in the biogenesis of LDs and localized at the ER, often at ER-LD junctions, or even on nascent LDs (Brasaemle and Wolins, 2012; Pol et al., 2014; Walther and Farese, 2009; Wang et al., 2016; Wilfling et al., 2014). One of the proteins that has been implicated in the sequestration of TAG for LD formation is the fat storage-inducing transmembrane protein (FIT) (Kadereit et al., 2008). There are two isoforms of the FIT protein in mammals, namely FIT1 and FIT2. FIT1 is found in skeletal muscle and is less evolutionarily conserved than FIT2. FIT2 homologues, on the other hand, are found in organisms ranging from Saccharomyces cerevisiae to humans and in mammals are expressed predominantly in adipose tissue (Gross et al., 2010; Miranda et al., 2014). Ectopic overexpression of FIT2 in mammals induced the accumulation of large-sized LDs at higher levels than are normally found and in tissues that are not normally LD-rich (Gross et al., 2010; Kadereit et al., 2008). FIT2 overexpression did not, however, increase the synthesis of TAGs directly, but rather promoted the more efficient partitioning of neutral lipids from the ER membrane into nascent LDs (Gross et al., 2011; Kadereit et al., 2008). Notably, modifications of the FIT2 polypeptide sequence that reduced or enhanced TAG-binding affinity resulted in the production of less or more LDs, respectively (Gross et al., 2011). Further, knock-down of FIT2 expression in adipose cells, zebrafish and mice reduced the accumulation of LDs (Goh et al., 2015; Gross et al., 2011; Kadereit et al., 2008; Miranda et al., 2014).

FIT2 was shown recently to be essential for the budding of nascent LDs from the ER into the cytosol in yeast and mammalian cells (Choudhary et al., 2015), and disruption of FIT2 in Caenorhabditis elegans was lethal (Choudhary et al., 2015). Interestingly, no homologues of FIT2 have been detected in the genomes of plants (Chapman et al., 2012). The importance of FIT2 in promoting LD formation in other eukaryotes prompted us to ask whether mammalian FIT2 would/could function to promote LD biogenesis in a plant cell context. Towards that end, we expressed the FIT2 protein from mouse (Mus musculus) in two transient plant cell systems, Nicotiana tabacum

suspension-cultured cells and Nicotiana benthamiana leaves, as well as via stable expression in transgenic Arabidopsis thaliana plants. Overall, our results indicate that FIT2, as it does in other eukaryotes, localizes to the ER in plant cells, including regions of the ER that appear to be involved in TAG biosynthesis and LD formation. FIT2 also promotes the accumulation of LDs in plant cells that normally do not accumulate substantial amounts of LDs, like those in leaves. We conclude that FIT2 likely interacts with the native LD biosynthetic machinery in plant cells to promote enhanced neutral lipid compartmentalization and the budding of LDs from ER, and its ectopic expression represents a novel strategy to elaborate the neutral lipid compartment in plants for various biotechnology-oriented applications.

Results

Mouse FIT2 induces the accumulation of cytosolic LDs in transiently transformed tobacco leaves

The FIT2 coding sequence from mouse was expressed in Agrobacterium-infiltrated tobacco leaves to evaluate the effect of FIT2 expression on LD proliferation in a plant cell context (Figure 1). As shown in Figure 1a, confocal images of BODIPY 493/503-stained LDs in mock (infiltration buffer only)-transformed leaves or leaves that were expressing only the P19 viral protein (serving as an enhancer of transgene expression and included in all infiltrations [Petrie et al., 2010]) showed few LDs distributed throughout the cytosol (Figure 1a), which is typical of mesophyll cells in N. benthamiana leaves (Cai et al., 2015; Gidda et al., 2016). Similarly, chlorophyll autofluorescence in mock- and P19-transformed cells revealed the expected distribution of chloroplasts, which mostly surrounded the cell's large central vacuole (Figure 1a). By contrast, expression of FIT2 in leaf cells resulted in a marked increase in both the numbers and sizes of LDs (Figure 1a, quantified in Figure 1b), all of which appeared to be localized within the cytosol and not in chloroplasts, indicating they were distinct from plastoglobuli. Further, expression of a FIT2 variant (FIT2-N[80]A) known to have reduced TAG-binding affinity compared to native FIT2 (Gross et al., 2011) resulted in an increase in the number of LDs in comparison with mock or P19 controls, but relatively fewer medium (3–6 μm^2), large (6–10 μm^2) or so-called super-sized (>10 μm^2) LDs in comparison with cells expressing the native FIT2 protein. Expression of a FIT2 variant with enhanced TAG-binding activity (i.e. FIT2-FLL[157-159]AAA; Gross et al., 2011), on the other hand, resulted in a further increase in super-sized LDs in comparison with native FIT2 (Figure 1a and b).

As shown also in Figure 1a, ectopic expression of the LEC2 transcription factor, which induces genes for seed oil like synthesis in leaves (Santos Mendoza et al., 2005), resulted in an increase in LD proliferation, as expected (Cai et al., 2015; Gidda et al., 2016). While the total numbers of LDs induced by LEC2 expression were similar to that of FIT2, the numbers of large and super-sized LDs were conspicuously greater with FIT2 in comparison with LEC2 (Figure 1b). These data suggest that FIT2 promotes the formation of larger-sized LDs in plant cells. Indeed, combined expression of LEC2 and FIT2, particularly with the 'enhanced' FIT2 variant (FLL[157-159]AAA), exaggerated further the total numbers of LDs in leaves, with a greater proportion of large and super-sized LDs, as well as medium-sized LDs, than with either FIT2 alone or LEC2 alone (Figure 1a and b). Moreover, FIT2 cooperated with the LEC2-induced machinery to package ~2.5 times more neutral lipids (i.e. TAGs and some SEs) in these leaves

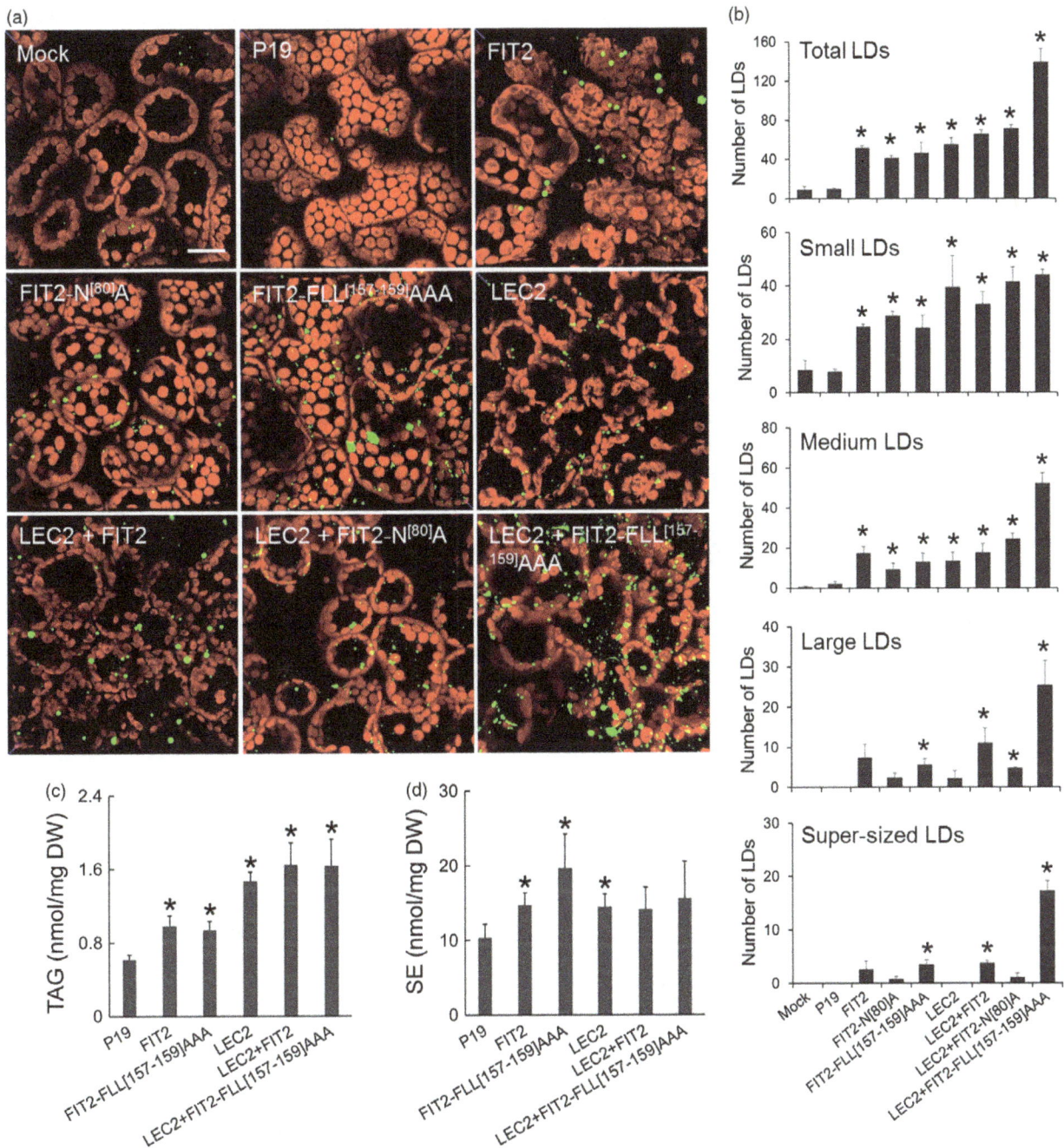

Figure 1 FIT2 expression in *Nicotiana benthamiana* leaves elaborates a neutral lipid compartment. (a) Representative confocal images of *N. benthamiana* leaves expressing FIT2 or mutated FIT2 proteins (i.e. FIT2-N[80]A or FIT2-FLL[157-159]AAA) in the presence or absence of LEC2. Cells were (co)transformed as indicated by the panel label, and LDs (green) were visualized by staining with BODIPY 493/503. Red colour corresponds to chloroplast autofluorescence. Images are 2D projections of Z-stacks of 50 optical sections. Bar = 20 μm. (b) Number of total LDs and LDs in different size categories per image area. The numbers and sizes of LDs were quantified by ImageJ as BODIPY-stained lipid area in 2D projections of Z-stacks (see Experimental procedures for details). Small LDs: BODIPY-stained lipid area <3 μm². Medium LDs: 3–6 μm². Large LDs: 6–10 μm². Super-sized LDs: >10 μm². Averages and SDs were calculated from three biological replicates (three Z-stacks from each replicate). Asterisks indicate significant difference relative to the mock control at $P \leq 0.05$ determined by Student's *t*-test. (c) and (d) TAG and SE content in infiltrated *N. benthamiana* leaves. Averages and SDs were plotted from four biological replicates. Asterisks indicate significant difference in comparison with the P19 control at $P \leq 0.05$ determined by Student's *t*-test.

(Figure 1c and d). Detailed analysis of the TAG and SE molecular species by conventional, direct-infusion ESI-MS/MS (electrospray ionization tandem mass spectrometry) (Figure S1) indicated that, generally, the TAGs and SEs that are normally present in leaves were all increased upon FIT2 expression. These same molecular species of TAG and SE were exaggerated further by co-expression of FIT2 and LEC2, making the overall increase in TAGs and SE even more obvious (Figure S1). Notably, there was no further increase in TAG and SE in leaves co-expressing LEC2 and the enhanced FIT2 variant (FLL[157-159]AAA) in comparison with amounts observed in leaves co-expressing LEC2 and native FIT2 (Figure 1c), despite a nearly twofold difference in the total number of LDs (Figure 1b). These observations would be consistent, however, with a primary role for FIT2 in the binding and partitioning of TAG into LDs, rather than a role in stimulating TAG synthesis (Gross *et al.*, 2011; Kadereit *et al.*, 2008).

FIT2 localizes to the ER in plant cells and is enriched in ER subdomains involved in TAG assembly and LD biogenesis

The proliferation of LDs in plant cells following FIT2 expression suggested that it was functioning at the subcellular level in a manner similar to that in animal and yeast cells. Indeed, as shown in Figure 2, confocal imaging revealed that when GFP-tagged FIT2 was expressed in *N. benthamiana* leaves, it localized exclusively to the ER, as evidenced by colocalization with a co-expressed ER marker protein, CFP-HDEL (Szymanski *et al.*, 2007). Notably, GFP-FIT2 and CFP-HDEL co-expressing cells also displayed a marked increase in the abundance of Nile red-stained LDs when compared to cells expressing CFP-HDEL alone (Figure 2), consistent with the induction of LDs in tobacco leaves transiently expressing FIT2 (Figure 1a and b). GFP-tagged versions of FIT2-N[80]A or FIT2-FLL[157-159]AAA also localized to the ER in tobacco leaf cells (Figure 2), and the elaboration of LDs in these cells was distinct from each other and that of wild-type GFP-FIT2, similar to results obtained using their non-GFP-tagged counterparts (Figure 1). That is, leaf cells expressing GFP-FIT2-N[80]A or GFP-FIT2-FLL[157-159]AAA, relative to those expressing GFP-FIT2, appeared to possess fewer or more larger-sized LDs, respectively (Figure 2). Cells expressing GFP-FIT2-FLL[157-159]AAA also displayed several conspicuous alterations in ER morphology, whereby the fusion protein (and CFP-HDEL) localized not only throughout the normal, reticular-like ER network, but also accumulated in specific regions of the ER that appeared globular in shape and were not observed in cells expressing CFP-HDEL alone (Figure 2; refer to arrowheads in bottom row). These ER structures typically colocalized with or next to Nile red-stained LDs, suggesting that these might be ER-LD junction sites that formed as a result of excessive TAG binding and/or inefficient release of the nascent LDs from the ER, or simply might be a reflection of a normal process that was exaggerated by ectopic expression. It is possible also that the LDs with TAG may be trapped in the ER lumen at these locations (Mishra *et al.*, 2016) or wrapped in ER membranes (Choudhary *et al.*, 2015), but the resolution of light microscopy prevents more conclusive statements to be made about the nature of these structures.

We tested next whether mouse FIT2 could colocalize with other proteins known to be involved in TAG synthesis and/or LD formation in plant cells. For this purpose, we employed tobacco (*N. tabacum*) Bright Yellow 2 (BY-2) suspension-cultured cells, which serve as a well-established model system for protein localization studies (Brandizzi *et al.*, 2003; Miao and Jiang, 2007) and, given their relatively large size, are amenable to microscopic analysis of ER subdomains (Gidda *et al.*, 2011; Hanton *et al.*, 2007; Shockey *et al.*, 2006). The proteins tested for colocalization included *Vernicia fordii* (tung tree) diacylglycerol acyltransferase 2 (DGAT2), which is known to be involved in TAG synthesis and localized to ER subdomains (Shockey *et al.*, 2006), and *Arabidopsis* SEIPIN1, which localizes to ER-LD sites where it is thought to promote the compartmentalization of neutral lipids into nascent, large-sized LDs from the ER (Cai *et al.*, 2015). As shown in Figures 3a and S2, and similar to the results presented above for GFP-FIT2 in *N. benthamiana* leaf cells (Figure 2), transient

expression of Cherry-FIT2 or GFP-FIT2 in BY-2 cells resulted in a marked increase in monodansylpentane (MDH)-stained LDs in comparison with mock-transformed control cells, and the fusion protein localized throughout concanavalin A (ConA)-stained ER. Unlike control cells, however, the ER of cells expressing FIT2 was generally more punctate and globular in nature (Figure 3a), which is often observed for transiently expressed proteins that target to ER subdomains. Co-expression of Cherry-tagged FIT2 with either myc-epitope-tagged DGAT2 or GFP-tagged SEIPIN1 showed clear colocalizations in specific regions of ER (Figure 3b). These associations appeared to be subdomain specific, as Cherry-FIT2 did not colocalize with co-expressed GFP-tagged *Arabidopsis* SEC24, a soluble protein that localizes to subdomains of ER that serve as transport-vesicle exit sites (Gidda *et al.*, 2011; Hanton *et al.*, 2007), or with a GFP-tagged isoform of tung tree cytochrome b_5 (Cb5), which is known to induce ER membrane 'whorls' or 'karmallae' when ectopically overexpressed (Hwang *et al.*, 2004; Snapp *et al.*, 2003) (Figure 3b). Control experiments are shown to help illustrate that each of the respective proteins is targeted to ConA-stained ER subdomains, regardless of FIT2 co-expression (Figure 3c). While we recognize that caution should be exercised in the interpretation of these results, the data suggest that mouse FIT2 colocalizes specifically with machinery in plant cells that is devoted to TAG biosynthesis and LD formation, and not likely indiscriminately localized to regions involved in other ER processes (e.g. vesicle export) or due to nonspecific membrane protein aggregation.

Stable expression of FIT2 in *Arabidopsis* increases neutral lipid accumulation in both leaf and seed tissues

While mouse FIT2 appeared to localize to ER subdomains and promote LD formation in transient plant expression assays (Figures 1–3), we next determined how this protein would function in stably transformed plants. Towards this end, we generated several independent transgenic *Arabidopsis* lines ectopically expressing GFP-tagged or nontagged mouse FIT2. Transgene expression was verified using reverse transcriptase (RT)-PCR (Figure S3a). As shown in Figure 4, leaves of mature, 24-day-old transgenic plants showed significant increases in the numbers of LDs compared to nontransformed, wild-type plants (Figure 4a and b), consistent with results obtained for transient expression in tobacco leaves and suspension cells (Figures 1–3). Further, in one line (i.e. FIT2-OE-C6), in addition to increased numbers of LDs, the LDs were significantly larger than those in wild-type plants or other FIT2 lines (Figure 4c).

Figure 4 shows also that the neutral lipid content was increased in FIT2 transgenic *Arabidopsis* plant leaves (with or without GFP), with significant increases in both TAG and SE quantified by ESI-MS/MS, in comparison with wild-type plants (Figure 4d and e; and Figure S4; to 0.1% TAG and 3% SE on a dry weight basis). Recently, it has become possible to visualize lipid species directly in tissues by MALDI-MS (Horn and Chapman, 2014a; Sturtevant *et al.*, 2016). Hence, we confirmed *in situ* by MALDI-MS (matrix-assisted laser desorption/ionization MS) imaging of cryosectioned leaves that two major neutral lipid molecular species, one TAG (TAG 52:6, m/z 889.668) and one SE (Campesterol [CA] 18:3, m/z 699.547), were in fact elevated and distributed in the mesophyll tissues of FIT2 transgenic leaves, that is FIT2-OE-C6 (Figure 4f). Curiously, there seemed to be a somewhat heterogeneous distribution of the TAG species relative to the SE species in the leaves of this line (Figure 4g), suggesting that there is a difference in the location in leaves for the synthesis

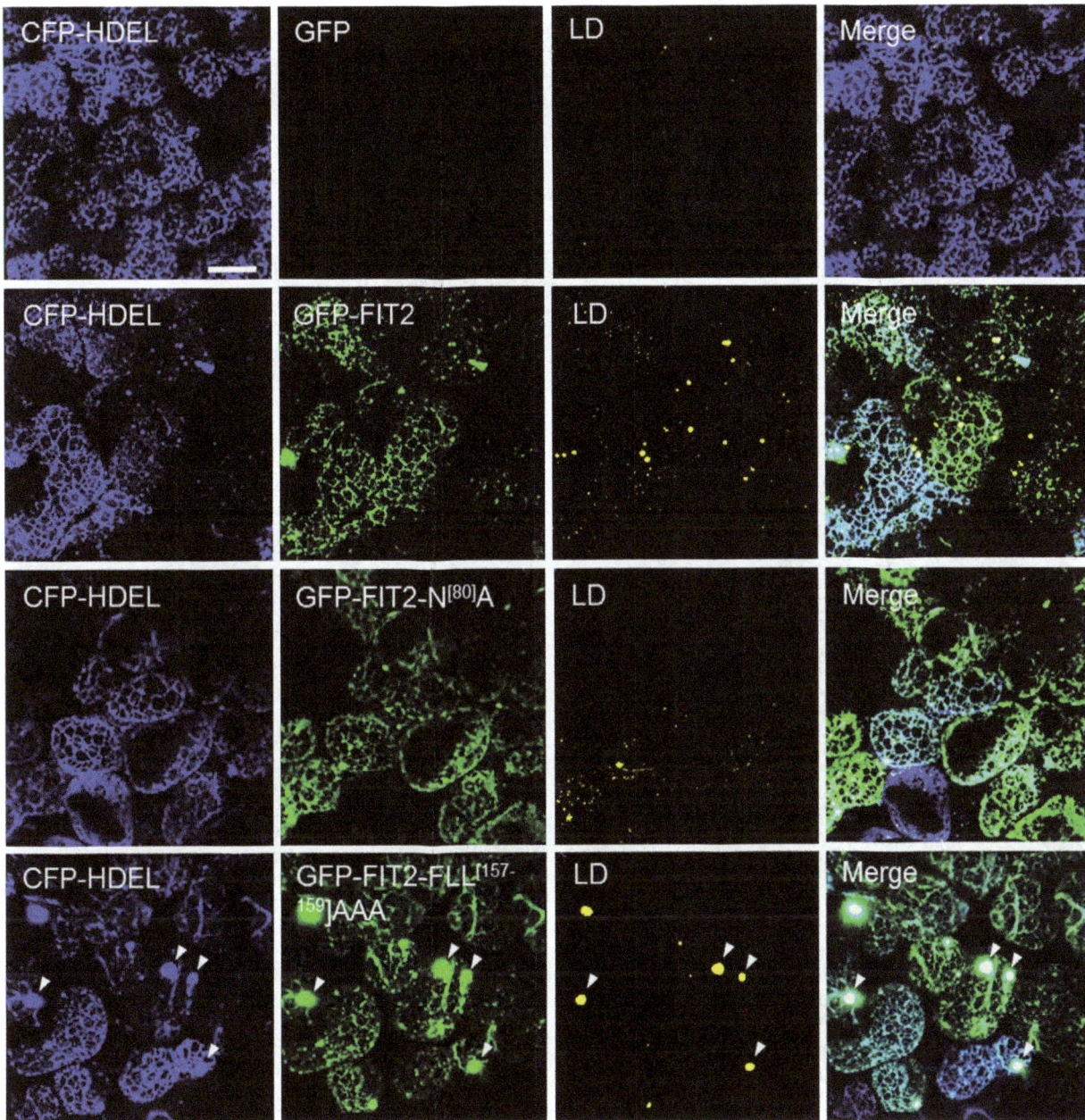

Figure 2 FIT2 and mutated FIT2 GFP-tagged fusion proteins localize to the ER in *Nicotiana benthamiana* leaf mesophyll cells. As indicated by panel labels, ER, GFP-FIT2/FIT2 variants and LDs were visualized (via confocal microscopy) in mesophyll cells based on fluorescence signals of CFP-HDEL (blue), N-terminal-GFP-tagged FIT2 or FIT2 variants (i.e. FIT2-N[80]A or FIT2-FLL[157-159]AAA) (green) and Nile red (yellow), respectively. Shown also are the corresponding merged images. Note that no GFP fluorescence signal was detectable in the cell expressing CFP-HDEL alone (top row), as expected. Arrowheads in the bottom row indicate examples of regions of the ER with altered morphology, where both GFP-FIT2-FLL[157-159]AAA and Nile red-stained LDs appear to accumulate. All images are 2D projections of Z-stacks of ~30 slices. Bar = 20 μm.

and/or accumulation of these two neutral lipid classes, but this possibility, and how FIT2 might participate in this process, needs to be examined in more detail. Nevertheless, this general distribution of elevated amounts of neutral lipids throughout the mesophyll tissues of leaves, compared to wild-type leaves, was consistent with an increase in BODIPY 493/503-stained LDs (Figure 4f, left) or Nile red-stained LDs (Figure 4a) observed in mesophyll cells expressing FIT2 in *Arabidopsis* transgenic lines. We observed also that LDs were more prevalent in young and mature leaves (i.e. the 7th true leaf and the 5th true leaf from the bottom of the plant, respectively) compared with older leaves (i.e. the 3rd

true leaf from the bottom of the plant) (Figure S3b), perhaps reflecting an influence of FIT2 in more metabolically active tissues where neutral lipids are likely to be synthesized. Further, transgenic leaves exhibited quantifiable increases in both TAGs and SEs, and most major molecular species of these neutral lipids increased significantly (Figure S4), indicating that FIT2 promoted a general elevation in neutral lipids that were already present in leaves, and not a specific accumulation of new or selected species. The composition of PC molecular species was not altered in a substantial way by stable, ectopic expression of FIT2 (Figure S5).

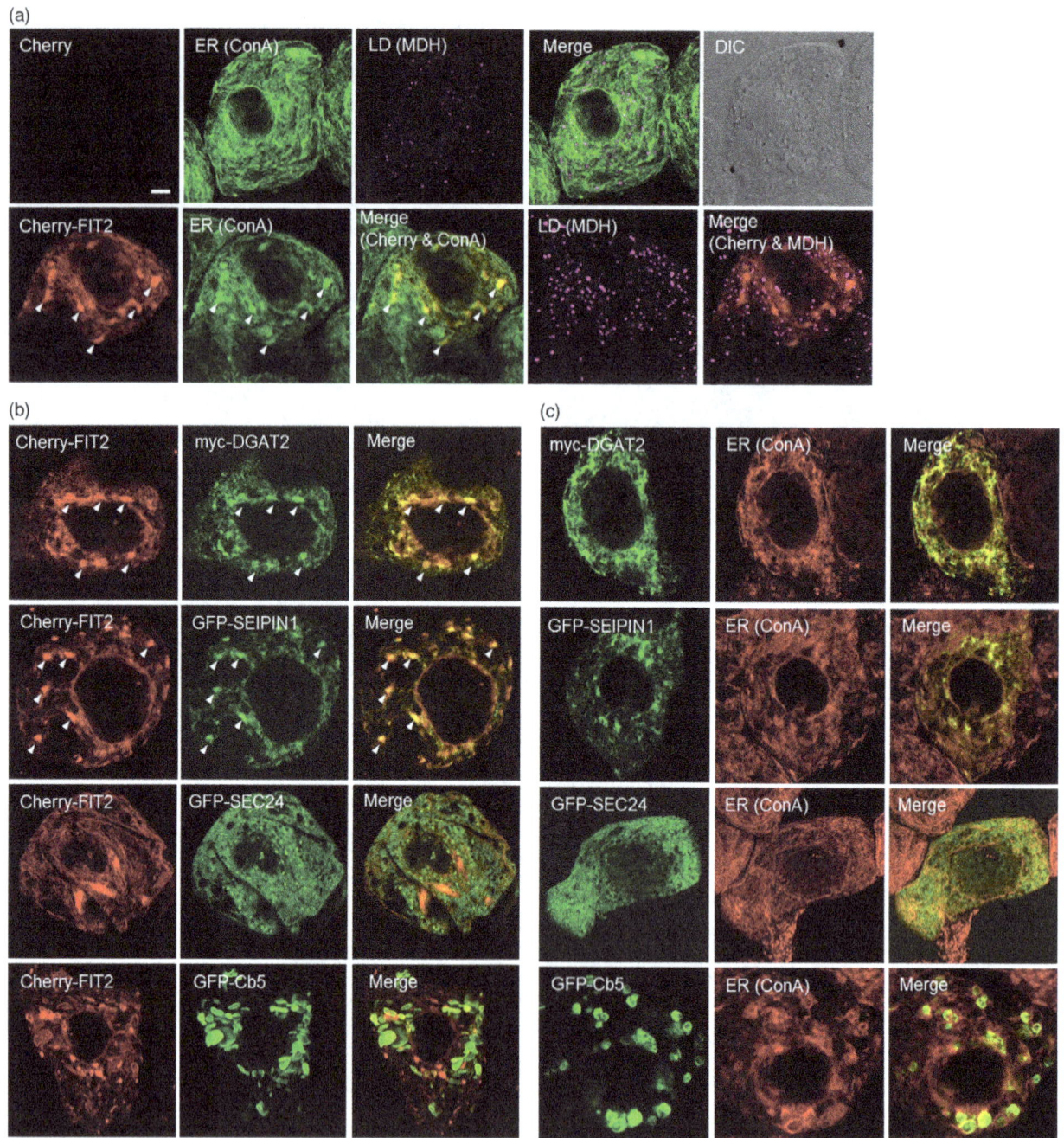

Figure 3 FIT2 expressed in *Nicotiana tabacum* suspension-cultured BY-2 cells is enriched in ER domains that also accumulate proteins involved in TAG biosynthesis and LD biogenesis. BY-2 cells were transiently (co)transformed with the indicated proteins, and then, cells were visualized using confocal microscopy. (a) Representative confocal images showing the ER and LDs in mock-transformed BY-2 cells (top row), whereby the ER was stained using fluor-conjugated ConA and LDs were stained using MDH (false-coloured magenta). Shown also is the corresponding merged and differential interference contrast (DIC) images. The bottom row shows the fluorescence-staining pattern of transiently expressed Cherry-FIT2 in comparison with ConA-stained ER and MDH-stained LDs. Arrowheads in the bottom row indicate examples of Cherry-FIT2 enriched in specific regions of ConA-stained ER. Note that no Cherry fluorescence signal was detected in the mock-transformed cell (top row), as expected. Bar = 5 μm. (b) Confocal images showing the localizations of Cherry-FIT2 and various other co-expressed ER-localized proteins, including myc-tagged DGAT2 or GFP-tagged SEIPIN1, SEC24 or Cb5. Shown also are the corresponding merged images for each pair of co-expressed proteins. Arrowheads illustrate examples of colocalization of FIT2 with DGAT2 and SEIPIN1. (c) Confocal images showing the localization of individually expressed Myc-DGAT2 or GFP-tagged SEIPIN1, SEC24 or Cb5 at the ConA-stained ER. Shown also are the corresponding merged images for each protein and ConA-stained ER.

While there was a significant increase in the numbers of LDs and contents of neutral lipids in vegetative tissues of *Arabidopsis* plants stably expressing mouse FIT2 (Figure 4), we also asked whether FIT2 might enhance the accumulation of TAGs in lipid-rich seed tissues in these same lines. As shown in Figure 5a, confocal microscopy of parenchyma cells in embryo cotyledons suggested that the size and distribution of Nile red-stained LDs was somewhat different in transgenic seeds compared to wild-

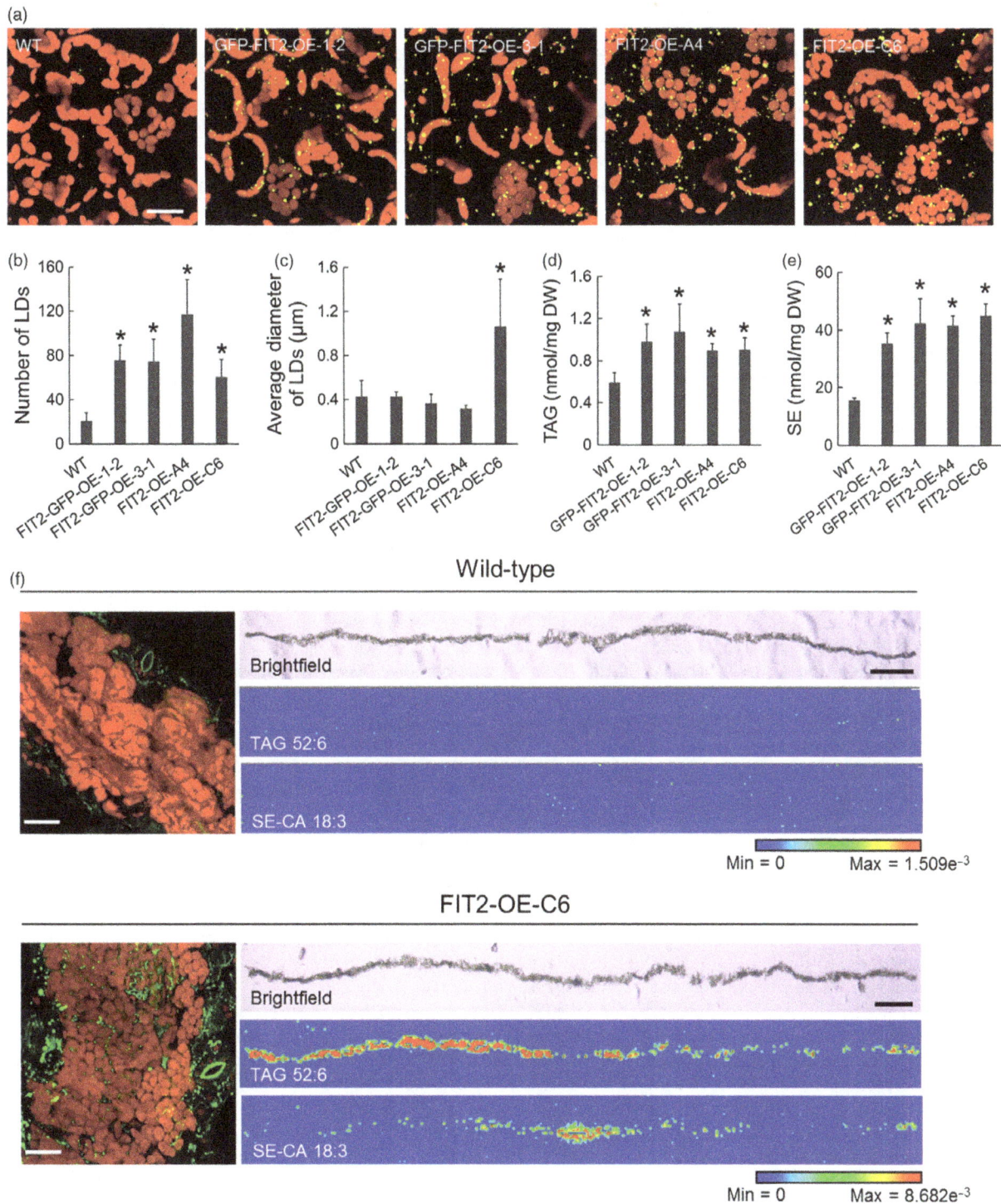

Figure 4 Stable expression of FIT2 in *Arabidopsis* increases LD number and neutral lipid content in leaf tissues. (a) Representative confocal images of wild-type and FIT2 transgenic *Arabidopsis* leaves. Red colour indicates autofluorescence attributable to chloroplasts. LDs were stained with Nile red (false-coloured yellow). Images are projections of Z-stacks of 30 optical sections. (b) and (c) Numbers and sizes of LDs per view (image area) of *Arabidopsis* leaf mesophyll. LDs were quantified as Nile red-stained lipid area in 2D images (as single optical slices). Averages and SDs were calculated based on nine images from at least three *Arabidopsis* plants. Asterisks indicate significant difference at $P \leq 0.05$ determined by Student's t-test. (d) and (e) TAG and SE content in wild-type and FIT2 transgenic *Arabidopsis* leaves. Averages and SDs are calculated from four biological replicates. Asterisks indicate significant difference at $P \leq 0.05$ determined by Student's t-test. (f) Spatial distribution of BODIPY-stained LDs (green), one major TAG species (TAG 52:6, m/z 889.668) and one major SE species (campesterol [CA] 18:3, m/z 699.547) in wild-type and transgenic (i.e. FIT2-OE-C6) *Arabidopsis* leaf cross sections. Left panels are representative confocal images of BODIPY-stained LDs (green) in wild-type and transgenic *Arabidopsis* leaf tissue. Autofluorescence of chloroplasts in confocal images are shown in red colour. Confocal images are projections of Z-stacks. Bar = 20 μm. Right panels are bright field microscopy images of leaf sections and corresponding spatial distribution of TAG 52:6 and CA 18:3 detected by MALDI. Bar = 0.5 mm. The blue-to-red scale bar is used to indicate minimum and maximum ion intensity of potassiated TAG 52:6 and potassiated CA 18:3 molecules.

type, with larger-sized LDs often observed within the interior of these cells (Figure 5a). Furthermore, while seed size and dry weight were not significantly altered by FIT2 expression (Figure 5b and c), there was a modest but significant increase in seed oil content in some, but not all, FIT2 transgenic lines (measured by time-domain ^{1}H-NMR), and also in the amounts of total fatty acids per seed measured by gas chromatography-flame ionization detection (GC-FID) (Figure 5d and e). More specifically, when plants were grown under identical conditions, some FIT2-expressing lines displayed an increase in seed oil content over wild-type by as much as 13% (e.g. 34% oil versus 30% oil on a dry weight basis for FIT2-OE-C6 versus WT, respectively) (Figure 5d), suggesting that FIT2 expression can enhance overall oil content in seed tissues, as well as in nonseed (vegetative) tissues of plants. Furthermore, fatty acid composition of the seed oil was mostly similar between wild-type and FIT2 lines, with only some subtle differences due to an increase in 18:1 and decrease in 20:1 acyl moieties (Figure 5f).

GFP-tagged FIT2 was also visualized in *Arabidopsis* leaves and seeds to confirm protein expression and evaluate the localization of FIT2 relative to LDs in these tissues (Figure S6). Similar to results in tobacco leaves and suspension cells (Figures 2 and 3), in *Arabidopsis* leaves, GFP-tagged FIT2 showed accumulations of concentrated fluorescence often in the proximity of Nile red-stained LDs (Figure S6). In seeds, the GFP fluorescence signal was relatively weak but indicated that GFP-FIT2 was associated with the LDs in places (Figure S6); however, the ER organization is rearranged in desiccated seed tissues (Hsieh and Huang, 2004; Wang *et al.*, 2010), and perhaps the residual FIT2 persisting through seed desiccation was localized to LDs or regions of ER intimately surrounding the LDs.

Discussion

FIT2 has been described as part of an evolutionarily conserved family of proteins that are important for the subcellular compartmentalization of neutral lipids in mammals, yeast, insects and *C. elegans* (Choudhary *et al.*, 2015; Goh *et al.*, 2015; Gross *et al.*, 2010; Kadereit *et al.*, 2008). It is curious, therefore, that this essential protein would not be present in all eukaryotes that form LDs from the ER, as no obvious homologues have been identified in plant genomes. Cytosolic LD biogenesis from the ER membrane shares conserved features at the cellular level across kingdoms, and there are likely overlapping sets of proteins that cooperate in this process, perhaps encoded by distantly related genes. Indeed, SEIPIN genes were recently identified in plant genomes, and three isoforms in *Arabidopsis* were shown to participate in the biogenesis of LDs in plant cells, similar to their function in other eukaryotes (Cai *et al.*, 2015). In some cases, a shared cellular process might also be carried out by distinctly different proteins, such as the oleosins and LDAPs in plants and the perilipins in mammals, all of which bind the LD surface and stabilize LDs in the cytosol, yet share no obvious sequence similarity (Chapman *et al.*, 2012; Gidda *et al.*, 2016).

Here, we have asked whether a protein without any known homologues in plants might still function in a plant cell context to influence LD biogenesis. FIT2 functions, at least in part, by binding to TAG (Gross *et al.*, 2011), and thus, its role in LD biogenesis is likely dependent on biophysical properties of cellular membranes that might also be present in plant cells. Notably, similar approaches were used to explore the targeting and function of oleosins, caleosins and perilipins in LD formation in yeast cells,

which otherwise lack these proteins endogenously (Beaudoin and Napier, 2002; Froissard *et al.*, 2009; Jacquier *et al.*, 2013; Mishra and Schneiter, 2015; Rowe *et al.*, 2016). Clearly, ectopic expression of mouse FIT2 in plant cells, similar to results observed in yeast and animal cells (Choudhary *et al.*, 2015; Gross *et al.*, 2010; Kadereit *et al.*, 2008), increases production of LDs and associated TAGs (Figures 1, 2, 4 and 5). From the results presented here, it also appears that FIT2, due to its inherent ability to bind TAGs (Gross *et al.*, 2011), localizes to TAG-forming domains of the ER in plant cells where it functions to promote LD formation. SEs were also elevated in leaves expressing FIT2 (Figures 1 and 4; Figures S1 and S6), however, suggesting that FIT2 engages in and enhances a general coordination of neutral lipid synthesis and LD formation. Indeed, when co-expressed with DGAT2 or SEIPIN1, FIT2 colocalized with both proteins (Figure 3), consistent with the premise that FIT2 associates with domains involved in both TAG synthesis and nascent LD biogenesis (Choudhary *et al.*, 2015). The activity of FIT2 in plant cells suggests it might be useful as a probe for further studying the molecular nature of LD-forming domains in plants, possibly identifying other, as yet unidentified proteins involved in LD biogenesis.

The ectopic expression of a seed-specific transcription factor such as LEC2 in leaves has been shown by us and others to promote the accumulation of TAGs in leaf tissues (Mu *et al.*, 2008; Sanjaya *et al.*, 2013; Slocombe *et al.*, 2009), and this is accentuated when additional proteins that promote TAG assembly, like DGAT, or proteins that bind to cytosolic LDs, like oleosins, are coproduced (Andrianov *et al.*, 2010; Vanhercke *et al.*, 2013, 2014; Winichayakul *et al.*, 2013). Together, these engineering approaches have been described in a 'push, pull and protect' strategy to overproduce TAG in vegetative tissues (Vanhercke *et al.*, 2014; Xu and Shanklin, 2016). In this model, seed-specific transcription factors up-regulate the fatty acid biosynthetic machinery (Grimberg *et al.*, 2015), effectively 'pushing' reduced photosynthetic carbon into lipid, up-regulation of DGAT promotes the synthesis of TAGs at the ER, 'pulling' reduced carbon into TAG, and oleosins 'protect' the sequestered TAG from the active metabolic pool and prevent the hydrolysis of TAGs and turnover of LDs. In fact, remarkable amounts of TAG have been accumulated in leaves of transgenic tobacco plants—up to more than 15% dry weight of leaves—using a combination of these engineering processes (Vanhercke *et al.*, 2014). The increase in the energy density of leaves (i.e. increased lipid content relative to carbohydrate) will likely find multiple applications in agriculture and bioenergy, such as higher calorie and nutritionally balanced feeds, or as a source of oleochemicals and biodiesel feedstock (Horn and Benning, 2016). Indeed, it has been suggested that an increase in vegetative biomass of lipid up to 10% dry weight would result in a 30% increase in usable liquid fuels, through a combination of both biodiesel and lignocellulosic-derived fuels (Ohlrogge and Chapman, 2011). The production of large amounts of lipids in vegetative tissues makes possible the separation of calories in edible oilseeds needed for human nutrition from a potential bioenergy feedstock derived from vegetative biomass, preventing renewable sources of energy from competing with food production. What remains to be explored is how proteins involved in packaging TAG into LDs, such as SEIPIN, FIT2 (or its yet-to-be identified functionally equivalent plant orthologue), might complement ongoing approaches to enhance the energy content of plant vegetative tissues. Certainly, it seems that co-expression of LEC2 with

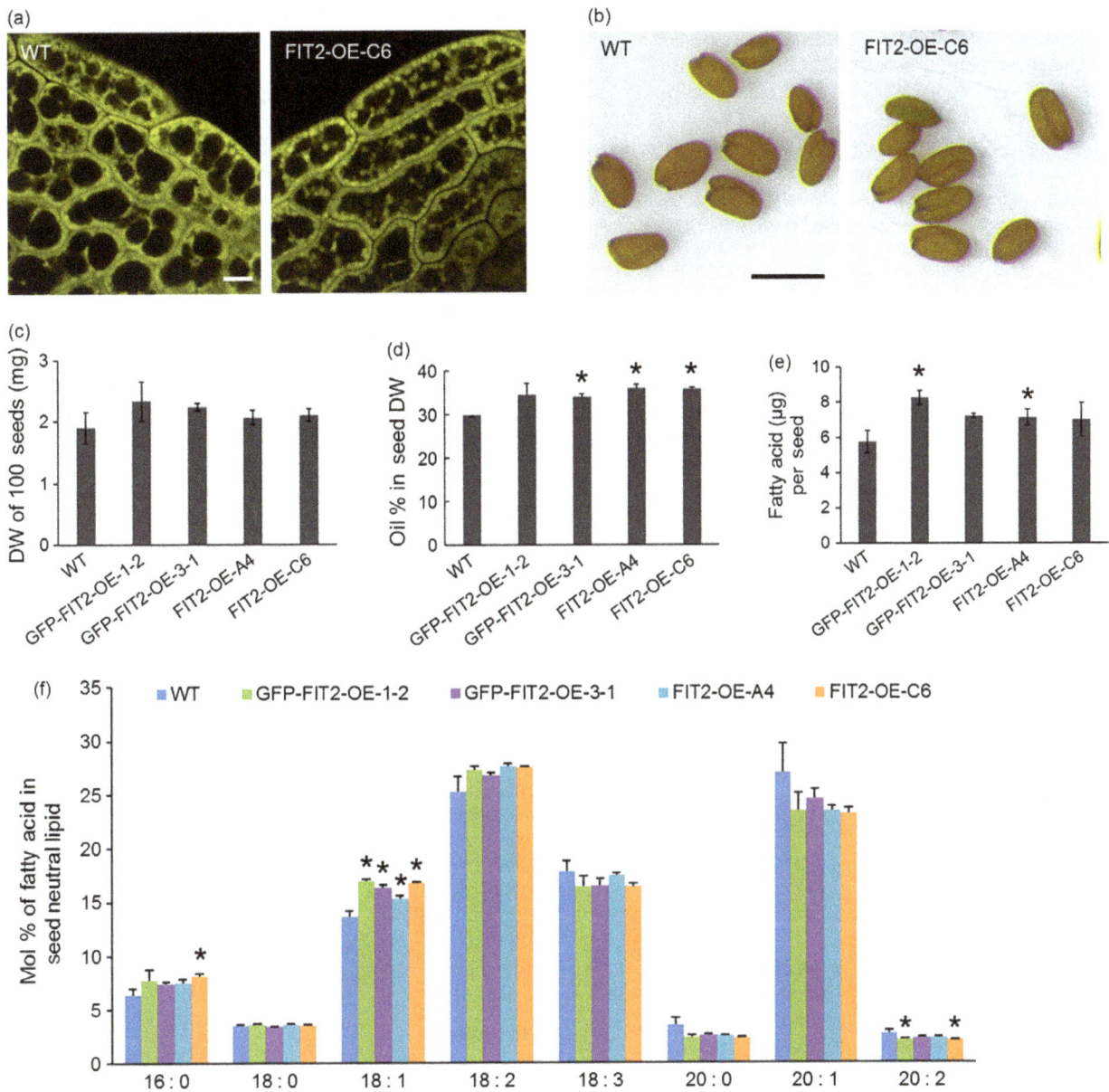

Figure 5 Stable FIT2 expression increases oil content in *Arabidopsis* seeds. (a) Representative confocal images of LDs in wild-type and FIT2 transgenic (i.e. FIT2-OE-C6 line) *Arabidopsis* seeds. LDs were stained with Nile red (false-coloured yellow). Bar = 5 μm. (b) Mature seeds from wild-type and FIT2 transgenic *Arabidopsis* plants. Bar = 0.5 mm. (c) to (f) Seed size and seed oil content of wild-type and FIT2 transgenic *Arabidopsis*. Averages and SDs were calculated from three biological replicates. Seeds were collected from *Arabidopsis* plants grown under the same conditions. Asterisks indicate significant difference at $P \leq 0.05$ determined by Student's *t*-test. (c) Weight of 100 dry seeds. (d) Oil content in dry seeds on a weight basis measured by NMR. (e) Oil content per seed on a fatty acid basis quantified by GC-FID. (f) Fatty acid composition in mature seeds.

variants of FIT2 in plant leaves can result in marked increase of compartmentalized LDs and neutral lipid content above that seen with either LEC2 or FIT2 alone (Figure 1), and so perhaps some component of LD 'release' would add significantly to the 'push–pull–protect' theory of increasing lipid accumulation in plant vegetative tissues.

Experimental procedures

Plant material, growth conditions and transformation

A. thaliana (Columbia-0) plants were grown in growth chambers on either one-half-strength MS-media-containing plates (Murashige and Skoog, 1962) or in soil at 21 °C under a 16/8-h light/dark cycle with 50 μE/m²/s light intensity. *Nicotiana benthamiana* plants were grown in soil at 28 °C under a 14/10-h light/dark cycle. Plants used in the same experiments were grown concurrently, in the same conditions. Plasmids were transformed into *Agrobacterium tumefaciens* (GV3101) by electroporation (Bio-Rad), and stable transformation of *Arabidopsis* plants was carried out using the floral dip method (Clough and Bent, 1998). Gene expression was evaluated in transgenic *Arabidopsis* seedlings (progeny) using RT-PCR, and independent lines were selected for further study. Infiltration of *N. benthamiana* leaves was carried out as described previously (Cai *et al.*, 2015; Petrie *et al.*, 2010). Tomato Bushy Stunt Virus (TBSV) *P19* was included in all infiltrations to suppress transgene silencing and intensify transgene expression (Petrie *et al.*, 2010); *Arabidopsis LEC2* was included in some co-infiltrations to promote seed-like lipid synthesis in leaves (Cai *et al.*, 2015; Gidda *et al.*, 2016; Santos Mendoza *et al.*, 2005). *Nicotiana tabacum* Bright Yellow (BY)-2 suspension-cultured cells were maintained and prepared for (co)transformation via biolistic particle bombardment using a Bio-Rad PDS system 1000/HE, as described previously (Lingard *et al.*, 2008).

Plasmid construction

The cDNA of *M. musculus FIT2* (EMBL Accession No. BAE37420.1) was kindly donated by David Silver (Duke—National University of Singapore). The *FIT2* ORF was subcloned into the plant binary expression plasmids pMDC32 and pMDC43 [Curtis and Gross-niklaus, 2003;] using restriction enzymes *Asc*I and *Pac*I, yielding pMDC32/FIT2 and pMDC43/GFP-FIT2, respectively. Both pMDC32 and pMDC43, as well as all other plant expression plasmids used in this study, contain the CaMV 35S constitutive promoter (Benfey *et al.*, 1989). Mutated versions of *FIT2* (i.e. FIT2-N[80]A and FIT2-FLL[157-159]AAA) were generated using PCR-fusion cloning procedures (Atanassov *et al.*, 2009) and then subcloned into pMDC32 and pMDC43. pRTL2/Cherry-FIT2 was constructed by subcloning the *FIT2* ORF into the transient expression plasmid pRTL2/Cherry-MCS, which contains the Cherry fluorescent protein coding sequence (Gidda *et al.*, 2011). Plant binary expression plasmids containing *Arabidopsis LEC2* (pORE04-*LEC2*) and *Tomato Bushy Stunt Virus P19* (pORE04-*P19*) were provided by Qing Liu (CSIRO) (Petrie *et al.*, 2010). Other plant expression plasmids, including pMDC32/CFP-HDEL, pMDC43/GFP-SEIPIN1, pRTL2/GFP-SEC24, pRTL2/Myc-DGAT2 and pRTL2/GFP-Cb5, have been described elsewhere (Cai *et al.*, 2015; Gidda *et al.*, 2011; Hwang *et al.*, 2004; Shockey *et al.*, 2006).

Rt-pcr

RNA was extracted from 4-week-old *Arabidopsis* plants using the RNeasy Plant Mini Kit (Qiagen, Germantown, MD, USA), and DNA contamination was eliminated using DNase (Promega, Madison, WI, USA). Approximately 100 ng of total RNA was used for RT-PCRs, which were performed with TaKaRa One Step RT-PCR Kit following the manufacturer's instructions. The PCR program was as follows: 42 °C for 15 min, 95 °C for 5 min, 35 amplification cycles (94 °C for 30 s, 55 °C for 30 s, 72 °C 30–60 s), and 72 °C for 7 min. Primers used to amplify *Arabidopsis EF1α* were those used elsewhere (Cai *et al.*, 2015). Primers used to amplify mouse *FIT2* were FIT2RTF 5′-ATGGAGCACCTGGAGCGC-3′ and FIT2RTR 5′- TCATTTCTTGTAAGTATCTCGCTTCAAAG-3′.

Lipid analysis

For neutral lipid content and compositional analysis, ~45 mg dry weight of *N. benthamiana* leaves, 15 mg of *Arabidopsis* leaves (dry weight) or 10 mg of *Arabidopsis* seeds were used in each replicate. Sample tissues were homogenized with glass beads in a bead beater (BioSpec Mini-Beadbeater-16) with 70 °C iso-propanol. TAG (tri-17:0) standard (Avanti) was added to *Arabidopsis* seed samples, while TAG (tri-15:0) standard and cholesterol ester (C13:0) standard (Nu-Chek Prep) were added to leaf samples. Total lipid was extracted as described by Cai *et al.* (2015). The neutral lipid fraction and polar lipid fraction were separated from total leaf lipid extract by solid-phase extraction using 6-mL silica columns (Sigma-Aldrich, St. Louis, MO, USA).

Hexane and diethyl ether (4:1) were used to elute neutral lipids, and chloroform and methanol (1:2) were used to elute polar lipids. PC (di-14:0) (Avanti) standard was added to the polar lipid fraction of *Arabidopsis* leaf samples. Lipid content in *Arabidopsis* (TAG, SE and PC) and *N. benthamiana* (TAG and SE) leaves was analysed by electrospray ionization mass spectrometry (ESI-MS) on an API3000 triple quadrupole mass spectrometer (Applied Biosystems/Sciex, Framingham, MA), and data were processed using Metabolite Imager (v.1.0) to quantify the TAG and SE content (Horn and Chapman, 2014b; Li *et al.*, 2014) and using LipidomeDB Data Calculation Environment to quantify the PC content (Welti *et al.*, 2002; Zhou *et al.*, 2011). Lipid extract from *Arabidopsis* seeds was transesterified with methanolic HCl, and the fatty acid methyl esters were analysed using gas chromatog-raphy–flame ionization detection (GC-FID, Hewlett Packard, HP 5890 II plus GC). Conditions for both ESI-MS and GC-FID were as described in James *et al.* (2010). Total oil content of *Arabidopsis* seeds was quantified using time-domain, pulse-field ^{1}H-NMR with a Bruker Minispec mq20 (Bruker optics), as described by Chapman *et al.* (2008), except calibrated for small volumes of *Arabidopsis* seeds.

Spatial distributions of TAG and SE in *Arabidopsis* leaves were determined *in situ* by matrix-assisted laser desorption/ionization mass spectrometry imaging (MALDI-MSI). *Arabidopsis* leaves were embedded in porcine gelatin and then frozen to −80 °C, before equilibration to −20 °C. Tissues were cut into 40-μm-thick cross sections on a cryostat (Leica CM1950) at −18 °C and lyophilized overnight before application of matrix (2,5-dihydrox-ybenzoic acid) by sublimation. A MALDI Orbitrap hybrid mass spectrometer (MALDI LTQ Orbitrap-XL; Thermo Fisher Scientific, Waltham, MA, USA) was used to scan the tissue sections at 40 micron steps, and the MS data were used to reconstruct false-colour MALDI-MSI images based on ion counts of selected lipid species (Horn and Chapman, 2014b). Lipid species were identified by exact mass comparisons with the LIPID MAPS database (http://www.lipidmaps.org/). Detailed procedures for data acquisition and analyses were described previously (Horn *et al.*, 2012).

Microscopy

Confocal images of *Arabidopsis* leaves and seeds and *N. ben-thamiana* leaves were obtained using a Zeiss LSM 710 confocal laser-scanning microscope (CLSM). Images of *N. tabacum* (BY-2) cells were obtained using either a Leica DM RBE microscope equipped with a Leica TCS SP2 scanning head or a Leica SP5 CLSM system equipped with a Radius 405-nm laser. *Arabidopsis* leaves and seeds, as well as tobacco leaves, were processed for CLSM imaging, including staining of LDs with the neutral lipid stains BODIPY 493/503 (Invitrogen) or Nile red (Sigma-Aldrich), as described previously (Cai *et al.*, 2015; Gidda *et al.*, 2016; Park *et al.*, 2013). The fifth leaf from the bottom of 24-day-old *Arabidopsis* plants was collected for analysis. *Arabidopsis* leaves used in comparison with young, mature and older mature leaves were the seventh, fifth and third true leaf from the bottom of the plant, respectively. Three fully expanded leaves from the top of the *N. benthamiana* plants (4-week-old) were used in the transient expression experiments. BY-2 cells were processed for CSLM as described previously (Gidda *et al.*, 2011, 2016; Lingard *et al.*, 2008). Briefly, cells ~4–6 h following biolistic bombard-ment were fixed in paraformaldehyde (Electron Microscopy Sciences), permeabilized and then incubated with the ER stain fluor-conjugated concanavalin A (ConA) (Sigma-Aldrich) and/or the neutral lipid stain MDH (Abgent). Cherry was excited with a 543-nm laser, and emission signals were collected from 590 to 640 nm. BODIPY 493/503, GFP and Nile red, as well as chlorophyll, were excited with a 488-nm laser, and emission signals were acquired from 500 to 540 nm (BODIPY and GFP), 560 to 620 nm (Nile red) and 640 to 720 nm (chlorophyll), respectively. CFP and MDH were excited with a 405-nm laser, and the emission signal was collected from 450 to 490 nm and 420 to 480 nm, respectively. All fluorophore emissions were collected sequentially in double- or triple-labelling experiments; single-labelling experiments showed no detectable crossover at the settings used for image acquisition. Depending on the plant material, images were acquired as a Z-series (leaves of *Arabidopsis* and *N. benthamiana*) or single optical sections (*Arabidopsis* seeds and BY-2 cells). All fluorescence images of cells shown are representative of at least two separate experiments, including at least three independently infiltrated leaves from three *N. benthamiana* plants or 20 transiently transformed BY-2 cells. The numbers and sizes of LDs in *Arabidopsis* and *N. benthamiana* leaves were quantified as BODIPY- or Nile red-stained lipid area according to Cai *et al.* (2015) using the Analyze Particles function at ImageJ (version 1.43). All the significance assessments in this study were performed using Student's *t*-test.

Acknowledgements

Research with FIT2 and LD formation in plants was supported initially by a grant from the U.S. Department of Energy (DOE), BER Division, DE-FG02-09ER64812, and completed with funds from the U.S. DOE, Office of Science, BES-Physical Biosciences program (DE-SC0016536). Support was also provided from the Natural Sciences and Engineering Research Council of Canada (NSERC), including an NSERC Undergraduate Student Research Award to Thuy Nguyen. Some of the experimental results reported here were initiated as part of a research-based undergraduate course, BIOL 3900, at UNT with the support of HHMI grant #52006955. Students deserving special recognition include Kevin Mutore, Charles McDaniel, Tucker Burke, Kathryn Haydon, Fancine Mascarenhas, Kyle Current and Reed Archer Killingsworth. Dr. Charlene Case assisted with manuscript preparation. Chris James assembled two of the binary vectors and generated some *Arabidopsis* transgenic lines for proof of concept. The authors declare no conflict of interest.

References

Andrianov, V., Borisjuk, N., Pogrebnyak, N., Brinker, A., Dixon, J., Spitsin, S., Flynn, J. *et al.* (2010) Tobacco as a production platform for biofuel: overexpression of Arabidopsis DGAT and LEC2 genes increases accumulation and shifts the composition of lipids in green biomass. *Plant Biotechnol. J.* **8**, 277–287.

Atanassov, I.I., Etchells, J.P. and Turner, S.R. (2009) A simple, flexible and efficient PCR-fusion/Gateway cloning procedure for gene fusion, site-directed mutagenesis, short sequence insertion and domain deletions and swaps. *Plant Meth.* **5**, 14.

Austin, J.R., Frost, E., Vidi, P.-A., Kessler, F. and Staehelin, L.A. (2006) Plastoglobules are lipoprotein subcompartments of the chloroplast that are permanently coupled to thylakoid membranes and contain biosynthetic enzymes. *Plant Cell*, **18**, 1693–1703.

Beaudoin, F. and Napier, J.A. (2002) Targeting and membrane-insertion of a sunflower oleosin in vitro and in Saccharomyces cerevisiae: the central hydrophobic domain contains more than one signal sequence, and directs oleosin insertion into the endoplasmic reticulum membrane using a signal anchor sequence mechanism. *Planta*, **215**, 293–303.

Beller, M., Bulankina, A.V., Hsiao, H.H., Urlaub, H., Jäckle, H. and Kühnlein, R.P. (2010) PERILIPIN-dependent control of lipid droplet structure and fat storage in *Drosophila*. *Cell Metab.* **12**, 521–532.

Benfey, P.N., Ren, L. and Chua, N.H. (1989) The CaMV 35S enhancer contains at least two domains which can confer different developmental and tissue-specific expression patterns. *EMBO J.* **8**, 2195–2202.

Brandizzi, F., Irons, S., Kearns, A. and Hawes, C. (2003) BY-2 cells: culture and transformation for live cell imaging. *Curr. Protoc. Cell Biol.*, Chapter **1**, Unit. 1.7.

Brasaemle, D.L. and Wolins, N.E. (2012) Packaging of fat: an evolving model of lipid droplet assembly and expansion. *J. Biol. Chem.* **287**, 2273–2279.

Cai, Y., Goodman, J.M., Pyc, M., Mullen, R.T., Dyer, J.M. and Chapman, K.D. (2015) Arabidopsis SEIPIN proteins modulate triacylglycerol accumulation and influence lipid droplet proliferation. *Plant Cell*, **27**, 2616–2636.

Chapman, K.D. and Ohlrogge, J.B. (2012) Compartmentation of triacylglycerol accumulation in plants. *J. Biol. Chem.* **287**, 2288–2294.

Chapman, K.D., Neogi, P.B., Hake, K.D., Stawska, A.A., Speed, T.R., Cotter, M.Q., Garrett, D.C. *et al.* (2008) Reduced oil accumulation in cottonseeds transformed with a Brassica non-functional allele of a Delta-12 Fatty Acid Desaturase (FAD2). *Crop Sci.* **48**, 1470–1481.

Chapman, K.D., Dyer, J.M. and Mullen, R.T. (2012) Biogenesis and functions of lipid droplets in plants. *J. Lipid Res.* **53**, 215–226.

Choudhary, V., Ojha, N., Golden, A. and Prinz, W.A. (2015) A conserved family of proteins facilitates nascent lipid droplet budding from the ER. *J. Cell Biol.* **211**, 261–271.

Clough, S.J. and Bent, A.F. (1998) Floral dip: a simplified method for Agrobacterium-mediated transformation of *Arabidopsis thaliana*. *Plant J.* **16**, 735–743.

Curtis, M.D. and Grossniklaus, U. (2003) A gateway cloning vector set for high-throughput functional analysis of genes in planta. *Plant Physiol.* **133**, 462–469.

Farese, R.V. and Walther, T.C. (2009) Lipid droplets finally get a little R-E-S-P-E-C-T. *Cell* **139**, 855–860.

Frandsen, G.I., Mundy, J. and Tzen, J.T. (2001) Oil bodies and their associated proteins, oleosin and caleosin. *Physiol. Plant.* **112**, 301–307.

Froissard, M., D'andréa, S., Boulard, C. and Chardot, T. (2009) Heterologous expression of AtClo1, a plant oil body protein, induces lipid accumulation in yeast. *FEMS Yeast Res.* **9**, 428–438.

Gao, Q. and Goodman, J.M. (2015) The lipid droplet-a well-connected organelle. *Front Cell Dev. Biol.* **3**, 49.

Gidda, S.K., Shockey, J.M., Falcone, M., Kim, P.K., Rothstein, S.J., Andrews, D.W., Dyer, J.M. *et al.* (2011) Hydrophobic-domain-dependent protein-protein interactions mediate the localization of GPAT enzymes to ER subdomains. *Traffic* **12**, 452–472.

Gidda, S.K., Park, S., Pyc, M., Yurchenko, O., Cai, Y., Wu, P., Andrews, D.W. *et al.* (2016) Lipid droplet-associated proteins (LDAPs) are required for the dynamic regulation of neutral lipid compartmentation in plant cells. *Plant Physiol.* **170**, 2052–2071.

Goh, V.J., Tan, J.S., Tan, B.C., Seow, C., Ong, W.Y., Lim, Y.C., Sun, L. *et al.* (2015) Postnatal deletion of fat storage-inducing transmembrane protein 2 (FIT2/FITM2) causes lethal enteropathy. *J. Biol. Chem.* **290**, 25686–25699.

Grimberg, Å., Carlsson, A.S., Marttila, S., Bhalerao, R. and Hofvander, P. (2015) Transcriptional transitions in *Nicotiana benthamiana* leaves upon induction of oil synthesis by WRINKLED1 homologs from diverse species and tissues. *BMC Plant Biol.* **15**, 192.

Gross, D.A., Snapp, E.L. and Silver, D.L. (2010) Structural insights into triglyceride storage mediated by fat storage-inducing transmembrane (FIT) protein 2. *PLoS ONE*, **5**, e10796.

Gross, D.A., Zhan, C. and Silver, D.L. (2011) Direct binding of triglyceride to fat storage-inducing transmembrane proteins 1 and 2 is important for lipid droplet formation. *Proc. Natl Acad. Sci. USA*, **108**, 19581–19586.

Hanton, S.L., Matheson, L.A., Chatre, L. and Brandizzi, F. (2007) Dynamic organization of COPII coat proteins at endoplasmic reticulum export sites in plant cells. *Plant J.* **57**, 963–974.

Horn, P.J. and Benning, C. (2016) The plant lipidome in human and environmental health. *Science* **353**, 1228–1232.

Horn, P.J. and Chapman, K.D. (2014a) Lipidomics in situ: insights into plant lipid metabolism from high resolution spatial maps of metabolites. *Prog. Lipid Res.* **54**, 32–52.

Horn, P.J. and Chapman, K.D. (2014b) Metabolite imager: customized spatial analysis of metabolite distributions in mass spectrometry imaging. *Metabolomics*, **10**, 337–348.

Horn, P.J., Korte, A.R., Neogi, P.B., Love, E., Fuchs, J., Strupat, K., Borisjuk, L. et al. (2012) Spatial mapping of lipids at cellular resolution in embryos of cotton. *Plant Cell*, **24**, 622–636.

Horn, P.J., James, C.N., Gidda, S.K., Kilaru, A., Dyer, J.M., Mullen, R.T., Ohlrogge, J.B. et al. (2013) Identification of a new class of lipid droplet-associated proteins in plants. *Plant Physiol.* **162**, 1926–1936.

Hsieh, K. and Huang, A.H. (2004) Endoplasmic reticulum, oleosins, and oils in seeds and tapetum cells. *Plant Physiol.* **136**, 3427–3434.

Huang, A. (1996) Oleosins and oil bodies in seeds and other organs. *Plant Physiol.* **110**, 1055–1061.

Hwang, Y.T., Pelitire, S.M., Henderson, M.P., Andrews, D.W., Dyer, J.M. and Mullen, R.T. (2004) Novel targeting signals mediate the sorting of different isoforms of the tail-anchored membrane protein cytochrome b5 to either endoplasmic reticulum or mitochondria. *Plant Cell*, **16**, 3002–3019.

Jacquier, N., Choudhary, V., Mari, M., Toulmay, A., Reggiori, F. and Schneiter, R. (2011) Lipid droplets are functionally connected to the endoplasmic reticulum in *Saccharomyces cerevisiae*. *J. Cell Sci.* **124**, 2424–2437.

Jacquier, N., Mishra, S., Choudhary, V. and Schneiter, R. (2013) Expression of oleosin and perilipins in yeast promotes formation of lipid droplets from the endoplasmic reticulum. *J. Cell Sci.* **126**, 5198–5209.

James, C.N., Horn, P.J., Case, C.R., Gidda, S.K., Zhang, D., Mullen, R.T., Dyer, J.M. et al. (2010) Disruption of the Arabidopsis CGI-58 homologue produces Chanarin-Dorfman-like lipid droplet accumulation in plants. *Proc. Natl Acad. Sci. USA*, **107**, 17833–17838.

Kadereit, B., Kumar, P., Wang, W.J., Miranda, D., Snapp, E.L., Severina, N., Torregroza, I. et al. (2008) Evolutionarily conserved gene family important for fat storage. *Proc. Natl Acad. Sci. USA*, **105**, 94–99.

Kim, E.Y., Park, K.Y., Seo, Y.S. and Kim, W.T. (2016) Arabidopsis small rubber particle protein homolog SRPs play dual roles as positive factors for tissue growth and development and in drought stress responses. *Plant Physiol.* **170**, 2494–2510.

Li, M., Baughman, E., Roth, M.R., Han, X., Welti, R. and Wang, X. (2014) Quantitative profiling and pattern analysis of triacylglycerol species in Arabidopsis seeds by electrospray ionization mass spectrometry. *Plant J.* **77**, 160–172.

Lingard, M.J., Gidda, S.K., Bingham, S., Rothstein, S.J., Mullen, R.T. and Trelease, R.N. (2008) Arabidopsis PEROXIN11c-e, FISSION1b, and DYNAMIN-RELATED PROTEIN3A cooperate in cell cycle-associated replication of peroxisomes. *Plant Cell* **20**, 1567–1585.

Miao, Y. and Jiang, L. (2007) Transient expression of fluorescent fusion proteins in protoplasts of suspension cultured cells. *Nat. Protoc.* **2**, 2348–2353.

Miranda, D.A., Kim, J.H., Nguyen, L.N., Cheng, W., Tan, B.C., Goh, V.J., Tan, J.S. et al. (2014) Fat storage-inducing transmembrane protein 2 is required for normal fat storage in adipose tissue. *J. Biol. Chem.* **289**, 9560–9572.

Mishra, S. and Schneiter, R. (2015) Expression of perilipin 5 promotes lipid droplet formation in yeast. *Comm. Integr. Biol.* **8**, e1071728.

Mishra, S., Khaddaj, R., Cottier, S., Stradalova, V., Jacob, C. and Schneiter, R. (2016) Mature lipid droplets are accessible to ER luminal proteins. *J. Cell Sci.* **129**, 3803–3815.

Mu, J., Tan, H., Zheng, Q., Fu, F., Liang, Y., Zhang, J., Yang, X. et al. (2008) LEAFY COTYLEDON1 is a key regulator of fatty acid biosynthesis in Arabidopsis. *Plant Physiol.* **148**, 1042–1054.

Murashige, T. and Skoog, F. (1962) A revised medium for rapid growth and bioassays with tobacco tissue cultures. *Plant Physiol.* **15**, 473–497.

Murphy, D.J. (2001) The biogenesis and functions of lipid bodies in animals, plants and microorganisms. *Prog. Lipid Res.* **40**, 325–438.

Murphy, D.J. (2012) The dynamic roles of intracellular lipid droplets: from archaea to mammals. *Protoplasma* **249**, 541–585.

Nacir, H. and Bréhélin, C. (2013) When proteomics reveals unsuspected roles: the plastoglobule example. *Front Plant Sci.* **4**, 114.

Ohlrogge, J. and Chapman, K. (2011) The seeds of green energy: expanding the contribution of plant oils as biofuels. *Biochemist* **33**, 34–38.

Park, S., Gidda, S.K., James, C.N., Horn, P.J., Khuu, N., Seay, D.C., Keereetaweep, J. et al. (2013) The α/β Hydrolase CGI-58 and Peroxisomal Transport Protein PXA1 Coregulate Lipid Homeostasis and Signaling in Arabidopsis. *Plant Cell*, **25**, 1726–1739.

Petrie, J.R., Shrestha, P., Liu, Q., Mansour, M.P., Wood, C.C., Zhou, X.R., Nichols, P.D. et al. (2010) Rapid expression of transgenes driven by seed-specific constructs in leaf tissue: DHA production. *Plant Meth.* **6**, 8.

Pol, A., Gross, S.P. and Parton, R.G. (2014) Biogenesis of the multifunctional lipid droplet: Lipids, proteins, and sites. *J. Cell Biol.* **204**, 635–646.

Robenek, H., Hofnagel, O., Buers, I., Robenek, M.J., Troyer, D. and Severs, N.J. (2006) Adipophilin-enriched domains in the ER membrane are sites of lipid droplet biogenesis. *J. Cell Sci.* **119**, 4215–4224.

Rowe, E.R., Mimmack, M.L., Barbosa, A.D., Haider, A., Isaac, I., Ouberai, M.M., Thiam, A.R. et al. (2016) Conserved amphipathic helices mediate lipid droplet targeting of perilipins 1-3. *J. Biol. Chem.* **291**, 6664–6678.

Sanjaya, Miller, R., Durrett, T.P., Kosma, D.K., Lydic, T.A., Muthan, B., Koo, A.J. et al. (2013) Altered lipid composition and enhanced nutritional value of Arabidopsis leaves following introduction of an algal diacylglycerol acyltransferase 2. *Plant Cell*, **25**, 677–693.

Santos Mendoza, M., Dubreucq, B., Miquel, M., Caboche, M. and Lepiniec, L. (2005) LEAFY COTYLEDON 2 activation is sufficient to trigger the accumulation of oil and seed specific mRNAs in Arabidopsis leaves. *FEBS Lett.* **579**, 4666–4670.

Shockey, J.M., Gidda, S.K., Chapital, D.C., Kuan, J.C., Dhanoa, P.K., Bland, J.M., Rothstein, S.J. et al. (2006) Tung tree DGAT1 and DGAT2 have nonredundant functions in triacylglycerol biosynthesis and are localized to different subdomains of the endoplasmic reticulum. *Plant Cell*, **18**, 2294–2313.

Slocombe, S.P., Cornah, J., Pinfield-Wells, H., Soady, K., Zhang, Q., Gilday, A., Dyer, J.M. et al. (2009) Oil accumulation in leaves directed by modification of fatty acid breakdown and lipid synthesis pathways. *Plant Biotechnol. J.* **7**, 694–703.

Snapp, E.L., Hegde, R.S., Francolini, M., Lombardo, F., Colombo, S., Pedrazzini, E., Borgese, N. et al. (2003) Formation of stacked ER cisternae by low affinity protein interactions. *J. Cell Biol.* **163**, 257–269.

Sreenivasulu, N. and Wobus, U. (2013) Seed-development programs: a systems biology-based comparison between dicots and monocots. *Annu. Rev. Plant Biol.* **64**, 189–217.

Sturtevant, D., Lee, Y.J. and Chapman, K.D. (2016) Matrix assisted laser desorption/ionization-mass spectrometry imaging (MALDI-MSI) for direct visualization of plant metabolites in situ. *Curr. Opin. Biotechnol.* **37**, 53–60.

Szymanski, K.M., Binns, D., Bartz, R., Grishin, N.V., Li, W.P., Agarwal, A.K., Garg, A. et al. (2007) The lipodystrophy protein seipin is found at endoplasmic reticulum lipid droplet junctions and is important for droplet morphology. *Proc. Natl Acad. Sci. USA*, **104**, 20890–20895.

Vanhercke, T., El Tahchy, A., Shrestha, P., Zhou, X.R., Singh, S.P. and Petrie, J.R. (2013) Synergistic effect of WRI1 and DGAT1 coexpression on triacylglycerol biosynthesis in plants. *FEBS Lett.* **587**, 364–369.

Vanhercke, T., El Tahchy, A., Liu, Q., Zhou, X.R., Shrestha, P., Divi, U.K., Ral, J.P. et al. (2014) Metabolic engineering of biomass for high energy density: oilseed-like triacylglycerol yields from plant leaves. *Plant Biotechnol. J.* **12**, 231–239.

Walther, T.C. and Farese, R.V. (2009) The life of lipid droplets. *Biochim. Biophys. Acta* **1791**, 459–466.

Wang, Y., Ren, Y., Liu, X., Jiang, L., Chen, L., Han, X., Jin, M. et al. (2010) OsRab5a regulates endomembrane organization and storage protein trafficking in rice endosperm cells. *Plant J.* **64**, 812–824.

Wang, H., Becuwe, M., Housden, B.E., Chitraju, C., Porras, A.J., Graham, M.M., Liu, X.N. et al. (2016) Seipin is required for converting nascent to mature lipid droplets. *Elife*, **5**, e16582. doi: 10.7554/eLife.16582.

Wanner, G., Formanek, H. and Theimer, R.R. (1981) The ontogeny of lipid bodies (spherosomes) in plant cells: Ultrastructural evidence. *Planta*, **151**, 109–123.

Welti, R., Li, W., Li, M., Sang, Y., Biesiada, H., Zhou, H.E., Rajashekar, C.B. et al. (2002) Profiling membrane lipids in plant stress responses. Role of phospholipase D alpha in freezing-induced lipid changes in Arabidopsis. *J. Biol. Chem.* **277**, 31994–32002.

Wilfling, F., Haas, J.T., Walther, T.C. and Farese, R.V. (2014) Lipid droplet biogenesis. *Curr. Opin. Cell Biol.* **29**, 39–45.

Winichayakul, S., Scott, R.W., Roldan, M., Hatier, J.H., Livingston, S., Cookson, R., Curran, A.C. *et al.* (2013) In vivo packaging of triacylglycerols enhances Arabidopsis leaf biomass and energy density. *Plant Physiol.* **162**, 626–639.

Xu, C. and Shanklin, J. (2016) Triacylglycerol metabolism, function, and accumulation in plant vegetative tissues. *Annu. Rev. Plant Biol.* **67**, 179–206.

Zhou, Z., Marepally, S.R., Nune, D.S., Pallakollu, P., Ragan, G., Roth, M.R., Wang, L. *et al.* (2011) LipidomeDB data calculation environment: online processing of direct-infusion mass spectral data for lipid profiles. *Lipids* **46**, 879–884.

PbrMYB21, a novel MYB protein of *Pyrus betulaefolia*, functions in drought tolerance and modulates polyamine levels by regulating arginine decarboxylase gene

Kongqing Li[1†], Caihua Xing[2†], Zhenghong Yao[2] and Xiaosan Huang[2]*

[1]Department of Rural Development, Nanjing Agricultural University, Nanjing, China
[2]College of Horticulture, State Key Laboratory of Crop Genetics and Germplasm Enhancement, Nanjing Agricultural University, Nanjing, China

*Correspondence
email huangxs@njau.edu.cn
†These authors have contributed equally to this work.

Summary

MYB comprises a large family of transcription factors that play significant roles in plant development and stress response in plants. However, knowledge concerning the functions of MYBs and the target genes remains poorly understood. Here, we report the identification and functional characterization of a novel stress-responsive MYB gene from *Pyrus betulaefolia*. The MYB gene, designated as *PbrMYB21*, belongs to the R2R3-type and shares high degree of sequence similarity to *MdMYB21*. The transcript levels of *PbrMYB21* were up-regulated under various abiotic stresses, particularly dehydration. PbrMYB21 was localized in the nucleus with transactivation activity. Overexpression of *PbrMYB21* in tobacco conferred enhanced tolerance to dehydration and drought stresses, whereas down-regulation of *PbrMYB21* in *Pyrus betulaefolia* by virus-induced gene silencing (VIGS) resulted in elevated drought sensitivity. Transgenic tobacco exhibited higher expression levels of *ADC* (arginine decarboxylase) and accumulated larger amount of polyamine in comparison with wild type (WT). VIGS of *PbrMYB21* in *Pyrus betulaefolia* down-regulated ADC abundance and decreased polyamine level, accompanied by compromised drought tolerance. The promoter region of *PbrADC* contains one MYB-recognizing cis-element, which was shown to be interacted with PbrMYB21, indicating the ADC may be a target gene of *PbrMYB21*. Take together, these results demonstrated that *PbrMYB21* plays a positive role in drought tolerance, which may be, at least in part, due to the modulation of polyamine synthesis by regulating the *ADC* expression.

Keywords: abiotic stress, polyamine biosynthesis, MYB, *Pyrus betulaefolia*, transcriptional regulation.

Introduction

Drought is one of the major abiotic stresses affecting plant growth, development, productivity and geographic distribution (Farooq et al., 2009). To cope with this stress, plants have evolved sophisticated mechanisms to adapt to drought stress, ranging from perception of stress signals to modifications of physiological and biochemical responses (Ingram and Bartels, 1996; Pastori and Foyer, 2002). In these regulating processes, plants undergo tremendous molecular changes by reprogramming an array of stress-responsive genes (Hirayama and Shinozaki, 2010; Liu et al., 2014; Osakabe et al., 2014; Seki et al., 2001). These genes can be generally classified into two main groups based on their products, effector molecules and regulator molecules. Among various regulatory genes, transcription factors (TFs) act as significant coordinators to transduce stress signals and to orchestrate expression of functional genes that play direct role in preventing plants from stress-associated damages (Ren et al., 2010). Therefore, genetic engineering of stress-associated TFs has been proposed to be a robust strategy for improving the stress tolerance of crop plants (Sreenivasulu et al., 2007; Thomashow et al., 2001; Yu et al., 2012; Zhao et al., 2016).

Plants modulate a battery of functional genes involved in the synthesis of various metabolites that resist the abiotic stresses. The most common polyamines (PAs), primarily diamine putrescine (Put), triamine spermidine (Spd) and tetramine spermine (Spm),

are low molecular weight aliphatic polycations that are ubiquitously present in living organisms, which are a group of aliphatic polycations that are ubiquitously distributed in higher plant. Accumulation of PAs serves as a metabolic hallmark under drought stresses in a large number of plants (Groppa and Benavides, 2008; Yang et al., 2007). It has also been suggested that polyamines act as osmotic regulators and scavengers of ROS (Drolet et al., 1986; Evans and Malmberg, 1989). Thus, it is conceivable that higher accumulation of polyamines through stimulated de novo biosynthesis can be conducive to ameliorating stress-induced damage. This implies that ADC pathway plays a key role in stress tolerance, as revealed by following experimental data. For example, previous studies demonstrated that overexpression of the *ADC2* gene led to enhance drought tolerance in *Arabidopsis thaliana* (Alcazar et al., 2010). In an earlier work, elevation of *ADC* transcript and endogenous polyamine levels by overexpression of polyamine biosynthetic genes has been demonstrated to enhance abiotic stresses tolerance in various plants (Huang et al., 2015a,b; Shi et al., 2010; Wang et al., 2011). In another study, PtADC was reported to be a potential target of PtrABF, which can specifically recognize the abscisic acid (ABA) response element within the promoter of PtADC (Zhang et al., 2015). WRKY70 was demonstrated to interact with W-box elements within the promoter of a Fortunella crassifolia ADC gene (Gong et al., 2015). Recently, a stress-responsive trifoliate orange NAC family TF, PtrNAC72, was reported to be a repressor of putrescine biosynthesis and may negatively regulate the drought stress response, via the modulation of putrescine-associated reactive oxygen species homoeostasis (Wu et al., 2016). All of these results suggest that the *ADC* gene is a promising candidate for enhancing stress tolerance and may be an important gene for molecular breeding of stress-tolerant plants.

The MYB transcription factors comprise a large gene family in plants. Proteins of this family contain a highly conserved MYB domain at the N-terminus, comprising one to four imperfect repeats (Lipsick, 1996). Each repeat contains 50–53 amino acid residues, which form each domain into three α-helices. Analysis of the three-dimensional structure of the MYB domain has shown that the second and third helices of each repeat have three regularly spaced tryptophan (W) residues forming a helix–turn–helix (HTH) structure with a hydrophobic core, while the third helix directly interacts with the major groove of the target DNA. Based on the number of imperfect adjacent repeats (R1, R2, R3) in the MYB domain, the MYB proteins are primarily classified into four major subgroups, R1R2R3-type MYB (MYB3R, three repeats), R2R3-type MYB (two repeats), MYB1R and 4R-like MYB protein (4R-MYB). MYB proteins appear to be widespread in plants, but no homologue has been identified thus far in other eukaryotes. The majority of plant MYB genes belong to the R2R3 types, although more than 100 members of this family have been suggested in both rice and *Arabidopsis* genomes (Chen et al., 2006). The R2R3-type MYB domain is the minimum DNA-binding domain, and an activation/repression domain, the balance between activators and repressors in this TFs family, may provide extra flexibility in terms of transcriptional regulation, while R1 appears to lose interaction with DNA (Ogata et al., 1994, 1995). However, C-terminus regions of MYB genes are quite variable, allowing a wide range of regulation roles in different biological processes (Kranz et al., 1998).

Previous studies have provided extensive data revealing that the plant MYBs play central roles in an array of physiological and biological processes, such as organ growth and development (Mu et al., 2009; Volpe et al., 2013), controlling stomatal aperture (Liang et al., 2005), primary and secondary metabolism (Czemmel et al., 2009; Mellway et al., 2009; Wang et al., 2010), cold and hormonal signalling (Lee and Seo, 2015; Seo et al., 2009; Shim et al., 2013), phenylpropanoid biosynthesis (Borevitz et al., 2000), anthocyanin biosynthesis (Matsui et al., 2008; Park et al., 2008), signal transduction (Borevitz et al., 2000) and cell division (Muller et al., 2006; Ryu et al., 2005). In addition, many evidences indicate that the MYB genes are closely associated with plant responses to abiotic stresses (Agarwal et al., 2006; Dubos et al., 2010; Kim et al., 2013; Lee and Seo, 2015). So far, well-characterized MYBs involved in abiotic stress response include *AtMYB1*, *AtMYB2*, *AtMYB15*, *AtMYB41*, *AtMYB44*, *AtMYB60*, *AtMYB74* and *MYB96* of *Arabidopsis* (Baek et al., 2013; Jung et al., 2008; Lee and Seo, 2015; Lippold et al., 2009; Oh et al., 2011; Seo et al., 2009, 2011; Wang et al., 2015; Xu et al., 2015); *OsMYB2*, *OsMYB3R-2*, *OsMYB511* and *MYBS3* of rice (Huang et al., 2015a,b; Ma et al., 2009; Su et al., 2010; Yang et al., 2012); *GmMYB76*, *GmMYB92* and *GmMYB100GmMYB177* of soya bean (Liao et al., 2008; Yan et al., 2015).

It is conceivable that the MYBs may achieve their function in stress tolerance via regulating a variety of stress-responsive genes, either regulatory or functional ones, which are considered as the potential target genes of the MYBs. Although a myriad of stress-responsive genes have been tested so far, knowledge concerning the target genes of MYBs is still poorly understood. So, it is necessary to investigate how a MYB gene can affect the expression of genes involved in other metabolic pathways which have been less examined. In this study, we report the functional identification of a R2R3-type MYB gene from *Pyrus betulaefolia*, designated as *PbrMYB21* in drought tolerance, belongs to the R2R3-type and shares high degree of sequence similarity to *MdMYB21*. In addition, a connection between plant MYBs and ADC-mediated polyamine accumulation has never been established although they have been separately verified as key players in abiotic stress tolerance. These results showed that *PbrMYB21* acts as a positive regulator of drought tolerance, which may be partly ascribed to its role in modulating polyamine levels through regulating arginine decarboxylase, suggesting that *ADC* might be a potential target gene of *PbrMYB21*.

Results

Isolation and analysis of *PbrMYB21*

We previously obtained a dehydration-induced MYB TF (Pbr028812.1) from a transcriptome of *Pyrus betulaefolia* (Li et al., 2016). Sequence analysis demonstrated that it was a full-length sequence with a complete open reading frame (ORF), which encodes a protein of 296 amino acid residues with a calculated molecular mass of 33.2 kDa and an isoelectric point of 5.98. Homology search showed that the gene shared the highest identity with *MdMYB21*, so it was designated as *PbrMYB21* (*Pyrus betulaefolia* MYB21). Multiple sequence alignment showed that PbrMYB21 protein had a conserved R2R3 region in its N-terminal region, but the C-terminal regions are divergent. The R2R3 domain contained highly conserved tryptophan residues, such as three tryptophans in the R2 repeat and two tryptophans in the R3 repeat (Figure 1). In addition, a phylogenetic tree was constructed using amino acid of PbrMYB21 and MYBs of other plants, such as *Malus domestica*, *Pyrus communis*, *Citrus sinensis* and *Glycine max*. In the phylogenetic tree, PbrMYB21 was most closely related to MdMYB21 as they shared 82% sequence identity (Figure S1).

Expression patterns of *PbrMYB21* under various treatments

Real-time quantitative PCR (qPCR) was used to examine the expression profiles of *PbrMYB21* under various treatments, including dehydration, salt and cold. *PbrMYB21* mRNA was quickly and sharply accumulated within 0.5 h of dehydration, and maintained stable at 1 h, but decreased thereafter (Figure 2a). Salt treatment for 1 h up-regulated *PbrMYB21*, which rose quickly to the peak value at 12 h, and then decreased within the 24 h (but still 3.4-fold higher than the initial level) (Figure 2b). When subjected to cold treatment, the transcript of *PbrMYB21* did not change notably except a slight increase at 48 h (Figure 2c), indicating that *PbrMYB21* was not cold-inducible.

PbrMYB21 was localized to nucleus

To find out the subcellular localization of PbrMYB21, full-length ORF of *PbrMYB21* was fused to N-terminal of GFP reporter protein driven by CaMV 35S promoter, generating a fusion protein PbrMYB21: GFP, using the plasmid containing GFP alone as control. The fusion protein (PbrMYB21) and the control (GFP) were analysed in tobacco leaf epidermis via *Agrobacterium-*

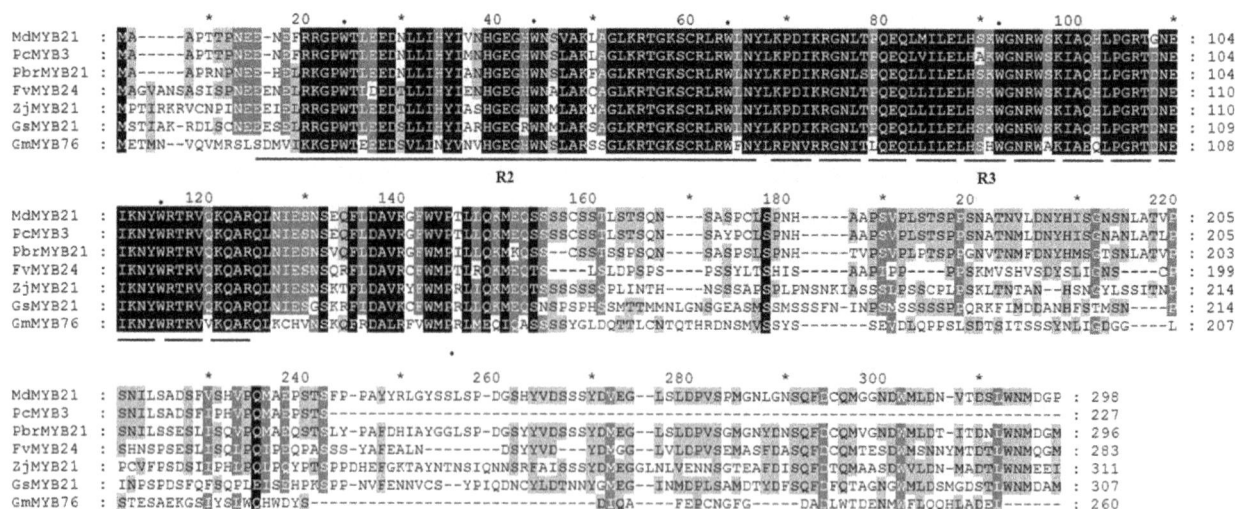

Figure 1 Multiple sequence alignment of PbrMYB21 and MYBs from other plant species, including *Malus domestica* (MdMYB21, NP_001280981), *Pyrus communis* (PcMYB3, AGL81356), *Fragaria vesca* subsp. *vesca* (FVMYB24, XP_004299054), *Ziziphus jujuba* (ZjMYB21, XP_015895640), *Glycine soja* (GsMYB21, KHN11407), *Glycine max* (GmMYB76, NP_001235761). The R2 and R3 domains are indicated by solid line and dotted line, respectively. Solid circles indicate the conserved residues, such as tryptophan (W).

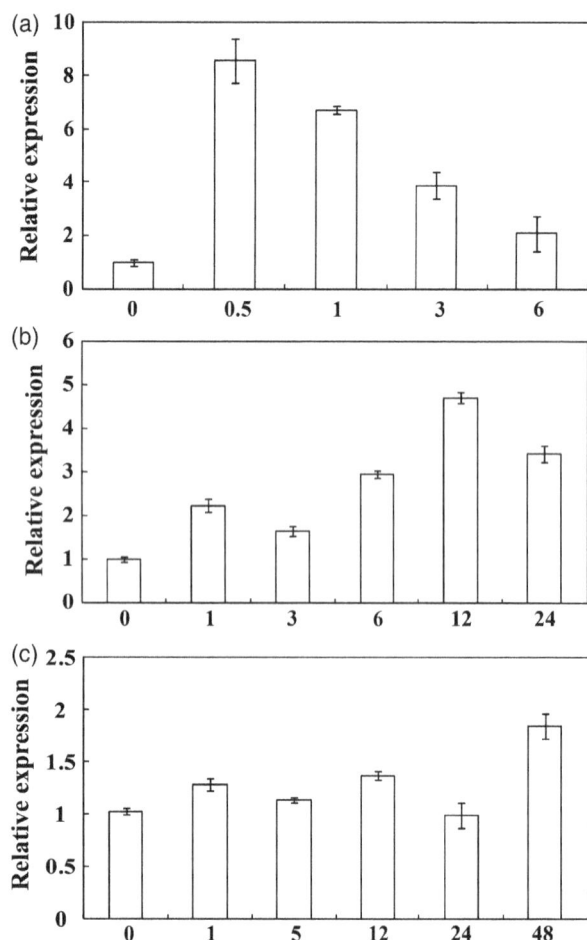

Figure 2 Time-course expression levels of *PbrMYB21* in *Pyrus betulaefolia* under abiotic stresses. (a–c) expression of *PbrMYB21* in response to dehydration (a), salt (b) and cold stress (c). The samples were collected at the designated time points and analysed by qPCR. Error bars stand for SD based on four replicates.

mediated transformation. Microscopic visualization showed that the control GFP was uniformly distributed throughout the whole cell (Figure 3a), whereas the PbrMYB21-GFP fusion protein was observed exclusively in the nucleus (Figure 3b). These results indicated that PbrMYB21 was a nuclear protein.

PbrMYB21 had transactivation activity

Transactivation activity is another defining feature for a transcription factor in addition to nuclear localization. The Y2H system was used to investigate whether PbrMYB21 functioned as a transcriptional activator. For this purpose, the full-length PbrMYB21 coding region was fused to the DNA-binding domain of GAL4 to generate a fusion plasmid, which was transformed into yeast AH109 and growth of the cells was compared with those transformed with the control plasmid (pGBKT7) on the same selection medium synthetic dropout (SD)/-Leu/-Trp. The yeast cells transformed with either vector could grow on the synthetic dropout medium (SD/-Leu/-Trp), indicating that the analysis system was reliable. On the SD/-Leu/-Trp/-His medium, the cells transformed with the control plasmid could not grow, whereas only the cells transformed with the recombinant vector survived on the selection medium alone or supplemented with 15 mM 3-amino-1,2,4-triazole (3-AT) (Figure 4, upper panels). In addition, when the yeast cells were cultured on the SD/-Leu/-Trp/-His medium added with both 15 mM 3-AT and 20 mM X-a-Gal, only those transformed with the fusion plasmid turned blue (Figure 4, bottom panels). Taken together, these results demonstrate PbrMYB21 had transactivation activity in yeast.

Overexpression of *PbrMYB21* confers enhanced dehydration and drought tolerance

Given that dehydration stress resulted in the strongest induction of *PbrMYB21* transcript, we speculated that *PbrMYB21* may play an important role in drought stress response. To test whether this hypothesis is true, transgenic tobacco overexpressing *PbrMYB21* were generated by *Agrobacterium*-mediated transformation,

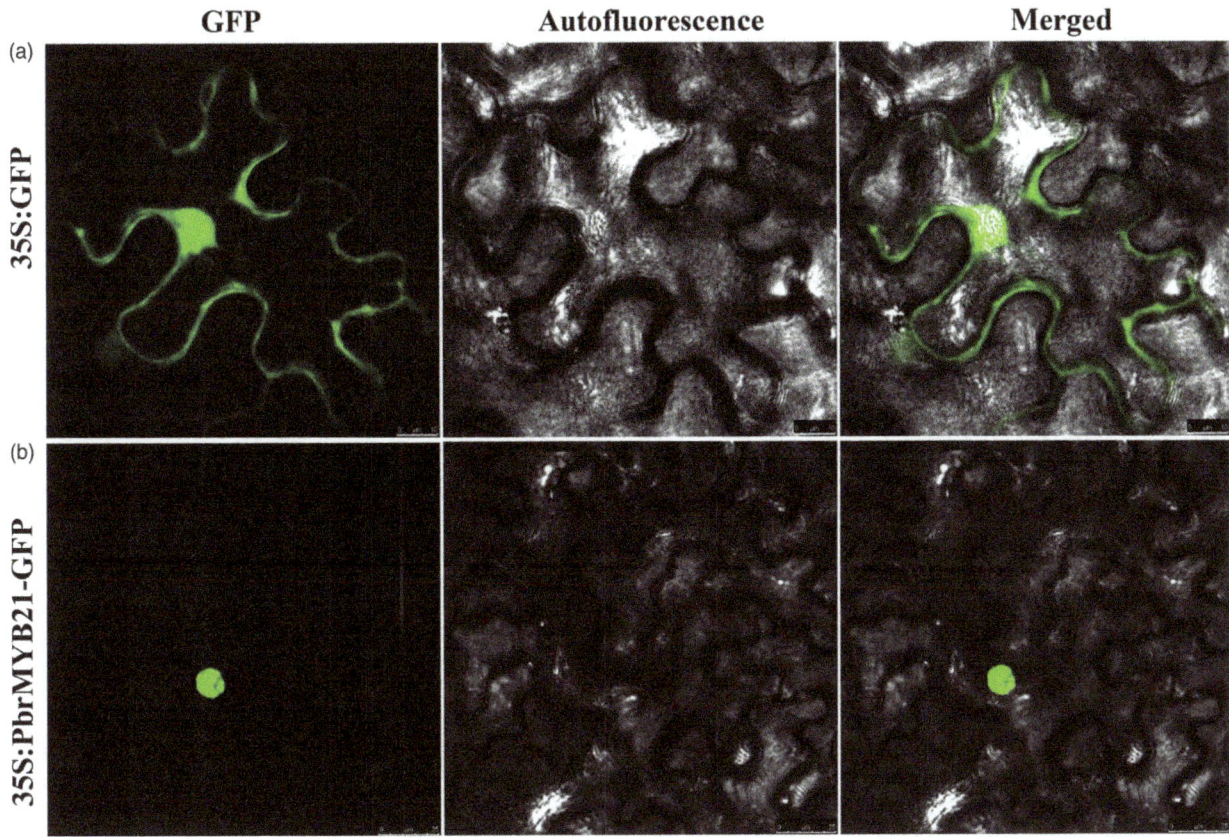

Figure 3 Subcellular localization of PbrMYB21. Tobacco epidermal cells were transiently transformed with constructs containing either control (GFP alone, (a) or fusion plasmid (PbrMYB21: GFP, (b)). Images under blight field (middle), fluorescence (left) and the merged images (right) are shown on the right.

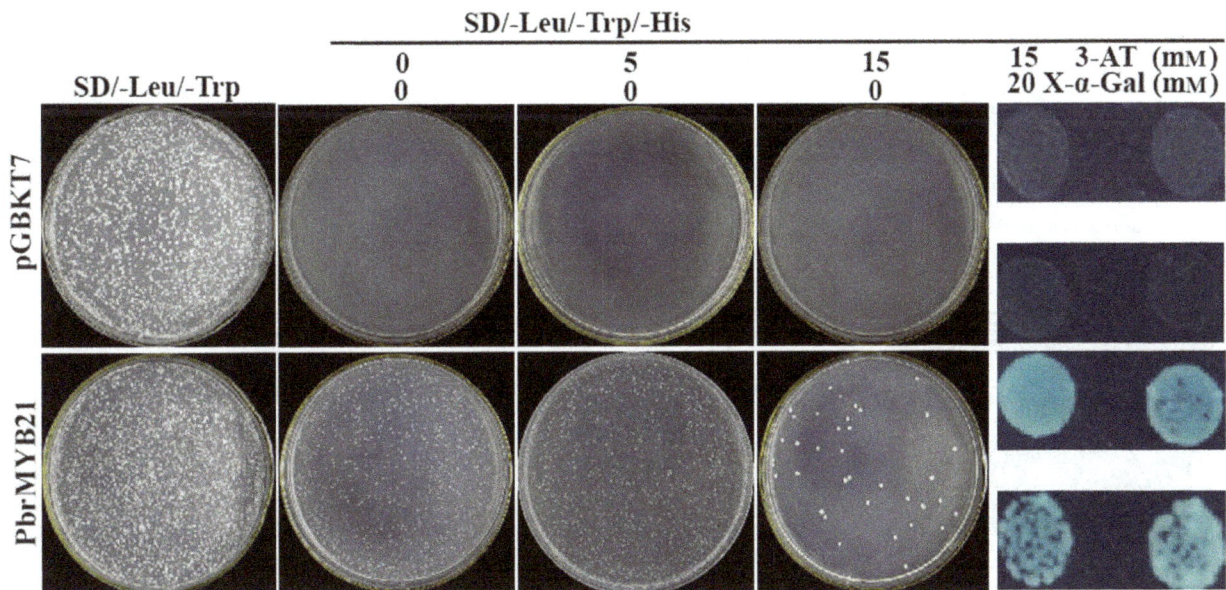

Figure 4 Transcriptional activation assay of PbrMYB21 in yeast. Growth of yeast cells (strain AH109) transformed with either control vector (upper panels) or fusion vector harbouring PbrMYB21 (bottom panels) on SD/–Leu/–Trp or SD/–Leu/–Trp/–His with or without 3-AT and 20 mM X-a-gal.

under the control of CAMV 35S promoter. Totally, twelve transformants (T_0 generation) were identified as positive lines by genomic PCR analysis. Semi-quantitative RT-PCR analysis showed that *PbrMYB21* (GSP1, Table S1) was overexpressed in six tested lines (Figure S2), from which three independent overexpression lines of tobacco (designated as OE-20, OE-17 and OE-9 hereafter) with high transcript levels of *PbrMYB21* were selected for following experiments, along with the corresponding untransformed wild type. These six overexpression lines of tobacco were further used for Western blotting assays with an anti-His antibody (Figure S2G), consistent with the results of the semi-quantitative RT-PCR. Southern blotting was carried out to examine *PbrMYB21* gene copy number in the transgenic tobacco genome. As shown in Figure S3, hybridization of three transgenic lines (OE-20, OE-17 and OE-9) genomic DNA digested with one restriction enzyme *Kpn*I using a 200-bp partial fragment of *PbrMYB21* as probe produced only one visible signal under high stringency conditions. As the probe sequence did not contain any recognition sites of the *Kpn*I, the hybridization pattern suggested that the *PbrMYB21* gene exists as a single-copy gene in the three transgenic lines genome.

When the aerial parts of 5-week-old *in vitro* seedlings were dehydrated in an ambient, a steady decrease in FW (fresh weight) was observed in both the WT and the three transgenic lines. However, at any time point within 70 min of dehydration, the three transgenic lines lost remarkably less water compared with the WT (Figure 5a). When dehydration was completed, leaves of WT exhibited more serious wilting relative to the transgenic lines (Figure 5b, c). EL (Electrolyte leakage), an important indicator of membrane injury, was significantly higher in the WT (65.2%) than in OE-20 (29.7%), OE-17 (26.9%) or OE-9 (30.5%), indicating that the WT suffered from severe membrane damage (Figure 5d). At the end of dehydration, stomatal apertures of WT were significantly larger than those of the transgenic lines (Figure 5e, f), consistent with their water loss dynamics. The leaves of 7-week-old *in vitro* seedlings were also subjected to dehydration for 70 min, followed by EL measurement and MDA contents. After 70 min of dehydration, more severe leaf withering was observed

in the WT plants when compared with the transgenic plants (Figure 5g). In agreement with the enhanced dehydration tolerance, the transgenic lines had lower values of EL and MDA following exposure to dehydration conditions than did the wild type (Figure 5h, i). These findings demonstrate that the transgenic lines were more resistant to dehydration.

Drought tolerance of the transgenic lines was also assessed using 45-day-old plants grown in soil pots (Figure 6a). Morphological difference became apparent after watering was stopped for 7 days, leaf wilting was more evident in the WT relative to the three transgenic lines (Figure 6b). At the end of drought, stomatal apertures of WT were significantly larger than those of the transgenic lines (Figure 6c, d). To compare the physiological differences, electrolyte leakage of OE-20 (30.9%), OE-17 (31.3%) and OE-9 (42.1%) was significantly lower in comparison with 89.5% of WT (Figure 6c). These results suggest that overexpression of *PbrMYB21* in tobacco conferred pronounced enhancement of drought stress tolerance.

Silencing of *PbrMYB21* in *Pyrus betulaefolia* confers sensitivity to drought stress

To further elucidate the role of *PbrMYB21* in drought tolerance, a virus-induced gene silencing (VIGS) was used to suppress the expression of *PbrMYB21* in *Pyrus betulaefolia*. Transcript analysis of the leaflets revealed that the transcripts for *PbrMYB21* were suppressed in the respective silenced plants (Figure S2H). We next compared the drought tolerance of these three silenced plants grown under drought treatment. Upon exposure to drought, pTRV-PbrMYB21 plants (pTRV-1, pTRV-2 and pTRV-3) displayed more serious wilting (Figure 7a, b) and compared with the empty

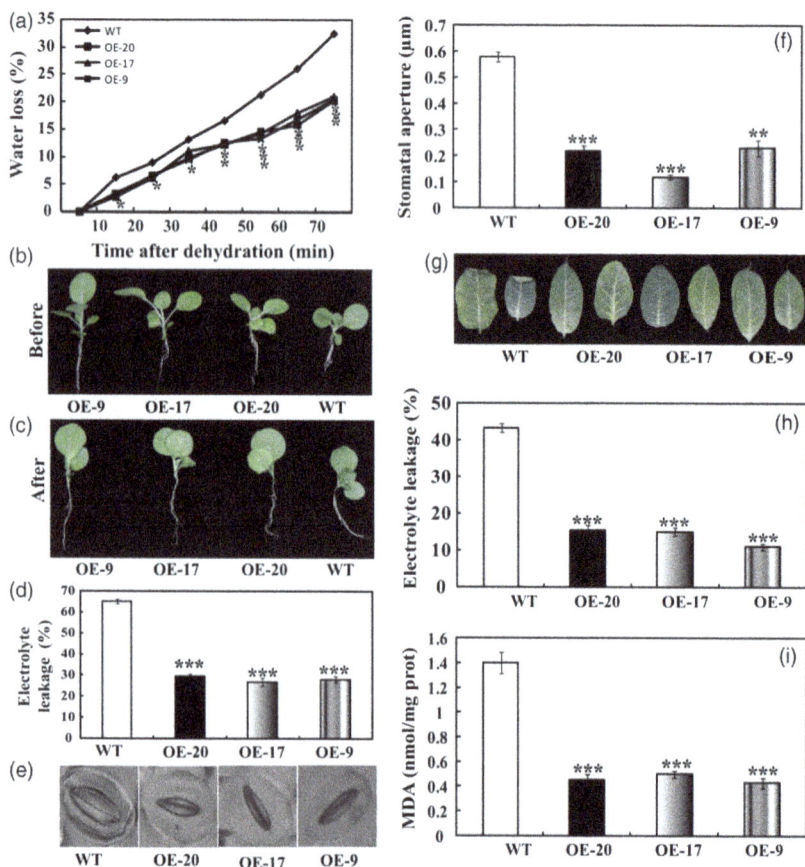

Figure 5 Overexpression of *PbrMYB21* conferred enhanced dehydration and drought tolerance in tobacco. (a) Time-course fresh water loss of WT, OE-20, OE-17 and OE-9 during 70-min dehydration. (b, c) Representative images of WT, OE-20, OE-17 and OE-9 after dehydration for 70 min. (d, e) Electrolyte leakage (d) and stomatal aperture size (e) of WT, OE-20, OE-17 and OE-9 after dehydration for 70 min. (g) Phenotypes of 7-week-old plants of transgenic lines (OE-20, OE-17 and OE-9) and WT after dehydration for 70 min. (h, i) Electrolyte leakage (h) and MDA contents (i) in the three lines measured after dehydration for 70 min. Asterisks indicate that the value is significantly different from that of the WT at the same time point (*$P < 0.05$; **$P < 0.01$; ***$P < 0.001$).

Figure 6 Overexpression of *PbrMYB21* conferred enhanced drought tolerance in tobacco. (a–b) Phenotypes of 7-week-old transgenic plants and WT before (a) and after (b) 7 days drought stress. (c–d) Representative images (c) and stomatal aperture size (d) of WT, OE-20, OE-17 and OE-9 after drought. (e) Electrolyte leakage of the WT, OE-20, OE-17 and OE-9 after drought treatment. Asterisks indicate that the value is significantly different from that of the WT at the same time point (**P < 0.01; ***P < 0.001).

vector (pTRV) transformed control plants (WT). At the end of drought, electrolyte leakage of the pTRV-PbrMYB21 plants was significantly higher than the control plants (Figure 7c). Water deprivation for 7 days, WT plants had significantly higher RWC (relative water content) relative to pTRV-PbrMYB21 plants (Figure 7d). The control plants displayed significantly lower malondialdehyde (MDA) contents compared with the pTRV-PbrMYB21 plants (Figure 7e).

Transgenic lines accumulate less ROS and the silencing lines accumulate more ROS

Histochemical staining with DAB (diaminobenzidine) and NBT (nitroblue tetrazolium) was carried out to reveal the levels of H_2O_2 and O_2^-, respectively, in the leaves subjected to dehydration and drought treatment. After the dehydration for 70 min, DAB staining showed that the intensity of brown precipitations in the WT leaves was stronger than that of the transgenic plants. Blue spots were distributed throughout the leaves of WT when NBT staining was applied, whereas leaves of the transgenic plants were only slightly stained (Figure 8a). When 45-day-old potted plants were subjected to drought treatment for 7 days, we also checked the ROS accumulation in the leaf discs of WT and the transgenic plants. WT leaf discs were stained to a deeper extent by DAB and NBT when compared with the transgenic counterparts (Figure 8b). Conversely, stronger DAB and NBT staining were detected in the pTRV-PbrMYB21 plants as compared with the control plants, suggesting that the accumulation of H_2O_2 and O_2^- was elevated when PbrMYB21 was down-regulated (Figure 8c). To verify the histochemical staining, we quantified the levels of H_2O_2 and O_2^- in the drought-treated samples using a detection kit. Consistent with the histochemical staining results, quantitative measurement showed that the levels of H_2O_2 and O_2^- in the transgenic lines were significantly lower than in WT during the drought treatment (Figure 8d, e), but the H_2O_2 and O_2^- contents in the pTRV-PbrMYB21 of *Pyrus betulaefolia* were significantly higher than in the WT (Figure 7f, g). Both histochemical staining and quantitative measurement indicated that the transgenic lines accumulated lower levels of ROS under the drought, whereas they were changed in an opposite way when *PbrMYB21* was silenced.

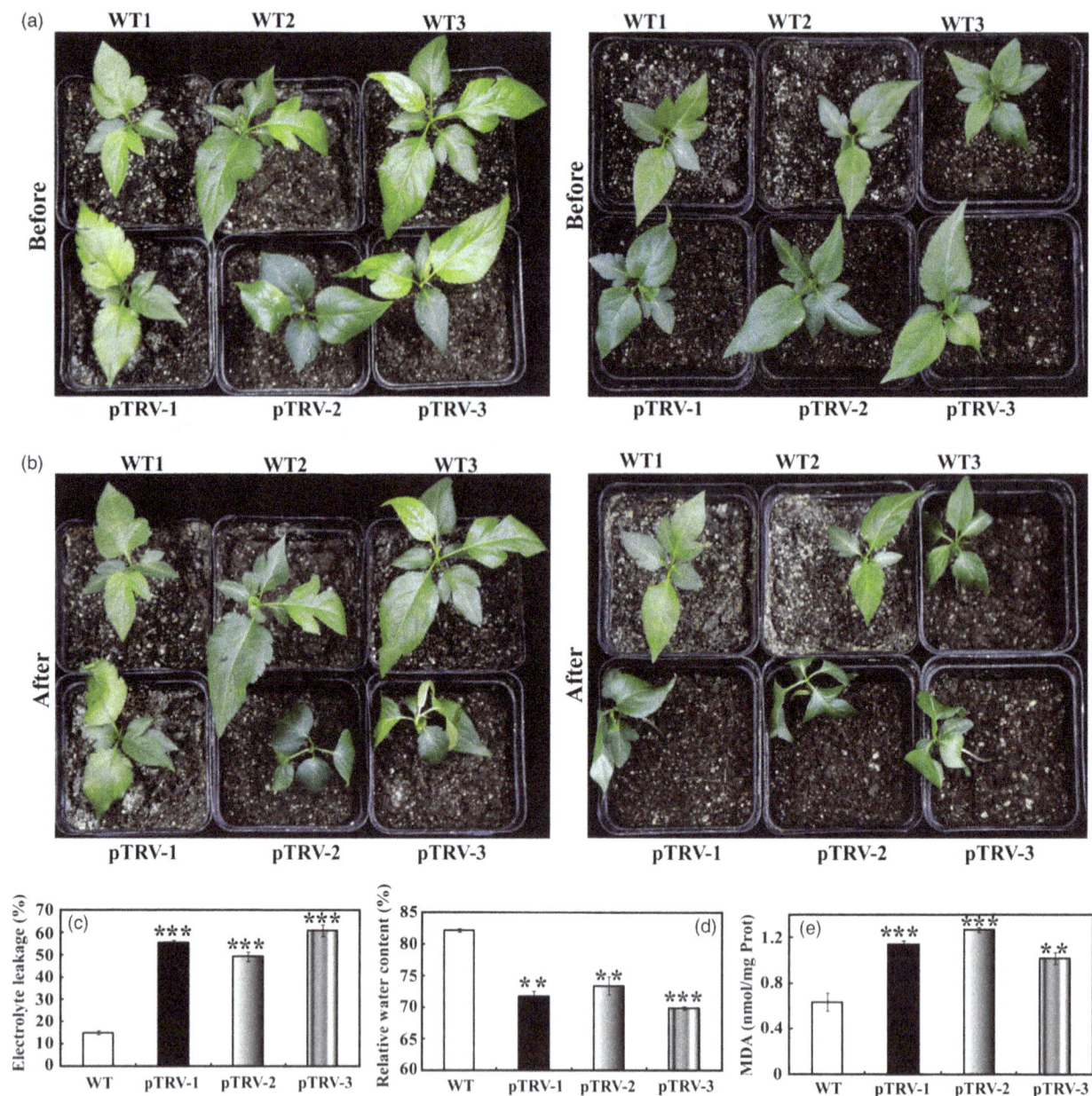

Figure 7 Silencing of the *PbrMYB21*i n *Pyrus betulaefolia* by virus-induced gene silencing (VIGS) resulted in elevated drought sensitivity. (a–b) Phenotypes of *PbrMYB21*-silenced plants before (a) and after drought stress for 7 days (b). (c–e) Electrolyte leakage (c), relative water content (d) and MDA levels (e) of the pTRV (WT) plants and pTRV-*PbrMYB21* silencing plants (pTRV-1, pTRV-2 and pTRV-3) after drought stress for 7 days. Asterisks indicate that the value is significantly different from that of the WT at the same time point (**$P < 0.01$; ***$P < 0.001$).

Analysis of antioxidant enzyme activities in the transgenic plants and the silencing plants

The crucial role of antioxidant enzymes in ROS scavenging (Gill and Tuteja, 2010) prompted us to examine the activities of three important enzymes (SOD, CAT and POD) in WT and the transgenic lines or the silencing plants during the drought treatment. In the normal condition, the enzyme activities of SOD, POD and CAT in tobacco transgenic lines were slightly higher than those of the WT, whereas activities of the three enzymes were significantly higher in the transgenic plants compared to WT when the plants experienced drought stress (Figure 9a–c). Conversely, the three enzyme activities in the three silencing plants (pTRV-1, pTRV-1 and pTRV-3) were significantly

lower than those of the WT before and after drought treatment (Figure 9d–f).

Expression analysis of stress-responsive genes in the WT and transgenic lines or the silencing plants before and after drought stress

To elucidate the molecular mechanism underlying the enhanced drought resistance in the transgenic lines, expression patterns of some genes involved stress response, including enzymes involved in biosynthesis of polyamine (*NtADC1*, *NtADC2* and *NtSAMDC*), osmoticum adjustment or water maintenance (*NtP5CS*, and *NtLEA5*) were analysed by qRT-PCR assay. In the absence of water, mRNA levels of all genes in three transgenic lines were significantly higher than those in the WT. Exposure to drought

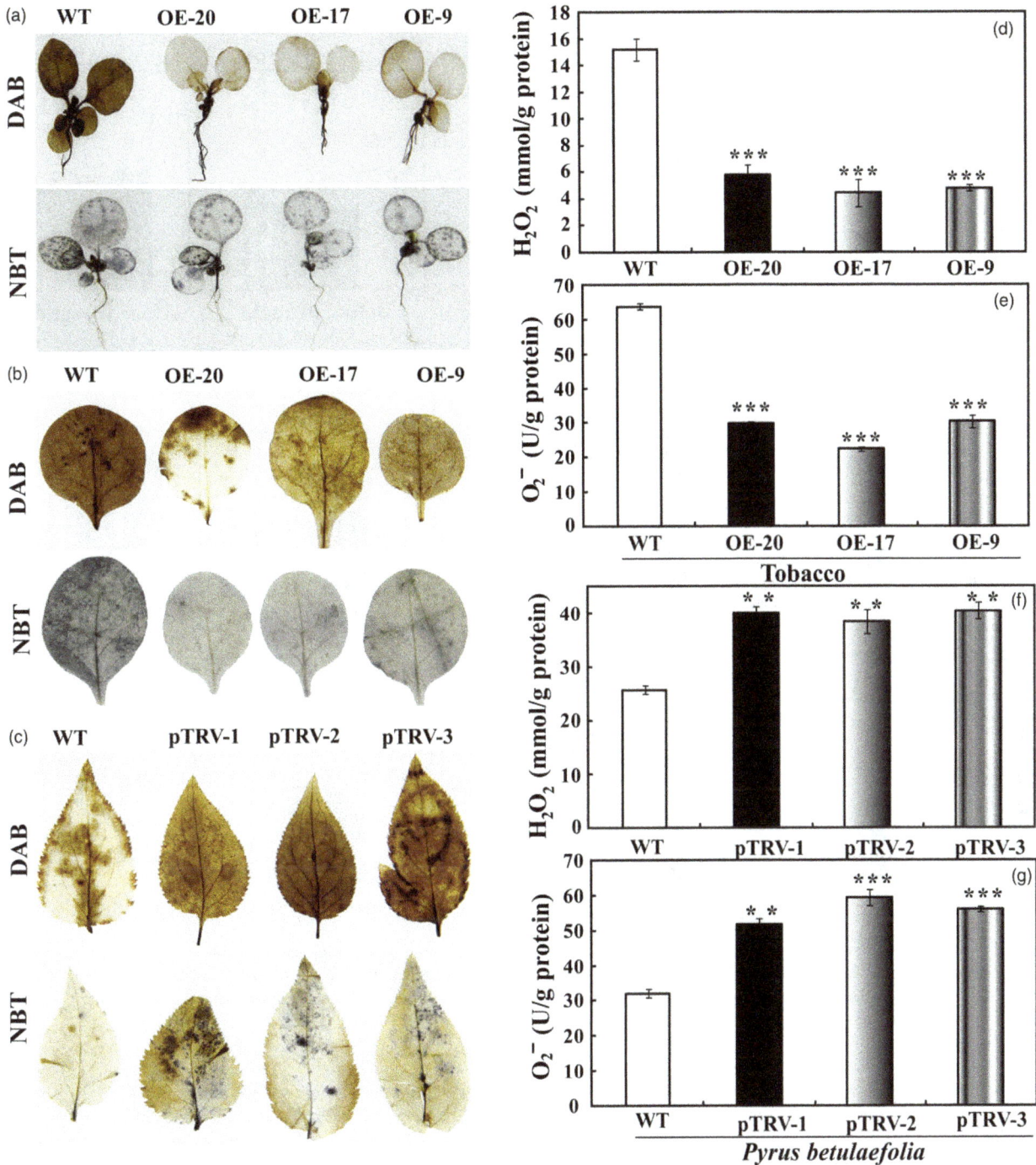

Figure 8 Analysis of H_2O_2 and anti-O_2^- in tobacco and *Pyrus betulaefolia* silenced plants after dehydration or drought stresses. (a) Histochemical staining with DAB and NBT for detection of H_2O_2 and O_2^-, respectively, in WT, OE-20, OE-17 and OE-9 after 70 min of dehydration. (b) Representative images indicating *in situ* accumulation of H_2O_2 and O_2^- in tobacco WT and transgenic lines (OE-20, OE-17 and OE-9) after drought stress for 7 days. (c) Representative images indicating *in situ* accumulation of H_2O_2 and anti-O_2^- in *Pyrus betulaefolia* WT and pTRV-*PbrMYB21* silencing plants (pTRV-1, pTRV-2 and pTRV-3) after drought for 7 days. (d–e) Levels of H_2O_2 (d) and O_2^- (e) in tobacco WT and transgenic lines (OE-20, OE-17 and OE-9) after drought stress. The insets in (d) and (e) are histochemical staining patterns using DAB and NBT, respectively. (f–g) Levels of H_2O_2 (f) and O_2^- (g) in *Pyrus betulaefolia* WT and pTRV-*PbrMYB21* silencing plants after drought treatment. Asterisks indicate that the value is significantly different from that of the WT at the same time point (**$P < 0.01$; ***$P < 0.001$).

caused up-regulation of the transcript levels of the examined genes in all lines, but the three transgenic lines still had a higher expression level in comparison with the WT (Figure 10). On the contrary, the mRNA abundance of these examined genes (*PbrLEA5*, *PbrADC*, *PbrSAMDC*, *PbrERD10C* and *PbrP5CS*) was prominently down-regulated in the silencing lines before and after drought treatment. Interestingly, exposure to drought stress caused minor up-regulation of *NtADC1* and *NtADC2* in the WT, but greater induction was observed in the transgenic lines. The transcript levels of *NtADC1* in three transgenic lines were 6.9 and

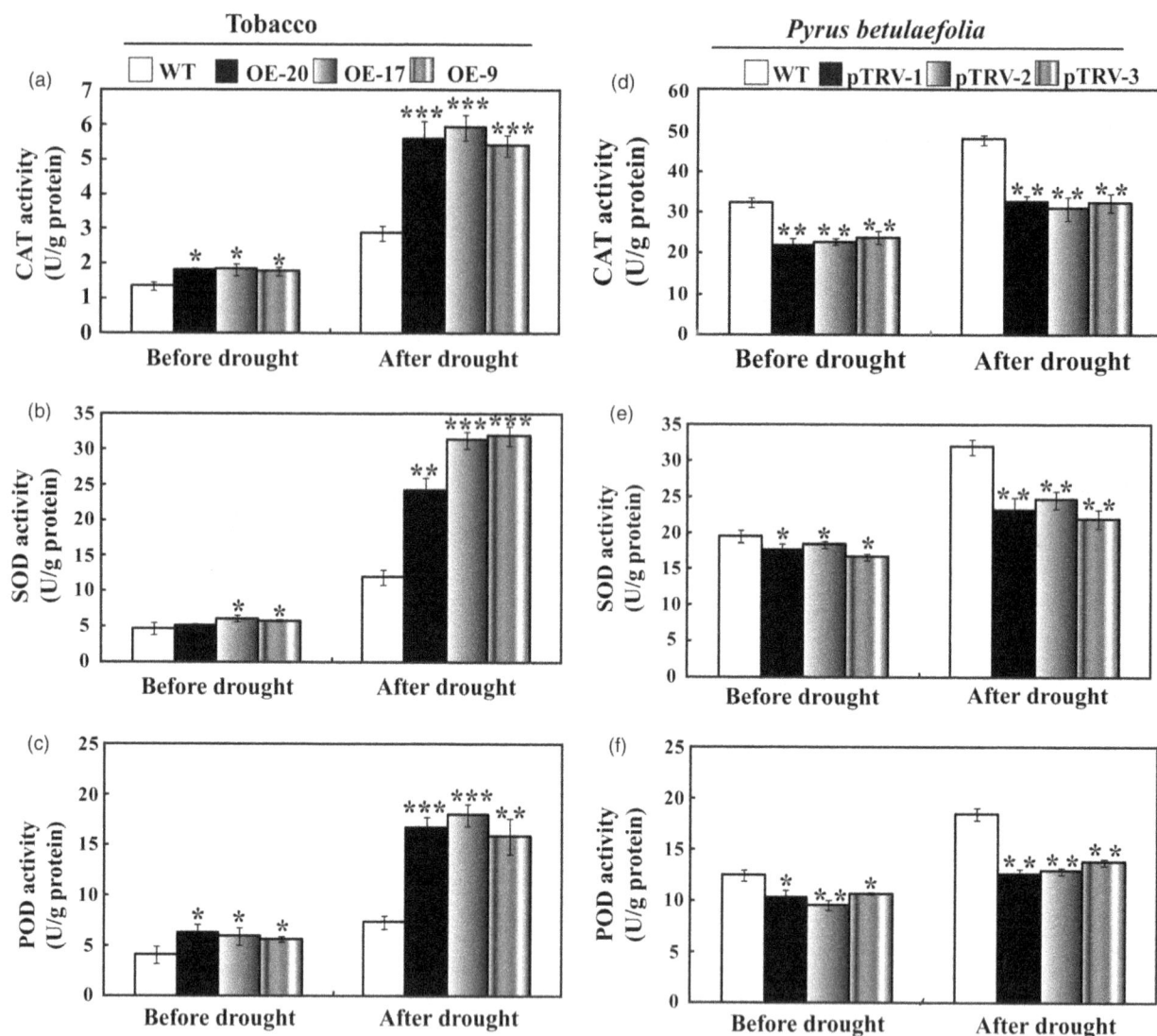

Figure 9 Analysis of antioxidant enzyme activities in tobacco and *Pyrus betulaefolia* silenced plants. (a–c) Activity of CAT (a), SOD (b) and POD (c) in tobacco WT and transgenic lines (OE-20, OE-17 and OE-9) before and after drought treatment. (d–f) Activity of CAT (d), SOD (e) and POD (f) in *Pyrus betulaefolia* WT and pTRV-*PbrMYB21* silencing plants (pTRV-1, pTRV-2 and pTRV-3) before and after drought stress. Asterisks indicate that the value is significantly different from that of the WT at the same time point (*$P < 0.05$; **$P < 0.01$; ***$P < 0.001$).

9.0 times that in the WT, respectively, while *NtADC2* levels in the three transgenic lines were 6.4–8.1 times that in the WT. The transcript levels of *PbrADC* in the silencing lines were significantly lower than in the WT without dehydration treatment. Drought increased the mRNA abundance of *PbrADC1* while the silencing lines had significantly lower levels than in WT (Figure 10). These results demonstrate that overexpression of *PbrMYB21* positively promoted the transcript level of *ADC* gene. In addition, *PtADC* of trifoliate orange has been shown previously to play a role in drought tolerance (Wang *et al.*, 2011), which prompted us to focus on this protein.

Alteration of levels of free polyamines in the transgenic plants and the silencing plants

Further work was carried out to elucidate physiological mechanism underlying the enhanced drought tolerance rendered by *PbrMYB21*. As polyamine is an important compound involved in drought tolerance (Wang *et al.*, 2011), we attempted to analyse

whether polyamine synthesis was altered in the tested plants. HPLC (high-performance liquid chromatography) was employed to analyse the free polyamine contents in the tested plants and WT. The polyamine contents in the three transgenic lines were higher than in the WT without drought stress (Figure 11a). Drought treatment led to an increase of putrescine, spermidine and spermine in all lines, whereas the elevation in the three transgenic lines was much higher than that of the WT (Figure 11b), which was consistent with the higher expression levels of *ADC* genes in the transgenic lines. Polyamine levels in the silencing plants were significantly lower than in the WT without drought treatment (Figure 11c). Drought increased the polyamine levels while the silencing plants had significantly lower levels than in WT (Figure 11d). These data demonstrate that overexpression of *PbrMYB21* elevated endogenous polyamine contents in the transgenic plants, whereas they were changed in an opposite way when *PbrMYB21* was knocked down.

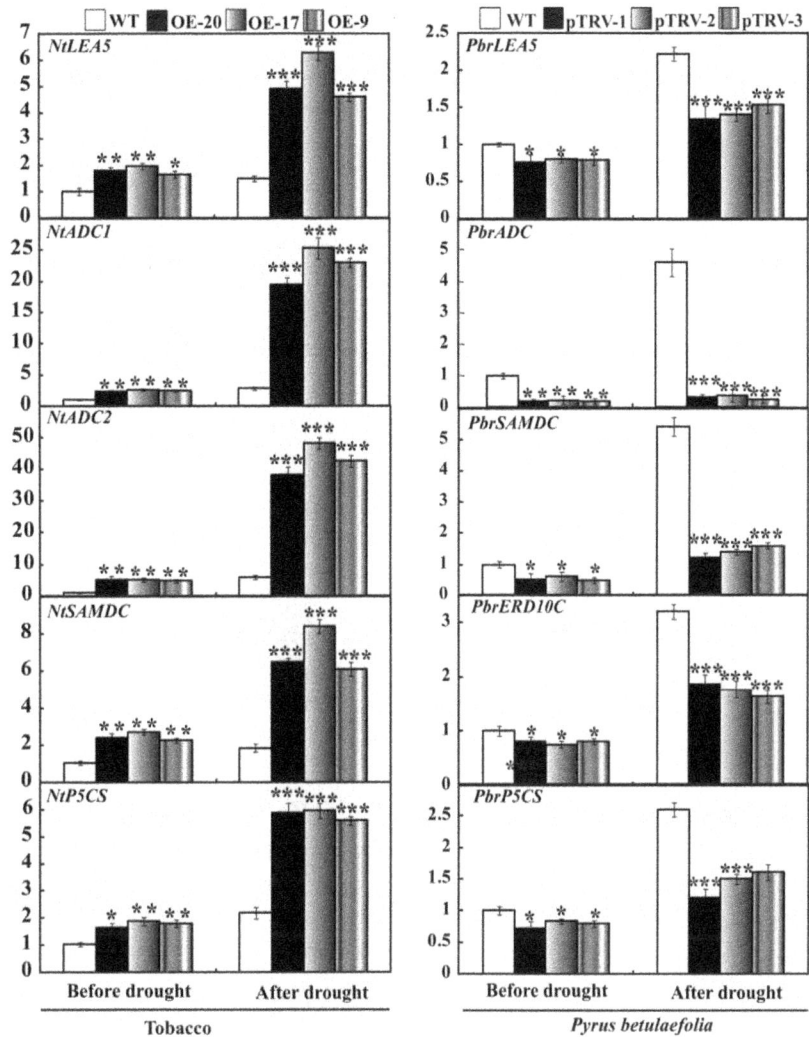

Figure 10 Quantitative real-time PCR analysis of expression levels of stress-responsive genes in tobacco and *Pyrus betulaefolia* silenced plants under normal and drought conditions. LEA5 (late embryogenesis abundant gene), ADC (arginine decarboxylase gene), SAMDC (S-Adenosyl methionine decarboxylase), ERD10C (late embryogenesis abundant gene), *P5CS* (Δ1-pyrroline-5-carboxylate synthetase). Data represent the means ± SE of four replicates. Asterisks indicate that the value is significantly different from that of the WT at the same time point (*$P < 0.05$; **$P < 0.01$; ***$P < 0.001$).

PbrMYB21 interacts with the promoters of PbrADC

The elevation or decrease of ADC genes in *PbrMYB21*-over-expressing or silencing plants, respectively, implied that ADC gene might be one of the potential target genes that are regulated by PbrMYB21. To verify this hypothesis, a 1980-bp promoter of *PbrADC* was isolated by genomic PCR. The putative transcriptional start site and the TATA-box were located 57 and 85–88 bp upstream of the translation start codon. Bioinformatics prediction demonstrated that the promoter of *PbrADC* contained putative a MYB-recognizing motif (CAACTG, -673 to -668 bp; Figure 12a). Therefore, we investigated whether PbrMYB21 could bind to this element using a yeast one-hybrid assay. A 246-bp fragment containing the MYB-recognizing element (P1) was used as bait and cloned into the reporter vector, while PbrMYB21 was used a prey. The yeast cells of positive and negative control and those co-transformed with effector and reporter grew well on the SD/-Ura/-Leu medium without the antibiotic (AbA). However, when 300 μM AbA was added, growth of negative control was fully inhibited, while the yeast cells of positive control and bait–prey survived (Figure 12c). These results indicate that PbrMYB21 interacted with the promoter of PbrADC in yeast. To confirm the results of the yeast one-hybrid assay, transient expression assay was further performed to confirm the interaction between PbrMYB21 and the promoter of PbrADC. The promoter activities, expressed as LUC/REN ratio, of the PbrMYB21 were significantly higher than those in the WT (Figure 12d, e). In contrast, when MYB *cis*-element in the promoter of PbrADC was mutated, there was no activity in the samples transformed with vectors of the negative control and mutated MYB element.

To further determine whether PbrMYB21 specifically binds to the MYB recognition site in the PbrADC promoter, electrophoretic mobility shift assay (EMSA) was performed using prokaryon-expressed and purified PbrMYB21-His fusion proteins. A specific DNA-PbrMYB21 protein complex was detected when the CAACTG-containing oligonucleotide was used as a labelled probe, and the same oligonucleotide that was unlabelled was used as a competitor (Figure 12f). The formation of these complexes was reduced with increasing amounts of the unlabelled competitor probe with the same sequence. In addition, when the cis-acting element was mutated from CAACTG to CGCTGG, a protein–DNA complex was not detected in the presence of the PbrMYB21 protein, indicating that the binding was specific for the CAACTG sequence (Figure 12g). The specificity of this competition confirms that PbrMYB21 recognized and bound specifically to the CAACTG motif within the PbrADC promoter.

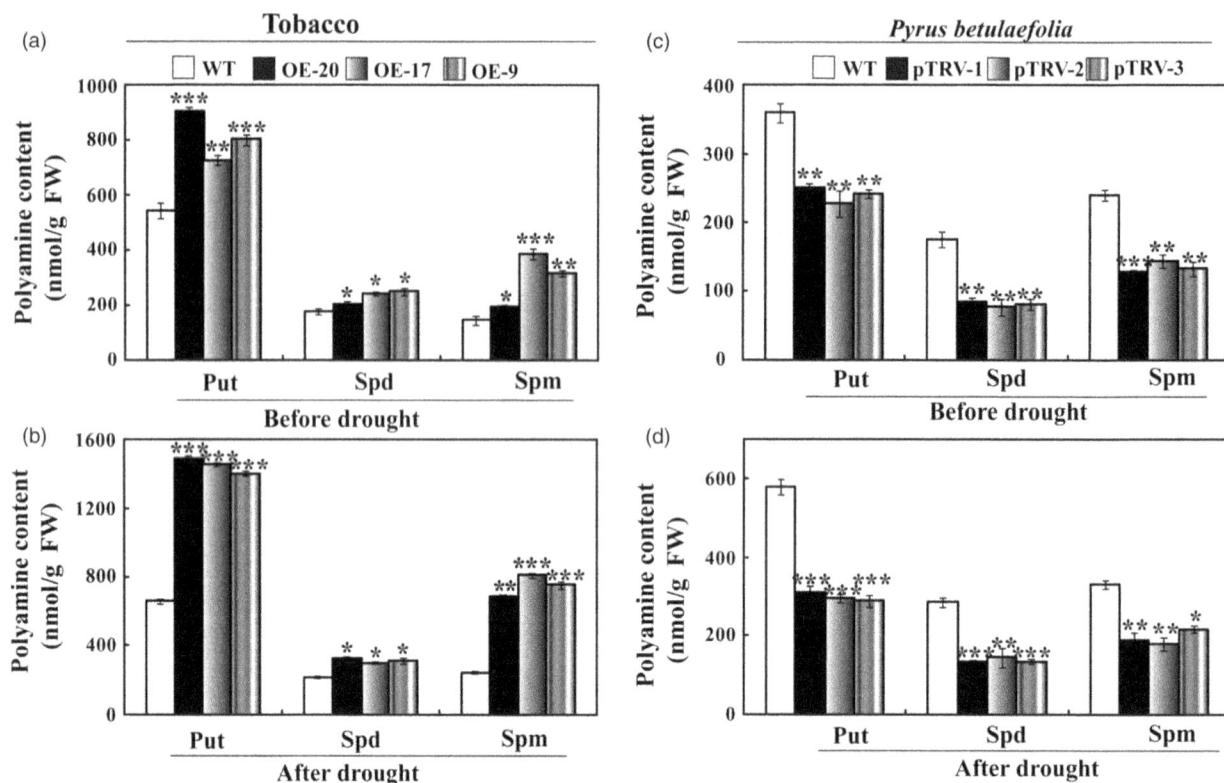

Figure 11 Analysis of free polyamine in tobacco and silencing *Pyrus betulaefolia* under normal and drought conditions. Free polyamine contents in WT and transgenic lines before (a) and after (b) drought stress. Put, putrescine; Spd, spermidine; Spm, spermine. (c–d) Free putrescine contents in *Pyrus betulaefolia* gene-silenced plants and the vector control plants before (c) and after (d) drought stress. Asterisks indicate that the value is significantly different from that of the WT at the same time point (*$P < 0.05$; **$P < 0.01$; ***$P < 0.001$).

The results above demonstrate that PbrMYB21 has transcriptional activity, but it remains to be verified whether it is an activator or a repressor. To address this, we performed a classical GAL4/UAS-based system assay to study the transcriptional activation or repression of a protein (Wang et al., 2014). As shown in Figure 12h, the GAL4 DNA-binding domain (G4DBD) that binds to the six copies of UAS to activate GUS expression. We then generated a fusion protein, G4DBD-PbrMYB21, which was co-transformed with 35S-UAS-GUS in *N. benthamiana* leaves to determine PbrMYB21 transcriptional activity. Histochemical staining showed that GUS expression was prominently activated in the *N. benthamiana* leaves compared with the leaves transformed with the empty vector control (Figure 12i). These results suggest that PbrMYB21 might act as a transcriptional activator of *PbrADC*.

Discussion

The MYB proteins constitute one of the largest transcription factor families in plants; 118 and 222 MYBs have been unravelled in the genomes of grape and apple. Numerous studies have provided extensive data revealing that plant MYB proteins play central roles in an array of physiological and biological processes, such as primary and secondary metabolism, cell cycle progression, hormonal signalling, and cell patterning and tissue differentiation, stomatal movement, signal transduction and cell division (Cominelli et al., 2005; Li et al., 2013; Mellway et al., 2009; Mu et al., 2009; Ryu et al., 2005; Shim et al., 2013). In addition, accumulating evidence shows that MYB proteins play pivotal roles in plant responses to various abiotic stresses, including salt (Guo

et al., 2016; Wang et al., 2015), drought (Guo et al., 2013), cold (Cominelli et al., 2005) and phosphate starvation (Baek et al., 2013). Although some MYB TFs have been characterized, the biological functions of most of the plant MYB TFs remain unclear, particularly in nonmodel plants, they are still poorly understood. Therefore, elucidation of the functions of MYBs in perennial plants, such as *Pyrus betulaefolia*, will advance understanding on the roles of MYBs. Of special note, the target genes in the identified MYBs, even those well-characterized ones, are still not clearly dissected so far. Therefore, target gene exploration is still one major challenge to gain a more comprehensive understanding on the roles and modes of action underlying the stress tolerance of the MYB genes.

MYB-containing genes have greatly diversified, being classified into four major subfamilies, namely 1R-MYB proteins, R2R3-MYB proteins, 3R-MYB proteins, and 4R-MYB proteins, containing one, two, three or four MYB domain repeats, respectively. R1R2R3-MYB proteins are commonly found in animals, but the majority of plant MYB genes belong to the R2R3-MYB types. In this study, we isolated a R2R3-type MYB from *Pyrus betulaefolia*. PbrMYB21 has the entire set of above-mentioned signature motifs required for defining a typical R2R3-type MYB, despite a low degree of sequence conservation outside the MYB DNA-binding domain. Multiple alignments revealed that PbrMYB21 shared a striking sequence similarity with MdMYB21 of *Malus domestica* and other plants, indicating that *PbrMYB21* is a putative MYB homologue of *Pyrus betulaefolia*.

The transcript levels of *PbrMYB21* were increased by dehydration and salt, in particular the former, indicating that *PbrMYB21* may be involved in the signal transduction of these two types of

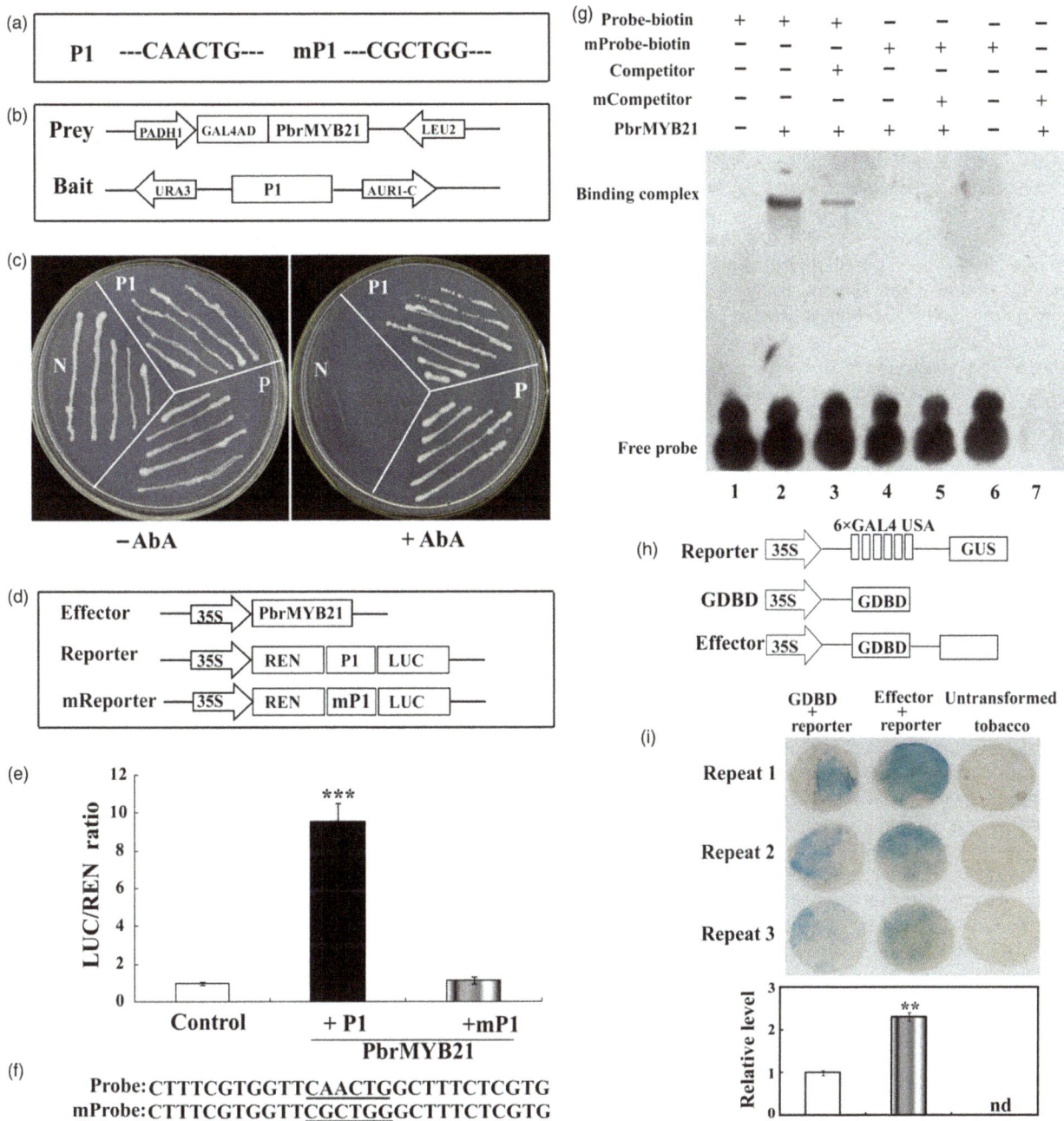

Figure 12 PbrMYB21 specifically binds to the promoter of *PbrADC* and acts as a transcriptional an activator. (a) Schematic diagrams of the promoter of *PbrADC*. The putative MYB binding site in the promoter fragment (P1) of *PbrADC*. Mutation of the MYB binding site sequence (mP1) is shown on the right. (b) Schematic diagrams of the prey and bait vectors used for yeast one-hybrid assay. (c) Growth of yeast (strain Y1H Gold) cells of positive control (P), negative control (N) and bait–prey (P1) co-transformation on SD/-Leu medium without (left) or with (right) 300 ng/mL AbA. (d) Schematic diagrams of the effector and reporter constructs used for transient expression assay. (e) Transient expression assay of the promoter activity in tobacco protoplasts co-transformed with effector and the reporter constructed using P1 containing normal MYB binding site element or mP1 containing mutation MYB binding site element. (f) Interaction of the PbrMYB21 protein with labelled DNA probes for the cis-elements of the *PbrADC* promoter in the EMSA. (g) The His-PbrMYB21 protein was incubated with the biotin-labelled promoter fragment containing the wild-type CAACTG or the mutated CGCTGG form; the nonlabelled fragment was used as a competitor; −, absence; +, presence. (h) Schematic diagrams of the three vectors used for the transient expression assay of the transcriptional activity of PbrMYB21 in *N. benthamiana* leaves using a GAL4/UAS-based system. 35S, the 35S promoter without the TATA-box; 6 × GAL4 UAS, six copies of the GAL4-binding site; G4DBD, the GAL4 DNA-binding domain; effector, PbrMYB21 was inserted downstream of GDBD. (i) GUS staining (top) and relative expression level (bottom) of the *N. benthamiana* leaves co-transformed with the indicated plasmids. Untransformed *N. benthamiana* leaves were used to show the original colour. The asterisks indicate a value that is significantly different (**$P < 0.01$). nd, Not detected. Experiments were performed three times, and each experiment contained at least three replicates.

stresses. Expression patterns of *PbrMYB21* were largely similar to *AtMYB41* that has been shown to be strongly and continuously induced by salt and drought (Lippold *et al.*, 2009). However, it has to be mentioned that *PbrMYB21* was not induced by abscisic acid treatment, and they also different from *AtMYB41*, as *PbrMYB21* was not induced under abscisic acid. Compared with salt stress, dehydration caused higher induction of *PbrMYB21* transcript levels, which forced us to do in-depth work and elucidate its function in drought tolerance. The three selected transgenic lines overexpressing *PbrMYB21* displayed better tolerance to dehydration and drought, while knockdown of *PbrMYB21* rendered the silencing plants more susceptible to drought, indicating that *PbrMYB21* acted as positive regulator of drought tolerance.

The overexpressing plants of tobacco displayed enhanced dehydration tolerance, as indicated by slower water loss rates and lower levels of MDA than the WT. As lipid peroxidation, represented by MDA levels and ELs, is detrimental to the membrane systems, it is conceivable that the transgenic plants are suffered from lower degree of membrane injury under the drought stress when compared with the WT. Lipid peroxidation largely results from oxidative stresses due to the excessive accumulation of ROS, in particular H_2O_2 and O_2^-. Histochemical staining by DAB and NBT clearly demonstrated that under drought stress the three transgenic lines accumulated remarkably less O_2^- and H_2O_2 than WT, implying they experienced milder oxidative stresses relative to the WT, consistent with the data on MDA and EL. As ROS accumulation under stressful situations largely relies on the homoeostasis between ROS generation and removal, accumulation of less ROS in the transgenic lines seems to indicate that scavenging systems in these plants might work more effectively compared with WT. To detoxify stress-induced ROS, plants have evolved a set of ROS detoxification system, in which several enzymes play essential roles, maintaining the ROS homoeostasis at a favourable situation so that the cells are less influence by the oxidative stress. ROS detoxification in higher plants is mainly achieved by ROS-scavenging enzymes, such as SOD, CAT and POD (Miller *et al.*, 2010). In this work, transgenic plants had higher activities of three antioxidant enzymes (CAT, POD and SOD), implying that they possess a more efficient enzymatic antioxidant system to eliminate ROS produced compared with the WT, which is consistent with the dramatic reduction of ROS level and ROS-associated membrane damage (lower MDA and EL). In this regard, it is seemingly likely that overexpression of *PbrMYB21* in the transgenic plants can alleviate ROS accumulation in a more powerful way. In contrast, the *PbrMYB21*-silenced *Pyrus betulaefolia* plants exhibited larger electrolyte leakage, higher MDA levels, higher H_2O_2 and O_2^- levels, and lower enzymes activities, indicating that suppression of PbrMYB21 rendered the silencing plants more susceptible to the drought stress. The plant phenotype or parameters of the silencing plants were opposite to what was seen with the overexpression lines.

To cope with unfavourable environmental constraints, plants modulate the expression of a large scale of stress-responsive genes, constituting an important molecular basis for the response and adaptation of plants to stresses (Umezawa *et al.*, 2006). To gain a deeper understanding of the regulatory function of *PbrMYB21* and to explain the enhanced the drought tolerance at molecular levels, transcript levels of some stress-responsive genes were monitored before and after drought treatment, including enzymes involved in biosynthesis of polyamine (*NtADC1, NtADC2*

and *NtSAMDC*), osmoticum adjustment or water maintenance (*NtP5CS, NtLEA5*), which or whose homologues in other plants have been shown to be involved in abiotic stress response. QRT-PCR analysis showed that the expression level of these genes was higher in the transgenic plants compared with those of WT under drought stress, consistent with earlier reports in which overexpression of a TF may active a group of target genes that function in a concerted manner to counteract the adverse effects of abiotic stresses (Huang *et al.*, 2013). Although expression levels of all of the tested genes were up-regulated by drought, they were still higher in the transgenic plants than in WT, indicating that these genes were more intensely induced in the transgenic lines. Interestingly, two genes (*NtADC1* and *NtADC1*) involved in polyamine synthesis displayed significantly higher mRNA abundance compared with the other stress-related genes. In additional, we also found that suppression of *PbrMYB21* in the silencing plants was accompanied by a noticeable reduction of these stress-responsive genes, particularly *PbrADC*, was lower in the silencing plants than in WT before and after drought treatment. *LEA5* and *ERD10C* encode hydrophilic late embryogenesis abundant (LEA) proteins that are assumed to play critical roles in combating cellular dehydration tolerance by binding water, and protecting cellular and membrane damage (Hundertmark and Hincha, 2008). *P5CS*, a key enzyme of proline biosynthesis, the transcript levels of this gene of the transgenic plants were higher than that of the WT before and after drought stress. *ADC1* and *ADC2* are genes involved in biosynthesis of polyamines, which are low molecular weight polycations and have been shown to be important stress molecules (Liu *et al.*, 2007). A number of earlier studies showed that more drastic induction of these genes implied that the transgenic plants might synthesize higher levels of polyamines to prevent them from lethal injury and maintain better growth under water stress (Shi *et al.*, 2010). In this study, we showed that increased *PbrMYB21* expression led to higher polyamines accumulation, which might contribute to the improved dehydration or drought in the transgenic plants on the basis of the following two arguments. Firstly, higher levels of *ADC* transcripts and free polyamines were detected in the transgenic lines, concomitant with accumulation of less ROS, which was positively consistent with their stress tolerance capacity. Secondly, *PbrMYB21*-silenced decreased the transcript level of *ADC* gene and polyamine contents and resulted in substantial impairment of drought tolerance. All of these results demonstrate that promotion of polyamine accumulation constituted one of the physiological or metabolic bases responsible for the enhanced drought tolerance in the transgenic plants overexpressing *PbrMYB21*.

Taken together, we isolated a R2R3-type MYB from *Pyrus betulaefolia*, which acts as a positive regulator of drought tolerance. The higher levels of *ADC* transcripts and free polyamines were detected in the *PbrMYB21* transgenic lines, whereas VIGS of *PbrMYB21* in *Pyrus betulaefolia* down-regulated *ADC* abundance and decreased free polyamine level, in agreement with these two genes common up-regulated from *Pyrus betulaefolia* drought transcriptome data set. Bioinformatics prediction revealed the presence of one MYBR element on the promoter of *PbrADC* gene, which were shown to be interacted by *PbrMYB21*, indicating that *ADC* might serve as a direct target of *PbrMYB21*. Y1H, transient expression assays and EMSA provided evidence further supporting a direct and specific interaction between *PbrMYB21* and the *PbrADC* promoter. In summary, these results indicated that *PbrMYB21* functioned in mediating drought tolerance by elevation of PAs levels via regulating *ADC* gene, which may explain the higher levels of *NtADC1*, *NtADC2* and free polyamines in the transgenic plants, concomitant with overexpression of *PbrMYB21* elevated *ADC* expression levels and polyamine content, whereas they were changed in an opposite way when *PbrMYB21* was silenced. Establishment of the MYB-ADC network provides new valuable knowledge of the function and underlying mechanism of MYB and expands our understanding of the complex drought signalling network. However, it has to be pointed out that *PbrMYB21* also may regulate other stress-responsive genes, extra efforts are required in the future to unravel other components related to *PbrMYB21* so as to gain a clear-cut silhouette of the major hub in the network.

Materials and methods

Plant materials and stress treatment

The leaves were obtained from 45-day-old *Pyrus betulaefolia* seedlings which grown at Nanjing Agricultural University for analysis of expression patterns of *PbrMYB21* under various stresses. The shoots were washed and cultured for 2 days in water in a growth chamber (25 °C) to minimize the mechanical stress on the tissues, followed by exposure to corresponding stress treatments, which were carried out as follows. The seedlings were used for gene cloning and analysis of expression patterns under stresses. For cold stress, the seedlings were placed in the chamber set at 4 °C for 0, 1, 5, 12, 24 and 48 h. For dehydration treatment, the seedlings were placed on dry filter papers for 0, 0.5, 1, 3 and 6 h under ambient environment. For salt stress, the seedlings were treated with 200 mM NaCl solution for 0, 1, 3, 6, 12 and 24 h. For each treatment, at least 60 seedlings were used, and the leaves were sampled from three randomly collected seedlings at designated time point and frozen immediately in liquid nitrogen and stored at −80 °C for further analysis.

RNA extraction and quantitative real-time PCR analysis

Total RNA was extracted from frozen leaves using RNAiso Plus RNA (TaKaRa, China) and treated with RNase-free DNase I (Thermo, USA) to eliminate the potentially contaminating DNA. First-strand cDNA was synthesized using RevertAid Reverse transcriptase (Thermo, USA) according to manufacturer's instructions. Quantitative real-time PCR was carried out with SYBR-Green PCR kit (TakaRa, China) using 20 μL of reaction mixture consisting 10 μL of 2× SYBR-PreMix EX Taq, 50 ng of cDNA, and 0.25 μM of each primer (GSP1, Table S1). The program of the qPCR was as follows: denaturation at 94 °C for 5 min, followed by 40 cycles of 94 °C for 10 s, 60 °C for 30 s, 72 °C for 30 s and a final extension 3 min at 72 °C. Each sample was analysed in four replicates, and the $2^{-\Delta\Delta Ct}$ method (Livak and Schmittgen, 2001) was applied to calculate the relative expression levels of each gene. *Tubulin* and *Ubiquitin* were analysed in parallel as reference controls for *Pyrus betulaefolia* and tobacco, respectively, to normalize expression levels.

Gene isolation and bioinformatics analysis

In an earlier study, a dehydration-induced MYB TF (Pbr028812.1) showing high sequence homology to MYB21 was found to be up-regulated in the transcriptome. RNA extraction and cDNA synthesis of the relevant samples were carried out as elaborated above. Based on the above sequence, a pair of gene-specific primers (GSP2, Table S1) was used for RT-PCR to amplify

PbrMYB21. The RT-PCR, in a total volume of 50 μL, consists of 250 ng of cDNA, 1× TransStart FastPfu buffer, 0.25 mM dNTP, 2.5 U of TransStart FastPfu DNA polymerase (TRANS), 0.4 μM of each primer and nuclease-free water. PCR was performed in a thermocycler with a programme consisting of 2 min at 95 °C, followed by 40 cycles of 20 s at 95 °C, 20 s at 55 °C, 60 s at 72 °C and a 5-min extension at 72 °C. The PCR product was purified, subcloned into pMD18-T vector (Takara, China) and sequenced in Invitrogen (Shanghai, China). The multiple alignments of the amino acid sequence in different species were used by ClustalW and GeneDoc software, and the phylogenetic tree was constructed by the maximum-likelihood (ML) method in the MEGA 6.0 program, theoretical isoelectric point (*pI*) and molecular weight were predicted by Expert Protein Analysis System (http://web.expasy.org/compute_pi/).

Transactivitional activity and subcellular localization of PbrMYB21

For the transactivation assay, the complete ORF of *PbrMYB21* was amplified by PCR using primers (GSP3, Table S1) containing *BamHI* and *NcoI* restriction sites, and the amplicon was double digested by *BamHI* and *NcoI*. The resultant fragment containing PbrMYB21 was then fused downstream of the yeast GAL4 DNA-binding domain in pGBKT7 by recombination reactions. The fusion vector and the negative control (pGBKT7) were independently expressed in yeast strain AH109 according to the manufacturer's instructions.

The full-length cDNA of the *PbrMYB21* was PCR amplified using primers (GSP4, Table S1) containing *BglII* and *SpeI* restriction sites and inserted into a polylinker site of the binary vector pCAMBIA1302 to generate a fusion construct (p35S-PbrMYB21-GFP). After identified the sequence, the fusion construct (p35S-PbrMYB21-GFP) and the control vector (pCAMBIA1302) were transferred into *Agrobacterium tumefaciens* strain GV3101 by heat shock. The abaxial surfaces of 5-week-old *N. benthamiana* leaves were agroinfiltration (Kumar and Kirti, 2010) with the bacterial suspension (OD600 = 0.5) and then kept in an incubator for 48 h, followed by live cell imaging under an inverted confocal microscopy (Zeiss LSM 780, Germany).

Vector construction and plant transformation

The full-length coding region of *PbrMYB21* was PCR amplified with GSP5 (Table S1) containing a His-tag (HHHHHH) and inserted into a polylinker site of the binary vector pCAMBIA1301. The tobacco rattle virus (TRV)-based vectors (pTRV1/2) construct was used for the VIGS (Wang et al., 2016; Xie et al., 2012). For the construction of pTRV-PbrMYB21, a fragment of the PbrMYB21 ORF (292, 91–382 bp) was amplified by PCR using primers of GSP6, which was then introduced into the pTRV2 vector. Empty pTRV2 vector was used as a control. All recombinant vectors were introduced into *Agrobacterium tumefaciens* strain GV3101 by the freeze–thaw method. The overexpression vector was used to transform tobacco (*Nicotiana tabacum*), and VIGS was performed by infiltration into the leaves of 30-day-old *Pyrus betulaefolia* seedling with *A. tumefaciens* harbouring a mixture of pTRV1 and pTRV2-target gene in a 1 : 1 ration (Ramegowda et al., 2013; Wang et al., 2016). Tobacco transformation was performed based on a leaf disc method (Horsch et al., 1985). The hygromycin-resistant overexpression plants were PCR verified. Transcript levels of PbrMYB21 in the positive transgenic plants were analysed by RT-PCR. Semi-quantitative RT-PCR was also performed to determine the gene-silencing efficiency. T2 homozygous plants of tobacco transgenic were used for the subsequent experiments. The reference genes, *Tublin* and *Ubiquitin*, were used as internal controls for *Pyrus betulaefolia* and tobacco, respectively.

Stress tolerance assay of the transgenic plants

Seeds of T_2-generation transgenic plants and wild type were sterilized for 30 s in 70% (v/v) ethanol and incubated in 2.5% (v/v) H_2O_2 for 5 min, followed by rinse with sterile distilled water for three times before they were sown on germination medium (GM) containing MS salts, 30 g/L sucrose and 0.8% agar (pH 5.7). For dehydration stress, two sets of experiments were designed. First, 30-day-old *in vitro* seedlings of the three lines removed from germination medium as mentioned above were allowed to dry for up to 70 min on the filter papers. The fresh weight (FW) of the seedlings was measured every 10 min using a scale, and the percentage FW loss was determined relative to the initial weight. For drought experiment, 5-day-old transgenic tobacco lines and WT were first transplanted into plastic containers filled with a mixture of soil and sand (1 : 1) where they were regularly watered for 7-week-old until the drought treatment. At the onset of and after 7 days of withholding water, morphological changes of the drought plants were inspected during the drought treatment. Photographs were taken when the differences were noticeable. The leaves were collected from some plants for analysis of antioxidant enzyme activity, relative water content (RWC), metabolite levels and expression of stress-responsive genes.

Physiological measurement and histochemical staining

MDA, H_2O_2 and O_2^- were measured using specific detection kits following the manufacturer's instructions (Nanjing Jiancheng Bioengineering Institute, Jiangsu, China). Histochemical staining with DAB and NBT was used to analyse the *in situ* accumulation of H_2O_2 and O_2^- according to Huang *et al.* (2011). Antioxidant enzyme (POD, SOD and CAT) activity, H_2O_2 level and antisuperoxide anion activity were quantified using the relevant detection kits (Nanjing Jiancheng Bioengineering Institute) based on the manufacturer's instructions. For microscopic observation of stomata, the epidermis was stripped and examined under an Olympus BH-2 microscope (Olympus, Tokyo, Japan) equipped with an Olympus DP70CCD camera (Olympus, Tokyo, Japan). The pictures were captured as JPEG digital files, and stomatal apertures were measured using Image Pro Plus 5.0.

Quantification of free polyamine levels by high-performance liquid chromatography (HPLC)

Free PAs were extracted and derived according to Liu *et al.* (2009) and Shi *et al.* (2010) with slight modification. In brief, nearly 0.1 g of sample powder was extracted in 1 mL of 5% cold perchloric acid on ice for 30 min. After centrifugation at 10 800*g* (4 °C) for 15 min, the supernatant was shifted to a new tube, and the pellet was extracted in 1 mL of 5% perchloric acid again. The supernatant from two centrifugations was mixed, and 200 μL of mixture was derived with 400 μL dansyl chloride (10 mg/mL in acetone) plus 200 μL saturated sodium bicarbonate; meanwhile, 10 μL 1, 6-hexanediamine (100 μM stock) was added as an internal standard. The derivatization was performed at 60 °C for 60 min, and then 100 μL proline (100 mg/mL) was added and the mixture was incubated for an additional 30 min at 60 °C. The dansylated polyamines were measured using a high-performance liquid chromatography (Waters, USA) equipped with a C^{18} reversed-phase column (4.6 mm × 150 mm, particle size 5 μM) and a fluorescence spectrophotometer with excitation and emission wavelength of 365 and 510 nm, respectively. Standard substances of PUT, SPD and SPM (sigma) were used to determine the retention times of three polyamines in the HPLC chromatograms. Three replicates were performed for each treatment.

Yeast one-hybrid assays in yeast

Putative MYB *cis*-element was identified in the promoter region of the ADC gene (Pbr022368.1) based on the pear genome sequence (http://peargenome.njau.edu.cn/). Yeast one-hybrid assay was performed to investigate whether PbrMYB21 could interact with the MYB *cis*-element. The full-length ORF of *PbrMYB21* was amplified by PCR using primers (GSP7, Table S1) and integrated into the *BamH*I and *Nco*I sites of pGADT7-Rec (Clontech) to create a prey vector (pGADT7-PbrMYB21). Based on the distribution of the MYB *cis*-element, one fragment (P1, -673 to -668 bp) was PCR amplified using primers (GSP8, Table S1) containing *Sma*I and *Xho*I restriction sites (ADC-P1) and cloned into the pAbAi vector to construct the bait. Both the effector vector and reporter vector were co-transformed into yeast strain Y1H Gold following the manufacturer's instructions (Clontech). The transformed cells were spread on SD/-Ura/-Leu medium added with (0 or 300 ng/mL) AbA, and incubated for 3 days at 30 °C. Both positive (pGAD-p53+ p53-AbAi) and negative (pGADT7-AD + P1) controls were included and processed in the same way.

Transient expression analysis

For transient expression assays using *Arabidopsis* protoplasts, The *PbrADC* promoter region was amplified using primers (GSP9, Table S1) containing *Pst*I and *BamH*I restriction sites and inserted into pGreenII 0800-LUC to generate the reporter plasmids. The MYB *cis*-element in P1 was mutated by PCR (GSP10, Table S1). The coding region of *PbrMYB21* (GSP11, Table S1) was digested with *Sma*I and *Xho*I from the prey vector and inserted into pGreenII 62-SK linearized with the same enzymes, generating an effector plasmid. The assays for transient expression in protoplasts were performed as described (Agarwal *et al.*, 2006). All the plasmids used in this assay were extracted with QIA–GEN plasmid Midi Kit. The promoter activity was expressed as the ratio of LUC/REN. The protoplasts co-transformed with normal reporters and pGreenII 62-SK vector without PbrMYB21 was considered as control; LUC/REN ratio of which was set as 1.

For transcriptional activity analysis, a 35S-UAS-GUS reporter construct was obtained from Xia Li. For this purpose, the coding sequence of PbrMYB21 was amplified and ligated to the pYF503 vector, generating an effector GDBD-PbrMYB21 construct. The effector construct, empty control (pYF503; designated as GDBD), and reporter plasmid 35S-UAS-GUS were transformed into *A. tumefaciens* strain GV3101 cells, and the two effectors were co-infiltrated with the reporter 35S-UAS-GUS into *N. benthamiana* leaves as described previously (Voinnet *et al.*, 2003). The abaxial surfaces of tobacco leaves were then kept in an incubator for 2 days, followed by histochemical GUS staining. For GUS staining, the tobacco leaves were submersed in GUS reaction mix [0.05 mM sodium phosphate buffer, pH 7.0, 1 mM X-gluc and 0.1% (v/v) Triton X-100], and were incubated at 37 °C overnight, followed by a 70% (v/v) ethanol wash.

Southern and Western blot analysis

Genomic DNA was extracted from transgenic tobacco leaves using a DNeasy Plant Mini Kit (Qiagen GmbH, Hilden, Germany). For Southern blotting analysis, 15 µg of DNA was digested overnight with the restriction enzyme KpnI without cleavage sites in PbrMYB21. The PbrMYB21 overexpressing vector was used as a positive control. The digested products were fractionated on a 0.8% agarose gel, followed by transfer to Hybond-N+ nylon membrane (Amersham Pharmacia Biotech, NJ, USA). Hybridization was carried out using a DIG-High Prime DNA Labeling and Detection Starter Kit II according to the method of manufacturer's instruction (Roche, Germany).

A total of 1 g of transgenic tobacco plants for each sample were ground in the buffer containing 20 mM Tris (pH 7.4), 100 mM NaCl, 0.5% NonidetP-40, 0.5 mM EDTA, 0.5 mM phenylmethylsulfonyl fluoride and 0.5% Protease Inhibitor Cocktail (Sigma-Aldrich). PbrMYB21 protein levels were determined by protein gel blotting using an anti-His antibody, and the protein extracts were separated on a 10% SDS-PAGE gel and transferred to polyvinylidene difluoride membranes (Millipore) using an electrotransfer apparatus (Bio-Rad). The membranes were incubated with anti-His (Sigma-Aldrich) primary antibodies and then peroxidase-conjugated secondary antibodies (Abcam) before visualization of immunoreactive proteins using an ECL detection kit (Millipore). Actin served as a protein loading control.

Emsa

EMSAs were performed using the LightShift Chemiluminescent EMSA Kit (Pierce) according to the manufacturer's protocol and as described by Hu et al. (2016). For the His-PbrMYB21 construct, the CDS of PbrMYB21 was amplified and inserted into pCzn1-His vector to generate the recombinant His-PbrMYB21 protein. The recombinant protein was expressed and purified using Ni-IDA resin according to the manufacturer's instructions. The binding activity of the protein was analysed using an oligonucleotide containing CAACTG motif present in the PbrADC promoter, labelled using the biotin 3′ End DNA Labeling Kit (Thermo Fisher Scientific). The same fragment without biotin labelling was used as a competitor. In addition, a mutated fragment of the probe was also labelled, in which CAACTG was replaced by CGCTGG. The binding reaction was incubated with the fusion protein in a 20 µL reaction solution at room temperature for 20 min. The DNA–protein complexes were separated on 6.5% nondenaturing polyacrylamide gels and electrophoretically transferred to a nylon membrane (GE Healthcare), and detected following the manufacturer's instructions. After UV cross-linking, migration of the biotin-labelled probes on the membrane was detected using streptavidin–horseradish peroxidase conjugates that bind to biotin and the chemiluminescent substrate according to the manufacturer's instructions. This experiment was performed three times.

Statistical analysis

The data were statistically processed using the SAS software package (SAS Institute); analysis of variance (ANOVA) was used to compare the statistical difference based on t-test at the significance levels of $P < 0.05$ (*), $P < 0.01$ (**), and $P < 0.001$ (***). Three technical replicates were used for each sample, and the data are shown as means ± standard errors (SE; $n = 3$). Three biological replicates were used for each of the genotypes, the wild type, OE-9, OE-17, OE-20, pTRV-PbrMYB21 plants and pTRV plants.

Acknowledgements

This work was supported by the National High-tech R&D Program (863) of China (2013AA102606-02), National Natural Science Foundation of China (31301758), the Jiangsu Provincial Natural Science Foundation (BK20150681, BK20130689), the Research Fund for the Doctoral Program of Higher Education (20130097120020), the Fundamental Research Funds for the Central Universities (KYZ201607, KYTZ201401, SK2014007), the Ministry of Education of Humanities and Social Science project (14YJC630058), the National Postdoctoral Fund (2014M551615, 2013T60545, 2012M521092) and the Jiangsu Provincial Postdoctoral Fund (1401125C, 1201019B).

References

Agarwal, M., Hao, Y.J., Kapoor, A., Dong, C.H., Fujii, H., Zheng, X.W. and Zhu, J.K. (2006) A R2R3 type MYB transcription factor is involved in the cold regulation of CBF genes and in acquired freezing tolerance. J. Biol. Chem. **281**, 37636–37645.

Alcazar, R., Altabella, T., Marco, F., Bortolotti, C., Reymond, M., Koncz, C., Carrasco, P. et al. (2010) Polyamines: molecules with regulatory functions in plant abiotic stress tolerance. Planta, **231**, 1237–1249.

Baek, D., Kim, M.C., Chun, H.J., Kang, S., Park, H.C., Shen, M. et al. (2013) Regulation of miR399f transcription by AtMYB2 affects phosphate starvation responses in Arabidopsis. Plant Physiol. **161**, 362–373.

Borevitz, J.O., Xia, Y., Blount, J., Dixon, R.A. and Lamb, C. (2000) Activation tagging identifies a conserved MYB regulator of phenylpropanoid biosynthesis. Plant Cell. **12**, 2383–2394.

Chen, Y.H., Yang, X.Y., He, K., Liu, M.H., Li, J.G., Li, Z.Q. et al. (2006) The MYB transcription factor superfamily of Arabidopsis: expression analysis and phylogenetic comparison with the rice MYB family. Plant Mol. Biol. **60**, 107–124.

Cominelli, E., Galbiati, M., Vavasseur, A., Conti, L., Sala, T., Vuylsteke, M., Leonhardt, N. et al. (2005) A guard-cell-specific MYB transcription factor regulates stomatal movements and plant drought tolerance. Curr. Biol. **15**, 1196–1200.

Czemmel, S., Stracke, R., Weisshaar, B., Cordon, N., Harris, N.N., Walker, A.R., Robinson, S.P. et al. (2009) The grapevine R2R3-MYB transcription factor VvMYBF1 regulates flavonol synthesis in developing grape berries. Plant Physiol. **151**, 1513–1530.

Drolet, G., Dumbroff, E.B., Legge, R.L. and Thompson, J.E. (1986) Radical scavenging properties of polyamines. Phytochemistry, **25**, 367–371.

Dubos, C., Stracke, R., Grotewold, E., Weisshaar, B., Martin, C. and Lepiniec, L. (2010) MYB transcription factors in Arabidopsis. Trends Plant Sci. **15**, 573–581.

Evans, P.T. and Malmberg, R.L. (1989) Do polyamines have roles in plant development. Annu. Rev. Plant Physiol. Plant Mol. Biol. **40**, 235–269.

Farooq, M., Wahid, A., Kobayashi, N., Fujita, D. and Basra, S.M.A. (2009) Plant drought stress: effects, mechanisms and management. AgronSustain Dev. **29**, 185–212.

Gill, S.S. and Tuteja, N. (2010) Reactive oxygen species and antioxidant machinery in abiotic stress tolerance in crop plants. Plant Physiol. Biochem. **48**, 909–930.

Gong, X., Zhang, J., Hu, J., Wang, W., Wu, H., Zhang, Q. and Liu, J.H. (2015) FcWRKY70, a WRKY protein of Fortunella crassifolia, functions in drought tolerance and modulates putrescine synthesis by regulating arginine decarboxylase gene. Plant Cell Environ. **38**, 2248–2262.

Groppa, M.D. and Benavides, M.P. (2008) Polyamines and abiotic stress: recent advances. *Amino Acids*, **34**, 35–45.

Guo, L., Yang, H., Zhang, X. and Yang, S. (2013) Lipid transfer protein 3 as a target of MYB96 mediates freezing and drought stress in Arabidopsis. *J. Exp. Bot.* **64**, 1755–1767.

Guo, H., Wang, Y., Wang, L., Hu, P., Wang, Y., Jia, Y., Zhang, C. *et al.* (2016) Expression of the MYB transcription factor gene BplMYB46 affects abiotic stress tolerance and secondary cell wall deposition in Betula platyphylla. *Plant Biotechnol. J.* **15**, 107–121. doi:10.1111/pbi.12595.

Hirayama, T. and Shinozaki, K. (2010) Research on plant abiotic stress responses in the post-genome era: past, present and future. *Plant J.* **61**, 1041–1052.

Horsch, R.B., Fry, J.E., Hoffmann, N.L., Eichholtz, D., Rogers, S.G. and Fraley, R.T. (1985) A simple and general-method for transferring genes into plants. *Science*, **227**, 1229–1231.

Hu, D.G., Sun, C.H., Ma, Q.J., You, C.X., Cheng, L.L. and Hao, Y.J. (2016) MdMYB1 regulates anthocyanin and malate accumulation by directly facilitating their transport into vacuoles in apples. *Plant Physiol.* **170**, 1315–1330.

Huang, X.S., Luo, T., Fu, X.Z., Fan, Q.J. and Liu, J.H. (2011) Cloning and molecular characterization of a mitogen-activated protein kinase gene from *Poncirus trifoliata* whose ectopic expression confers dehydration/drought tolerance in transgenic tobacco. *J. Exp. Bot.* **62**, 5191–5206.

Huang, X.S., Wang, W., Zhang, Q. and Liu, J.H. (2013) A basic helix-loop-helix transcription factor, PtrbHLH, of Poncirus trifoliata confers cold tolerance and modulates peroxidase-mediated scavenging of hydrogen peroxide. *Plant Physiol.* **162**, 1178–1194.

Huang, P., Chen, H., Mu, R., Yuan, X., Zhang, H.S. and Huang, J. (2015a) OsMYB511 encodes a MYB domain transcription activator early regulated by abiotic stress in rice. *Genet. Mol. Res.* **14**, 9506–9517.

Huang, X.S., Zhang, Q.H., Zhu, D.X., Fu, X.Z., Wang, M., Zhang, Q., Moriguchi, T. *et al.* (2015b) ICE1 of *Poncirus trifoliata* functions in cold tolerance by modulating polyamine levels through interacting with arginine decarboxylase. *J. Exp. Bot.* **66**, 3259–3274.

Hundertmark, M. and Hincha, D.K. (2008) LEA (Late Embryogenesis Abundant) proteins and their encoding genes in *Arabidopsis thaliana*. *BMC Genomics* **9**, 118.

Ingram, J. and Bartels, D. (1996) The molecular basis of dehydration tolerance in plants. *Annu. Rev. Plant Physiol. Plant Mol. Biol.* **47**, 377–403.

Jung, C., Seo, J.S., Han, S.W., Koo, Y.J., Kim, C.H., Song, S.I., Nahm, B.H. *et al.* (2008) Overexpression of AtMYB44 enhances stomatal closure to confer abiotic stress tolerance in transgenic *Arabidopsis*. *Plant Physiol.* **146**, 623–635.

Kim, J.H., Nguyen, N.H., Jeong, C.Y., Nguyen, N.T., Hong, S.W. and Lee, H. (2013) Loss of the R2R3 MYB, AtMyb73, causes hyper-induction of the SOS1 and SOS3 genes in response to high salinity in Arabidopsis. *J. Plant Physiol.* **170**, 1461–1465.

Kranz, H.D., Denekamp, M., Greco, R., Jin, H., Leyva, A., Meissner, R.C., Petroni, K. *et al.* (1998) Towards functional characterisation of the members of the R2R3-MYB gene family from *Arabidopsis thaliana*. *Plant J.* **16**, 263–276.

Kumar, K.R.R. and Kirti, P.B. (2010) A mitogen-activated protein kinase, AhMPK6 from peanut localizes to the nucleus and also induces defense responses upon transient expression in tobacco. *Plant Physiol. Biochem.* **48**, 481–486.

Lee, H.G. and Seo, P.J. (2015) The MYB96-HHP module integrates cold and abscisic acid signaling to activate the CBF-COR pathway in Arabidopsis. *Plant J.* **82**, 962–977.

Li, Y., Jiang, J., Du, M.L., Li, L., Wang, X.L. and Li, X.B. (2013) A cotton gene encoding MYB-like transcription factor is specifically expressed in pollen and is involved in regulation of late anther/pollen development. *Plant Cell Physiol.* **54**, 893–906.

Li, K.Q., Xu, X.Y. and Huang, X.S. (2016) Identification of differentially expressed genes related to dehydration resistance in a highly drought-tolerant pear, Pyrus betulaefolia, as through RNA-Seq. *PLoS ONE*, **11**, e0149352. doi: 10.1371/journal.pone.0149352.

Liang, Y.K., Dubos, C., Dodd, I.C., Holroyd, G.H., Hetherington, A.M. and Campbell, M.M. (2005) AtMYB61, an R2R3-MYB transcription factor controlling stomatal aperture in *Arabidopsis thaliana*. *Curr. Biol.* **15**, 1201–1206.

Liao, Y., Zou, H.F., Wang, H.W., Zhang, W.K., Ma, B., Zhang, J.S. and Chen, S.Y. (2008) Soybean GmMYB76, GmMYB92, and GmMYB177 genes confer stress tolerance in transgenic Arabidopsis plants. *Cell Res.* **18**, 1047–1060.

Lippold, F., Sanchez, D.H., Musialak, M., Schlereth, A., Scheible, W.R., Hincha, D.K. and Udvardi, M.K. (2009) AtMyb41 regulates transcriptional and metabolic responses to osmotic stress in Arabidopsis. *Plant Physiol.* **149**, 1761–1772.

Lipsick, J.S. (1996) One billion years of Myb. *Oncogene*, **13**, 223–235.

Liu, J.H., Kitashiba, H., Wang, J., Ban, Y. and Moriguchi, T. (2007) Polyamines and their ability to provide environmental stress tolerance to plants. *Plant Biotechnol.* **24**, 117–126.

Liu, J.H., Ban, Y., Wen, X.P., Nakajima, I. and Moriguchi, T. (2009) Molecular cloning and expression analysis of an arginine decarboxylase gene from peach (*Prunus persica*). *Gene*, **429**, 10–17.

Liu, J.H., Peng, T. and Dai, W.S. (2014) Critical cis-acting elements and interacting transcription factors: key players associated with abiotic stress responses in plants. *Plant Mol. Biol. Rep.* **32**, 303–317.

Livak, K.J. and Schmittgen, T.D. (2001) Analysis of relative gene expression data using real-time quantitative PCR and the 2(−Delta Delta C(T)) Method. *Methods*, **25**, 402–408.

Ma, Q.B., Dai, X.Y., Xu, Y.Y., Guo, J., Liu, Y.J., Chen, N., Xiao, J. *et al.* (2009) Enhanced tolerance to chilling stress in OsMYB3R-2 transgenic rice is mediated by alteration in cell cycle and ectopic expression of stress genes. *Plant Physiol.* **150**, 244–256.

Matsui, K., Umemura, Y. and Ohme-Takagi, M. (2008) AtMYBL2, a protein with a single MYB domain, acts as a negative regulator of anthocyanin biosynthesis in Arabidopsis. *Plant J.* **55**, 954–967.

Mellway, R.D., Tran, L.T., Prouse, M.B., Campbell, M.M. and Constabel, C.P. (2009) The wound-, pathogen-, and ultraviolet B-responsive MYB134 gene encodes an R2R3 MYB transcription factor that regulates proanthocyanidin synthesis in poplar. *Plant Physiol.* **150**, 924–941.

Miller, G., Suzuki, N., Ciftci-Yilmaz, S. and Mittler, R. (2010) Reactive oxygen species homeostasis and signalling during drought and salinity stresses. *Plant Cell Environ.* **33**, 453–467.

Mu, R.L., Cao, Y.R., Liu, Y.F., Lei, G., Zou, H.F., Liao, Y., Wang, H.W. *et al.* (2009) An R2R3-type transcription factor gene AtMYB59 regulates root growth and cell cycle progression in Arabidopsis. *Cell Res.* **19**, 1291–1304.

Muller, D., Schmitz, G. and Theres, K. (2006) Blind homologous R2R3 Myb genes control the pattern of lateral meristem initiation in Arabidopsis. *Plant Cell*, **18**, 586–597.

Ogata, K., Morikawa, S., Nakamura, H., Sekikawa, A., Inoue, T., Kanai, H., Sarai, A. *et al.* (1994) Solution structure of a specific DNA complex of the Myb DNA-binding domain with cooperative recognition helices. *Cell*, **79**, 639–648.

Ogata, K., Morikawa, S., Nakamura, H., Hojo, H., Yoshimura, S., Zhang, R., Aimoto, S. *et al.* (1995) Comparison of the free and DNA-complexed forms of the DNA-binding domain from c-Myb. *Nat. Struct. Biol.* **2**, 309–320.

Oh, J.E., Kwon, Y., Kim, J.H., Noh, H., Hong, S.W. and Lee, H. (2011) A dual role for MYB60 in stomatal regulation and root growth of *Arabidopsis thaliana* under drought stress. *Plant Mol. Biol.* **77**, 91–103.

Osakabe, Y., Osakabe, K., Shinozaki, K. and Tran, L.S.P. (2014) Response of plants to water stress. *Front Plant Sci.* **5**, 86.

Park, J.S., Kim, J.B., Cho, K.J., Cheon, C.I., Sung, M.K., Choung, M.G. and Roh, K.H. (2008) Arabidopsis R2R3-MYB transcription factor AtMYB60 functions as a transcriptional repressor of anthocyanin biosynthesis in lettuce (*Lactuca sativa*). *Plant Cell Rep.* **27**, 985–994.

Pastori, G.M. and Foyer, C.H. (2002) Common components, networks, and pathways of cross-tolerance to stress. The central role of "redox" and abscisic acid-mediated controls. *Plant Physiol.* **129**, 460–468.

Ramegowda, V., Senthil-Kumar, M., Udayakumar, M. and Mysore, K.S. (2013) A high-throughput virus-induced gene silencing protocol identifies genes involved in multi-stress tolerance. *BMC Plant Biol.* **13**, 193.

Ren, X.Z., Chen, Z.Z., Liu, Y., Zhang, H.R., Zhang, M., Liu, Q.A., Hong, X.H. *et al.* (2010) ABO3, a WRKY transcription factor, mediates plant responses to abscisic acid and drought tolerance in Arabidopsis. *Plant J.* **63**, 417–429.

Ryu, K.H., Kang, Y.H., Park, Y.H., Hwang, D., Schiefelbein, J. and Lee, M.M. (2005) The WEREWOLF MYB protein directly regulates CAPRICE transcription during cell fate specification in the Arabidopsis root epidermis. *Development*, **132**, 4765–4775.

Seki, M., Narusaka, M., Abe, H., Kasuga, M., Yamaguchi-Shinozaki, K., Carninci, P., Hayashizaki, Y. *et al.* (2001) Monitoring the expression pattern of 1300 Arabidopsis genes under drought and cold stresses by using a full-length cDNA microarray. *Plant Cell*, **13**, 61–72.

Seo, P.J., Xiang, F., Qiao, M., Park, J.Y., Lee, Y.N., Kim, S.G., Lee, Y.H. *et al.* (2009) The MYB96 transcription factor mediates abscisic acid signaling during drought stress response in Arabidopsis. *Plant Physiol.* **151**, 275–289.

Seo, P.J., Lee, S.B., Suh, M.C., Park, M.J., Go, Y.S. and Park, C.M. (2011) The MYB96 transcription factor regulates cuticular wax biosynthesis under drought conditions in *Arabidopsis*. *Plant Cell*, **23**, 1138–1152.

Shi, J., Fu, X.Z., Peng, T., Huang, X.S., Fan, Q.J. and Liu, J.H. (2010) Spermine pretreatment confers dehydration tolerance of citrus in vitro plants via modulation of antioxidative capacity and stomatal response. *Tree Physiol.* **30**, 914–922.

Shim, J.S., Jung, C., Lee, S., Min, K., Lee, Y.W., Choi, Y., Lee, J.S. *et al.* (2013) AtMYB44 regulates WRKY70 expression and modulates antagonistic interaction between salicylic acid and jasmonic acid signaling. *Plant J.* **73**, 483–495.

Sreenivasulu, N., Sopory, S.K. and Kishor, P.B.K. (2007) Deciphering the regulatory mechanisms of abiotic stress tolerance in plants by genomic approaches. *Gene*, **388**, 1–13.

Su, C.F., Wang, Y.C., Hsieh, T.H., Lu, C.A., Tseng, T.H. and Yu, S.M. (2010) A novel MYBS3-dependent pathway confers cold tolerance in rice. *Plant Physiol.* **153**, 145–158.

Thomashow, M.F., Gilmour, S.J., Stockinger, E.J., Jaglo-Ottosen, K.R. and Zarka, D.G. (2001) Role of the Arabidopsis CBF transcriptional activators in cold acclimation. *Physiol Plantarum.* **112**, 171–175.

Umezawa, T., Fujita, M., Fujita, Y., Yamaguchi-Shinozaki, K. and Shinozaki, K. (2006) Engineering drought tolerance in plants: discovering and tailoring genes to unlock the future. *Curr. Opin. Biotechnol.* **17**, 113–122.

Voinnet, O., Rivas, S., Mestre, P. and Baulcombe, D. (2003) An enhanced transient expression system in plants based on suppression of gene silencing by the p19 protein of tomato bushy stunt virus. *Plant J.* **33**, 949–956.

Volpe, V., Dell'Aglio, E., Giovannetti, M., Ruberti, C., Costa, A., Genre, A., Guether, M. *et al.* (2013) An AM-induced, MYB-family gene of *Lotus japonicus* (LjMAMI) affects root growth in an AM-independent manner. *Plant J.* **73**, 442–455.

Wang, K., Bolitho, K., Grafton, K., Kortstee, A., Karunairetnam, S., McGhie, T.K., Espley, R.V. *et al.* (2010) An R2R3 MYB transcription factor associated with regulation of the anthocyanin biosynthetic pathway in Rosaceae. *BMC Plant Biol.* **10**, 50.

Wang, J., Sun, P.P., Chen, C.L., Wang, Y., Fu, X.Z. and Liu, J.H. (2011) An arginine decarboxylase gene PtADC from *Poncirus trifoliata* confers abiotic stress tolerance and promotes primary root growth in *Arabidopsis*. *J. Exp. Bot.* **62**, 2899–2914.

Wang, Y., Wang, L., Zou, Y., Chen, L., Cai, Z., Zhang, S., Zhao, F. *et al.* (2014) Soybean miR172c targets the repressive AP2 transcription factor NNC1 to activate ENOD40 expression and regulate nodule initiation. *Plant Cell*, **26**, 4782–4801.

Wang, T., Tohge, T., Ivakov, A., Mueller-Roeber, B., Fernie, A.R., Mutwil, M., Schippers, J.H. *et al.* (2015) Salt-related MYB1 coordinates abscisic acid biosynthesis and signaling during salt stress in Arabidopsis. *Plant Physiol.* **169**, 1027–1041.

Wang, F., Guo, Z.X., Li, H.Z., Wang, M.M., Onac, E., Zhou, J., Xia, X.J. *et al.* (2016) Phytochrome A and B function antagonistically to regulate cold tolerance via abscisic acid-dependent jasmonate signaling. *Plant Physiol.* **170**, 459–471.

Wu, H., Fu, B., Sun, P., Xiao, C. and Liu, J.H. (2016) A NAC transcription factor represses putrescine biosynthesis and affects drought tolerance. *Plant Physiol.* **172**, 1532–1547.

Xie, X.B., Li, S., Zhang, R.F., Zhao, J., Chen, Y.C., Zhao, Q., Yao, Y.X. *et al.* (2012) The bHLH transcription factor MdbHLH3 promotes anthocyanin accumulation and fruit colouration in response to low temperature in apples. *Plant Cell Environ.* **35**, 1884–1897.

Xu, R., Wang, Y.H., Zheng, H., Lu, W., Wu, C.G., Huang, J.G., Yan, K. *et al.* (2015) Salt-induced transcription factor MYB74 is regulated by the RNA-directed DNA methylation pathway in *Arabidopsis*. *J. Exp. Bot.* **66**, 5997–6008.

Yan, J.H., Wang, B., Zhong, Y.P., Yao, L.M., Cheng, L.J. and Wu, T.L. (2015) The soybean R2R3 MYB transcription factor GmMYB100 negatively regulates plant flavonoid biosynthesis. *Plant Mol. Biol.* **89**, 35–48.

Yang, J., Zhang, J., Liu, K., Wang, Z. and Liu, L. (2007) Involvement of polyamines in the drought resistance of rice. *J. Exp. Bot.* **58**, 1545–1555.

Yang, A., Dai, X.Y. and Zhang, W.H. (2012) A R2R3-type MYB gene, OsMYB2, is involved in salt, cold, and dehydration tolerance in rice. *J. Exp. Bot.* **63**, 2541–2556.

Yu, Z.X., Li, J.X., Yang, C.Q., Hu, W.L., Wang, L.J. and Chen, X.Y. (2012) The jasmonate-responsive AP2/ERF transcription factors AaERF1 and AaERF2 positively regulate artemisinin biosynthesis in *Artemisia annua* L. *Mol. Plant*, **5**, 353–365.

Zhang, Q.H., Wang, M., Hu, J.B., Wang, W., Fu, X.Z. and Liu, J.H. (2015) PtrABF of *Poncirus trifoliata* functions in dehydration tolerance by reducing stomatal density and maintaining reactive oxygen species homeostasis. *J. Exp. Bot.* **66**, 5911–5927.

Zhao, Q., Ren, Y.R., Wang, Q.J., Yao, Y.X., You, C.X. and Hao, Y.J. (2016) Overexpression of MdbHLH104 gene enhances the tolerance to iron deficiency in apple. *Plant Biotechnol. J.* **14**, 1633–1645.

3

Eugenol specialty chemical production in transgenic poplar (*Populus tremula* × *P. alba*) field trials

Da Lu[1,†], Xianghe Yuan[1,†], Sung-Jin Kim[1,†], Joaquim V. Marques[1], P. Pawan Chakravarthy[1], Syed G. A. Moinuddin[1], Randi Luchterhand[2], Barri Herman[1,2], Laurence B. Davin[1] and Norman G. Lewis[1,*]

[1]Institute of Biological Chemistry, Washington State University, Pullman, WA, USA
[2]Puyallup Research and Extension Center, Washington State University, Puyallup, WA, USA

*Correspondence
email lewisn@wsu.edu
[†]These authors contributed equally to the article.

Summary

A foundational study assessed effects of biochemical pathway introduction into poplar to produce eugenol, chavicol, *p*-anol, isoeugenol and their sequestered storage products, from potentially available substrates, coniferyl and *p*-coumaryl alcohols. At the onset, it was unknown whether significant carbon flux to monolignols vs. other phenylpropanoid (acetate) pathway metabolites would be kinetically favoured. Various transgenic poplar lines generated eugenol and chavicol glucosides in *ca.* 0.45% (~0.35 and ~0.1%, respectively) of dry weight foliage tissue in field trials, as well as their corresponding aglycones in trace amounts. There were only traces of any of these metabolites in branch tissues, even after ~4-year field trials. Levels of bioproduct accumulation in foliage plateaued, even at the lowest introduced gene expression levels, suggesting limited monolignol substrate availability. Nevertheless, this level still allows foliage collection for platform chemical production, with the remaining (stem) biomass available for wood, pulp/paper and bioenergy product purposes. Several transformed lines displayed unexpected precocious flowering after 4-year field trial growth. This necessitated terminating (felling) these particular plants, as USDA APHIS prohibits the possibility of their interacting (cross-pollination, etc.) with wild-type (native plant) lines. In future, additional biotechnological approaches can be employed (e.g. gene editing) to produce sterile plant lines, to avoid such complications. While increased gene expression did not increase target bioproduct accumulation, the exciting possibility now exists of significantly increasing their amounts (e.g. 10- to 40-fold plus) in foliage and stems via systematic deployment of numerous 'omics', systems biology, synthetic biology and metabolic flux modelling approaches.

Keywords: Eugenol, hybrid poplar, phytochemical factories, phenylpropanoid metabolism.

Introduction

Naturally occurring tree species worldwide have long been used by humanity as major sources of wood products, pulp, paper, other polymers (e.g. natural rubber), resins and oils, spices, fragrances, perfumes, various specialty chemicals, medicinals and numerous foodstuffs, such as fruits and nuts (Patten *et al.*, 2010). More recently, through genetic engineering, hybrid poplar (*Populus tremula* × *P. alba*) was also demonstrated able to produce the characteristic rose oil constituent, phenylethanol (**1**, Figure 1a), as a specialty chemical in amounts potentially suitable for commercial production. The phenylethanol (**1**) accumulating was largely sequestered *in vivo* as its phenylethanol glucoside (**2**), rather than in its aglycone form (Costa *et al.*, 2013a,b; Doughton, 2014).

The U.S. Congress has stipulated that, by 2030, 30% of petroleum-derived gasoline is to be replaced by renewables from lignocellulosic plant/algae sources (Houghton *et al.*, 2006). Yet, currently, the key technical barriers to massively utilize lignocellulosic feedstocks for liquid fuels have not been overcome so as to be economically competitive. Much attention is thus now being increasingly given to developing the means to economically produce value-added bioenergy crops with valuable bioproducts, in order to meet—if not exceed—congressional targets. That is, there is an increasing emphasis on producing bioproducts, as well

as lignocellulosic biomass, to enable future biorefineries to be economically competitive.

Among various plant species considered for such purposes, hybrid poplar has many advantages including: rapid growth and development; ease of biological transformation (e.g. for introducing new biochemical pathways as above or for reducing lignocellulosic recalcitrance); facile harvesting of plant material through coppicing (e.g. perhaps three harvests per 2 years growth in ultrashort rotations), as well as for its considerable potential as a bioenergy crop suitable for deployment in marginal lands (i.e. land not used for food production). An additional potential advantage for designing future genetic manipulations (including gene editing) with hybrid poplar is that its close relative *P. trichocarpa* has had its genome sequenced (Tuskan *et al.*, 2006).

In the current study, genetic engineering of poplar was envisaged to mainly produce bioproducts, eugenol (**4**), isoeugenol (**8**) or homologs/derivatives thereof via heterologous introduction of genes encoding their pathways from coniferyl alcohol (**12**) (Figure 1b) (Kim *et al.*, 2014). Eugenol (**4**) was of particular interest, as it is a major constituent of the clove tree (*Syzygium aromaticum*, or *Eugenia caryophyllata*), where it largely accumulates in cloves and clove oil [with the latter up to 89% eugenol (**4**)] (Chaieb *et al.*, 2007). Moreover, cloves and clove oil have been highly valued in the West since the Age of Exploration,

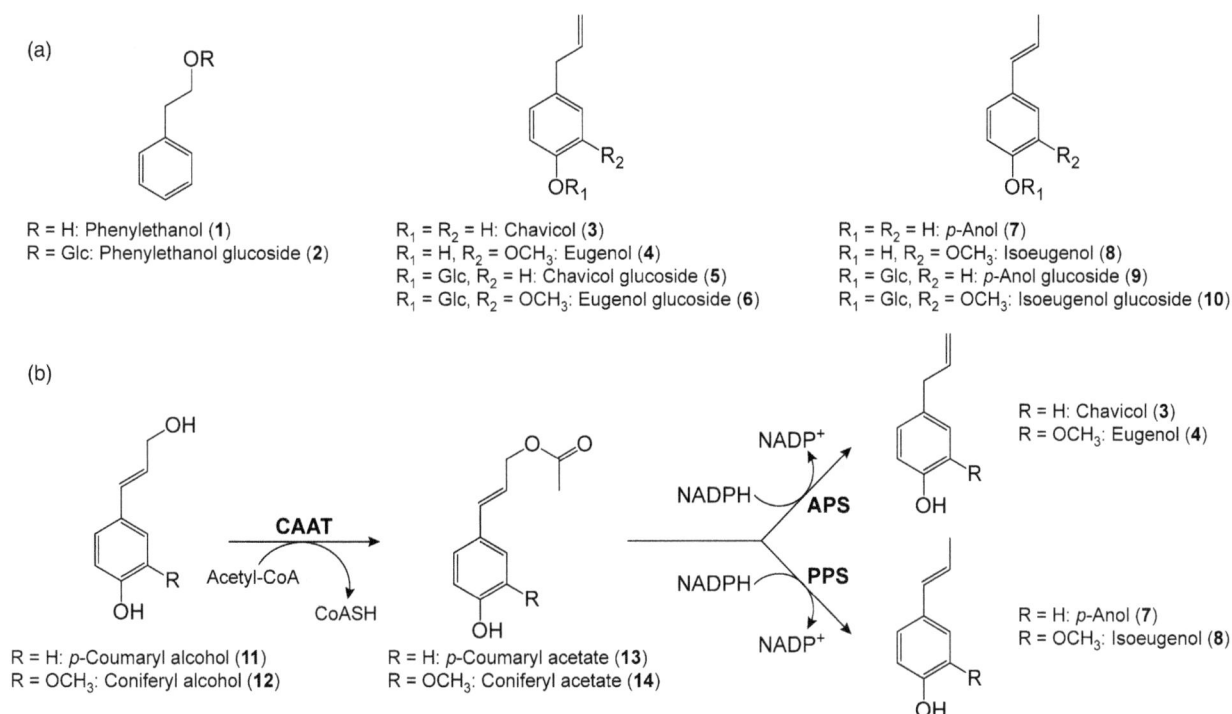

Figure 1 Phenyl derivatives and monolignol conversions to allyl/propenylphenols. (a) Phenylethanol (**1**), phenylethanol glucoside (**2**), allyl/propenylphenols **3**, **4**, **7** and **8** and allyl/propenylphenol glucosides **5**, **6**, **9** and **10**. (b) Biosynthetic pathway to allyl/propenylphenols **3**, **4**, **7** and **8**. APS, allylphenol synthase; CAAT, cinnamyl alcohol acyltransferase; PPS, propenylphenol synthase.

when maritime trade routes connected European consumers to spice producers in Asia and Eastern Africa. Clove oils are currently obtained from dried flower buds, as well as leaves and stems, with the bud oil being most highly valued. Estimated clove oil production for 2016 was 3500–4000 tonnes, with prices ranging from ~$12/kg (clove leaf oil) to ~$80/kg (clove bud oil) (Caiger, 2016).

Eugenol (**4**) has well-known antioxidant, antimicrobial and anti-inflammatory properties. It is commonly used as a pesticide and fumigant, as well as to protect foods from micro-organisms during storage. It also has the property of enhancing skin permeation of various drugs (for a review see Kamatou et al., 2012) and is widely used as an analgesic in dentistry (Skinner, 1940; Vassão et al., 2010; Weinberg et al., 1972).

Isoeugenol (**8**), by contrast, is one of the main volatiles emitted from petunia (Petunia hybrida) flowers (Verdonk et al., 2003) and is used somewhat as a fragrance/flavour additive in beverages, baked goods, chewing gum and other products such as perfumes and soaps (Badger et al., 2002). It also acts as a defence compound in planta (Koeduka et al., 2008).

It was, therefore, of interest to determine to what extent eugenol (**4**) and isoeugenol (**8**) (or derivatives thereof) could be produced in transgenic poplar in initial (foundational) field trials, as a potential chemical platform.

We describe the genetic engineering of P. tremula × P. alba for (i) potentially producing specialty chemicals eugenol (**4**), isoeugenol (**8**) and/or sequestered derivatives thereof, and (ii) establishing what, if any the effects are of such transformations on P. tremula × P. alba growth/development and (unexpected) precocious flowering in a field trial environment over a 4-year time frame. The genes required for the transformations were

originally obtained from the creosote bush (Larrea tridentata), with proof of principle for eugenol (**4**)/isoeugenol (**8**) production being first demonstrated in Escherichia coli (Kim et al., 2014). Results obtained are described below.

Results and Discussion

The biosynthetic pathway to eugenol (**4**)/isoeugenol (**8**) from coniferyl alcohol (**12**) involves an initial acylation step by a substrate versatile monolignol acyl transferase (trivially called coniferyl alcohol acyl transferase (CAAT)) to form, for example, acetate **14** which is converted into eugenol (**4**)/isoeugenol (**8**) via action of either the corresponding allylphenol synthase (APS) or propenylphenol synthase (PPS, Figure 1b) (Kim et al., 2014; Koeduka et al., 2006; Lewis et al., 2015; Vassão et al., 2006, 2007). The corresponding genes, namely LtCAAT, LtAPS1 and LtPPS1, were cloned from the creosote bush (Kim et al., 2014; Lewis et al., 2015; Vassão et al., 2010) as described in the Supporting Information.

LtAPS1 and LtPPS1 are members of a vascular plant phenyl-propanoid oxidoreductase family, which includes pinoresinol–lariciresinol reductase (PLR) in lignan biosynthesis, all of which have very similar biochemical mechanisms (Vassão et al., 2010). LtAPS1 and LtPPS1, however, only share moderate homology (49/42% identity and 64/60% similarity, respectively) to PLR from Forsythia intermedia (Dinkova-Kostova et al., 1996; Min et al., 2003).

Hybrid poplar transformations

For potential production of allyl/propenylphenols **4** and **8** in transgenic hybrid poplar [as previously carried out for E. coli (Kim et al., 2014)], p35S::LtAPS1::T35S and p35S::LtPPS1::T35S gene cassettes were individually excised from their pK2GW7 gateway vector constructs. These were separately fused together with the pK2GW7/LtCAAT1 vector construct, with constructs harbouring LtCAAT1/LtAPS1 and LtCAAT1/LtPPS1 (Figure S1) individually subjected to an Agrobacterium EHA105 transformation procedure (see Costa et al., 2013b and Supporting Information).

Transformed Agrobacterium EHA105 lines harbouring each pK2GW7 vector construct were then individually incubated with leaves, petioles and internodal stem segments harvested from 50-day-old P. tremula × P. alba (INRA 717-1B4) for the poplar transformation (see Table S1 and Figure S2).

Quantitative real-time PCR analyses were next carried out on leaf tissue (see Table S2) using first-strand cDNA synthesized from total RNA from each plant as a template. For each line, the LtCAAT1 and LtAPS1 or LtPPS1 mRNA expression levels were individually determined, with WT hybrid poplar (P. tremula × P. alba) and L. tridentata employed as comparative controls. (There was, as expected, no detectable expression under the conditions employed with qPCR in the WT lines, when using specific primers for L. tridentata LtCAAT1, LtAPS1 and LtPPS1, hence zero values).

Gene expression levels of LtCAAT1 and LtAPS1 in 212 transformed poplar lines growing in the greenhouse ranged from 3 to 5000 and 0.2 to 50 times higher, respectively, as compared to the L. tridentata control. Analogously, LtCAAT1 and LtPPS1 expression levels were 10–450 and 0.1–18 times higher in the 178 LtCAAT1::LtPPS1 transgenic lines. Representative transformants (57 APS lines in Figure S3a,b and 46 PPS lines in Figure S4a, b) were selected for subsequent field trials, to cover a broad range of relative gene expression levels.

Transgenic hybrid poplar field trials

Selected transgenic lines (Figures S3a,b and S4a,b) were analysed for eugenol (4), isoeugenol (8) and derivatives thereof (6 and 10) produced under natural outdoor settings, together with their overall growth/development characteristics relative to WT controls. Growth/development under field trial conditions was particularly important, as one of our other earlier collaborative studies (Voelker et al., 2010) had found significant deleterious effects on transgenic poplar growth when 4-coumarate CoA ligase (4CL) was down-regulated using comparable approaches as here. That is, some of the 4CL lines in the previous field trials had bushy/stunted growth forms, whereas they were tree-like under greenhouse conditions. This was not the case though for the poplar producing phenylethanol glucoside (2), as the transformants grew normally in field trials (Costa et al., 2013b).

Transgenic poplar parent plants above were delivered for field trial evaluation in September and October of 2011, with each parent being ca. 30–60 cm tall. These were grown in 1-gallon containers in greenhouses, under 16-h light and 24.5 °C day temperature and 20 °C night temperature conditions. Fertilizer (Wilgro, Wilbur-Ellis, 20:20:20) was supplied for every watering at 600-μs conductivity (incoming H_2O 208 μs), with parents flushed every 10 days for 3 days with only incoming H_2O to reduce pot salt load. The first cuttings from the parents were placed in rooting trays in December 2011, then transplanted in Beaver Plastics 45 cavity trays in February 2012 and further grown in the greenhouse to the end of March 2012. Thereafter, transplanted cuttings with a 3 to 4-mm caliper above the original cutting stem were moved to cold frames and grown until planting specification sizes of 45–60 cm in height were obtained. These thus provided the biological replicates needed for the field trial described below.

The field trial with these lines was planted over 4 days in June (June 4th to 8th) 2012 (see Experimental procedures for details). The complete trial (together with other transgenic lines) was planted at 1.52 × 1.52 m spacing to fit a total of 5019 trial trees inside the area allocated by the USDA APHIS permit. There were 20 replicates, generated as described in the paragraph above, for each WT and transgenic line, these being placed in a statistically random design.

In 2012, stem height measurements were made, whereas in 2013 and 2014 (see Figure S5), plant volume production was estimated by measuring stem heights and diameters. There was no obvious massive deleterious effect on height measurements through increasing (introduced) gene expression levels, even though some variations in growth were noted between different lines.

The trial design had very good statistical power, where 20 replicates provided 93% power at a 95% confidence limit; that is, designed so that the results supported the null hypothesis that there was a significant difference of 30 cm of growth between clones. This is a standard clonal forestry trial layout design, which would generally be used to select for the best growing clones across a trial site. The variation observed, however, was not in fact due to the site but instead competition between plant lines.

The aim of the trial here was to find the largest tree with the highest desired metabolite of interest level—in contrast to a production setting, which would select for the biggest tree with most wood product. With these trees, some lines grew slower. However, as the trial closes canopy (about year 2.5), then shorter trees cannot compete well for light and because of that also water and nutrients. To try and compete, they grow thinner and taller to try to capture some extra canopy growing space. Consequently, some cannot keep up and their diameters become more and more variable across the trial, an expected result in such trials. The key here, however, was in trying to select for the biggest volume tree with highest levels of metabolites of interest, with the smaller trees becoming more variable as they attempt to compete and are outgrown.

Field trial evaluations from 2012 to July 2016 were carried out under USDA APHIS approval, which allows for growth/development periods of 3–5 years, provided no flowering/pollination occurs with either transgenic lines or any adjacent/nearby WT poplars. This current requirement is to ensure that there is no 'escape' of the transgenic lines.

Unexpectedly, during Spring 2016 (4th growing season), 114 transgenic trees from 36 of the lines carrying the LtCAAT1/LtAPS1 genes (Figure S3c) and 23 transgenic trees from 16 of the lines carrying the LtCAAT1/LtPPS1 genes (Figure S4c) flowered precociously and were felled according to USDA APHIS requirements. (The USDA APHIS permit does not allow flowering, and the trees were felled as soon as catkins were noted.) Flowering was heavier either in the outer edges of the trial or where LtAPS and LtPPS lines were in close proximity to each other in the inner areas. However, as for stem height and diameter data, there was no direct correlation noted between expression level of introduced genes and trees that flowered (Figures S3 and S4). The reason(s) for precocious flowering is (are) not known at this time and could not have been anticipated. Precocious flowering has, however, been very recently reported with transformed Eucalyptus lines, although in that case, this was expected as they were specifically engineered to shorten flowering time (Klocko et al., 2016b).

Precocious flowering has two possible ramifications for commercialization: (i) plant lines are coppiced within 3 years or (ii) emerging gene editing procedures are further developed to prevent flowering onset (Klocko et al., 2016a).

Metabolomics

In the field trial, eleven of each of the different LtAPS and LtPPS transgenic poplar plant lines were selected for both targeted and nontargeted metabolomics analyses, with the remainder assessed for growth/development parameters. As indicated earlier, each line had 20 replicates in the trial randomly configured. Selected lines for metabolomics analyses were chosen to cover a broad range of LtCAAT, LtAPS and LtPPS gene expression levels, with six of the 20 replicates being analysed. In particular, it was of interest to establish what, if any, correlations could be made between gene expression and bioproduct accumulation level.

Targeted metabolomics analyses

Leaf and branch tissues from selected lines were individually harvested (August 2014, after ca. 2 years growth), extracted and analysed (see Experimental procedures). Two complementary approaches were utilized: (i) GCMS analysis of t-butyl methyl ether (containing 0.5 mM benzyl methyl ether internal standard (IS)) extracts of fresh tissue to determine amounts of 'free' eugenol (4), chavicol (3), isoeugenol (8) or p-anol (7) present in foliage or branch tissues, and (ii) LC/MS analyses of aqueous methanolic extracts of freeze-dried tissue to establish amounts, if any, of sequestered allyl/propenylphenols, such as the glucosides (5, 6, 9 and 10) or as other derivatives. [In this context, glycosylation is a very well-known mechanism in planta to sequester phenolic and other alcohol functionality storage products, for example phenylethanol glucoside (2) in transgenic poplars (Costa et al., 2013b).]

For assessments of eugenol (4), chavicol (3), isoeugenol (8) and p-anol (7) accumulating in the various transformed poplar lines, relative to WT, analyses (including identification and quantification) were on a fresh weight basis. In this regard, while eugenol (4), chavicol (3) and isoeugenol (8) were commercially available as standards, p-anol (7) was synthesized from p-hydroxybenzaldehyde via reaction with ethyl triphenyl phosphonium iodide in the presence of n-BuLi in ca. 86% yield (Kim et al., 2014). Conditions for their chromatographic resolution, and establishing their mass spectroscopic fragmentation patterns, were developed, as well as for GC/MS calibration and recovery. Only data for eugenol (4), chavicol (3) and isoeugenol (8) standards are shown in Figures 2a–f, as p-anol (7) was not detected in the plant sample analysed.

For plant analyses (harvesting, storage, extraction and recovery), authentic standard calibrations and spiking of tissues with standards individually (see Supporting Information) were carried out, with results normalized by use of the benzyl methyl ether IS, to account for recovery losses, etc. In WT leaf tissues, eugenol (4), chavicol (3) and isoeugenol (8) were in trace amounts (near detection limits), these being ≤0.0006, 0.0001, and 0.00005% fresh weight, respectively, with p-anol (7) not detected under the conditions employed. By contrast, in the LtAPS lines, amounts of eugenol (4), chavicol (3) and isoeugenol (8) in leaf tissue increased somewhat in amounts of up to 0.01, 0.0007 and 0.00009% fresh weight, that is 167, 70 and 1.8 times higher, respectively (data not shown). For LtPPS lines, amounts of eugenol (4) in leaf tissue increased to somewhat similar levels (up to 0.009% fresh weight), with chavicol (3) and isoeugenol (8) being present in

amounts of ca. 0.0005% and 0.0004%, respectively [All metabolites were identified based on same retention times and mass spectral fragmentation patterns (Figures 2g–l) as compared to authentic standards (Figures 2a–2f)]. Formation of chavicol (3) was not unexpected, given the known substrate versatility of the cloned enzymes, which can utilize other monolignols such as p-coumaryl alcohol (11) (Kim et al., 2014).

It was next instructive to ascertain whether allyl/propenylphenols 5, 6, 9, 10 were accumulating in sequestered form in leaf tissues, such as noted for phenylethanol glucoside (2) (Costa et al., 2013b). Thus, eugenol glucoside (6), isoeugenol glucoside (10), chavicol glucoside (5) and p-anol glucoside (9) were individually synthesized (see Supporting Information), with general synthetic protocols employed being described using eugenol glucoside (6) as an example. That is, α-acetobromoglucose was reacted with sodium eugenolate (obtained from eugenol (4) and sodium ethoxide) to generate eugenol-β-glucoside (6). After crystallization, its β-configuration was determined by analysis of its one- and two-dimensional homo- and heteronuclear ^{1}H spectra in dimethylsulphoxide-d6 solutions, with the β-glucopyranosyl linkage deduced from the higher coupling constant (J = 7.32 Hz) and lower δ (4.78 ppm) value of the anomeric proton (Mastelic et al., 2004).

Similarly, glucosides 5, 9 and 10 were synthesized, and all four glucosides were used for calibration curve standardization and normalization to the naringenin IS. Both eugenol glucoside (6) and chavicol glucoside (5) were detected as [M + HCOO]$^-$ ions. Conditions for their separation and calibration of extraction protocols were then developed (Figures 3a–d) in the negative ion mode, following which analyses of aqueous methanolic extracts of WT, LtAPS and LtPPS leaf tissues were carried out. For WT leaf tissue, allyl/propenylphenol glucosides 5, 6, 9 and 10 were not unambiguously detected/identified, as all were below detection limits with the methodologies employed (data not shown).

By contrast, with LtAPS lines, eugenol glucoside (6) and chavicol glucoside (5) were unambiguously identified in the leaf tissues, based on elution time, UV and mass spectral fragmentation patterns (see Figures 3e,f), whereas neither isoeugenol glucoside (10) nor p-anol glucoside (9) were detected. Quantification of glucoside 6 ranged from 0.25 to 0.35% dry weight relative to both the standard glucoside 6 calibration and the naringenin IS recovery (see Supporting Information). For the eugenol glucoside (6) standard, the ion at m/z 371.1344 corresponded to [M + HCOO]$^-$ (Figure 3c) with the same ion observed in the LtAPS line (Figure 3e). The amounts of chavicol glucoside (5) at m/z 341.1235 (Figure 3f) ranged from 0.07 to 0.1% dry weight.

With LtPPS leaf tissues, both eugenol glucoside (6) and chavicol glucoside (5) were present (see Figures 3g,h), but considerably lower in amounts ranging from 0.04 to 0.09 and 0.006 to 0.015% dry weight, respectively. Eugenol glucoside (6) had a m/z 371.1335 [M + HCOO]$^-$ ion, whereas chavicol glucoside (5) had a m/z 341.1231 [M + HCOO]$^-$ ion. Expected products, isoeugenol glucoside (10) and p-anol glucoside (9) were not detected.

It was next instructive to compare the 11 selected LtAPS and LtPPS lines as regards their differential levels of gene expression (Figures 4a and 5a) and accumulation levels of eugenol glucoside (6) (Figures 4b and 5b) and chavicol glucoside (5) (Figures 4c and 5c), respectively. In the LtAPS lines, while gene expression levels ranged from 1 to 30, there was essentially no effect on eugenol glucoside (6) accumulation levels, with these reaching a maximum plateau value even at the lowest gene expression level. The

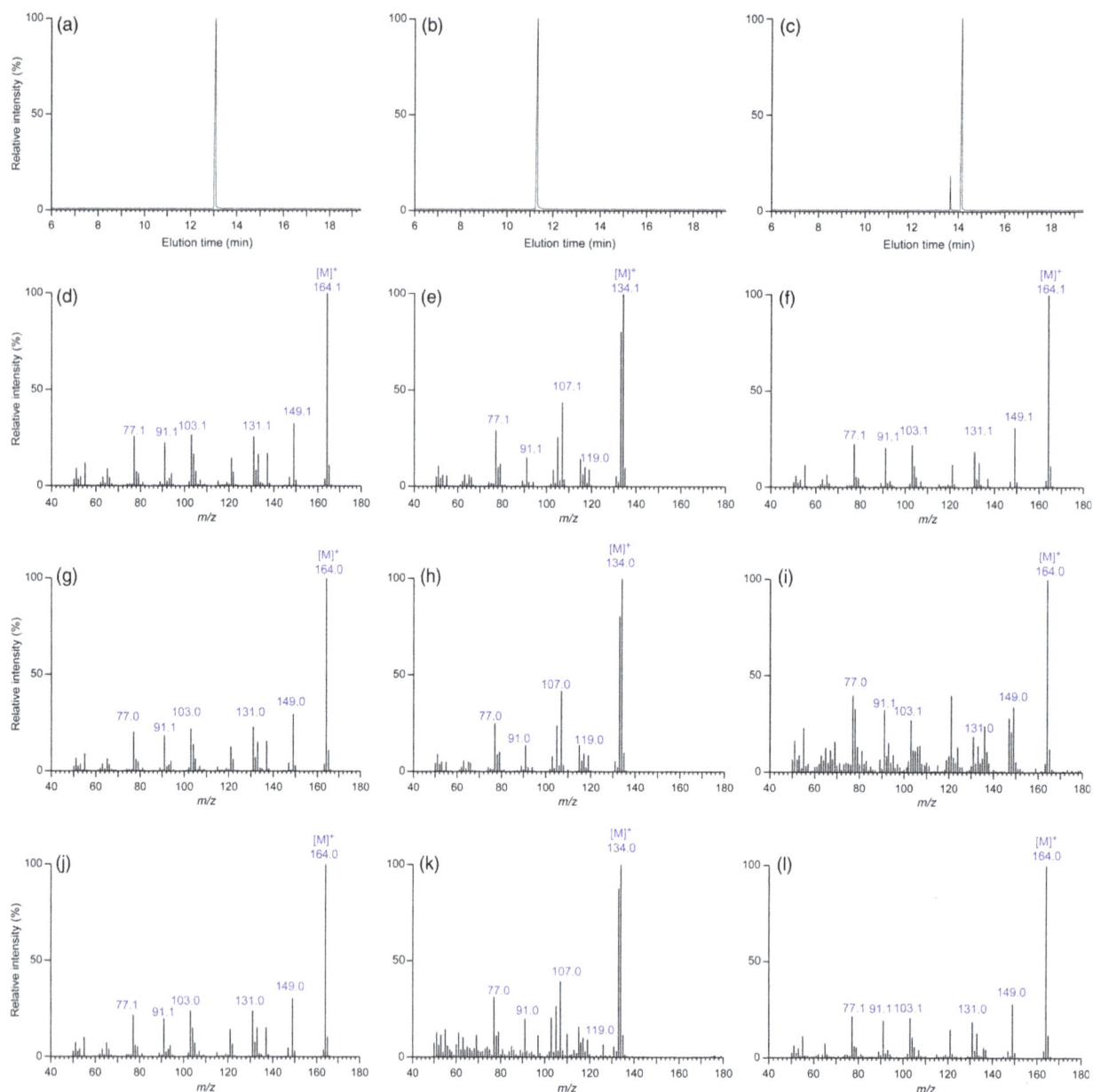

Figure 2 Eugenol (**4**)-, chavicol (**3**)- and isoeugenol (**8**)-extracted ion chromatograms and their mass spectra. Extracted ion chromatograms (a–c) of eugenol (**4**), chavicol (**3**) and isoeugenol (**8**) standards correlated with elution times, and their respective mass spectra (d–f). Mass spectra of eugenol (**4**), chavicol (**3**) and isoeugenol (**8**) in LtCAAT1/LtAPS1 (g–i) and LtCAAT1/LtPPS1 (j–l) transgenic hybrid poplars, respectively.

same trend was noted for chavicol glucoside (**5**) levels, except they were about one-third of eugenol glucoside (**6**) amounts (Costa et al., 2013a; Kim et al., 2013).

Provisionally, these data suggest that there was limited substrate monomer supply, or that some other form of metabolic block (including pathway inhibition) was operative, that is, as metabolite levels did not increase with increasing level of gene expression. This result thus contrasted with our findings (Costa et al., 2013b) of much higher accumulation levels of phenylethanol glucoside (**2**), these being ca. >10-fold higher than for the combined amounts of compounds **5** and **6**, respectively, even though they were all derived from Phe. However, phenylethanol glucoside (**2**) is directly produced from available Phe, whereas eugenol (**6**) and chavicol (**5**) glucosides are much further downstream in terms of metabolic pathway placement. Indeed,

in other studies with transgenic Arabidopsis, a three-step pathway from tyrosine to dhurrin (Kristensen et al., 2005; Morant et al., 2007) gave somewhat similar levels of metabolite accumulation to that observed for phenylethanol glucoside (**2**) (Costa et al., 2013b). An assumption here, though, is also that the accumulation levels of **5** and **6** represent terminal storage products.

Analogously, comparison of LtPPS gene expression levels was also very informative, in terms of accumulation of eugenol glucoside (**6**), chavicol glucoside (**5**) and p-anol (**9**) glucoside. Similar trends were noted, in terms of metabolite levels, except that amounts of (**6**) were much lower (by a ca. 3-fold reduction), whereas those of (**5**) remained about the same, relative to the LtAPS observation. This was unexpected, as in E. coli, LtPPS produces isoeugenol (**8**), but not eugenol (**4**) (Kim et al., 2014).

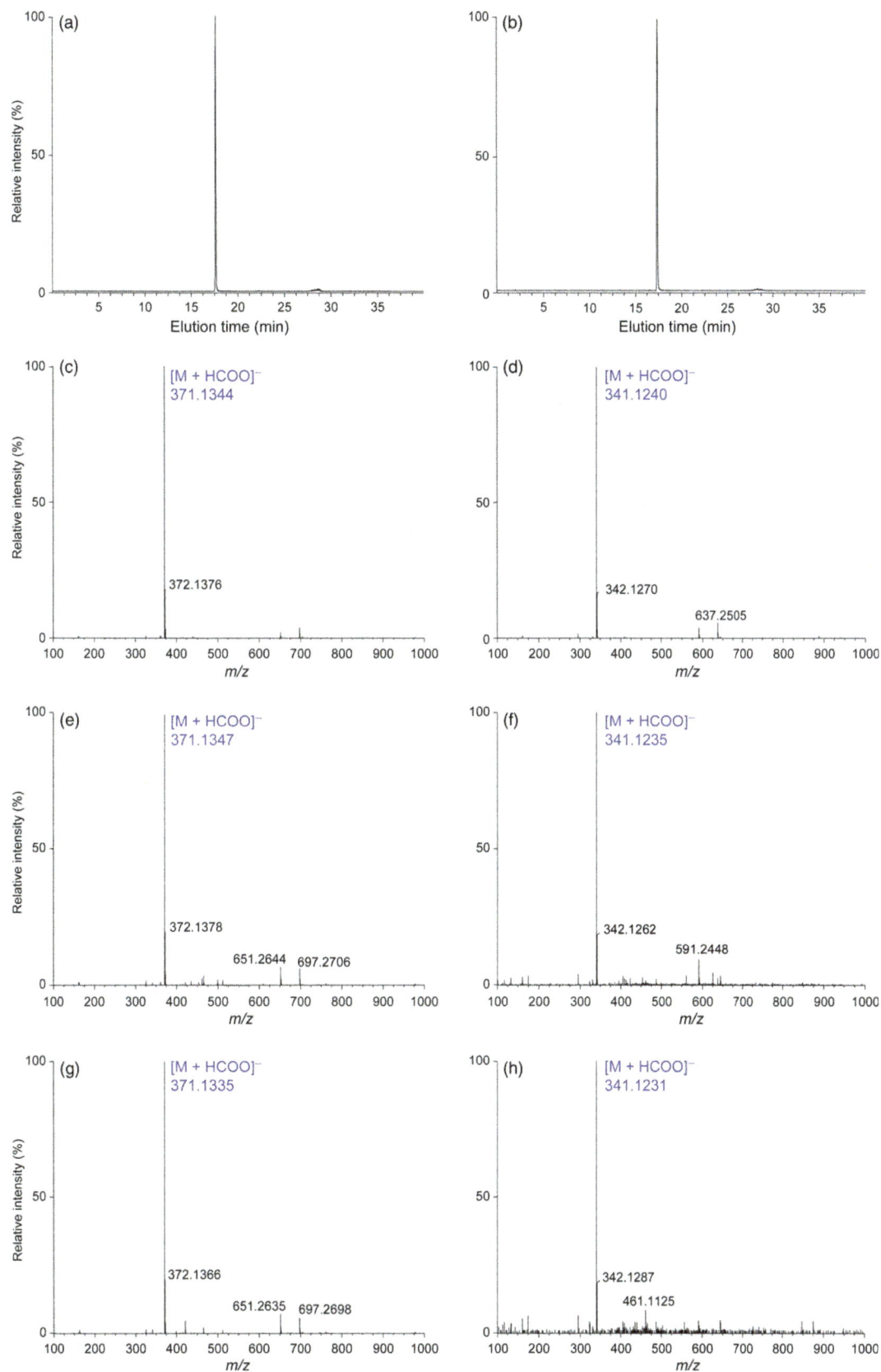

Figure 3 Eugenol glucoside (**6**)- and chavicol glucoside (**5**)-extracted ion chromatograms and their [M + HCOO]⁻ molecular ion mass spectra. Extracted ion chromatograms of (a) eugenol glucoside (**6**) and (b) chavicol glucoside (**5**) standards correlated with elution times, and their single ion patterns (c and d), respectively. Single ion patterns of eugenol glucoside (**6**) and chavicol glucoside (**5**) in LtCAAT1/LtAPS1 (e, f) and LtCAAT1/LtPPS1 (g, h) transgenic hybrid poplars, respectively.

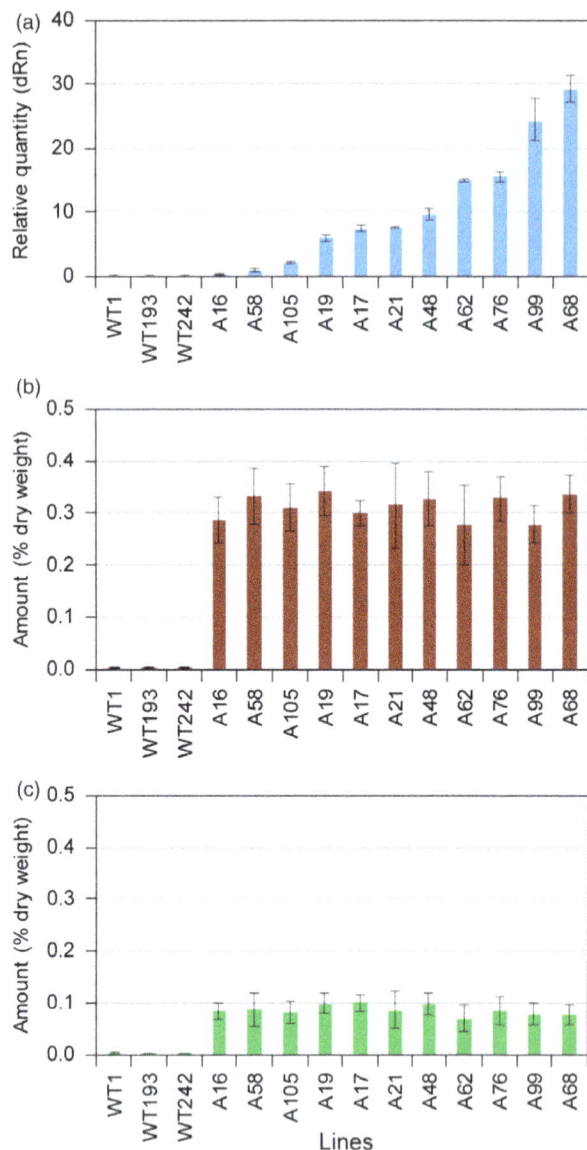

Figure 4 Correlation of engineered pathway expression levels with eugenol glucoside (**6**) and chavicol glucoside (**5**) amounts in transformed hybrid poplar. (a) Expression levels of *LtAPS1* in hybrid poplar transformed with the *p35S::LtCAAT1::T35S::p35S::LtAPS1::T35S* construct. Accumulation levels of (b) eugenol glucoside (**6**) and (c) chavicol glucoside (**5**) in leaves of WT and transgenic hybrid poplars harvested in August 2014.

Figure 5 Correlation of engineered pathway expression levels with eugenol glucoside (**6**) and chavicol glucoside (**5**) amounts in transformed hybrid poplar. (a) Expression levels of *LtPPS1* in hybrid poplar transformed with the *p35S::LtCAAT1::T35S::p35S::LtPPS1::T35S* construct. Accumulation levels of (b) eugenol glucoside (**6**) and (c) chavicol glucoside (**5**) in leaves of WT and transgenic hybrid poplars harvested in August 2014.

Interestingly, very small amounts of one of the possible products, *p*-anol glucoside (**9**), were observed, but its levels did not increase with increasing gene expression level. To account for the absence of **8**, perhaps its fully conjugated (more reactive) nature enabled it to undergo phenoxy radical coupling more readily *in situ* (i.e. thus reflecting a potential increased susceptibility of the more highly conjugated (**8**) as compared to eugenol (**4**)).

The aqueous methanol extracts of branch tissues from WT, LtAPS and LtPPS lines were also analysed to determine whether sequestered products **5**, **6**, **9** and **10** were present. None were, however, detected in WT branches, although both LtAPS and LtPPS lines contained eugenol glucoside (**6**) at very low but similar levels, with amounts ranging from 0.00008 to 0.003% dry weight (Figures 6a,b). By contrast, chavicol glucoside (**5**) was only

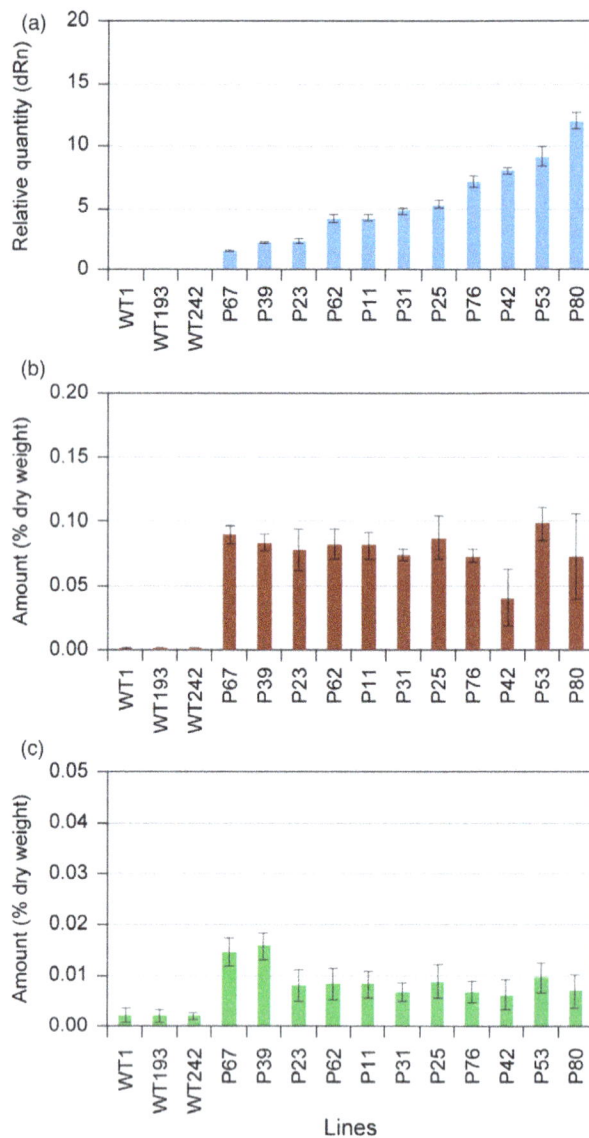

detected in LtAPS lines in very low amounts of *ca.* 0.0002% dry weight (Figure 6a), whereas isoeugenol glucoside (**10**) was again not detected; *p*-anol glucoside (**9**) was potentially detected in trace amounts (*ca.* 0.00004% dry weight) in the LtPPS lines (Figure 6b).

Nontargeted metabolomics analyses

In addition to the above metabolic products, an untargeted metabolomic analysis of the UPLC-qTOF data was performed to establish whether other unforeseen metabolic effects occurred, with a particular focus on phenylpropanoid and/or upstream shikimate–chorismate pathway metabolite/metabolic pools. Negative ion mode data were thus processed using XCMS and CAMERA R packages for the untargeted metabolomic analyses,

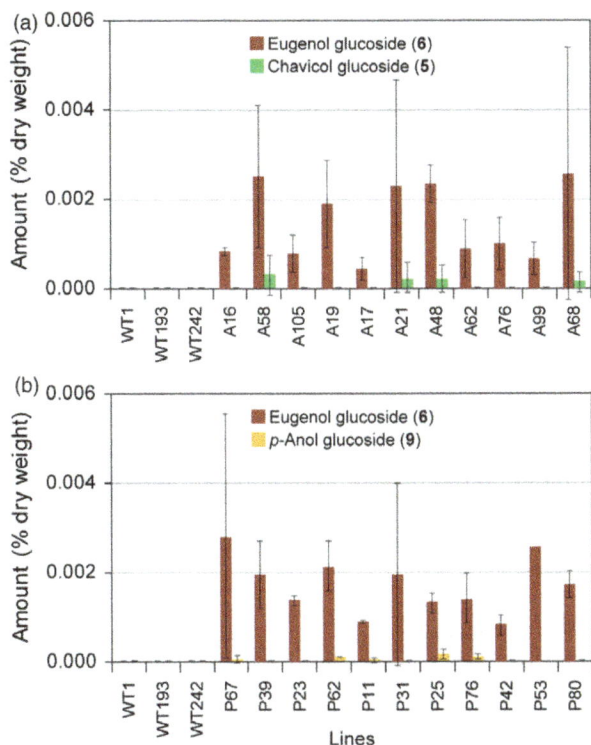

Figure 6 Accumulation levels of eugenol glucoside (**6**), chavicol glucoside (**5**) and p-anol glucoside (**9**) in branches of WT and transgenic hybrid poplar expressing *LtAPS* (a) and *LtPPS* (b).

as this mode works quite well for phenolic/aromatic ring compounds in *Populus* species. Such metabolites are mainly derived from phenylpropanoid and/or shikimate–chorismate pathways. Indeed, several peaks in HPLC chromatograms of polar extracts have been putatively annotated as flavonoid- and salicylate-derived compounds (Morreel *et al.*, 2006). However, data for unambiguous identification of these metabolites was lacking. In any event, our approach led to detection of 1102 features (i.e. unique *m/z* vs. RT signals integrated during XCMS processing) in 24 plant lines, of which 280 were significantly different in comparison of transgenic and control plants ($P \leq 0.05$). [This approach, however, depends on careful analysis of the corresponding pseudo MS2 spectra extrapolated by the CAMERA tool, that is based on chromatographic behaviour of the different features that were grouped as belonging to the same metabolite.]

In addition to eugenol glucoside (**6**) and chavicol glucoside (**5**), another eugenol (**4**) derivative, reported as being similar (Koeduka *et al.*, 2013; Tikunov *et al.*, 2013) but not identical to eugenol-xylosyl-glucoside (Straubinger et al., 1999) was detected (Tables S3 and S4). While this metabolite was not fully identified, its presence underscores the importance and diversity of metabolism that can occur in engineered pathways, thereby contributing to increasing the sink and productivity in such plant biofactories.

There were also some 14 other metabolites whose levels showed differences (up to <1.67-fold) between transgenic and WT lines. Supplemental Tables S3 and S4 and Figure S6 summarize the detection of these molecular species, their fold changes between WT and transgenic lines, and their tentative structural types. These were not explored further in this study given the

relatively small (fold) changes noted in their amounts. The transgenic plant lines also had reduced levels of a syringic acid glucose derivative relative to WT, that is by *ca.* 0.83- and 0.71-fold reductions (p values of 0.1431, 0.0046, respectively; Tables S3 and S4). However, whether this metabolite was phenolically linked (Wolfram *et al.*, 2010) or ester-linked (Yue *et al.*, 1994) was not further investigated.

Cell wall analyses

Analyses of polymeric carbohydrates (cellulose, hemicelluloses), as well as lignin and matrix sugars, were also carried out for branch samples (see Figures S7–S10); differences noted, however, were provisionally considered being due to variability in branch growth/development on sampling.

Lessons learned and future directions

This foundational study assessed whether introduction of bio-chemical pathways into poplar to potentially produce eugenol (**4**), chavicol (**3**), p-anol (**7**) and/or isoeugenol (**8**), as well as corresponding sequestered storage products **5**, **6**, **9** and **10**, could be achieved. This approach differed from an earlier successful undertaking to produce phenylethanol (**1**), via biotech-nologically introducing a two-enzyme step process from Phe in poplar foliage tissue (Costa *et al.*, 2013b). This was because the study here required availability of much farther downstream phenylpropanoid pathway substrates, namely the monolignols coniferyl (**12**) and p-coumaryl (**11**) alcohols, respectively. As poplar foliage produces many Phe-derived phenylpropanoid (acetate) products, such as the aforementioned flavonoids and other pathway-related metabolites, it was by no means certain that carbon flux to monolignol substrates vs. other phenyl-propanoid (acetate) pathway metabolites would be favoured with these manipulations.

The data obtained established that the various poplar lines generated were not only transformed with the biochemical pathways of interest, but that eugenol glucoside (**6**) and chavicol glucoside (**5**) accumulated in foliage tissue in ~0.35 and ~0.1% dry weight, as well as traces of corresponding aglycones. By contrast, they were only traces of these metabolites in branch tissues.

On the other hand, and unlike the earlier *E. coli* study (Kim *et al.*, 2014), neither isoeugenol (**8**) nor p-anol (**7**) accumulated, perhaps due to either further downstream metabolism of these metabolites or the enzyme kinetics being such that they could not favourably compete for monolignols **12** and **11** in the foliage tissues.

While various LtAPS and LtPPS lines accumulated some of the desired products in their foliage tissues, their accumulation levels reached a plateau even at lowest gene expression levels achieved. Their formation also had no obvious massive effect on plant productivity (volume and diameter), in contrast to previous field trial analyses (Voelker *et al.*, 2010). As increased gene expression in the various lines did not result in increased accumulation levels of the targeted bioproducts, this raises the exciting possibility of, in future, significantly increasing the metabolite levels through further iterations.

The eugenol glucoside (**6**) bioproducts production, harvesting and processing to this point in poplar foliage tissue allow for considering potential entry into economically displacing the current annual worldwide production of around 4000 tonnes of natural clove eugenol (**4**) (with pricing of *ca.* $12-80/kg, or $48-320M revenue annually). Clove production in 2013 in Tanzania, for example, was about 1.14 tonnes/ha, and whose clove oil content can range between 2 and 2.5% (22.8 -28.5 kg/ha) (FAOSTAT; World Agriculture). This, in turn, gives maximum eugenol (**4**) amounts of ca.18.2–22.4 kg/ha at an estimated 80% eugenol (**4**).

In this regard, full-size trees of commercial poplar (*Populus deltoides* × *P. trichocarpa)* species can produce ca. 2.2 tonnes/ ha of leaf biomass which could give approx. 8 kg/ha of eugenol glucoside (**6**), or ca. 4 kg/ha eugenol (**4**). Thus, to meet the 4000 tonnes current production levels of eugenol (**4**) from clove, some 800 000–1 M ha of land would be required. Increasing levels of their amounts by up to ca. 40-fold or so would reduce the land need down to 20–25 000 ha, and potentially produce value-added bioproducts (in addition to wood and pulp) of $48–320 M additional revenue annually. Currently, clove production worldwide uses in excess of 400 000 ha (World Agriculture).

Accordingly, in future, studies will employ numerous 'omics', systems biology, synthetic biology and metabolic modelling approaches to next identify and overcome metabolic blocks/ restricted carbon flux to the desired products in both foliage and stems. Furthermore, while some LtAPS and LtPPS lines displayed precocious flowering after a 4-year growth period, there are biotechnological approaches available (e.g. gene editing) to produce sterile plant lines which would potentially prevent this unexpected finding from occurring.

Experimental procedures

Field trial

For isolation/cloning of *LtCAAT1-*, *LtAPS1-* and *LtPPS1-*coding genes and vector construction for heterologous expression, hybrid poplar transformation, total RNA isolation and quantitative real-time PCR for selection of transgenic poplar, see Supporting Information.

The field trial, planted in June 2012, consisted of a total of 5019 poplar trees (*P. trichocarpa* and *P. tremula* × *P. alba*) derived from 245 mixed transgenic lines (4883 trees) and 136 WT (20 *P. trichocarpa* and 116 *P. tremula* × *P. alba*). This included the 57 *P. tremula* × *P. alba* lines (1126 trees) transformed with the LtCAAT-LtAPS vector and the 46 *P. tremula* × *P. alba* lines (905 trees) transformed with the LtCAAT-LtPPS vector, as well as the 116 WT *P. tremula* × *P. alba* as controls. The layout was a completely randomized single-tree plot design with 20 biological replications (Jayaraman, 1999). The trial was planted as a 1.52 m × 1.52 m square lattice in June 2012. Irrigation was by means of drippers (2 L/h) installed in 17-mm LDPE piping (Netafim, Fresno, CA, USA) through Triton X irrigation, with watering set for 1 h twice a day so that each tree received approx. 4 L H_2O per day. After the initial settling period, fertilizer (Wil-Sol® Pro-Balance 20-20-20; Wilbur Ellis, Aurora, CO, USA) was supplied at an 1:120 dilution ratio via a proportional dosing pump (D45RE15-20 GPM; Dosatron, Clearwater, FL, USA) as follows: 14.3, 16.1, and 12.4 g of concentrate per tree was applied in years 1, 2 and 3, respectively. Once the trees began to go dormant, a chemical hardener fertilizer (Wil-Sol® Pro-Finisher 4-25-35; Wilbur Ellis) was applied at the same dilution giving 4.4, 5.6 and 7.4 g of concentrate per tree in years 1, 2 and 3, respectively. Watering ceased after completion of the third growing season.

Tree heights and diameters were measured, tree heights only for the first year (2012) and then heights and diameters yearly (e.g. 2013 and 2014). Heights were measured for the first 3 years

using a height pole, and then, a Nikon Laser Forestry Pro Rangefinder 8381 was used to determine heights using three-point measurement. Diameters were measured with a 0-25 cm tree caliper rotated around the stem at the standard nominal breast height (1.4 m)—diameter at breast height (DBH).

Plant harvesting

The bottom branches and their leaves were individually collected from selected lines (replicates 10–15) as per USDA APHIS required protocols: Branches, cut into 10-cm sections, and leaves from each tree were separately flash-frozen in liquid N_2, kept on dry ice until reaching destination, and stored at −80 °C until sample processing.

Metabolite extractions

For allyl/propenylphenol determinations, frozen leaf tissues (0.2–0.5 g) were individually ground to a fine powder using a TissueLyser II (Qiagen, Germantown, MD, USA; 30 s at a frequency of 30 Hz) equipped with one 5-mm steel ball, with resulting ground tissues stored at −80 °C until needed. Lower branch tissues, by comparison, were first chopped into small pieces, ball-milled into a fine powder for 6 h and sieved (600 μm).

To each tissue sample type, 300–500 mg was added *tert*-butyl methyl ether (2 μL per mg tissue) containing 0.5 mM benzyl methyl ether as internal standard (IS). Each tissue was vortexed for a few seconds, sonicated in cold H_2O for 15 min, vortexed again and centrifuged (20 000 g, 15 min, 4 °C). Each supernatant was subjected to GC-MS analysis.

For allyl/propenylphenol glucoside determinations, freeze-dried leaves and branch samples from each line were prepared as above. Each sample was individually extracted with MeOH–H_2O (70:30, v/v) containing 0.1 mM naringenin IS at a ratio of 1 mL to 50 mg of tissue. After adding extraction solvent, each solution was vortexed for a few seconds, sonicated in cold H_2O for 15 min, vortexed again and centrifuged (20 000 g, 15 min, 4 °C). Each supernatant was transferred individually to UPLC vials and subjected to UPLC-QTOF-MS analyses.

GC-MS analyses

These were carried out on a HP 6890 Series GC System equipped with a HP 5973 MS detector (EI mode, 70 eV). Separations of metabolites were achieved using a RESTEK-5Sil-MS (30 m × 250 mm × 0.25 mm) column, with column conditions as follows: 40 °C maintained for 2 min, then 40–150 °C at 10 °C/min and 150–250 °C at 20 °C/min with a final holding time of 2 min; total run time was 20 min, with a detector temperature of 250 °C. Amounts of eugenol (**4**)/isoeugenol (**8**) and chavicol (**3**)/*p*-anol (**7**) were estimated based on *m/z* 164 and 134 extracted ion traces, respectively. Peak areas were normalized to the benzyl methyl ether IS peak area and quantified using calibration with authentic standards.

UPLC-QTOF-MS analyses

UPLC-QTOF-MS analyses employed a Waters Acquity™ Ultra Performance LC system (Waters, Milford, MA) equipped with a photodiode array (PDA) eλ detector (Waters) coupled to a Xevo™ G2 QTof mass spectrometer (Waters MS Technologies, Manchester, UK) using MassLynx (V4.1) software. Separations of metabolites used a ACQUITY UPLC® BEH C_{18} column (Waters, 2.1 × 150 mm, 1.7 μm particle size), with a linear gradient for separation: 100% A (0.1% HCO_2H in H_2O) over 0.5 min, 0–45%

B (0.1% HCO_2H in CH_3CN) over 25.0 min, 45–100% B over 1.0 min, held at 100% B for a further 2.0 min and then re-equilibrated at 100% A for 11.5 min. Flow rate was 0.2 mL/min, with column and sample temperatures kept at 25 and 5 °C, respectively. Injection volumes were 2 μL for LC-MS analysis and 4 μL for LC-MS/MS, with UV–visible spectra recorded between 200 and 500 nm (1.2 nm resolution). An electrospray ionization (ESI) source was used to detect masses of eluted compounds (m/z range: 50–1000 Da) and Ar was the collision gas. Detection settings were as follows: Negative ion mode: capillary voltage at 2.0 kV, cone voltage at 30 eV, collision energy at 6 eV and at 27 eV. Positive ion mode: capillary voltage at 3.0 kV, cone voltage at 30 eV, collision energy at 6 eV and at 18 eV. For LC-MS/MS analyses, 5–50, 30 and 30–60 eV collision energies were used. Sodium formate (5 mM in 2-propanol-H_2O, 90:10, v/v) was used for calibrating the mass spectrometer, with leucine enkephalin (2 ng/μL in CH_3CN − H_2O + 0.1% HCO_2H, 50:50, v/v) employed as lock mass. For extraction protocol standardization and calibration, see Supporting Information.

Metabolomics data processing and normalization

After UPLC-QTOF-MS analysis, chromatograms were reviewed using MassLynx software (Waters) with obvious outliers excluded. To conduct subsequent data processing, RAW data files were converted to NetCDF using MassLynx's Databridge software. NetCDF files were processed in RStudio (V 0.9.332) using the package XCMS (Smith et al., 2006) for feature (unique mass/retention time signals, MxxxTxxx) detection and retention time correction, followed by the CAMERA package (Kuhl et al., 2012) for features grouping and partial annotation (https://biocond uctor.org/packages/release/bioc/html/CAMERA.html). XCMS-based feature detection and integration were performed using the Centwave method (Tautenhahn et al., 2008) with the following parameters: prefilter = c(0,0), snthr = 2, ppm = 15, peakwidth = c(5,20), and nSlaves = 4. Retention time correction was performed twice via the obiwarp method as follows: bw = 2, minfrac = 0.5, mzwid = 0.015 and plottype = c ('deviation'). The CAMERA package was also used as follows: perfwhm = 0.6 and groupCorr, findIsotopes and findAdducts with polarity = 'negative'.

Pairwise comparison in XCMS was conducted after setting the two groups, one including all WT samples and the other samples from all transformed lines (both LtAPS and LtPPS). After initial XCMS and CAMERA processing, TSV and CSV files generated were combined and further analysed manually. Feature lists were normalized to the naringenin IS (integrated in XCMS results as feature M271T1315; retention time (RT) 21.92; m/z 271.0607, calculated 271.0606 for $C_{15}H_{11}O_5$) for statistical analysis. Comparisons between WT and transgenic lines (either from LtAPS or LtPPS plants) were performed separately for selection of statistically significant features, including being subjected to provisional annotation efforts.

Nontargeted metabolite provisional annotation

Following computational processing, information obtained was used for manual peak identification of known and putatively annotated poplar metabolites. Targeted known metabolites were identified by comparison with retention time, UV spectrum and MS/MS fragmentation of authentic standards. Nontargeted metabolites were putatively annotated as follows: (i) molecular formulae calculated from accurate mass and isotopic pattern recognition were used for screening Metlin (https://metlin.sc ripps.edu/index.php) and SciFinder Scholar metabolite databases (SciFinder Scholar™ 2015), as well as other sources including KEGG (http://www.genome.jp/kegg/pathway.html) and Plant Metabolic Network (PMN, http://www.plantcyc.org/) databases. (ii) MS/MS fragmentations of nontargeted metabolites were compared with candidate molecules found in databases [ReSpect for Phytochemicals (http://spectra.psc.riken.jp/) and MassBank (http://www.massbank.jp/)] and verified with literature reports, especially for metabolites (known and/or putative) previously reported in poplar (see supporting information).

Acknowledgements

This work was mainly funded by U.S. Department of Agriculture National Institutes for Food and Agriculture (USDA NIFA, #683A757612), the Chemical Sciences, Geosciences and Biosciences Division, Office of Basic Energy Sciences, Office of Science, U.S. Department of Energy (DE-FG-0397ER20259) and, in part, by USDA NIFA (#2011-68005-30416) and McIntire Stennis (WNP00202). The DOE funding enabled development/application of all the chromatographic and metabolomics approaches used in this study. Ms. Xianghe Yuan was supported by the China Scholarship Council. Thanks are also extended to Drs. Daryl J. Petrey (Plant Health Safeguarding Specialist) and Gary W. Brown (Biotechnology Compliance Specialist) for USDA APHIS field trial inspections and compliance of same. The authors declare no conflict of interest.

Author contributions

Da Lu performed plant harvesting, metabolite extractions, GC-MS analyses, UPLC-QTOF-MS analyses, metabolomics, nontargeted metabolite identification and annotation, cell wall residue preparation, lignin monomer composition and content assessments, matrix polysaccharide composition and crystalline cellulose contents, and draft manuscript writing and research discussions. Xianghe Yuan performed nontargeted metabolite detection and provisional annotation, and draft manuscript section writing. Joaquim V. Marques performed metabolite provisional annotation and plant metabolomics, and draft manuscript writing. Sung-Jin Kim performed hybrid poplar transformation and molecular biology analyses, and draft manuscript section writing. Randi Luchterhand and Barri Herman performed design, implementation, oversight and monitoring of all transgenic hybrid poplar field trials and plant harvesting, as well as in related draft manuscript section preparation. P. Pawan Chakravarthy performed synthesis of allyl/propenylphenol glucosides, and draft manuscript section writing. Laurence B. Davin and Norman G. Lewis responsibilities included research conception and discussions/guidance, completion of draft paper sections integration, multiple revisions and finalizing manuscript.

References

Badger, D.A., Smith, R.L., Bao, J., Kuester, R.K. and Sipes, I.G. (2002) Disposition and metabolism of isoeugenol in the male Fischer 344 rat. Food Chem. Toxicol. **40**, 1757–1765.

Caiger, S. (2016) Market Insider: Essential oils and oleoresins. http://www. intracen.org/uploadedFiles/intracenorg/Content/Exporters/Market_Data_

and_Information/Market_information/Market_Insider/Essential_Oils/Monthly%
20Report%20April%20%202016.pdf.

Chaieb, K., Hajlaoui, H., Zmantar, T., Kahla-Nakbi, A.B., Rouabhia, M.,
Mahdouani, K. and Bakhrouf, A. (2007) The chemical composition and
biological activity of clove essential oil, *Eugenia caryophyllata* (*Syzigium
aromaticum* L. Myrtaceae): a short review. *Phytother. Res.* **21**, 501–506.

Costa, M.A., Davin, L.B. and Lewis, N.G. (2013a) *Modified plants for producing
aroma/fine/specialty chemicals.* Provisional Patent Application No.
61,849,839 filed April 11, 2013.

Costa, M.A., Marques, J.V., Dalisay, D.S., Herman, B., Bedgar, D.L., Davin, L.B.
and Lewis, N.G. (2013b) Transgenic hybrid poplar for sustainable and scalable
production of the commodity/specialty chemical, 2-phenylethanol. *PLoS One*,
8, e83169.

Dinkova-Kostova, A.T., Gang, D.R., Davin, L.B., Bedgar, D.L., Chu, A. and
Lewis, N.G. (1996) (+)-Pinoresinol/(+)-lariciresinol reductase from *Forsythia
intermedia*: protein purification, cDNA cloning, heterologous expression and
comparison to isoflavone reductase. *J. Biol. Chem.* **271**, 29473–29482.

Doughton, S. (2014) Rose scent in poplar trees? WSU turns to genetic
engineering. *The Seattle Times*, February 9, 2014.

FAOSTAT. http://www.factfish.com/statistic/cloves%2C%20yield.

Houghton, J., Weatherwax, S. and Ferrell, J. (2006) *Breaking the biological
barriers to cellulosic ethanol: A joint research agenda.* A Research Roadmap
Resulting from the Biomass to Biofuels Workshop Sponsored by the U.S.
Department of Energy (http://www.doegenomestolife.org/biofuels/
2005workshop/2005low_strategy.pdf).

Jayaraman, K. (1999) *A Statistical Manual For Forestry Research* (ftp://ftp.fao.
org/docrep/fao/003/X6831E/X6831E00.pdf). Bangkok: Food and Agriculture
Organization of the United Nations Regional Office for Asia and The Pacific.

Kamatou, G.P., Vermaak, I. and Viljoen, A.M. (2012) Eugenol—From the
remote Maluku Islands to the international market place: a review of a
remarkable and versatile molecule. *Molecules*, **17**, 6953–6981.

Kim, S.-J., Marques, J.V., Lu, D., Davin, L.B. and Lewis, N.G. (2013) *Modified
poplar producing allyl/propenyl phenols.* 52nd Annual Meeting of the
Phytochemical Society of North America, Corvallis, OR.

Kim, S.-J., Vassão, D.G., Moinuddin, S.G.A., Bedgar, D.L., Davin, L.B. and Lewis,
N.G. (2014) Allyl/propenyl phenol synthases from the creosote bush and
engineering production of specialty/commodity chemicals, eugenol/
isoeugenol, in *Escherichia coli. Arch. Biochem. Biophys.* **541**, 37–46.

Klocko, A.L., Brunner, A.M., Huang, J., Meilan, R., Lu, H., Ma, C., Morel, A.
et al. (2016a) Containment of transgenic trees by suppression of *LEAFY. Nat.
Biotechnol.* **34**, 918–922.

Klocko, A.L., Ma, C., Robertson, S., Esfandiari, E., Nilsson, O. and Strauss, S.H.
(2016b) *FT* overexpression induces precocious flowering and
normal reproductive development in *Eucalyptus. Plant Biotechnol. J.* **14**,
808–819.

Koeduka, T., Fridman, E., Gang, D.R., Vassão, D.G., Jackson, B.L., Kish, C.M.,
Orlova, I. *et al.* (2006) Eugenol and isoeugenol, characteristic aromatic
constituents of spices, are biosynthesized via reduction of a coniferyl alcohol
ester. *Proc. Natl Acad. Sci. USA*, **103**, 10128–10133.

Koeduka, T., Louie, G.V., Orlova, I., Kish, C.M., Ibdah, M., Wilkerson, C.G.,
Bowman, M.E. *et al.* (2008) The multiple phenylpropene synthases in both
Clarkia breweri and *Petunia hybrida* represent two distinct protein lineages.
Plant J. **54**, 362–374.

Koeduka, T., Suzuki, S., Iijima, Y., Ohnishi, T., Suzuki, H., Watanabe, B.,
Shibata, D. *et al.* (2013) Enhancement of production of eugenol and its
glycosides in transgenic aspen plants *via* genetic engineering. *Biochem.
Biophys. Res. Commun.* **436**, 73–78.

Kristensen, C., Morant, M., Olsen, C.E., Ekstrøm, C.T., Galbraith, D.W., Møller,
B.L. and Bak, S. (2005) Metabolic engineering of dhurrin in transgenic
Arabidopsis plants with marginal inadvertent effects on the metabolome and
transcriptome. *Proc. Natl Acad. Sci. USA*, **102**, 1779–1784.

Kuhl, C., Tautenhahn, R., Böttcher, C., Larson, T.R. and Neumann, S. (2012)
CAMERA: an integrated strategy for compound spectra extraction and
annotation of liquid chromatography/mass spectrometry data sets. *Anal.
Chem.* **84**, 283–289.

Lewis, N.G., Davin, L.B., Kim, S.-J., Vassão, D.G., Patten, A.M. and Eichinger, D.
(2015) *Genes encoding chavicol/eugenol synthase from the creosote bush
Larrea tridentata.* US Patent No. 9,131,648, issued 2015-09-15.

Mastelic, J., Jerkovic, I., Vinkovic, M., Dzolic, Z. and Vikic-Topic, D. (2004)
Synthesis of selected naturally occurring glucosides of volatile compounds.
Their chromatographic and spectroscopic properties. *Croat. Chem. Acta*, **77**,
491–500.

Min, T., Kasahara, H., Bedgar, D.L., Youn, B., Lawrence, P.K., Gang, D.R., Halls,
S.C. *et al.* (2003) Crystal structures of pinoresinol-lariciresinol and
phenylcoumaran benzylic ether reductases and their relationship to
isoflavone reductases. *J. Biol. Chem.* **278**, 50714–50723.

Morant, A.V., Jørgensen, K., Jørgensen, B., Dam, W., Olsen, C.E., Møller, B.L.
and Bak, S. (2007) Lessons learned from metabolic engineering of cyanogenic
glucosides. *Metabolomics*, **3**, 383–398.

Morreel, K., Goeminne, G., Storme, V., Sterck, L., Ralph, J., Coppieters, W.,
Breyne, P. *et al.* (2006) Genetical metabolomics of flavonoid biosynthesis in
Populus: a case study. *Plant J.* **47**, 224–237.

Patten, A.M., Vassão, D.G., Wolcott, M.P., Davin, L.B. and Lewis, N.G. (2010)
Trees: a remarkable biochemical bounty. In *Comprehensive Natural Products
II. Chemistry and Biology* (Mander, L. and Liu, H.-W. eds), Vol. 3. Discovery,
Development and Modification of Bioactivity, pp. 1173–1296. Oxford, UK:
Elsevier.

Skinner, E.W. (1940) *The Science of Dental Materials.* Philadelphia: W.B.
Saunders Company.

Smith, C.A., Want, E.J., O'Maille, G., Abagyan, R. and Siuzdak, G. (2006)
XCMS: processing mass spectrometry data for metabolite profiling using
nonlinear peak alignment, matching, and identification. *Anal. Chem.* **78**,
779–787.

Straubinger, M., Knapp, H., Watanabe, N., Oka, N., Washio, H. and
Winterhalter, P. (1999) Three novel eugenol glycosides from rose flowers,
Rosa damascena Mill. *Nat. Prod. Lett.* **13**, 5–10.

Tautenhahn, R., Böttcher, C. and Neumann, S. (2008) Highly sensitive feature
detection for high resolution LC/MS. *BMC Bioinformatics*, **9**, 504.

Tikunov, Y.M., Molthoff, J., de Vos, R.C.H., Beekwilder, J., van Houwelingen,
A., van der Hooft, J.J.J., Nijenhuis-de Vries, M. *et al.* (2013) NON-SMOKY
GLYCOSYLTRANSFERASE1 prevents the release of smoky aroma from tomato
fruit. *Plant Cell*, **25**, 3067–3078.

Tuskan, G.A., Difazio, S., Jansson, S., Bohlmann, J., Grigoriev, I., Hellsten, U.,
Putnam, N. *et al.* (2006) The genome of black cottonwood, *Populus
trichocarpa* (Torr. & Gray). *Science*, **313**, 1596–1604.

Vassão, D.G., Gang, D.R., Koeduka, T., Jackson, B., Pichersky, E., Davin, L.B.
and Lewis, N.G. (2006) Chavicol formation in sweet basil (*Ocimum basilicum*):
cleavage of an esterified C9 hydroxyl group with NAD(P)H-dependent
reduction. *Org. Biomol. Chem.* **4**, 2733–2744.

Vassão, D.G., Kim, S.-J., Milhollan, J.K., Eichinger, D., Davin, L.B. and Lewis,
N.G. (2007) A pinoresinol-lariciresinol reductase homologue from the
creosote bush (*Larrea tridentata*) catalyzes the efficient *in vitro* conversion
of *p*-coumaryl/coniferyl alcohol esters into the allylphenols chavicol/eugenol,
but not the propenylphenols *p*-anol/isoeugenol. *Arch. Biochem. Biophys.*
465, 209–218.

Vassão, D.G., Kim, K.-W., Davin, L.B. and Lewis, N.G. (2010) Lignans
(neolignans) and allyl/propenyl phenols: biogenesis, structural biology, and
biological/human health considerations. In: *Comprehensive Natural Products
II. Chemistry and Biology* (Mander, L. and Liu, H.-W. eds), Vol. 1: Structural
Diversity I, pp. 815–928. Oxford, UK: Elsevier.

Verdonk, J.C., Ric de Vos, C.H., Verhoeven, H.A., Haring, M.A., van Tunen,
A.J. and Schuurink, R.C. (2003) Regulation of floral scent production in
petunia revealed by targeted metabolomics. *Phytochemistry*, **62**, 997–
1008.

Voelker, S.L., Lachenbruch, B., Meinzer, F.C., Jourdes, M., Ki, C., Patten, A.M.,
Davin, L.B. *et al.* (2010) Antisense down-regulation of *4CL* expression alters
lignification, tree growth, and saccharification potential of field-grown
poplar. *Plant Physiol.* **154**, 874–886.

Weinberg, J.E., Rabinowitz, J.L., Zanger, M. and Gennaro, A.R. (1972) [14]C-
Eugenol: I. Synthesis, polymerization, and use. *J. Dent. Res.* **51**, 1055–1061.

Wolfram, K., Schmidt, J., Wray, V., Milkowski, C., Schliemann, W. and Strack,
D. (2010) Profiling of phenylpropanoids in transgenic low-sinapine oilseed
rape (*Brassica napus*). *Phytochemistry*, **71**, 1076–1084.

World Agriculture. http://www.agrotechnomarket.com/2011/07/clove-oil.html.

Yue, J., Lin, Z., Wang, D. and Sun, H. (1994) A sesquiterpene and other
constituents from *Erigeron breviscapus. Phytochemistry*, **36**, 717–719.

4

Degradation of lignin β-aryl ether units in *Arabidopsis thaliana* expressing *LigD*, *LigF* and *LigG* from *Sphingomonas paucimobilis* SYK-6

Ewelina Mnich[1], Ruben Vanholme[2,3], Paula Oyarce[2,3], Sarah Liu[4], Fachuang Lu[4], Geert Goeminne[3], Bodil Jørgensen[5], Mohammed S. Motawie[1], Wout Boerjan[2,3], John Ralph[4], Peter Ulvskov[5], Birger L. Møller[1,6], Nanna Bjarnholt[1,*] and Jesper Harholt[5,6,*]

[1]*Plant Biochemistry Laboratory, Department of Plant Biology and Environmental Sciences, University of Copenhagen, Frederiksberg C, Denmark*

[2]*Department of Plant Biotechnology and Bioinformatics, Ghent University, Ghent, Belgium*

[3]*Department of Plant Systems Biology, VIB, Ghent, Belgium*

[4]*Department of Biochemistry and DOE Great Lakes Bioenergy Research Center, Wisconsin Energy Institute, Madison, WI, USA*

[5]*Section for Plant Glycobiology, Department of Plant Biology and Environmental Sciences, University of Copenhagen, Frederiksberg C, Denmark*

[6]*Carlsberg Research Laboratory, Copenhagen, Denmark*

*Correspondence
email nnb@plen.ku.dk and
email jesper.harholt@carlsberg.com

Keywords: biofuel, lignin
modification, bacteria, *Sphingomonas
paucimobilis*, Ligβ-aryl ether,
saccharification yield.

Summary

Lignin is a major polymer in the secondary plant cell wall and composed of hydrophobic interlinked hydroxyphenylpropanoid units. The presence of lignin hampers conversion of plant biomass into biofuels; plants with modified lignin are therefore being investigated for increased digestibility. The bacterium *Sphingomonas paucimobilis* produces lignin-degrading enzymes including LigD, LigF and LigG involved in cleaving the most abundant lignin interunit linkage, the β-aryl ether bond. In this study, we expressed the *LigD*, *LigF* and *LigG* (*LigDFG*) genes in *Arabidopsis thaliana* to introduce postlignification modifications into the lignin structure. The three enzymes were targeted to the secretory pathway. Phenolic metabolite profiling and 2D HSQC NMR of the transgenic lines showed an increase in oxidized guaiacyl and syringyl units without concomitant increase in oxidized β-aryl ether units, showing lignin bond cleavage. Saccharification yield increased significantly in transgenic lines expressing *LigDFG*, showing the applicability of our approach. Additional new information on substrate specificity of the LigDFG enzymes is also provided.

Introduction

Lignin is one of the most abundant biopolymers in the world. Together with cellulose, hemicelluloses, pectin and additional minor components that constitute the plant cell wall, lignin offers mechanical strength and affords protection against pathogens. The hydrophobic properties of lignin make it a crucial polymer controlling water conduction (Bonawitz and Chapple, 2010). During the polymerization process, lignin fills up spaces between polysaccharides in the secondary plant cell wall and may be covalently cross-linked to some of these. The properties of lignin impede enzymatic lignocellulose deconstruction and thus constitute a major bottleneck in biofuel production (Pauly and Keegstra, 2010; Van Acker et al., 2013; Vanholme et al., 2013a). Energy-demanding and expensive pretreatment is required to facilitate the access of hydrolytic enzymes to the polysaccharides in the cell wall. Modification of the cell wall structure in ways that ease deconstruction without compromising the agronomic performance of the crop plant is therefore an important research topic. However, plants with large reductions in lignin content exhibit reduced growth, fitness and development. Altering lignin composition and structure might therefore be a better strategy to improve the efficiency of biomass processing (Bonawitz and Chapple, 2010; Li et al., 2008; Ralph, 2007; Sederoff et al., 1999; Vanholme et al., 2008, 2012).

Lignin is a complex aromatic polymer derived mainly from three monolignols: *p*-coumaryl alcohol, coniferyl alcohol and sinapyl alcohol. When incorporated into lignin, these monolignols give rise to the *p*-hydroxyphenyl (H), guaiacyl (G) and syringyl (S) units, respectively. Once monolignols are synthesized in the cytoplasm, they are exported to the apoplast, activated to form radicals by the action of peroxidases or laccases and thereby subjected to radical coupling polymerization reactions. The most abundant interunit linkage type, formed primarily by the coupling of a monolignol (at its β-position) to the phenolic end (at its 4–O-position) of a growing polymer chain, is β–O–4 referring to an ether linkage between the aliphatic side-chain of one monolignol and the phenolic moiety of another. Additional linkage types formed are β–5, 5–5, β–β and 4–O–5 (Boerjan et al., 2003; Ralph et al., 2004).

The ability of microorganisms to degrade lignin has been extensively studied. White-rot and brown-rot fungi are involved in high-molecular-mass lignin degradation whereas bacteria like *Sphingomonas paucimobilis* SYK-6 degrade low-molecular-mass lignin fragments and small oligomers (Kirk and Farrell, 1987; Masai et al., 2007). The metabolism of *S. paucimobilis* has been evolutionarily adapted to allow the bacterium to grow on lignin-derived compounds. This involves development of enzymatic machinery to fully degrade different types of units. Reductive cleavage of β-aryl ether units has been demonstrated *in vitro* using guaiacylglycerol-β-guaiacyl ether (GGE) as a model sub-strate (Masai et al., 2003) (Figure 1). The β–O–4-linked moieties are reductively cleaved in the course of a three-step sequence that involves initial dehydrogenation (catalysed by LigD, LigL and LigN enzymes), followed by reductive cleavage of the ether linkage catalysed by a glutathione S-transferase (LigE or LigF), involving the formation of a glutathione conjugate, and finally reductive cleavage of this conjugate, producing the monomer and oxidized glutathione in a step using additional glutathione, catalysed by a glutathione lyase (LigG). The enzymes catalysing these processes are highly stereospecific. The GGE substrate exists as four enantiomers. The αR-GGE enantiomers are dehydrogenated by LigD, whereas the αS-GGE enantiomers are dehydrogenated by LigL or LigN. The LigF enzyme conjugates the βS enantiomer of α-(2-methoxyphenoxy)-β-hydroxypropio-vanil-lone (MPHPV) with glutathione (GSH) whereas the βR enantiomer is conjugated by LigE (Sato et al., 2009). The substrate specificity assays were mainly performed on guaiacyl dimers (Gall et al., 2014a; Masai et al., 2007), but in vitro the Lig enzymes are also capable of degrading synthetic high-molecular-mass polymers mimicking the structure of lignin (Sonoki et al., 2002).

A recent paper described an attempt to engineer LigD into the plant with the idea of creating benzylic-oxidized lignins that might chemically degrade more readily (Tsuji et al., 2015). In this study, the codon-optimized S. paucimobilis SYK-6 genes encod-ing the complete set of lignin-degrading enzymes LigD, LigF and LigG (henceforth abbreviated LigDFG) were inserted into the Arabidopsis genome by stable transformation using transit peptides targeting each of the heterologous proteins to the secretory pathway.

Results

Transient expression of LigDFG in tobacco

The S. paucimobilis LigD, LigF, LigG gene sequences were codon-optimized for expression in Arabidopsis. The bacterial Lig genes have a high GC content (approx. 60%), and as a result of codon optimization, the GC content was lowered from 64% to 54% in LigD, from 61% to 51% in LigF and from 63% to 54% in LigG. With the goal of modifying the cell wall, proteins involved in the LigDFG pathway had to be targeted to the secretory pathway to achieve localization in the apoplast. To achieve this, each separate protein was fused with the barley α-amylase signal peptide (amy) (Rogers, 1985) or potato proteinase inhibitor II (ppi) (Liu et al., 2004). These transit peptides have previously been used to target heterologous proteins to the secretory pathway (Borkhardt et al.,

2010; Buanafina et al., 2010; Herbers et al., 1995). For expres-sion of the entire pathway, LigD, LigF and LigG were expressed in tri-cistronic constructs in which the three coding sequences were linked by the 2A gene sequence coding for a self-processing polypeptide similar to the one found in the Foot-and-mouth disease virus (El Amrani et al., 2004). Successful expression of multiple proteins in plants from constructs and targeting them to different cellular compartments has previously been reported, including apoplastically localized cell wall degrading enzymes (El Amrani et al., 2004; de Felipe, 2002; Øbro et al., 2009). To test whether the tri-cistronic constructs amy-LigD-2A-amy-LigF-2A-amy-LigG (amyDFG) and ppi-LigD-2A-ppi-LigF-2A-ppi-LigG (ppiDFG) yielded active proteins in planta, they were transiently expressed in Nicotiana benthamiana (tobacco) by infiltration into the leaves. Enzymatic assays with proteins extracted from the infiltrated leaves using GGE or MPHPV as substrate showed formation of hydroxypropiovanillone (HPV) from both (Figure S1). HPV is the expected product formed from GGE by the action of LigDFG (Figure 1). This demonstrated that the heterologously expressed proteins were functional. The intermediate MPHPV was also formed from GGE in protein extracts from both constructs, whereas α-glutathionyl-β-hydroxypropiovanillone (GS-HPV) was only detected in the enzymatic assay with proteins from plants expressing amyDFG (Figure S1). Presumably, the steady-state pool size of GS-HPV in plants expressing ppiDFG was below the detection limit determined by the experimental design.

Stable Arabidopsis transformants expressing either LigD, LigF or LigG

In order to investigate the in vivo catalytic properties of the individual enzymes in Arabidopsis and to optimize the protein extraction method, Arabidopsis plants were transformed with constructs to separately express either LigD, LigF or LigG, each with the amy or ppi signal peptide code, under the control of the 35S promotor.

GGE, MPHPV and GS-HPV were used as a substrate to monitor the catalytic function of LigD, LigF and LigG, respectively, using crude protein extracted from leaves of the transformed plants (see pathway in Figure 1). Leaves were used as the enzyme source due to the ease of sufficient amounts of protein by extraction and their formation in the early part of the growth cycle thereby reducing the growth period. All genes were expressed under the control of the 35S promotor which would be expected to result in comparable protein accumulation in other tissues. Because GS-HPV was not available as a chemical standard, this reference compound was produced enzymatically

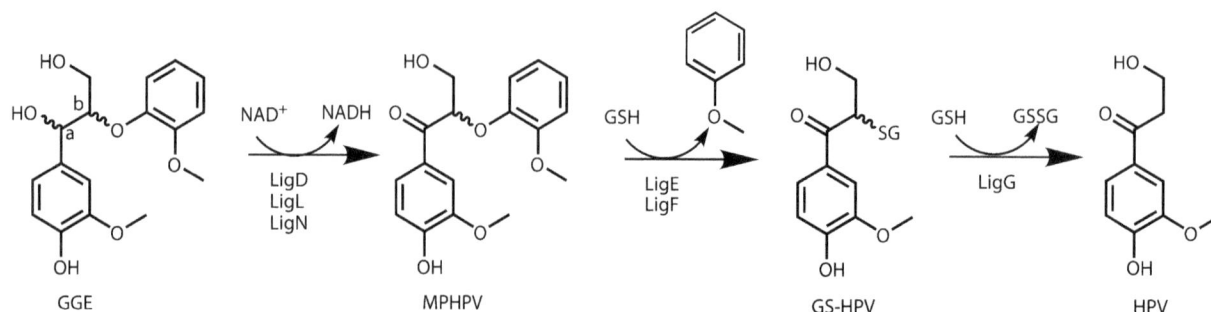

Figure 1 Pathway for degradation of β–O–4-linked units by Lig enzymes from S. paucimobilis exemplified with the model compound GGE (Masai et al., 2003). GGE, guaiacylglycerol-β-guaiacyl ether; MPHPV, α-(2-methoxyphenoxy)-β-hydroxypropiovanillone; GS-HPV, α-glutathionyl-β-hydroxypropiovanillone; HPV, β-hydroxypropiovanillone; GSH, glutathione.

by incubation of MPHPV with protein extracts of *Escherichia coli* expressing LigF. The reaction mixture obtained contained mainly GS-HPV and was used to test for LigG activity. The enzyme assays with the majority of the transformed lines indeed produced the expected products (Figure S2). The total number of lines identified that were positive for insertion of the transgene as detected by PCR and for the expected corresponding enzyme activity were three lines expressing *amy-LigD* (*amyD*), two lines expressing *amy-LigF* (*amyF*), two lines expressing *amy-LigG* (*amyG*), two lines expressing *ppi-LigD* (*ppiD*) and two lines expressing *ppi-LigF* (*ppiF*).

Interestingly, HPV was also detected in plants expressing only *LigF* when assayed using MPHPV as substrate (Figure S2). In the control samples where MPHPV was incubated with protein extracted from wild-type plants, no formation of HPV was detected. This indicated that an endogenous LigG-like activity was present in wild-type Arabidopsis plants or that LigF possessed LigF as well as LigG activity. No HPV could be detected when wild-type protein extracts were incubated with GS-HPV, supporting the contention that LigF possesses additional LigG activity, but this needs to be studied in more detail.

The results showed that it was possible to monitor the activity of the individual enzymes when expressed with either the amy or ppi signal peptides and that the optimized protein extraction and assay conditions were adequate as a screening method for identification of plants successfully expressing the entire LigDFG

pathway after transformation of Arabidopsis with the tri-cistronic constructs.

Engineering the entire LigDFG pathway into Arabidopsis

To metabolically engineer lignin *in planta* using the LigDFG pathway, *amyDFG* and *ppiDFG* constructs were stably transformed into Arabidopsis. Initially, positive first-generation transformants obtained using the tri-cistronic constructs were identified by PCR. The positive transformants were then screened using the optimized assays for enzymatic activity using protein extracts from leaves. LigD activity was detected in seven and five independent transformants expressing *amyDFG* and *ppiDFG*, respectively, as demonstrated by the observed enzymatic conversion of GGE to MPHPV. Three of the transformants exhibiting LigD activity (*amyDFG10*, *amyDFG12* and *ppiDFG1*) also showed LigFG activity as demonstrated by formation of HPV (Figure 2). These independent transformed lines were selected for further characterization.

Polyclonal antibodies against LigD were raised in rabbits and used to investigate the abundance of LigD protein in the transgenic *amyDFG10*, *amyDFG12* and *ppiDFG1* plants. By use of the polyclonal antibodies, LigD was detected in crude protein extracts from rosette leaves of *amyDFG10*, *amyDFG12* and *ppiDFG1* plants and in *ppiD4* (*LigD*-expressing) plants that were used as a positive control (Figure 3). No bands corresponding to unsuccessful 2A proteolytic cleavage, between LigD and LigF

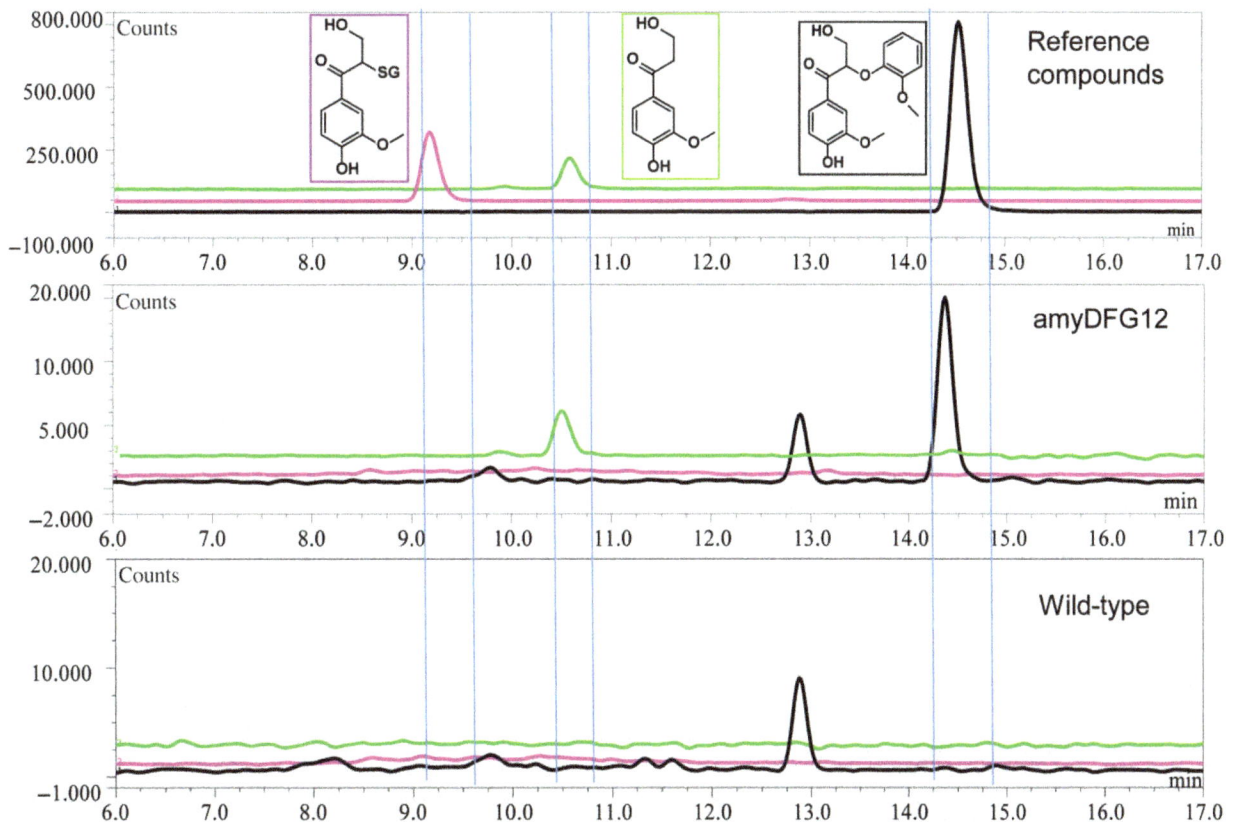

Figure 2 Detection of LigDFG catalysed products and intermediates formed in transgenic Arabidopsis plants expressing *LigDFG*. The chromatogram in the upper panel shows the retention times of the reference compounds. In the panels below, the metabolism of guaiacylglycerol-β-guaiacyl ether following incubation with leaf protein extracts from a transgenic line and wild-type plants is shown. Black line: Extracted ion chromatogram (EIC) for *m/z* 319 recorded to detect formation of α-(2-methoxyphenoxy)-β-hydroxypropio-vanillone. Pink line: EIC for *m/z* 502 recorded to detect formation of GS-HPV. Green line: EIC *m/z* 197 recorded to detect formation of HPV. LigDFG, enzymes LigD, LigF and LigG.

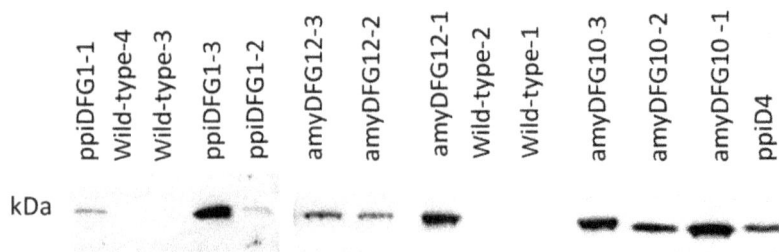

Figure 3 Detection of LigD protein in *LigDFG* transgenic lines (third-generation T3) as monitored by Western blotting with anti-LigD antibody. No bands corresponding to unsuccessful 2A proteolytic cleavage, between LigD and LigF could be detected, as only the shown 31.9 kDa band could be detected with anti-LigD antibody. The figure is a composite of several blots.

could be detected, as only a single band of 31.9 kDa could be detected with anti-LigD antibody. No LigD was detected in the protein extract from wild-type rosette leaves used as a negative control (Figure 3). The three independent biological replicates of *amyDFG10* showed a high and consistent abundance of LigD, whereas substantial variation in LigD abundance was observed among the three biological replicates of *ppiDFG1* and among the three biological replicates of *amyDFG12* (Figure 3).

Lignin characterization by 2D NMR

NMR is a powerful tool for qualitative analysis and for quantifying the relative amounts of the different linkage types and lignin building blocks present in lignin (Lu and Ralph, 2011; Mansfield et al., 2012; Morreel et al., 2010). The signals in the aliphatic region of 2D HSQC NMR spectra depict the linkage-type distribution, whereas the signals in the aromatic region can be used to deduce the monomeric lignin composition. Enzymatic lignins extracted from dried, mature but still green inflorescence stems of T3 *amyDFG10*, *amyDFG12*, *ppiDFG1* and wild-type plants were used for the analyses. The relative amounts of different lignin substructures between the wild-type and transgenic lines were obtained from volume integration of the HSQC contours (Table S3), and the summary of the results is shown in Figures 4 and 5. In general, the analyses revealed that the overall monomeric and linkage-type composition of the transgenic plants were similar to those in the wild type. However, the *LigDFG* lines showed a statistically significant twofold increase in α-keto-G (Gox) and α-keto-S (Sox) units. The Gox/G ratio was increased from 0.34% in wild type to 0.64%, 0.73% and 0.65% in the *amyDFG10*, *amyDFG12* and *ppiDFG1* lines, respectively (Figure 4). The Sox/S ratio was 1.09% in wild type and increased to 2.16%, 2.13% and 2.13% in the *amyDFG10*, *amyDFG12* and *ppiDFG1* lines, respectively. The observed higher abundance of Gox and Sox might be logically explained by LigD activity (Tsuji et al., 2015). However, the increased amounts of Gox and Sox units observed in the aromatic region were not reflected in the aliphatic region; there were no significant differences in oxidized β-aryl ether units (Aox) as deduced from the Aox/A ratio, showing that the additional oxidized G and S units (Gox and Sox) were not bound in such units (Figures 4 and 5).

Phenolic metabolite profiling

Phenolic metabolite profiling of inflorescence stems from seven individual *amyDFG10* plants and from ten individual *amyDFG12*, *ppiDFG1* and wild-type plants was carried out using UHPLC-MS analysis. As many as 1286 different mass signals were detected in all replicates of at least one genotype. ANOVA tests showed that 33 of these signals were present at significantly different intensities between the lines ($P < 0.05$). *Post hoc* analysis showed that all significant differences could be assigned to differences between the *amyDFG10* line and the wild type. No significant differences were observed in signal intensity between the other *LigDFG* lines and wild type. Further analyses demonstrated that the 33 mass signals were derived from a total of 26 different compounds (Table 1, Figure 6). The abundance of 24 of these compounds was higher, and the abundance of 2 compounds was lower in the *amyDFG10* line than in the wild-type plants (Table 1). Twenty-one of the 24 compounds that accumulated in the *amyDFG10* line were structurally characterized based on their MS/MS fragmentation patterns. The detailed structural characterization of compounds **1** to **3** is described in the Supporting Information (Figure S3) whereas structural characterization of compounds **4** to **21** was based on literature data (Tsuji et al., 2015).

Several phenolic compounds with an α-keto functionality were found in the *amyDFG10* plants. The majority of these compounds were dimers of an α-keto-G (Gox) unit β-O-4-linked to either a G unit (**4** and **6**), a G′ unit (derived from coniferaldehyde; **5**), a ferulic acid moiety (**7**–**19**) or to a sinapic acid moiety (**20** and **21**). The dimers were found to be further conjugated with hexose, malic acid, an unknown constituent of 224 Da, glutamate or a combination thereof. The oxidized dimers were also found in plants expressing only *LigD* (Tsuji et al., 2015), and their formation was therefore assigned as resulting from LigD activity. The putative LigD substrates giving rise to formation of the oxidized dimers detected in this study were previously identified in Arabidopsis inflorescence stems (Matsuda et al., 2010; Morreel et al., 2014; Vanholme et al., 2012). G(β-O-4)ferulic acid and G (β-O-4)sinapic acid, and their conjugates are classified as neolignan-like compounds (Davin and Lewis, 2005) as they share the same structures seen in neolignans, but they are likely produced in the cytosol (Dima et al., 2015).

GS-HPV is the product formed by LigF, but it was not detected in any of the extracts from the transgenic lines. HPV is the final product of the LigDFG pathway and was indeed present as a γ-O-glucoside conjugate (**1**) and as a γ-O-acetyl hexoside conjugate (**3**) in extracts of the *amyDFG10* line. In addition, hydroxypropiosyringone (HPS) hexose (**2**) was detected in the *amyDFG10* line, hinting that LigDFG can also act on (β-O-4)-linked S units. Notably, the HPV γ-O-hexoside (**1**) was also increased in plants expressing only LigD without an apoplastic target signal peptide. This compound was marked as 'Unknown $C_{16}H_{21}O_9$' in a previous study involving expression of *LigD* in Arabidopsis (Tsuji et al., 2015). Re-evaluation of the data presented in the latter study showed that the amount of HPS γ-O-hexoside (**2**) in plants expressing only *LigD* also had a tendency of being increased whereas the level of HPV γ-O-acetyl hexoside was not significantly increased (Figure S4). In any case, the reported fold-change as compared to wild type of these HPV-like compounds was lower in the *LigD*-expressing lines compared to the *amyDFG10* line (Figure S4, Table 1). It is important to note that the α-keto-oxidized G(β-O-4)G units were detectable in wild-type plants as well, suggesting that wild-type plants do have LigD

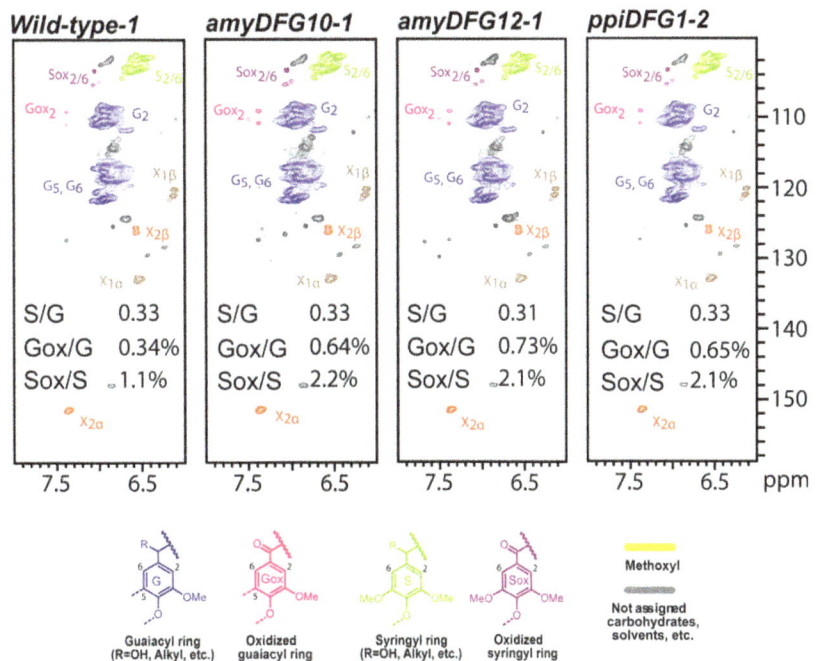

Figure 4 Aromatic region of partial 2D HSQC NMR spectra obtained by analysis of three biological replicates of the *LigDFG* lines and wild type. A single representative spectrum from each line is shown. The content of particular units is the average of the three replicates. See Table S3 for details.

activity. Furthermore, hexosylated HPV was likewise detected in wild-type stems, indicating that wild-type Arabidopsis also exhibits LigF and LigG activity.

Glucose release

The *LigDFG* transgenic Arabidopsis plants were tested for cell wall saccharification yield. To measure the effects of rather subtle cell wall compositional differences between transgenic and wild-type lines, partial glucose release was accomplished using a mild enzyme based saccharification procedure. Two growth cohorts of each eight biological replicates of ground dried stems from each of the transgenic Arabidopsis lines (*amyDFG10*, *amyDFG12* and *ppiDFG1*) and from wild type were analysed (Figure 7). Two-way ANOVA showed statistically significant increases in glucose release from *amyDFG10* ($P < 0.02$), *amyDFG12* ($P < 0.02$) and *ppiDFG1* ($P < 0.001$), with no interaction between genotype and growth cohort.

Discussion

Lignin embedded in the plant cell wall impedes enzymatic lignocellulose deconstruction for biofuel production. In this study, we targeted lignin-degrading Lig enzymes from the bacterium *S. paucimobilis* to the apoplast of Arabidopsis with the aim of modifying lignin structure in a manner that would improve the accessibility of polysaccharide-hydrolysing enzymes without influencing the overall content of lignin. The overall lignin composition, in terms of G and S units and linkage types, was not changed in the transgenic *LigDFG* Arabidopsis lines, except that the degree of α-oxidation of G and S units was approximately twofold increased, as deduced from the aromatic regions of the 2D HSQC NMR spectra (Figures 4 and S3). The detected changes in lignin structure were significant and displayed little variation between the individual transgenic plant lines obtained. In Arabidopsis plants expressing only *LigD*, an increase in oxidized β–O–4-ether structures (Aox structures in Figure 5) was observed consistent with the action of LigD on the Cα-OH of β–O–4-linked units (A structures in Figure 5) as exemplified in Figure 8 (Tsuji

et al., 2015). Upon expression of the entire LigDFG pathway, the LigD product serves as substrate for LigF, and therefore, the oxidized β–O–4-ether structures (Aox) are further metabolized. In the *LigDFG* lines, the observed increase in the content of oxidized structures (Gox and Sox units) in the absence of elevated levels of oxidized β–O–4-ether structures (Aox) (Figure 4) demonstrated that the entire LigDFG pathway was functional: the set of LigDFG enzymes reductively cleaved some of the β–O–4-units resulting in an increase in HPV-like end-units. Naturally occurring benzaldehyde and benzoic acid end-units will contribute to the intensity of Gox and Sox signals (Boerjan *et al.*, 2003; Rahimi *et al.*, 2013). However, LigDFG activity would not influence the content of, nor modify, these groups and therefore the increase in Gox and Sox in the transgenic *LigDFG* lines was derived from reductive cleavage of α-oxidized β–O–4-units. The unaltered Aox/A ratio demonstrated that under the given experimental conditions, the LigD-step was rate limiting, that is those G units with β–O–4-bonds that are oxidized to α-ketones by LigD are cleaved by the action of LigF and LigG.

The observed reductive cleavage of some of the β–O–4-linked lignin units in the transgenic *LigDFG* lines caused a moderate, but significant, increase in the digestibility of the cell wall by polysaccharide-hydrolysing enzymes. In a recently published study, Tsuji *et al.* (2015) did not detect increased saccharification yields in transgenic Arabidopsis plants expressing *LigD* alone, although it was expected that the increase in oxidized β–O–4-structures should render the lignin more easily degradable by alkaline pretreatment. Similarly to the present study, only minor changes in lignin structure were observed by 2D HSQC NMR analysis, although this might reflect issues with respect to achieving apoplastic localization of the enzyme. At the current stage, it may thus be concluded that expression of the entire LigDFG pathway with the amy and ppi signal peptides may be more efficient in terms of achieving increased enzymatic cell wall decomposition. The data seem to suggest that the cleavage of the oxidized β–O–4-structures (Aox) achieved by expressing LigDFG is needed to improve the saccharification, not merely the oxidation of the β–O–4-structures.

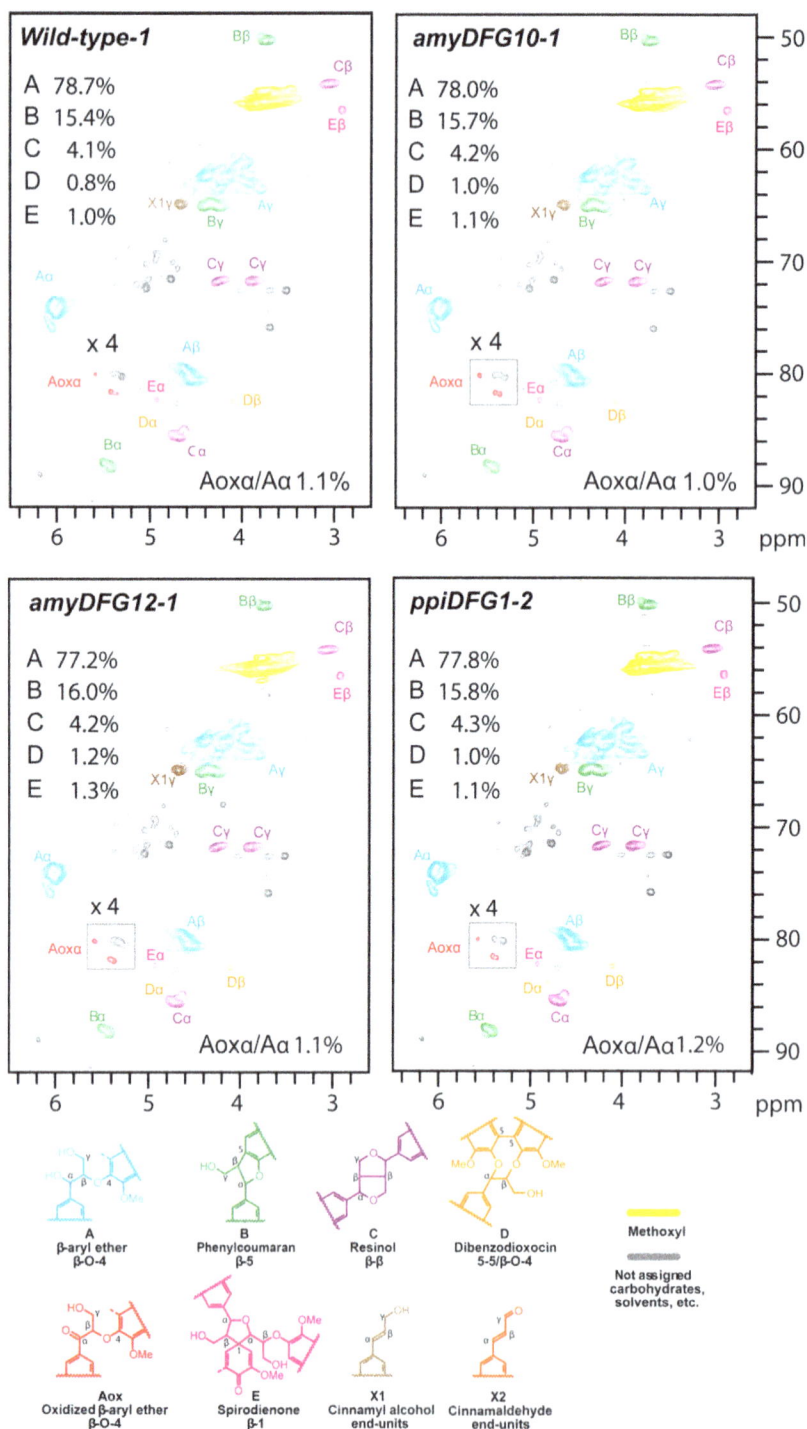

Figure 5 Aliphatic region of partial 2D HSQC NMR spectra obtained by analysis of three biological replicates of the *LigDFG* lines and wild type. A single representative spectrum from each line is shown. The content of particular units is the average of the three replicates. See Table S3 for details.

Phenolic profiling of the transgenic *amyDFG10* Arabidopsis line demonstrated accumulation of dimers oxidized at the Cα position. This is consistent with LigD activity. In agreement with previously reported observations (Tsuji *et al.*, 2015), the detected compounds were mainly oxidized G units linked with other G units or with ferulic acid and their hexose and malate derivatives. Such phenylpropanoid dimers are classified as neolignans or lignan-like and are thought to be involved in plant defence mechanisms rather than in the lignification process (Davin and Lewis, 2005). The cellular localization of this coupling reaction is not known. In lignin formation, oxidative coupling of monolignols takes place in the cell wall by radical reactions catalysed by the

action of oxidases (peroxidases or laccases) (Boerjan *et al.*, 2003). The G(β-O-4)ferulic acid is unlikely to be formed in the cell wall as it would mean that such units would subsequently be incorporated into lignin, but no or very little cell wall-bound ferulic acid can be detected in dicots (Vogel, 2008). It has recently been demonstrated that radical coupling may also proceed in the cytosol (Dima *et al.*, 2015; Niculaes *et al.*, 2014) and peroxidases are also found in the vacuole (Welinder *et al.*, 2002; Weng and Chapple, 2010). However, as glycosylation is a cytoplasmic process and the vacuole is a storage compartment for glycosylated monolignols (Dima *et al.*, 2015; Liu *et al.*, 2011), the radical coupling of monolignols to form neolignan-like compounds most

Table 1 Phenolic compounds which show significant differences between their abundance in *LigDFG* transgenic Arabidopsis plants and wild type

nr	m/z	RT (min)	Identity	WT	amyDFG10	amyDFG12	ppiDFG1	Ratio of amyDFG10/WT
Compounds with increased abundance in amyDFG10								
HPV-like compounds								
1	357.12	4.50	HPV + hexose	1569	8751	1502	2062	5.58
2	387.13	5.01	HPS + hexose	240	1944	187	196	8.10
3	399.16	6.45	HPV + hexose + acetate	2568	6277	2365	2000	2.44
Dilignols and dilignol hexosides								
4	373.13	11.85	Gox(β-O-4)G	1838	10 382	2079	2127	5.65
5	371.11	14.26	Gox(β-O-4)G′	246	2020	132	405	8.21
6	581.19	8.42	Gox(β-O-4)G 4-O-hexoside (formic acid adduct)	28	2616	32	19	93.43
Neolignan-like compounds								
7	549.16	8.68	Gox(β-O-4)FA + hexose	1126	8516	562	982	7.56
8	549.16	9.19	Gox(β-O-4)FA + hexose	5	1634	4	4	326.80
9	549.15	9.73	Gox(β-O-4)FA + hexose	470	4759	313	375	10.13
10	503.12	12.21	Gox(β-O-4)FA + malate	1218	11 972	1039	1351	9.83
11	503.12	12.85	Gox(β-O-4)FA + malate	639	6681	559	636	10.46
12	665.17	9.06	Gox(β-O-4)FA + malate + hexose	6	1358	7	0	226.33
13	665.17	9.21	Gox(β-O-4)FA + malate + hexose	354	6307	185	384	17.82
14	773.23	14.02	Gox(β-O-4)FA + hexose + sinapic acid	1897	21 354	2273	2520	11.26
15	773.23	15.36	Gox(β-O-4)FA + hexose + sinapic acid	43	5128	123	79	119.26
16	935.29	11.14	Gox(β-O-4)FA + hexose + hexose + sinapic acid	94	5794	114	244	61.64
17	935.29	11.24	Gox(β-O-4)FA + hexose + hexose + SA	63	4159	11	173	66.02
18	935.29	11.71	Gox(β-O-4)FA + hexose + hexose + SA	17	1982	0	75	116.59
19	516.15	10.13	Gox(β-O-4)FA + glutamic acid	17	1459	40	18	85.82
20	417.12	12.29	Gox(β-O-4)SA + malate (-malate)	419	1620	320	382	3.87
21	417.12	13.22	Gox(β-O-4)SA + malate (-malate)	218	1367	193	181	6.27
Compounds with unknown identity								
22	225.04	3.47	Unknown	393	1010	333	417	2.57
23	385.12	12.17	Unknown	49 607	86 594	43 207	55 871	1.75
24	387.15	8.74	Unknown	2275	6362	2198	2226	2.80
Compounds with decreased abundance in amyDFG10								
Compounds with unknown identity								
25	419.14	8.57	Unknown	1583	394	1319	1264	0.25
26	389.13	8.88	Unknown	4931	1970	3762	3823	0.40

All significant differences in abundance were observed between the *amyDFG10* line and wild type. The ratio represents the fold-change in abundance between the listed transgenic line as compared to wild type. s.d., standard deviation; HPV, hydroxypropiovanillone; HPS, hydroxypropiosyringone; G, guaiacyl unit; Gox, α-keto-oxidized guaiacyl unit. The chemical structures of compounds 1–21 are shown in Figure 6.

likely takes place in the cytosol before glycosylation and transport into the vacuole. This is corroborated by the observation that the phenolic alterations observed by Tsuji and co-workers in both purely cytoplasmic and mixed cytoplasmic/apoplastic localized LigD-expressing transformants are also neolignan-like compounds (Tsuji *et al.*, 2015). Malate (trans-) esterification occurs inside the vacuole (Hause *et al.*, 2002; Strack and Sharma, 1985).

In this study, the LigDFG proteins were targeted to the secretory pathway using well-proven signal peptides (Liu *et al.*, 2004; Rogers, 1985), although leakage or mis-targeting cannot be fully excluded. The presence of oxidized neolignan-like compounds such as Gox(β-O-4)FA in the transgenic *amyDFG10* Arabidopsis line indicates that the LigD enzyme and G(β-O-4) ferulic acid have at some point been present in the same cellular compartment. This could either be caused by mis-targeting of LigD to the cytoplasm in the *amyDFG10* line, or by co-occurrence of the LigD enzyme and its substrate G(β-O-4)ferulic acid as the protein travels through the secretory pathway to the apoplast. The former explanation is the most likely due to similarities in phenolic profiling with the cytoplasmically localized LigD (Tsuji

et al., 2015). Quantification of the LigD enzyme in soluble protein extracts from all three *LigDFG*-expressing lines using Western blot analysis (Figure 3) demonstrated that the expression level of *LigD* was strongly variable between replicate plants from the two homozygous transgenic lines (*amyDFG12* and *ppiDFG1*), but high and more stable in the transgenic *amyDFG10* line, potentially due to the mixed localization of LigD in *amyDFG10* lines. All three transgenic Arabidopsis lines possessed LigDFG activity in the *in vitro* assay, expressed the LigD enzyme to a level detectable by Western blot and displayed benzylic oxidation in β-O-4-structures. However, two of the lines (*amyDFG12* and *ppiDFG1*) did not display a significant change in soluble phenolics, supporting the notion that the absolute levels of intracellular enzyme accumulation in these lines were negligible. Furthermore, the required cofactors [nicotinamide adenine dinucleotide (NAD$^+$) and glutathione (GSH)] are expected to be nonlimiting in the cytosol, meaning that if LigDFG had been present in this compartment in the *amyDFG12* and *ppiDFG1* lines, we would have detected changes in soluble phenolics. The fact that we did not strongly implies that the detected oxidation of G and S lignin

1 R=Hex
3 R=acetyl Hex

2

4 R=H
6 R=Hex

5

7-9 R₁=H, R₂=Hex
10, 11 R₁=H, R₂=malate
12, 13 R₁=H, R₂=malate + Hex
14, 15 R₁=H, R₂=malate + 224 Da
16, 17 R₁=H, R₂=Hex + Hex + 224 Da
18 R₁=Hex, R₂=Hex +224 Da
19 R₁=H, R₂=glutamate

20, 21

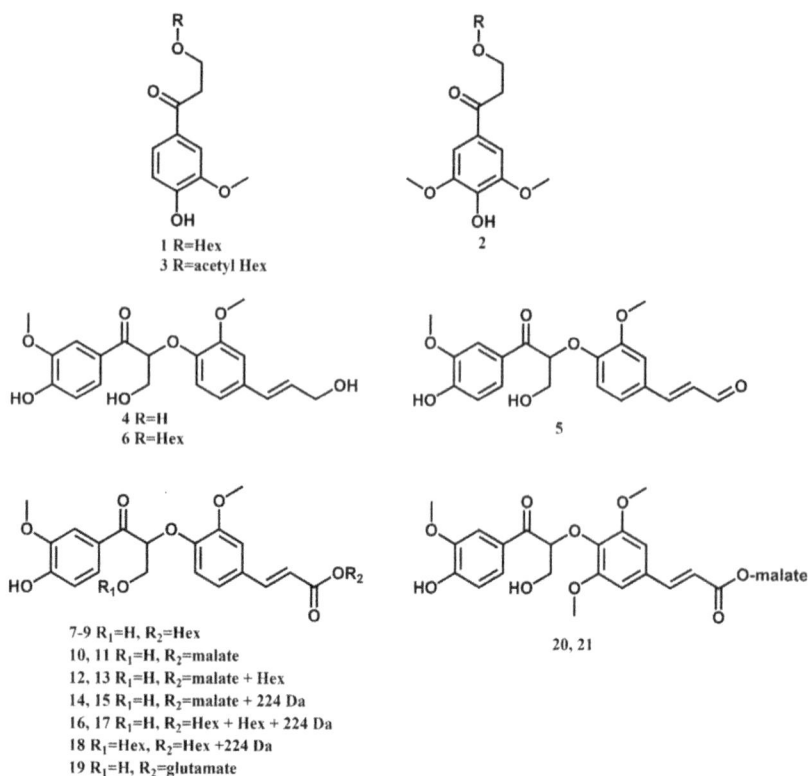

Figure 6 Structures of phenolic compounds which show significant differences between their abundance in *LigDFG* transgenic Arabidopsis plants and wild type. Structural identifications of the compounds were essentially based on MS/MS fragmentation data. The structural elucidation of compounds 1–3 is reported in SM, whereas structural characterization of compounds 4–21 has previously been reported (Tsuji *et al.*, 2015).

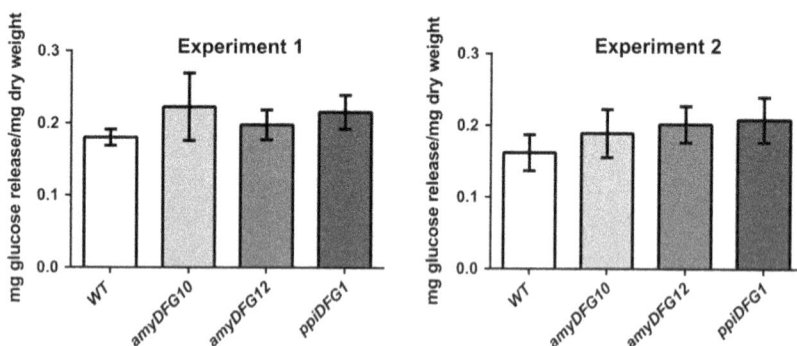

Figure 7 Analysis of saccharification yield in *LigDFG* transgenic Arabidopsis plants. Two-way ANOVA showed statistically significant increases in glucose release from *amyDFG10* ($P < 0.02$), *amyDFG12* ($P < 0.02$) and *ppiDFG1* ($P < 0.001$), with no interaction between genotype and growth cohort (two cohorts, $n = 8$ in each cohort).

units (Gox and Sox in the NMR) and the proposed cleavage of β–O–4-bonds had indeed taken place in the apoplast.

Two cofactors are required for the ability of LigDFG enzymes from *S. paucimobilis* to catalyse cleavage of the β-aryl ether units in lignin. LigD activity is dependent on the presence of NAD^+ whereas LigF and LigG activity requires the presence of GSH. The concentration of these two cofactors is expected to be low in the apoplast although NAD^+ is detected in the apoplastic fluid and associated with peroxidase activity (Otter and Polle, 1997) and GSH is involved in detoxification processes and small amounts are found in the apoplast (Foyer *et al.*, 2001). Low concentrations of cofactors are therefore considered to be the limiting factor for LigDFG activity in the apoplast rather than the level of the LigDFG proteins. A completely separate factor that may also restrict the level of modifications introduced into lignin by the expression of *LigDFG* is that LigD and LigF show high substrate specificity with only one of four stereoisomers of G(β-O-4)G being accepted as substrate and cleaved.

The results obtained from the characterization of the transgenic Arabidopsis plants expressing *LigDFG* provided additional knowledge on the substrate specificity of LigDFG. The 2D HSQC NMR analysis showed an increase in the amount of Sox units formed. In addition, S(β-O-4)G was converted into HPS by the LigDFG enzymes. The phenolic analysis also revealed formation of Gox(β-O-4)sinapic acid. This demonstrated that coniferyl alcohol dimers as well as sinapyl alcohol and sinapic acid ether dimers were recognized by the LigDFG enzymes. These *in vivo* activities match recently reported *in vitro* activities of LigD on S(β-O-4) units (Gall *et al.*, 2014b; Tsuji *et al.*, 2015). The ability of the LigDFG enzymes to reductively cleavage not only G units expands the potential of using this approach to modify lignin structure in plants as this allows a greater variety of ether-linked units to be cleaved.

Expression of the individual *LigD*, *LigF* and *LigG* genes in transgenic Arabidopsis lines provided interesting insights into the ability of endogenous plant enzyme activities to metabolize intermediates in the LigD-LigF-LigG pathway. When protein extracts from *LigF*-expressing lines were incubated with MPHPV as substrate, HPV was formed in addition to the expected product GS-HPV (Figure S2). In the bacterial pathway, the

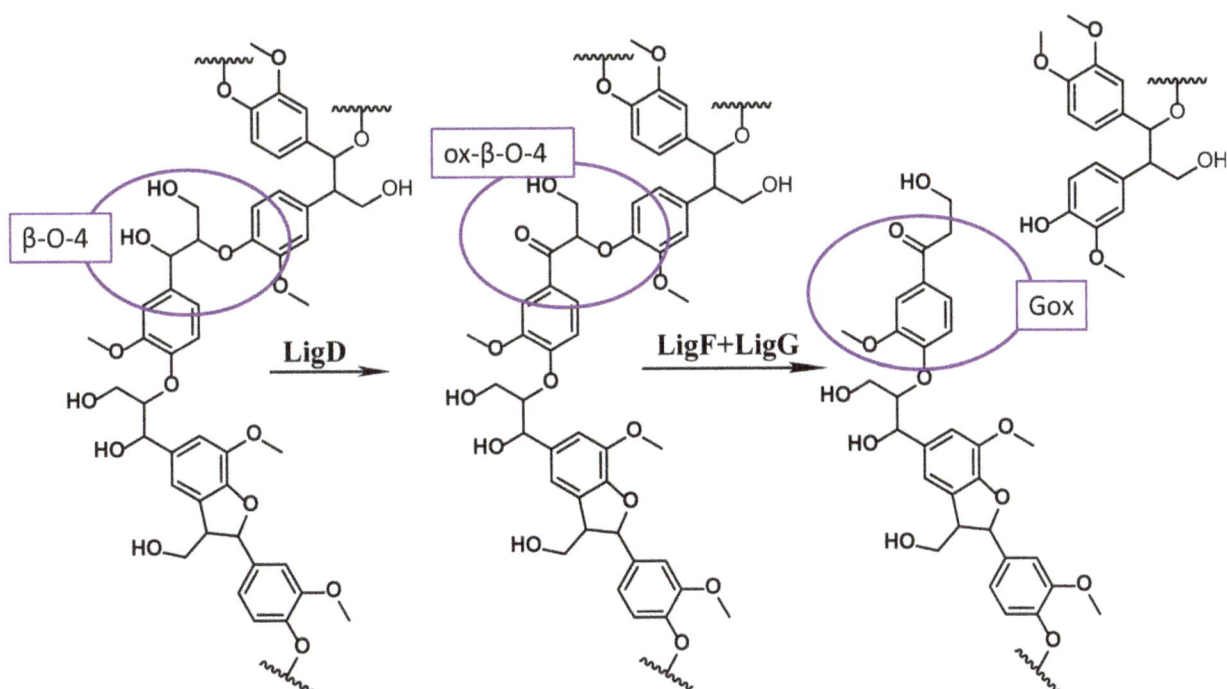

Figure 8 Cleavage of a β–O–4-bond in lignin by the LigDFG enzymes depicting their theoretical contribution to the observed increase in Gox units as revealed by the 2D HSQC NMR spectra.

conversion of GS-HPV to HPV is catalysed by LigG. We did not detect LigG activity in enzyme extracts from WT Arabidopsis leaves, even upon prolonged incubation, and based on this it can be suggested that the enzyme responsible for the conversion in the transgenic lines was LigF. However, the phenolic profiling demonstrated that HPV derivatives are indeed present in stems from WT plants, and re-examination of the results from Tsuji et al. (2015) demonstrated that the amounts of these compounds were also slightly increased in plants expressing only LigD. These results suggest that Arabidopsis does harbour enzymes with LigF and LigG activities. The discrepancy in enzyme activity between leaves and stems can be explained either by differential expression of endogenous LigF and LigG-like enzymes in different tissues or at different growth stages, or by the endogenous activities being collated into one enzyme with LigFG activity. LigG belongs to the omega class of GSTs (Meux et al., 2012) that are characterized by catalysing redox reactions and reductive deglutathionylations instead of a classical glutathionylation reactions (Board et al., 2008). A cysteine residue in the active site of the omega class of GSTs is essential for their different mode of action. In plants, GSTs of the lambda class are known to have cysteine in the active site as opposed to glutathionylating plant GSTs that harbour a serine in this position (Lallement et al., 2014b). The lambda class GSTs from wheat and poplar are able to reduce glutathionylquercetin to quercetin (Dixon and Edwards, 2010; Lallement et al., 2014a). A lambda class GST is therefore a good candidate for the LigG activity found in wild-type Arabidopsis, but no plant enzymes are known which catalyse a LigF-like reaction. In bacterial degradation of pentachlorophenols, a single enzyme is responsible for a presumably two-step reaction similar to the combined LigFG-reaction (Huang et al., 2008). The reaction consumes two GSH-molecules and, if the enzyme is damaged by oxidation at the

catalytic Cys-residue, it produces a GSH-conjugate that it is unable to cleave. It is therefore possible that under certain conditions, LigF harbours additional LigG activity, and that Arabidopsis produces an enzyme with both activities. Further analysis is required in order to identify the Arabidopsis enzymes and compare their activity to LigF and LigG to elucidate whether the introduction of these enzymes into the apoplast in concert with LigD is sufficient to introduce changes into lignin structure. As endogenous lambda GSTs in plants are targeted to either the cytosol, chloroplast or peroxisome (Dixon et al., 2009), such genetic engineering will still be required to obtain degradation of lignin.

In conclusion, we have shown that expression of *LigDFG* encoding the β-aryl ether-unit-degrading enzymes from *S. paucimobilis*, has an impact on lignin structure, thus providing a platform for saccharification yield improvement, a desired feature of plants for biofuel production. Even though there are limitations to the amount of change possible from this approach, we have shown that the changes observed are statistically significant and that it is possible to modify lignin structure, without significant alterations to neolignan-like compounds, by targeting bacteria-derived lignin-degrading enzymes to the secretory pathway. Fine-tuning of lignin cross-linking may be combined with other modest modifications of lignin structure offering improved properties with respect to biomass conversion into biofuels without compromising plant robustness.

Materials and methods

Vector construction

For expression of the various Lig genes in *E. coli* and *Arabidopsis thaliana*, genes were synthesized by GenScript (Hong Kong). For expression in *E. coli*, the following genes were synthesized: *LigD*

(DDBJ accession no. D11473–1), *LigF* (DDBJ accession no. D11473-2) and *LigG* (DDBJ accession no. AB026292) and subcloned into the pETSTREP vector (Dixon *et al.*, 2009). A list of the synthetic gene sequences and the primers used for amplification can be found in Tables S1 and S2.

Vectors for plant expression studies were constructed using the USER cloning technique (Geu-Flores *et al.*, 2007). The gene sequences were codon-optimized for plant expression and synthesized with added signal peptides, either from barley α-amylase (amy) (Genbank accession no. K02637) (Rogers, 1985) or from potato proteinase inhibitor II (ppi) (Liu *et al.*, 2004). Genes were amplified independently by PCR (Samalova *et al.*, 2006) linking the *LigD*, *LigF*, *LigG* sequences with 2A sequences. Each gene was provided with a signal peptide (amy or ppi). The genes were inserted into the PacI/Nt.BbvCI USER cloning site in pCAMBIA1300 USER compatible vectors and controlled by an enhanced 35S promotor (Geu-Flores *et al.*, 2009). Vectors harbouring a single gene (amy-LigD, amy-LigF, amy-LigG, ppi-LigD, ppi-LigF, ppi-LigG) were constructed by ligation of the PCR products into the PacI/Nt.BbvCI USER cassette in p1300 (for amy-LigF, amy-LigG, ppi-LigF and ppi-LigG) or p2300 (amy-LigD and ppi-LigD).

Plant cultivation

Arabidopsis thaliana Col-0 plants used for protein assays and screening experiments and *N. benthamiana* plants used for transient expression were grown at a long-day regime (16-h light) at a temperature of 21 °C and a relative humidity of 60%. For 2D-NMR, metabolite profiling and saccharification experiments, Arabidopsis plants were grown at a short-day regime (8 h light) for 10 weeks followed by growth in the long-day regime. For metabolite profiling, the stems were harvested when they reached a height of approximately 40 cm. For NMR and saccharification experiments, the plants were grown until senescence.

Transient and stable plant transformation

For plant transformation, expression vectors were introduced into *Agrobacterium tumefaciens* strain Agl1 (Lazo *et al.*, 1991) by electroporation (Shen and Forde, 1989). Arabidopsis transformation was performed using the floral dip method (Bent, 2006). Transient expression in tobacco was accomplished by agro-infiltration (Goodin *et al.*, 2002) using 4-week-old plants.

Selection of transformed Arabidopsis plants

Mature seeds were collected from fully senesced Arabidopsis plants. Sterilized seeds were plated on ½ Murashige and Skoog (pH 5.8) containing 30 μg/mL geneticin (G418) or 30 μg/mL hygromycin B. After cold-stratification (4 °C, 72 h), seeds on plates were incubated at 21 °C, long-day regime until the transformed seedlings had developed true leaves. The seedlings were transferred to soil and grown in a greenhouse. Transgenic plants were verified by PCR on gDNA isolated from their leaves. Primers used to amplify inserted *LigDFG* genes were LigDfw (5′-CGCTCACGGCATCGTTCTTGATA-3′) and LigGrev (5′-CCACC TTGTGTATAGTCATAG-3′) (Tm 54 °C, 30 cycles).

The transgenic plants were characterized by NMR, saccharification and phenolic profiling using plants grown from seeds of T3 plants of homozygous amyDFG12 and ppiDFG1 lines sown alongside segregating amyDFG10 seeded plants and wild type. In the phenolic profiling experiment, segregating amyDFG10 plants were genotyped by PCR for the presence of the amyDFG10

construct. For the experiments with plants expressing only *amyD*, *amyF*, *amyG*, *ppiD* and *ppiF*, T3 homozygous lines were used.

Expression in *E. coli*

LigD, LigF and LigG were expressed in the *E. coli* BL21DE3 strain (New England Biolabs, Ipswich, MA, USA). Bacteria were grown (28 °C) in LB medium (50 mL) containing 50 μg/mL of kanamycin. Gene expression was induced by addition of Isopropyl β-D-1-thiogalactopyranoside (IPTG) (final concentration 1 mM) at $OD_{600 \text{ nm}}$ 0.6. After the induction (5 h), cells were harvested by centrifugation and protein was isolated.

Extraction and metabolite profiling of phenylpropanoids

The lower part (8 cm) of Arabidopsis flowering stems (total height of around 40 cm) carrying multiple fully developed siliques was frozen and ground in liquid nitrogen. Sample preparation procedure and UHPLC-MS setting were previously reported (Vanholme *et al.*, 2013b). From the resulting chromatograms, 1286 de-isotoped constituent signals were integrated and aligned via Progenesis QI (Waters Corporation, Milford, MA), with each constituent being defined by its *m/z* value and retention time. Statistics (ANOVA with *post hoc t*-test) were performed in Progenesis QI extension EZinfo. The following stringent filters were used: abundance > 1000, *P*-value ANOVA < 0.05. Statistical analyses were performed on arcsinh-transformed ion intensities. For structural elucidation, MS/MS was used. For MS/MS, all settings were the same as in full MS, except the collision energy was ramped from 15 to 35 eV in the trap, and the scan time was set at 0.5 s.

Protein isolation

Plant soluble proteins were extracted from rosette leaves of T3 plants by homogenization in liquid nitrogen and extraction into 100 mM potassium phosphate buffer (pH 7.5) including 3 mM EDTA, 1 mM phenylmethyl-sulfonyl fluoride (PMSF), and 1 mM GSH in the presence of polyvinylpoly-pyrrolidone (PVPP) (100 mg/1 g of plant material). The homogenate was filtered through muslin gauze and centrifuged (30 min; 20 000 *g*), and the supernatant obtained used for Western blotting, protein assays, etc.

LigDFG enzyme assays

Appropriate substrates, GGE (0.5 mM final concentration) or MPHPV (0.5 mM final concentration), NAD$^+$ (1 mM final concentration) and GSH (1 mM final concentration) were mixed in 100 mM phosphate buffer (pH 7.5) (total volume: 100 μL), and proteins of bacterial or plant origin were added. Following incubation (16 h, room temperature), the enzyme reaction was stopped by adding MeOH (50% final concentration).

Chemical synthesis

The chemical synthesis of MPHPV S4 and HPV-glucopyranoside S14 is described in Data S1.

Immunodetection by Western blot

Antibodies against LigD were produced by Agrisera (Vännäs, Sweden) in rabbits using the LigD amino acid sequence NH2-CDIMDREAYARAADE-CONH2 (residues 63–76) as an antigen. The serum obtained (1 : 500 dilution) 15 weeks after the first immunization was used for Western blots to obtain a semi-quantitative assessment of the LigD protein levels in transgenic plants.

LC-MS analyses

LC-MS analyses were performed using a Dionex HPLC fitted with a UV detector (UVD340U with microflow cell) and connected to an MSQplus MS detector (Thermoscientific, Waltham, MA, USA). Chromatographic separation was obtained using a Zorbax SB C18 column (2.1 × 50 mm, Phenomex, Værlose, Denmark). The mobile phases were A: 50 μM NaCl, 0.1% HCOOH; B: 50 μM NaCl, 0.1% HCOOH, 80% MeCN. The gradient used was 0% B to 70% B (linear) in 30 min, 70% B to 100% B in 5 min (flow rate: 0.2 mL/min). The column was rinsed with 100% B and re-equilibrated with A for 6 min (flow rate: 0.3 mL/min). Mass spectrometry was performed in the positive ionization mode ESI (cone voltage: 75 V, needle voltage: 3.5 kV). Samples were analysed in full-scan mode (*m/z* 100–800).

2D HSQC NMR

The Arabidopsis samples (1 g) obtained from the 25 cm lower part of 10 flowering stems (total plant height around 40 cm and carrying multiple fully developed siliques) were used for NMR analyses, and acetylated lignin samples were prepared as described in Bonawitz *et al.* (2014), Lu and Ralph (2003), Tsuji *et al.* (2015).

For NMR investigations, the acetylated lignin sample (55 mg) was dissolved in deuterated chloroform ($CDCl_3$, 0.5 mL) in a 5 mm NMR tube. The 2D HSQC spectrum was acquired using a Bruker 700 MHz instrument with a cryogenically cooled QCI ($^{1}H/^{31}P/^{13}C/^{15}N$) 5 mm inverse ($^{1}H$ coils closest to the sample) probe using a standard HSQC program. Statistics (ANOVA with *post hoc t*-test) were performed.

Saccharification yield

Dried ground Arabidopsis stems (30 mg) obtained from lower 8 cm were resuspended in MilliQ water (600 μL, 70 °C, 2 h), centrifuged (12 000 ***g***, 10 min) and the pellet obtained resuspended in sodium acetate buffer (600 μL, 50 mM, pH 4.5) containing Cellic Ctec2 (5 FPU/g of dry matter; Novozymes, Bagsværd, Denmark) and xylanase BioFeed Wheat (Novozymes, 1% of biomass). Samples were incubated (24 h, 50 °C, shaking), centrifuged (16 000 ***g***, 5 min) and aliquots (20 μL) of the supernatant were hydrolysed (1 h, 120 °C) by addition of trifluoroacetic acid (TFA) (150 μL) dissolved in MilliQ water (830 μL). Solvent was removed under vacuum and the residue obtained dissolved in water (1 mL). Glucose concentration was determined by high-performance anion exchange chromatography with pulsed amperometric detection (HPAEC-PAD) using a PA20 column (Thermoscientific, Waltham, MA, USA) (Øbro *et al.*, 2004).

Acknowledgements

The Carlsberg Foundation and Villum Foundation are gratefully acknowledged for support to J.H. Financial support from the VILLUM Center of Excellence 'Plant Plasticity' and from the Center for Synthetic Biology 'bioSYNergy' funded by the UCPH Excellence Program for Interdisciplinary Research is gratefully acknowledged. Additionally, we acknowledge the European Commission's Directorate-General for Research within the 7th Framework Program (FP7/2007–2013) under the grant agreement No 270089 (MULTIBIOPRO), the Hercules program of Ghent University for the Synapt Q-Tof (grant no. AUGE/014), the Multidisciplinary Research Partnership 'Biotechnology for a Sustainable Economy' (01MRB510W) of Ghent University and the SBO-FISCH project 'ARBOREF', funded by the Agency for Innovation by Science and Technology (IWT) in Flanders. PO is indebted to the Agency for Innovation by Science and Technology of Chile for a predoctoral fellowship, and RV is indebted to the Research Foundation-Flanders (FWO) for a postdoctoral fellowship. JR, FL and SL were funded by the DOE Great Lakes Bioenergy Research Center (DOE BER Office of Science DE-FC02-07ER64494). Innovation Fund Denmark is acknowledged for support to BJ and PU (FLABBERGAST A, The Biological Track—4106-00037B).

References

Bent, A. (2006) *Arabidopsis thaliana* floral dip transformation method. In *Agrobacterium Protocols* (Wang, K., ed.), pp. 87–104. Totowa, NJ: Humana Press.

Board, P.G., Coggan, M., Cappello, J., Zhou, H., Oakley, A.J. and Anders, M.W. (2008) S-(4-Nitrophenacyl)glutathione is a specific substrate for glutathione transferase omega 1-1. *Anal. Biochem.* **374**, 25–30.

Boerjan, W., Ralph, J. and Baucher, M. (2003) Lignin biosynthesis. *Annu. Rev. Plant Biol.* **54**, 519–546.

Bonawitz, N.D. and Chapple, C. (2010) The genetics of lignin biosynthesis: connecting genotype to phenotype. *Annu. Rev. Genet.* **44**, 337–363.

Bonawitz, N.D., Kim, J.I., Tobimatsu, Y., Ciesielski, P.N., Anderson, N.A., Ximenes, E., Maeda, J. *et al.* (2014) Disruption of mediator rescues the stunted growth of a lignin-deficient *Arabidopsis* mutant. *Nature* **509**, 376–380.

Borkhardt, B., Harholt, J., Ulvskov, P., Ahring, B.K., Jørgensen, B. and Brinch-Pedersen, H. (2010) Autohydrolysis of plant xylans by apoplastic expression of thermophilic bacterial endo-xylanases. *Plant Biotechnol. J.* **8**, 363–374.

Buanafina, M.M.O., Langdon, T., Hauck, B., Dalton, S., Timms-Taravella, E. and Morris, P. (2010) Targeting expression of a fungal ferulic acid esterase to the apoplast, endoplasmic reticulum or golgi can disrupt feruloylation of the growing cell wall and increase the biodegradability of tall fescue (*Festuca arundinacea*). *Plant Biotechnol. J.* **8**, 316–331.

Davin, L.B. and Lewis, N.G. (2005) Dirigent phenoxy radical coupling: advances and challenges. *Curr. Opin. Biotechnol.* **16**, 398–406.

Dima, O., Morreel, K., Vanholme, B., Kim, H., Ralph, J. and Boerjan, W. (2015) Small glycosylated lignin oligomers are stored in *Arabidopsis* leaf vacuoles. *Plant Cell*, **27**, 695–710.

Dixon, D.P. and Edwards, R. (2010) Roles for stress-inducible lambda glutathione transferases in flavonoid metabolism in plants as identified by ligand fishing. *J. Biol. Chem.* **285**, 36322–36329.

Dixon, D.P., Hawkins, T., Hussey, P.J. and Edwards, R. (2009) Enzyme activities and subcellular localization of members of the *Arabidopsis* glutathione transferase superfamily. *J. Exp. Bot.* **60**, 1207–1218.

El Amrani, A., Barakate, A., Askari, B.M., Li, X., Roberts, A.G., Ryan, M.D. and Halpin, C. (2004) Coordinate expression and independent subcellular targeting of multiple proteins from a single transgene. *Plant Physiol.* **135**, 16–24.

de Felipe, P. (2002) Polycistronic viral vectors. *Curr. Gene Ther.* **2**, 355–378.

Foyer, C.H., Theodoulou, F.L. and Delrot, S. (2001) The functions of inter- and intracellular glutathione transport systems in plants. *Trends Plant Sci.* **6**, 486–492.

Gall, D.L., Kim, H., Lu, F., Donohue, T.J., Noguera, D.R. and Ralph, J. (2014a) Stereochemical features of glutathione-dependent enzymes in the *Sphingobium* sp. strain SYK-6 β-aryl etherase pathway. *J. Biol. Chem.* **289**, 8656–8667.

Gall, D.L., Ralph, J., Donohue, T.J. and Noguera, D.R. (2014b) A group of sequence-related sphingomonad enzymes catalyzes cleavage of β-aryl ether linkages in lignin β-guaiacyl and β-syringyl ether dimers. *Environ. Sci. Technol.* **48**, 12454–12463.

Geu-Flores, F., Nour-Eldin, H.H., Nielsen, M.T. and Halkier, B.A. (2007) USER fusion: a rapid and efficient method for simultaneous fusion and cloning of multiple PCR products. *Nucleic Acids Res.* **35**, e55.

Geu-Flores, F., Olsen, C.E. and Halkier, B.A. (2009) Towards engineering glucosinolates into non-cruciferous plants. *Planta* **229**, 261–270.

Goodin, M.M., Dietzgen, R.G., Schichnes, D., Ruzin, S. and Jackson, A.O. (2002) pGD vectors: versatile tools for the expression of green and red fluorescent protein fusions in agroinfiltrated plant leaves. *Plant J.* **31**, 375–383.

Hause, B., Meyer, K., Viitanen, P.V., Chapple, C. and Strack, D. (2002) Immunolocalization of 1-O-sinapoylglucose:malate sinapoyltransferase in *Arabidopsis thaliana*. *Planta* **215**, 26–32.

Herbers, K., Wilke, I. and Sonnewald, U. (1995) A thermostable xylanase from *Clostridium thermocellum* expressed at high levels in the apoplast of transgenic tobacco has no detrimental effects and is easily purified. *Bio/Technology* **13**, 63–66.

Huang, Y., Xun, R., Chen, G. and Xun, L. (2008) Maintenance role of a glutathionyl-hydroquinone lyase (PcpF) in pentachlorophenol degradation by *Sphingobium chlorophenolicum* ATCC 39723. *J. Bacteriol.* **190**, 7595–7600.

Kirk, T.K. and Farrell, R.L. (1987) Enzymatic "combustion": the microbial degradation of lignin. *Annu. Rev. Microbiol.* **41**, 465–505.

Lallement, P.-A., Brouwer, B., Keech, O., Hecker, A. and Rouhier, N. (2014a) The still mysterious roles of cysteine-containing glutathione transferases in plants. *Front. Pharmacol.* **5**, 192.

Lallement, P.-A., Meux, E., Gualberto, J.M., Prosper, P., Didierjean, C., Saul, F., Haouz, A. *et al.* (2014b) Structural and enzymatic insights into Lambda glutathione transferases from *Populus trichocarpa*, monomeric enzymes constituting an early divergent class specific to terrestrial plants. *Biochem. J.* **462**, 39–52.

Lazo, G.R., Stein, P.A. and Ludwig, R.A. (1991) A DNA transformation-competent *Arabidopsis* genomic library in *Agrobacterium*. *Biotechnology (N. Y.)*, **9**, 963–967.

Li, X., Weng, J.-K. and Chapple, C. (2008) Improvement of biomass through lignin modification. *Plant J.* **54**, 569–581.

Liu, Y.-J., Yuan, Y., Zheng, J., Tao, Y.-Z., Dong, Z.-G., Wang, J.-H. and Wang, G.-Y. (2004) Signal peptide of potato PinII enhances the expression of Cry1Ac in transgenic tobacco. *Acta Biochim. Biophys. Sin. (Shanghai)* **36**, 553–558.

Liu, C.-J., Miao, Y.-C. and Zhang, K.-W. (2011) Sequestration and transport of lignin monomeric precursors. *Molecules* **16**, 710–727.

Lu, F. and Ralph, J. (2003) Non-degradative dissolution and acetylation of ball-milled plant cell walls: high-resolution solution-state NMR. *Plant J.* **35**, 535–544.

Lu, F. and Ralph, J. (2011) Solution-state NMR of lignocellulosic biomass. *J. Biobased Mater. Bio.* **5**, 169–180.

Mansfield, S.D., Kim, H., Lu, F. and Ralph, J. (2012) Whole plant cell wall characterization using solution-state 2D NMR. *Nat. Protoc.* **7**, 1579–1589.

Masai, E., Ichimura, A., Sato, Y., Miyauchi, K., Katayama, Y. and Fukuda, M. (2003) Roles of the enantioselective glutathione S-transferases in cleavage of beta-aryl ether. *J. Bacteriol.* **185**, 1768–1775.

Masai, E., Katayama, Y. and Fukuda, M. (2007) Genetic and biochemical investigations on bacterial catabolic pathways for lignin-derived aromatic compounds. *Biosci. Biotechnol. Biochem.* **71**, 1–15.

Matsuda, F., Hirai, M.Y., Sasaki, E., Akiyama, K., Yonekura-Sakakibara, K., Provart, N.J., Sakurai, T. *et al.* (2010) AtMetExpress development: a phytochemical atlas of *Arabidopsis* development. *Plant Physiol.* **152**, 566–578.

Meux, E., Prosper, P., Masai, E., Mulliert, G., Dumarçay, S., Morel, M., Didierjean, C. *et al.* (2012) *Sphingobium* sp. SYK-6 LigG involved in lignin degradation is structurally and biochemically related to the glutathione transferase ω class. *FEBS Lett.* **586**, 3944–3950.

Morreel, K., Dima, O., Kim, H., Lu, F., Niculaes, C., Vanholme, R., Dauwe, R. *et al.* (2010) Mass spectrometry-based sequencing of lignin oligomers. *Plant Physiol.* **153**, 1464–1478.

Morreel, K., Saeys, Y., Dima, O., Lu, F., Van de Peer, Y., Vanholme, R., Ralph, J. *et al.* (2014) Systematic structural characterization of metabolites in

Arabidopsis via candidate substrate-product pair networks. *Plant Cell*, **26**, 929–945.

Niculaes, C., Morreel, K., Kim, H., Lu, F., McKee, L.S., Ivens, B., Haustraete, J. *et al.* (2014) Phenylcoumaran benzylic ether reductase prevents accumulation of compounds formed under oxidative conditions in poplar xylem. *Plant Cell*, **26**, 3775–3791.

Øbro, J., Harholt, J., Scheller, H.V. and Orfila, C. (2004) Rhamnogalacturonan I in *Solanum tuberosum* tubers contains complex arabinogalactan structures. *Phytochemistry* **65**, 1429–1438.

Øbro, J., Borkhardt, B., Harholt, J., Skjøt, M., Willats, W.G.T. and Ulvskov, P. (2009) Simultaneous in vivo truncation of pectic side chains. *Transgenic Res.* **18**, 961–969.

Otter, T. and Polle, A. (1997) Characterisation of acidic and basic apoplastic peroxidases from needles of Norway spruce (*Picea abies*, L., Karsten) with respect to lignifying substrates. *Plant Cell Physiol.* **38**, 595–602.

Pauly, M. and Keegstra, K. (2010) Plant cell wall polymers as precursors for biofuels. *Curr. Opin. Plant Biol.* **13**, 305–312.

Rahimi, A., Azarpira, A., Kim, H., Ralph, J. and Stahl, S.S. (2013) Chemoselective metal-free aerobic alcohol oxidation in lignin. *J. Am. Chem. Soc.* **135**, 6415–6418.

Ralph, J. (2007) The compromised wood workshop 2007. In *Perturbing Lignification* (Entwistle, K., Harris, P.J. and Walker, J., eds), pp. 85–112. Canterbury: Wood Technology Research Centre, University of Canterbury.

Ralph, J., Lundquist, K., Brunow, G., Lu, F., Kim, H., Schatz, P.F., Marita, J.M. *et al.* (2004) Lignins: natural polymers from oxidative coupling of 4-hydroxyphenyl-propanoids. *Phytochem. Rev.* **3**, 29–60.

Rogers, J.C. (1985) Two barley alpha-amylase gene families are regulated differently in aleurone cells. *J. Biol. Chem.* **260**, 3731–3738.

Samalova, M., Fricker, M. and Moore, I. (2006) Ratiometric fluorescence-imaging assays of plant membrane traffic using polyproteins. *Traffic* **7**, 1701–1723.

Sato, Y., Moriuchi, H., Hishiyama, S., Otsuka, Y., Oshima, K., Kasai, D., Nakamura, M. *et al.* (2009) Identification of three alcohol dehydrogenase genes involved in the stereospecific catabolism of arylglycerol-beta-aryl ether by *Sphingobium* sp. strain SYK-6. *Appl. Environ. Microbiol.* **75**, 5195–5201.

Sederoff, R.R., MacKay, J.J., Ralph, J. and Hatfield, R.D. (1999) Unexpected variation in lignin. *Curr. Opin. Plant Biol.* **2**, 145–152.

Shen, W.J. and Forde, B.G. (1989) Efficient transformation of *Agrobacterium* spp. by high voltage electroporation. *Nucleic Acids Res.* **17**, 8385.

Sonoki, T., Iimura, Y., Masai, E., Kajita, S. and Katayama, Y. (2002) Specific degradation of β-aryl ether linkage in synthetic lignin (dehydrogenative polymerizate) by bacterial enzymes of *Sphingomonas paucimobilis* SYK-6 produced in recombinant *Escherichia coli*. *J. Wood Sci.* **48**, 429–433.

Strack, D. and Sharma, V. (1985) Vacuolar localization of the enzymatic synthesis of hydroxycinnamic acid esters of malic acid in protoplasts from *Raphanus sativus* leaves. *Physiol. Plant.* **65**, 45–50.

Tsuji, Y., Vanholme, R., Tobimatsu, Y., Ishikawa, Y., Foster, C.E., Kamimura, N., Hishiyama, S. *et al.* (2015) Introduction of chemically labile substructures into *Arabidopsis* lignin through the use of LigD, the Cα-dehydrogenase from *Sphingobium* sp. strain SYK-6. *Plant Biotechnol. J.* **13**, 821–832.

Van Acker, R., Vanholme, R., Storme, V., Mortimer, J.C., Dupree, P. and Boerjan, W. (2013) Lignin biosynthesis perturbations affect secondary cell wall composition and saccharification yield in *Arabidopsis thaliana*. *Biotechnol. Biofuels* **6**, 46.

Vanholme, R., Morreel, K., Ralph, J. and Boerjan, W. (2008) Lignin engineering. *Curr. Opin. Plant Biol.* **11**, 278–285.

Vanholme, R., Morreel, K., Darrah, C., Oyarce, P., Grabber, J.H., Ralph, J. and Boerjan, W. (2012) Metabolic engineering of novel lignin in biomass crops. *New Phytol.* **196**, 978–1000.

Vanholme, B., Desmet, T., Ronsse, F., Rabaey, K., Van Breusegem, F., De Mey, M., Soetaert, W. *et al.* (2013a) Towards a carbon-negative sustainable bio-based economy. *Front. Plant Sci.* **4**, 174.

Vanholme, R., Cesarino, I., Rataj, K., Xiao, Y., Sundin, L., Goeminne, G., Kim, H. *et al.* (2013b) Caffeoyl shikimate esterase (CSE) is an enzyme in the lignin biosynthetic pathway in *Arabidopsis*. *Science* **341**, 1103–1106.

Vogel, J. (2008) Unique aspects of the grass cell wall. *Curr. Opin. Plant Biol.* **11**, 301–307.

Welinder, K.G., Justesen, A.F., Kjaersgård, I.V.H., Jensen, R.B., Rasmussen, S.K., Jespersen, H.M. and Duroux, L. (2002) Structural diversity and transcription of class III peroxidases from *Arabidopsis thaliana*. *Eur. J. Biochem.* **269**, 6063–6081.

Weng, J.-K. and Chapple, C. (2010) The origin and evolution of lignin biosynthesis. *New Phytol.* **187**, 273–285.

Enhanced transport of plant-produced rabies single-chain antibody-RVG peptide fusion protein across an *in cellulo* blood–brain barrier device

Waranyoo Phoolcharoen[1,2,]*, Christophe Prehaud[3], Craig J. van Dolleweerd[1], Leonard Both[1], Anaelle da Costa[3], Monique Lafon[3] and Julian K-C. Ma[1]

[1]Institute for Infection and Immunity, St. George's Hospital Medical School, University of London, London, UK
[2]Pharmacognosy and Pharmaceutical Botany, Faculty of Pharmaceutical Sciences, Chulalongkorn University, Bangkok, Thailand
[3]Unité de Neuroimmunologie Virale, Département de Virologie, Institut Pasteur, Paris, France

*Correspondence
email Waranyoo.P@chula.ac.th

Keywords: rabies virus, single-chain antibody, blood–brain barrier, antibody engineering, plant biotechnology.

Summary

The biomedical applications of antibody engineering are developing rapidly and have been expanded to plant expression platforms. In this study, we have generated a novel antibody molecule *in planta* for targeted delivery across the blood–brain barrier (BBB). Rabies virus (RABV) is a neurotropic virus for which there is no effective treatment after entry into the central nervous system. This study investigated the use of a RABV glycoprotein peptide sequence to assist delivery of a rabies neutralizing single-chain antibody (ScFv) across an *in cellulo* model of human BBB. The 29 amino acid rabies virus peptide (RVG) recognizes the nicotinic acetylcholine receptor (nAchR) at neuromuscular junctions and the BBB. ScFv and ScFv-RVG fusion proteins were produced in *Nicotiana benthamiana* by transient expression. Both molecules were successfully expressed and purified, but the ScFv expression level was significantly higher than that of ScFv-RVG fusion. Both ScFv and ScFv-RVG fusion molecules had potent neutralization activity against RABV *in cellulo*. The ScFv-RVG fusion demonstrated increased binding to nAchR and entry into neuronal cells, compared to ScFv alone. Additionally, a human brain endothelial cell line BBB model was used to demonstrate that plant-produced ScFv-RVG[P] fusion could translocate across the cells. This study indicates that the plant-produced ScFv-RVG[P] fusion protein was able to cross the *in cellulo* BBB and neutralize RABV.

Introduction

Rabies remains a major burden in resource-limited countries particularly in Asia and Africa, accounting for approximately 60 000 deaths per year, mainly in children (Fooks *et al.*, 2014). The most common source of infection is from an animal bite. After a period of replication in muscle, the virus gains access to the peripheral nervous system before entering the central nervous system (CNS) (Hemachudha *et al.*, 2002) by a process of retrograde axonal transport. The virus spreads rapidly to the brain, resulting in an overwhelming encephalitis that kills the host (Hemachudha *et al.*, 2002; Lewis *et al.*, 2000). Rabies is unique in that once a productive infection has been established in the CNS, the outcome is invariably fatal.

Rabies postexposure prophylaxis (PEP) is highly effective if correctly administered promptly after a potential exposure (Shantavasinkul and Wilde, 2011; Uwanyiligira *et al.*, 2012). However, in the case of delayed treatment and the onset of symptoms, PEP is ineffective. RABV antibodies are unlikely to offer therapeutic benefits once RABV has entered the CNS as they cannot cross the blood–brain barrier (BBB) (Pardridge, 2010).

Nicotinic acetylcholine receptors (nAchRs) are ligand-gated channels located in the neuromuscular junction and in the CNS (Lentz *et al.*, 1988). nAchRs facilitate RABV entry into both muscle and neuronal cells (Burrage *et al.*, 1985; Lentz *et al.*, 1982). The rabies glycoprotein, which forms spikes on the surface of the virus, contains a short motif which interacts with nAchR to mediate entry into cells (Lentz, 1990; Lentz *et al.*, 1987). Previous studies have shown that a linear 29 amino acid peptide derived from the rabies glycoprotein (RVG) binds to the alpha subunit of nAchR enabling the delivery of conjugated molecules into the CNS, including siRNA (Kumar *et al.*, 2007), nanoparticles (Hwang do *et al.*, 2011; Kim *et al.*, 2013) and enzymes (Fu *et al.*, 2012; Xiang *et al.*, 2011).

The objective of this study was to engineer a RABV-specific antibody that was capable of crossing the BBB to neutralize RABV infection in the CNS. Monoclonal antibody (mAb) 62-71-3 IgG is a potent rabies neutralizing antibody (Muller *et al.*, 2009). Recombinant IgG and single-chain antibody (ScFv) of 62-71-3 was recently expressed in plants and potent RABV neutralization was demonstrated (Both *et al.*, 2013). The ScFv was developed further here to link the RVG peptide using a gene encoding 62-71-3. ScFv genetically fused with RVG was cloned, expressed in *Nicotiana benthamiana* and purified by Ni-affinity chromatography. This molecule was investigated for RABV neutralization and binding to nAchR. The results demonstrate that the RVG peptide does not affect RABV neutralization, but does facilitate nAchR binding and transport of the rabies ScFv across an *in cellulo* BBB model.

Results

Expression of 62-71-3 ScFv and ScFv -RVG fusion

ScFv and ScFv-RVG fusion genes were cloned into the pEAQ vector (Peyret and Lomonossoff, 2013) as shown in Figure 1, and the proteins were expressed in *N. benthamiana*. A time-course of the protein expression between days 4–7 postinfiltration indicated day 6 was the optimal day to harvest (data not shown). The expression level of ScFv and ScFv-RVG was approximately 100 and 2 µg/g fresh leaf weight, respectively. The Ni-affinity purified ScFv fusion and ScFv-RVG fusion were assessed by Coomassie-stained SDS-PAGE gel (Figure 2a) or by immunoblotting with anti-E tag antiserum (Figure 2b). The amounts of purified proteins were quantified by comparing the band intensity with standard BSA protein (MW 66 kDa). Major bands were observed at the expected sizes for ScFv and ScFv-RVG fusion of 56 kDa (lane 1) and 61 kDa (lane 2), respectively. The identity of the bands was confirmed by Western blot (Figure 2b), which also demonstrated the presence of higher molecular weight bands (probably aggregates) and lower molecular weight bands (possibly degradation products). Of note, the ratio of full-length protein over degraded protein as shown in the immunoblotting (Figure 2b) is similar for ScFv and ScFv-RVG.

Neutralization of rabies virus

The two versions of 62-71-3 ScFv were tested to determine their ability to neutralize RABV (ERA strain) *in cellulo* using a plaque-inhibition assay. With a starting concentration of 0.5 mg/mL, the neutralizing activity of ScFv and ScFv-RVG fusion was identical to the neutralizing activity of 62-71-3 IgG (Figure 3). Statistical analysis by one-way ANOVA (GraphPad Prism, GraphPad Software, Inc. La Jolla, California, USA, version 7.0) confirmed that there was no significant difference among 62-71-3 IgG, ScFv and ScFv-RVG neutralizing activities.

Binding to nAchR

Binding and penetration of ScFv and ScFv-RVG fusion of 293 cells overexpressing nAchR were tested by flow cytometry. A greater proportion of ScFv-RVG fusion (dotted line) bound to the 293 cells as evidenced by the shift to the right of the dotted line compared to ScFv (solid line), shown in Figure 4a. A greater amount of total ScFv-RVG fusion (dotted line) was also found in the 293 cells overexpressing nAchR compared to ScFv (solid line, Figure 4b).

UV-inactivated RABV and α-bungarotoxin were used as competitive inhibitors for the interaction between the RVG peptide and nAchR. Cells pre-incubated with each inhibitor were tested for their ability to bind and to internalize ScFv and ScFv-RVG

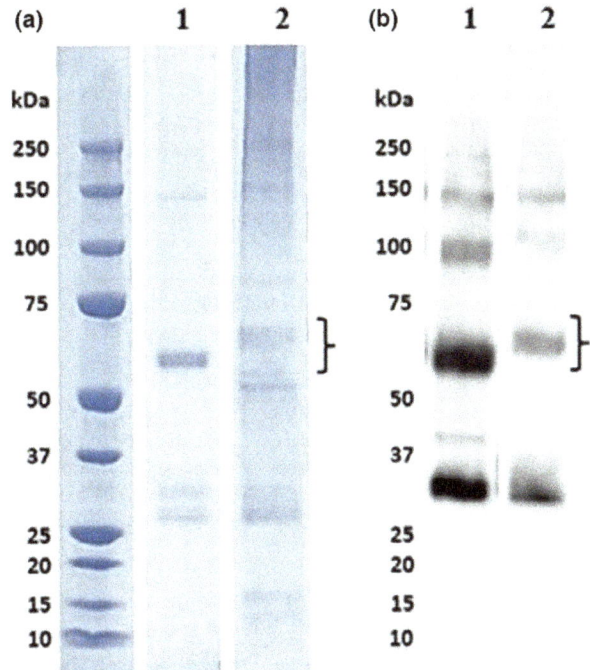

Figure 2 SDS-PAGE and Western blot analyses of ScFv and ScFv-RVG fusion proteins. The plant-produced ScFv^P (lane 1) and ScFv-RVG^P fusion proteins (lane 2) were purified by Ni-affinity chromatography. ScFv and ScFv-RVG fusion proteins were analysed by SDS-PAGE under reducing conditions, followed by (a) staining with Coomassie blue or (b) blotting onto nitrocellulose and probing with a mouse anti-E tag antiserum. The expected size of the ScFv and ScFv-RVG fusion is approximately 56 kDa and 61 kDa, respectively, which are indicated by curly braces.

fusion. There was a low-level background entry of ScFv into cells. This could not be inhibited by pre-incubation with either UV-inactivated RABV or α-bungarotoxin, indicating that its entry is mediated by a nonspecific mechanism (Figure 5a and c). In contrast, the presence of the UV-inactivated virus or α-bungarotoxin inhibited the entry of ScFv-RVG fusion as evidenced by the shift to the left of the dotted line compared to the absence of the competitor (solid line), shown in Figure 5b and d, respectively. These results confirmed that the entry of ScFv-RVG fusion protein into cells occurred via a nAchR-mediated pathway.

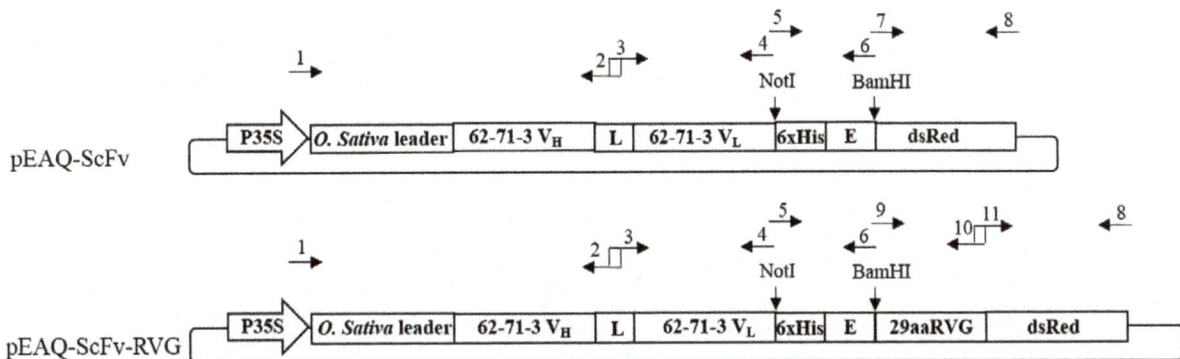

Figure 1 Schematic representation of the T-DNA regions of the vectors used in this study (Both *et al.*, 2013). P35S: CaMV 35S promoter, *Oryza sativa* leader: *O. sativa* leader sequence, 62-71-3 V$_H$: variable region of the heavy chain of 62-71-3 monoclonal antibody, L: the (Gly$_4$Ser)$_3$ linker, 62-71-3 V$_L$: variable region of the light chain of 62-71-3 monoclonal antibody, dsRed: red fluorescent protein from Discosoma sp., 29aaRVG: the 29 amino acid peptide (RVG) from RABV glycoprotein, 6xHis: 6 histidine residues, E: GAPVPYPDPLEPR peptide sequence, the sequences of primers number 1–11 were listed in Table S1.

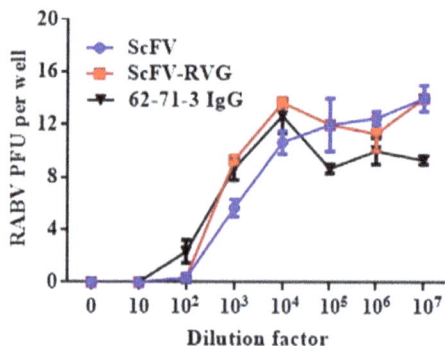

Figure 3 RABV neutralization of ScFv and ScFv-RVG fusion compared to 61-71-3 IgG. The neutralization assay was performed by the rapid fluorescent focus inhibition test on BSR cells. The starting concentration of antibodies was 0.5 mg/mL. Data presented are average values from three independent experiments, and the error bars indicate the standard deviation (SD). Statistical significance was determined by one-way ANOVA (GraphPad Prism, version 7.0).

These experiments were repeated with similar results using a second cell line, neuroscreen cells (Greene and Tischler, 1976), which are neuronal cells that express nAchRs (Figure 5e–h).

Passage of ScFv and ScFv-RVG fusion across an *in cellulo* model of the blood–brain barrier

The human hCMEC/D3 cell line, which retains morphological and functional characteristics of brain endothelium, is widely used as a human *in cellulo* BBB model (van der Helm et al., 2016). The *in cellulo* BBB transport experiment was conducted on the transwell device made with hCMEC/D3 cell monolayer as described in Figure 6 (Eigenmann et al., 2013). The barrier integrity of the human brain endothelial cell monolayer was assessed by transport of the small molecule Lucifer yellow and was determined to be 2.11×10^{-3} cm/min, attesting to the tightness of the junctions (Figure S1, Siflinger-Birnboim et al., 1987). The expression of nAchR alpha7 on hCMEC/D3 was also confirmed by real-time PCR (Figure S2). After antibodies were added to the upper chamber, the medium in the lower chamber was tested for RABV-neutralizing activity after 2 and 18 h (Figure 6a). These time points were chosen to eliminate the caveat of BBB alteration after adding the molecule (i.e. a transport after 2 h only is a very active transport across the endothelial cell barrier). The full-length 62-71-3 mAb did not cross the hCMEC/D3 monolayer, consistent with a previous report for an antibody molecule (Markoutsa et al., 2011). 62-71-3 IgG-RVG conjugate did not cross the endothelial cell barrier either (Figure 6b). Some ScFv was found to

cross the hCMEC/D3 cells as the 2 h medium sample had neutralizing activity (at dilution 1 : 100), but this did not increase by 18 h (Figure 6b). In contrast, ScFv-RVG fusion passed through the hCMEC/D3 cells to a much greater extent, and the neutralizing activity of the medium in the bottom well increased in a time-dependent manner (Figure 6b).

In a second assay, UV-inactivated RABV and α-bungarotoxin were used as competitive inhibitors (Figure 6c). Both are natural ligands of α7 nAchR. As before, the 62-71-3 IgG-RVG did not cross the hCMEC/D3 barrier, but the ScFv-RVG fusion did accumulate in the bottom well in a time-dependent manner. Pretreating cells with either UV-inactivated RABV or α-bungarotoxin reduced the passage of ScFv-RVG fusion at 2 and 18 h, resulting in at least 10-fold reduction in neutralizing activity found in the medium in the bottom well (Figure 6d). These inhibitors had no effect on the transport of 62-71-3 IgG-RVG across the barrier (Figure 6d).

Discussion

Several strategies for the transport across the BBB by drugs or antibodies have been proposed, including association with an antibody recognizing transferrin receptor as a carrier (Friden et al., 1991; Pardridge, 2015), targeting to the insulin receptor (Boado et al., 2010; Pardridge et al., 1985) and formulation with low-density lipoproteins to target the endothelial LDL-receptor (Alyautdin et al., 1997, 1998; Gulyaev et al., 1999). For plant-manufactured products, cholera toxin B subunit (CTB) was also used successfully to deliver proteins accross the BBB (Kohli et al., 2014; Kwon and Daniell, 2016) or to act as a strong mucosal adjuvant (Roy et al., 2010; Shahid and Daniell, 2016). Several proteins were used previously to target drugs to the brain, such as the human immunodeficiency virus TAT protein (Schwarze et al., 1999) and RVG peptide (Kumar et al., 2007; Liu et al., 2009). The RVG peptide constitutes part of the mature rabies viral glycoprotein (Kim et al., 2013) that can be visualized as trimeric peplomers on the surface of the virion and was previously shown to enable the transvascular delivery of siRNA to the brain (Kumar et al., 2007). The region of the viral G protein utilized here, as a linear peptide, has a similar amino acid composition to snake venom α-bungarotoxin (Lentz, 1991), which was previously shown to bind to nicotinic acetylcholine receptors (nAchR). These receptors are important as they occur in high density at the neuromuscular junction, and are present in the central nervous system and on endothelial cells. Thus, in the case of α-bungarotoxin, these receptors are also involved in penetration of the toxin into the brain (Bracci et al., 1988; Donnelly-Roberts and Lentz, 1989;

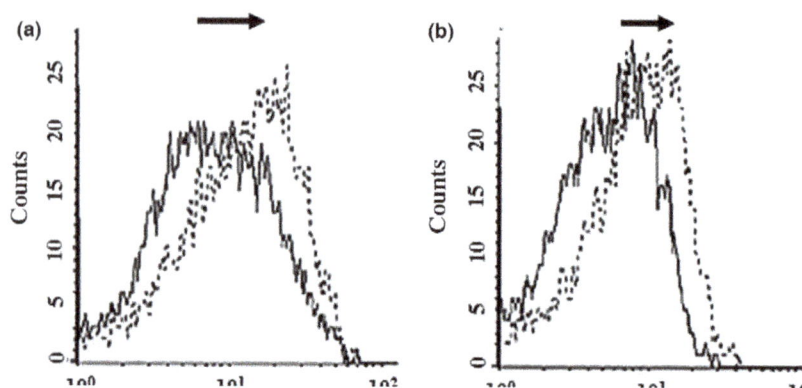

Figure 4 Binding and penetration of 62-71-3 ScFv to 293 cells overexpressing nAchR by flow cytometry. Binding (a) and entry (b) were detected with mouse anti-E antiserum and cy5-conjugated goat anti-mouse IgG antiserum. Solid line: ScFv, dotted line: ScFv-RVG fusion protein. The arrows represent the shift to the right of ScFv-RVG (dotted line) compared to ScFv (solid line).

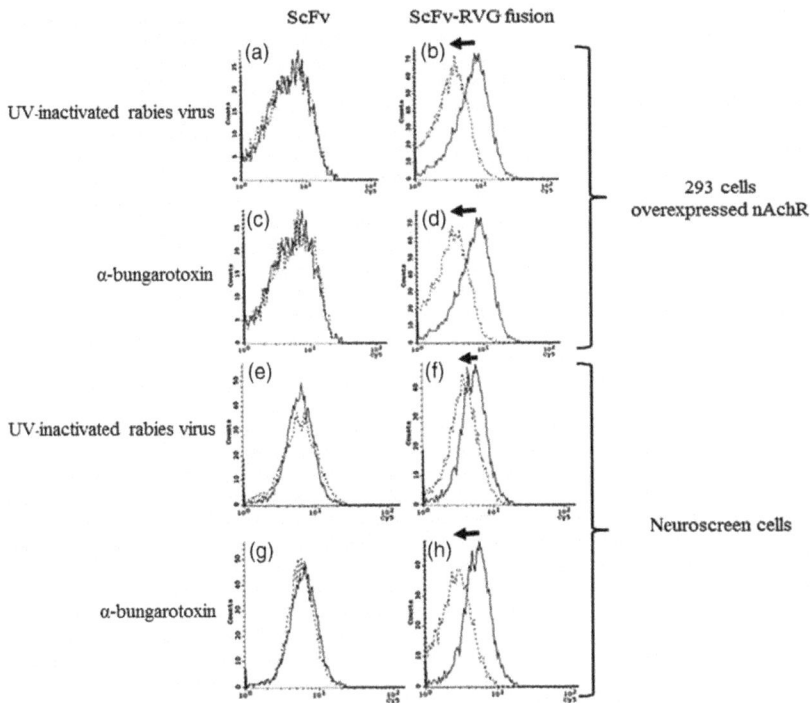

Figure 5 Inhibition of binding penetration of ScFv-RVG fusion into nAchR overexpressing 293 cells and neuroscreen cells by UV-inactivated RABV and α-bungarotoxin. Flow cytometry on nAchR overexpressing 293 cells pretreated with UV-inactivated RABV (a and b) and α-bungarotoxin (c and d) before incubation with ScFv (a and c) and ScFv-RVG fusion protein (b and d). Flow cytometry on neuroscreen cells pretreated with UV-inactivated RABV (e and f) and α-bungarotoxin (g and h) before incubation with ScFv (e and g) and ScFv-RVG fusion protein (f and h). Solid line: no inhibitor, dotted line: pretreated with UV-inactivated RABV or α-bungarotoxin. The arrows represent the shift to the left of ScFv-RVG (dotted line) compared to ScFv (solid line).

McQuarrie et al., 1976; Tzartos and Changeux, 1983). Similarly, the full-length RABV glycoprotein has been shown to interact with nAchR, allowing virus entry into the brain (Burrage et al., 1985; Lentz, 1990; Rustici et al., 1989).

Size is a key factor governing the ability of a molecule to pass the BBB (Jekic, 1979). 62-71-3 ScFv was used in this study because ScFvs are small molecules that retain the antigen specificity of the original immunoglobulin (Bird et al., 1988). The neutralization activity of the plant-produced 62-71-3 ScFv had previously been confirmed (Both et al., 2013). In this study, 62-71-3 ScFv was produced in N. benthamiana by transient expression at high yields whereas the ScFv-RVG fusion protein was expressed at significantly lower levels (Figure 2). Similar differences in expression levels between the two molecules were also observed in Escherichia coli (data not shown). Moreover, there is a degraded product in the purified protein, which is approximately half the size of the full protein. This degraded product appeared in the immunoblot, confirming the presence of the E tag. This fragment might be either the functional ScFv or the dsRed portion. However, only the band of full-length protein was used to quantify the amount of molecules used for the next studies for both ScFv and ScFv-RVG proteins.

Although there are several rabies vaccines and antibodies developed from plants (Hefferon, 2013; Rosales-Mendoza, 2015; Shahid and Daniell, 2016), here we show for the first time that a fusion protein with the RVG peptide can be produced in plants. Producing the RVG peptide fusion protein in this manner will remove the conjugation step and potentially reduce production costs. Both ScFv and ScFv-RVG fusion demonstrated equivalent neutralization of live RABV in cellulo, indicating that the ability of ScFv to neutralize the virus was not impaired by fusion to the RVG peptide.

To test nAchR binding, HEK293 cells overexpressing nAchR were used (Yamauchi et al., 2011). The ScFv-RVG fusion showed an increase in binding and penetration to cells overexpressing nAchR, compared to ScFv. To confirm that the increase in entry

was due to binding to nAchR, both UV-inactivated RABV and α-bungarotoxin were used independently as competitive inhibitors. α-bungarotoxin has a similar structure to RVG and binds to nAchR at the same site as rabies glycoprotein (Donnelly-Roberts and Lentz, 1989; Lentz, 1991; Lentz et al., 1984, 1987, 1988). This investigation demonstrated that entry of ScFv-RVG fusion into nAchR overexpressing cells decreased when the cells were pretreated with either UV-inactivated RABV or α-bungarotoxin, confirming the role of RVG peptide in mediating cell entry via the nAchR.

The BBB possesses specific characteristics that protect the brain from exposure to both endogenous and exogenous toxins. However, this protective barrier also limits the delivery of therapeutic molecules to the brain, a major constraint in developing suitable tools to neutralize RABV that is replicating in the CNS. The gold standard for studying transport across the BBB is to use in vivo animal models, but they are expensive, laborious, ethically contentious and often lack predictive data. Therefore, any researcher planning to use animals in their research must first show why there is no alternative to animal experimentation (European Commission, directive 201/63/EU) in order to fulfil the guiding principles underpinning the human use of animals in scientific research (i.e. the three Rs: **R**eplace, **R**educe and **R**efine). Previous study suggested that in cellulo models are robust, reproducible, easy to analyse and allow study of human cells and tissues (van der Helm et al., 2016) following the 3Rs rules. An in cellulo model was, therefore, used here to determine, in a first instance, the potential for the antibodies to cross the human BBB.

The hCMEC/D3 cell line has been developed as a model for the human BBB and has been used to test the permeability of several drugs (Al-Shehri et al., 2015; Ma et al., 2014). Here, the results indicated that 62-71-3 ScFv was able to pass across the hCMEC/D3 cells whilst the 62-71-3 IgG was not. Although the incubation time was increased from 2 to 18 h, the amount of ScFv crossing the cells did not increase. This might be due to the ScFv molecule crossing the cells by passive transport mechanisms and is probably

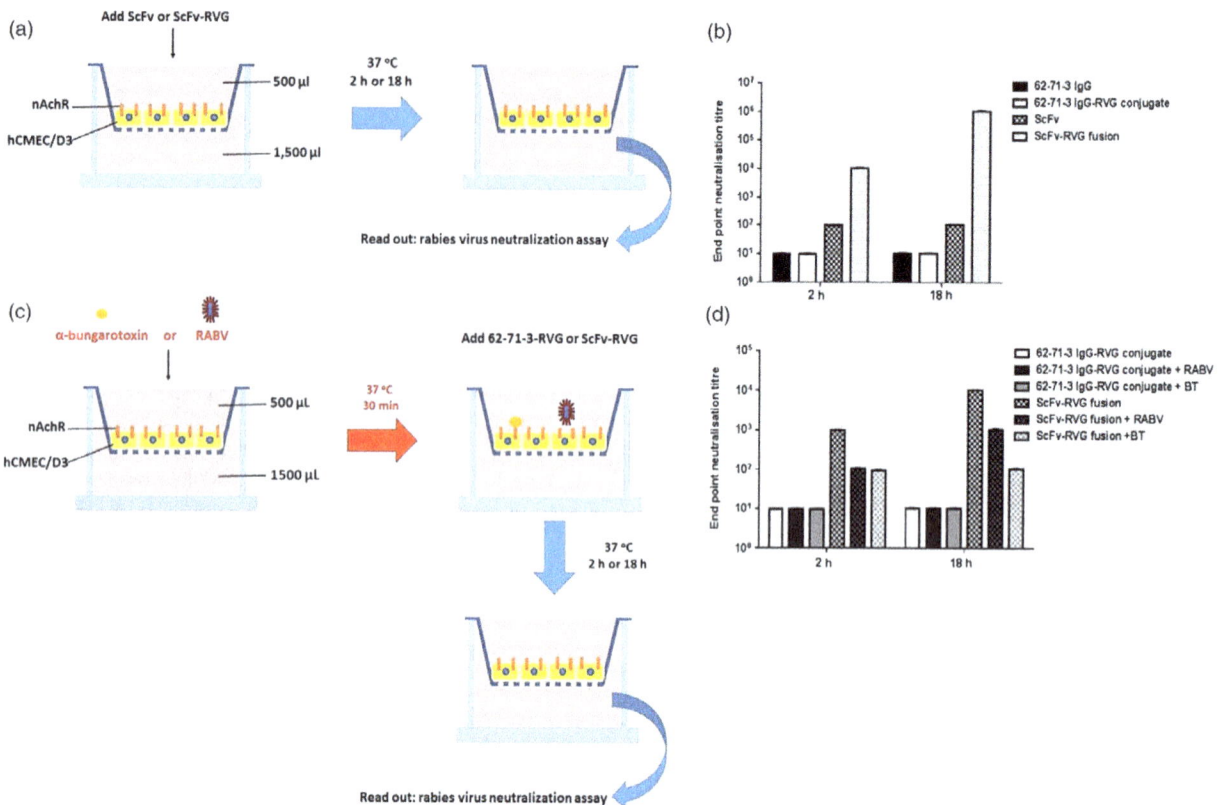

Figure 6 ScFv-RVG fusion transports across in cellulo BBB model. (a) A schematic diagram of the experiment. 10 μg of antibodies was added to the upper chamber of hCMEC/D3 cells in the transwell. Medium (collected after 2 or 18 h after adding the molecules) at the bottom of the well was tested for the presence of RABV-neutralizing antibodies by a RABV neutralization assay. (b) RABV neutralization titre of 62-71-3 IgG, 62-71-3 IgG-RVG conjugate, ScFv and ScFv-RVG fusion that crossed hCMEC/D3 cells. Each column represents the average values from three independent experiments, and the error bars indicate for the standard deviation (SD). (c) A schematic diagram of the inhibition experiment. hCMEC/D3 cells were pretreated with either UV-inactivated RABV or α-bungarotoxin (BT), and then, 10 μg of 62-71-3 IgG-RVG conjugate or ScFv-RVG fusion was added to the upper chamber. Medium at the bottom of the well was tested for the presence of RABV-neutralizing antibodies by a RABV neutralization assay after 2 and 18 h. (d) RABV neutralization titre of 62-71-3 IgG-RVG and ScFv-RVG fusion, which crossed hCMEC/D3 cells after the cells were pretreated with or without RABV or BT. Each column represents the average values from three independent experiments, and the error bars indicate the standard deviation (SD).

a reflection of the smaller size of this molecule compared to the IgG control. However, when the ScFv was fused with the RVG peptide, the penetration across the cells was significantly increased (Figure 6b) and occurred in a time-dependent manner indicating active penetration. When competitive inhibitors, UV-inactivated RABV and α-bungarotoxin, were used, transport across the in cellulo BBB decreased for the ScFv-RVG fusion protein (Figure 6c) suggesting that the ScFv-RVG fusion was transported across the in cellulo BBB by active transport mechanisms involving binding to nAchR.

Although postexposure prophylaxis in rabies is highly effective when correctly administered, significant challenges remain in treatment of infection, particularly when patient presentation is delayed. Alternative approaches to the treatment of late-stage rabies infection are still urgently required. The data presented here indicate a potential strategy to deliver potently neutralizing monoclonal antibody fragments across the BBB and into the CNS. Additional in vivo animal studies are required to assess pharmacokinetics of ScFv linked to RVG and efficacy of this form of postexposure tool following clinical presentation in an in vivo model. This approach may lead to a new mechanism by which postexposure tools can be administered to individuals exhibiting clinical rabies.

Experimental procedures

Genetic construct design

The 62-71-3 IgG was previously described (Both et al., 2013). For the cloning of pEAQ-ScFv, primers 1, 2, 3, 4, 5, 6, 7 and 8 were used (Figure 1; for the sequences see Table S1). Primer 1 was designed to introduce the attB recombination sites and the Oryza sativa signal peptide into the V_H domains of mAb 62-71-3. Primer 2 was used as a reverse primer for linking the V_H and V_L domains of mAb 62-71-3 with the $(Gly_4Ser)_3$ linker. Primers 3 and 4 were used as forward and reverse primers to amplify V_L domains of mAb 62-71-3 with NotI site at the 3' end. The V_H and V_L domains of mAb 62-71-3 with the $(Gly_4Ser)_3$ linker were linked using overlap PCR using primers 1 and 4. Primers 5 and 6 were used as forward and reverse primers, respectively, to amplify His tag- E tag fusion gene containing NotI and BamHI sites. Primer 7 was used as a forward primer to amplify dsRed gene containing BamHI site. Primer 8 was used as a reverse primer to amplify dsRed and also contained attB recombination sequence to the 3' end of dsRed gene. dsRed gene was included to monitor ScFv/ScFv-RVG expression in cells by immunofluorescence. The V_H and V_L domains of mAb 62-71-3 with the $(Gly_4Ser)_3$ linker were digested with NotI restriction enzyme. The fusion His tag - E tag portion was digested with NotI and BamHI restriction enzymes. The dsRed gene was digested with BamHI restriction enzyme. These three pieces were ligated, purified using the QIAquick PCR purification kit (Qiagen) and recombined into the Gateway entry vector pDONR/Zeo (all materials for Gateway recombination including enzymes, entry vector pDONR/Zeo, competent E. coli cells and zeocin, were obtained from Invitrogen). The E. coli cloning strain DH5α was heat-shocked with the plasmids and streaked on plates containing LB plus 50 µg/mL zeocin. Individual colonies were used for inoculating 5 mL LB medium containing 50 µg/mL zeocin and were shaken overnight (250 rpm, 37 °C). The plasmids were purified from a saturated overnight culture with the QIAprep Spin Miniprep Kit (Qiagen) and used for recombination with the Gateway destination vector pEAQ-HT-DEST3 (Sainsbury et al., 2009). For the cloning of pEAQ-ScFv-RVG (Figure 1), the V_H and V_L domains of mAb 62-71-3 with the $(Gly_4Ser)_3$ linker and the His tag – E tag portion were cloned using the same method as pEAQ-ScFv. Primers 9 and 10 were used as forward and reverse primers, respectively, to amplify RVG peptide with BamHI site at the 5' end. Primers 11 and 8 were used as forward and reverse primers, respectively, to amplify dsRed gene. The RVG peptide and dsRed genes were linked by overlap PCR using primers 8 and 9. After the three pieces were ligated, the Gateway recombination was performed using the same method as previously described for pEAQ-ScFv.

Plant inoculation and protein expression

Agrobacterium tumefaciens LBA4404 was transformed with the pEAQ-ScFv and the pEAQ-ScFv-RVG fusion vectors by electroporation. Recombinant bacterial strains were used to infiltrate leaves of N. benthamiana plants under vacuum. Leaves were harvested on days 4, 5, 6 or 7 postinfiltration for expression time-course experiments. For other experiments, the leaves were harvested on day 6 postinfiltration. Soluble proteins were extracted in 0.1M Tris-HCl pH 7.5 + 0.2% Triton X, using a blender before centrifugation at 18 000g for 10 min. The supernatant was retained for analysis.

SDS-PAGE and western blot

Plant extracts were denatured by boiling in NuPAGE® LDS Sample Buffer and separated on 4%–12% polyacrylamide gels (Life Technologies, Warrington, UK). Proteins were either visualized by Coomassie blue staining or transferred to a nitrocellulose membrane (Amersham Hybond-ECL; Amersham Biosciences, Little Chalfont, UK). The membrane was blocked with 5% nonfat dried milk, 0.1% Tween 20 in PBS. The membrane was probed with horseradish peroxidase (HRP)-conjugated mouse anti-His tag antiserum (Sigma) or HRP-conjugated mouse anti-E tag antiserum (Abcam, Cambridge, UK) diluted at 1 : 5000 in 1% nonfat dried milk in PBST. The membranes were developed by chemi-luminescence using ECL plus detection reagent (GE Healthcare, Buckinghamshire, UK).

Protein purification

Plant extract was filtered through Miracloth (EMD Millipore, Massachusetts, USA), centrifuged at 20 000 g for 15 min and passed through a 0.2-µm filter (Merck Millipore, Germany). Purification was by Ni-affinity chromatography with chelating SepharoseTM (GE healthcare) charged with NiSO4.6H2O. The antibody molecules were extensively purified from the crude

extract, but this affinity purification method does not reach a purification at homogeneity.

Cells and viruses

BSR cells (a clone of baby hamster kidney (BHK) cells) were grown in Dulbecco's modified Eagle's medium (DMEM)-Glutamax I (Life Technologies) supplemented with 10% foetal calf serum and penicillin/streptomycin. Neuroscreen cells (a subclone of PC12 cells, Cellomics, USA, which express α7-nicotinic acetylcholine receptor) were grown in RPMI medium (Sigma, Welwyn Garden City, UK) supplemented with 10% horse serum, 5% foetal calf serum and penicillin/steptomycin. Human Embryonic Kidney 293 (HEK) cells overexpressing human α7-nicotinic acetylcholine receptor (nAchR) (Yamauchi et al., 2011) were grown in DMEM supplemented with 10% foetal calf serum and Pen/Strep. The nAchR expression in this cell line was monitored by flow cytometry (data not shown). The hCMEC/D3 cells were grown in EndoGro™ medium (Millipore, Molsheim, France) according to the manufacturer's instruction. The nonpathogenic RABV laboratory strain ERA was propagated as previously described (Thoulouze et al., 1997).

In cellulo RABV neutralization assay

Neutralization of the ERA strain was performed on BSR cells using the rapid fluorescent focus inhibition test (Louie et al., 1975). The negative control consisted of medium without antibody. Dilutions of the test antibodies were incubated with RABV(<20 PFU) for 1 h at 37 °C before incubating with BSR cells at 37 °C with 5% CO2. After 48 h, the supernatant was removed and the cells were fixed with 80% acetone at 4 °C for 30 min. The cells were washed and incubated with 1 : 50 FITC-conjugated mouse anti-RABV nucleocapsid antiserum (Bio-Rad, Marnes-la-Coquette, France) at 37 °C for 30 min. After washing, RABV foci were counted using a fluorescent microscope. Assays were performed in triplicate.

nAchR binding and competition assay

HEK 293 cells expressing human α7-nicotinic acetylcholine receptor (nAchR) (Yamauchi et al., 2011) or Neuroscreen cells (entry assay) were seeded on six-well plates. After 24 h, cells were placed on ice and treated with ScFv preparations for 5 or 30 min, for the binding and entry assays, respectively. Of note, over a 5-min incubation on ice, it is expected that only a few single-chain antibody molecules are able to penetrate into the cell (Lim et al., 2013). After washing, the cells were harvested and incubated in cell fixation solution (BD Biosciences) for 15 min. For the binding assay, samples were washed with 1% inactivated foetal calf serum and 0.1% NaN3 in PBS, pH 7.4, whilst for the binding-penetration assay, samples were washed with 1% inactivated foetal calf serum, 0.1% NaN$_3$ and 0.1% saponin in PBS, pH 7.4, before incubation with 1 : 1000 mouse anti-E tag antiserum at 4 °C, overnight. The cells were washed and incubated with goat anti-mouse IgG antiserum conjugated with cy5 (Jackson laboratory, West Grove, Pennsylvania, USA) at 37 °C for 1 h. The absence of saponin in the binding assay allowed us to detect the ScFv cytoplasmic membrane bound molecule as the secondary antibody is not able to penetrate inside the cell. Again, the cells were washed, resuspended in staining buffer and analysed with FACS CellQuest software (BD, US). Alternatively for a competition assay, cells were pretreated with either 2x10^7 PFU of UV-inactivated RABV'Challenge Virus Strain' (CVS) (Megret et al., 2005) (i.e. still able to bind to RABV receptors but not replicative) or 16 µM α-bungarotoxin (Tocris Bioscience, Bristol,

UK) for 30 min on ice, before the ScFv or ScFv-RVG fusion was added.

In cellulo BBB transwell assay

The hCMEC/D3 cell line was prepared as described (Eigenmann et al., 2013) and seeded on the apical side of a Cultrex® Rat Collagen I (150 μg/mL-R&D Systems, Minneapolis, Minnesota, USA) coated 0.9 cm² polyethylene terephthalate filter insert with 3.0 μm porosity (BD Falcon, Loughborough, UK). 10 μg of each antibody preparation was added to the top chamber. The cells were incubated at 37 °C with 5% CO_2, and the medium was sampled after 2 h and 18 h from the bottom chamber for neutralizing antibody detection. For inhibition of in cellulo BBB penetration, UV-inactivated RABV or α-bungarotoxin was added to the top chamber for 30 min before 62-71-3 IgG-RVG conjugate and ScFv-RVG fusion protein were added. The medium in the bottom chamber was sampled as before at 2 h and 18 h.

Determination of the restrictive paracellular permeability with Lucifer Yellow

The restrictive paracellular permeability of hCMEC/D3 cells was assessed by their low permeability to the nonpermeant fluorescent marker Lucifer Yellow (LY) (Sigma-Aldrich, L0259). Briefly, after 5 days of culture on filters, hCMEC/D3 monolayers were transferred to 12-well plates containing 1.5 mL of transport medium (HBSS CaMg (Gibco, 14025-100) supplemented by 10 mM of hepes (Life technologies, 15630-080) and 1 mM of sodium pyruvate (Life technologies, 11360)) per well (abluminal compartment). 0.5 mL transport medium containing 50 μM of LY was then added to the luminal compartment. Incubations were performed at 37 °C, 5% CO_2 and 95% humidity. After 15, 25 and 45 min, the inserts were transferred into new wells, beforehand filled with 1.5 mL of transport medium. After 45 min, aliquots were taken for each time point, from both compartments and the concentration of LY determined using a fluorescence spectrophotometer (Tecan Infinite F500).

The endothelial permeability coefficient (P_e) of LY was calculated in centimetres/min (cm/min), as described previously (Siflinger-Birnboim et al., 1987). To obtain a concentration-independent transport parameter, the clearance principle was used. Briefly, the average volume cleared was plotted versus time, and the slope was estimated by linear regression. Both insert permeability (PS_f, for insert only coated with collagen) and insert plus endothelial cell permeability (PS_t, for insert with collagen and cells) were taken into consideration, according to the following formula: $1/PSe = 1/PSt - 1/PSf$.

The permeability value for the endothelial monolayer was then divided by the surface area of the porous membrane of the insert (Corning, 3460) to obtain the endothelial permeability coefficient (Pe) of the molecule (in cm/min).

Quantitative polymerase chain reaction

The procedure undertaken has been described in Chopy et al. (2011). Basically, cDNA synthesis was performed with 1 μg total RNA using SuperScript II reverse transcriptase (Life Technologies, France). Quantitative real-time RT-PCR (qRT-PCR) was performed in triplicate using an ABI Prism 7500 fast sequence detector system (primers 18S: F: CTT AGA GGG ACA AGT GGC G, R: ACG CTG AGC CAG TCA GTG TA; α7 AchR QT00074732, Qiagen, France) with GoTaq PCR master mix (Promega, Charbonniéres-les-Bains, France). After normalization to 18S

rRNA, the relative abundance of mRNA was obtained by calculation of the difference in threshold cycles of the test and control samples (mock value set to 1), commonly known as the $\Delta\Delta C_T$ method.

Acknowledgements

This work was supported by the Royal Society (Newton International fellowship to W. Phoolcharoen), a Wellcome Trust grant (WT093092MA), the Hotung Foundation and an ERC award (ERC-2010-AdG_20100317). We would like to thank Pierre-Jean Corringer for HEK 293 cells overexpressing nAchR, and Pierre-Emmanuel Ceccaldi, Philippe A. Afonso (Institut Pasteur) and Pierre Olivier Couraud (Inserm/Colchin) for hCMEC/D3 cells. The authors declare no conflict of interest.

References

Al-Shehri, A., Favretto, M.E., Ioannou, P.V., Romero, I.A., Couraud, P.O., Weksler, B.B., Parker, T.L. et al. (2015) Permeability of PEGylated immunoarsonoliposomes through in vitro blood brain barrier-medulloblastoma co-culture models for brain tumor therapy. Pharm. Res. **32**, 1072–1083.

Alyautdin, R.N., Petrov, V.E., Langer, K., Berthold, A., Kharkevich, D.A. and Kreuter, J. (1997) Delivery of loperamide across the blood-brain barrier with polysorbate 80-coated polybutylcyanoacrylate nanoparticles. Pharm. Res. **14**, 325–328.

Alyautdin, R.N., Tezikov, E.B., Ramge, P., Kharkevich, D.A., Begley, D.J. and Kreuter, J. (1998) Significant entry of tubocurarine into the brain of rats by adsorption to polysorbate 80-coated polybutylcyanoacrylate nanoparticles: an in situ brain perfusion study. J. Microencapsul. **15**, 67–74.

Bird, R.E., Hardman, K.D., Jacobson, J.W., Johnson, S., Kaufman, B.M., Lee, S.M., Lee, T. et al. (1988) Single-chain antigen-binding proteins. Science, **242**, 423–426.

Boado, R.J., Hui, E.K., Lu, J.Z., Zhou, Q.H. and Pardridge, W.M. (2010) Selective targeting of a TNFR decoy receptor pharmaceutical to the primate brain as a receptor-specific IgG fusion protein. J. Biotechnol. **146**, 84–91.

Both, L., van Dolleweerd, C., Wright, E., Banyard, A.C., Bulmer-Thomas, B., Selden, D., Altmann, F. et al. (2013) Production, characterization, and antigen specificity of recombinant 62-71-3, a candidate monoclonal antibody for rabies prophylaxis in humans. FASEB J. **27**, 2055–2065.

Bracci, L., Antoni, G., Cusi, M.G., Lozzi, L., Niccolai, N., Petreni, S., Rustici, M. et al. (1988) Antipeptide monoclonal antibodies inhibit the binding of rabies virus glycoprotein and alpha-bungarotoxin to the nicotinic acetylcholine receptor. Mol. Immunol. **25**, 881–888.

Burrage, T.G., Tignor, G.H. and Smith, A.L. (1985) Rabies virus binding at neuromuscular junctions. Virus Res. **2**, 273–289.

Chopy, D., Detje, C.N., Lafage, M., Kalinke, U. and Lafon, M. (2011) The type I interferon response bridles rabies virus infection and reduces pathogenicity. J. Neurovirol. **17**, 353–367.

Donnelly-Roberts, D.L. and Lentz, T.L. (1989) Synthetic peptides of neurotoxins and rabies virus glycoprotein behave as antagonists in a functional assay for the acetylcholine receptor. Pept. Res. **2**, 221–226.

Eigenmann, D.E., Xue, G., Kim, K.S., Moses, A.V., Hamburger, M. and Oufir, M. (2013) Comparative study of four immortalized human brain capillary endothelial cell lines, hCMEC/D3, hBMEC, TY10, and BB19, and optimization of culture conditions, for an in vitro blood-brain barrier model for drug permeability studies. Fluids Barriers CNS, **10**, 33.

Fooks, A.R., Banyard, A.C., Horton, D.L., Johnson, N., McElhinney, L.M. and Jackson, A.C. (2014) Current status of rabies and prospects for elimination. Lancet, **384**, 1389–1399.

Friden, P.M., Walus, L.R., Musso, G.F., Taylor, M.A., Malfroy, B. and Starzyk, R.M. (1991) Anti-transferrin receptor antibody and antibody-drug conjugates cross the blood-brain barrier. Proc. Natl Acad. Sci. USA, **88**, 4771–4775.

Fu, A., Wang, Y., Zhan, L. and Zhou, R. (2012) Targeted delivery of proteins into the central nervous system mediated by rabies virus glycoprotein-derived peptide. *Pharm. Res.* **29**, 1562–1569.

Greene, L.A. and Tischler, A.S. (1976) Establishment of a noradrenergic clonal line of rat adrenal pheochromocytoma cells which respond to nerve growth factor. *Proc. Natl Acad. Sci. USA*, **73**, 2424–2428.

Gulyaev, A.E., Gelperina, S.E., Skidan, I.N., Antropov, A.S., Kivman, G.Y. and Kreuter, J. (1999) Significant transport of doxorubicin into the brain with polysorbate 80-coated nanoparticles. *Pharm. Res.* **16**, 1564–1569.

Hefferon, K. (2013) Plant-derived pharmaceuticals for the developing world. *Biotechnol. J.* **8**, 1193–1202.

van der Helm, M.W., van der Meer, A.D., Eijkel, J.C., van den Berg, A. and Segerink, L.I. (2016) Microfluidic organ-on-chip technology for blood-brain barrier research. *Tissue Barriers*. **4**, e1142493.

Hemachudha, T., Laothamatas, J. and Rupprecht, C.E. (2002) Human rabies: a disease of complex neuropathogenetic mechanisms and diagnostic challenges. *Lancet Neurol.* **1**, 101–109.

Hwang do, W., Son, S., Jang, J., Youn, H., Lee, S., Lee, D., Lee, Y.S. *et al.* (2011) A brain-targeted rabies virus glycoprotein-disulfide linked PEI nanocarrier for delivery of neurogenic microRNA. *Biomaterials*, **32**, 4968–4975.

Jekic, M. (1979) Pathogenetic mechanisms of septic states in war injuries. *Acta Chir. Iugosl.* **26**, 71–72.

Kim, J.Y., Choi, W.I., Kim, Y.H. and Tae, G. (2013) Brain-targeted delivery of protein using chitosan- and RVG peptide-conjugated, pluronic-based nano-carrier. *Biomaterials*, **34**, 1170–1178.

Kohli, N., Westerveld, D.R., Ayache, A.C., Verma, A., Shil, P., Prasad, T., Zhu, P. *et al.* (2014) Oral delivery of bioencapsulated proteins across blood-brain and blood-retinal barriers. *Mol. Ther.* **22**, 535–546.

Kumar, P., Wu, H., McBride, J.L., Jung, K.E., Kim, M.H., Davidson, B.L., Lee, S.K. *et al.* (2007) Transvascular delivery of small interfering RNA to the central nervous system. *Nature*, **448**, 39–43.

Kwon, K.C. and Daniell, H. (2016) Oral Delivery of Protein Drugs Bioencapsulated in Plant Cells. *Mol. Ther.* **24**, 1342–1350.

Lentz, T.L. (1990) Rabies virus binding to an acetylcholine receptor alpha-subunit peptide. *J. Mol. Recognit.* **3**, 82–88.

Lentz, T.L. (1991) Structure-function relationships of curaremimetic neurotoxin loop 2 and of a structurally similar segment of rabies virus glycoprotein in their interaction with the nicotinic acetylcholine receptor. *Biochemistry* **30**, 10949–10957.

Lentz, T.L., Burrage, T.G., Smith, A.L., Crick, J. and Tignor, G.H. (1982) Is the acetylcholine receptor a rabies virus receptor? *Science*, **215**, 182–184.

Lentz, T.L., Wilson, P.T., Hawrot, E. and Speicher, D.W. (1984) Amino acid sequence similarity between rabies virus glycoprotein and snake venom curaremimetic neurotoxins. *Science*, **226**, 847–848.

Lentz, T.L., Hawrot, E. and Wilson, P.T. (1987) Synthetic peptides corresponding to sequences of snake venom neurotoxins and rabies virus glycoprotein bind to the nicotinic acetylcholine receptor. *Proteins*, **2**, 298–307.

Lentz, T.L., Hawrot, E., Donnelly-Roberts, D. and Wilson, P.T. (1988) Synthetic peptides in the study of the interaction of rabies virus and the acetylcholine receptor. *Adv. Biochem. Psychopharmacol.* **44**, 57–71.

Lewis, P., Fu, Y. and Lentz, T.L. (2000) Rabies virus entry at the neuromuscular junction in nerve-muscle cocultures. *Muscle Nerve*, **23**, 720–730.

Lim, K.J., Sung, B.H., Shin, J.R., Lee, Y.W., Kim, D.J., Yang, K.S. and Kim, S.C. (2013) A cancer specific cell-penetrating peptide, BR2, for the efficient delivery of an scFv into cancer cells. *PLoS ONE*, **8**, e66084.

Liu, Y., Huang, R., Han, L., Ke, W., Shao, K., Ye, L., Lou, J. *et al.* (2009) Brain-targeting gene delivery and cellular internalization mechanisms for modified rabies virus glycoprotein RVG29 nanoparticles. *Biomaterials*, **30**, 4195–4202.

Louie, R.E., Dobkin, M.B., Meyer, P., Chin, B., Roby, R.E., Hammar, A.H. and Cabasso, V.J. (1975) Measurement of rabies antibody: comparison of the mouse neutralization test (MNT) with the rapid fluorescent focus inhibition test (RFFIT). *J. Biol.Stand.* **3**, 365–373.

Ma, S., Liu, X., Xu, Q. and Zhang, X. (2014) Transport of ginkgolides with different lipophilicities based on an hCMEC/D3 cell monolayer as a blood-brain barrier cell model. *Life Sci.* **114**, 93–101.

Markoutsa, E., Pampalakis, G., Niarakis, A., Romero, I.A., Weksler, B., Couraud, P.O. and Antimisiaris, S.G. (2011) Uptake and permeability studies of BBB-targeting immunoliposomes using the hCMEC/D3 cell line. *Eur. J. Pharm. Biopharm.* **77**, 265–274.

McQuarrie, C., Salvaterra, P.M., De Blas, A., Routes, J. and Mahler, H.R. (1976) Studies on nicotinic acetylcholine receptors in mammalian brain. Preliminary characterization of membrane-bound alpha-bungarotoxin receptors in rat cerebral cortex. *J. Biol. Chem.* **251**, 6335–6339.

Megret, F., Prehaud, C., Lafage, M., Batejat, C., Escriou, N., Lay, S., Thoulouze, M.I. *et al.* (2005) Immunopotentiation of the antibody response against influenza HA with apoptotic bodies generated by rabies virus G-ERA protein-driven apoptosis. *Vaccine*, **23**, 5342–5350.

Muller, T., Dietzschold, B., Ertl, H., Fooks, A.R., Freuling, C., Fehlner-Gardiner, C., Kliemt, J. *et al.* (2009) Development of a mouse monoclonal antibody cocktail for post-exposure rabies prophylaxis in humans. *PLoS Negl. Trop Dis.* **3**, e542.

Pardridge, W.M. (2010) Biopharmaceutical drug targeting to the brain. *J. Drug Target.* **18**, 157–167.

Pardridge, W.M. (2015) Blood-brain barrier drug delivery of IgG fusion proteins with a transferrin receptor monoclonal antibody. *Expert. Opin. Drug. Deliv.* **12**, 207–222.

Pardridge, W.M., Eisenberg, J. and Yang, J. (1985) Human blood-brain barrier insulin receptor. *J. Neurochem.* **44**, 1771–1778.

Peyret, H. and Lomonossoff, G.P. (2013) The pEAQ vector series: the easy and quick way to produce recombinant proteins in plants. *Plant Mol. Biol.* **83**, 51–58.

Rosales-Mendoza, S. (2015) Current Developments and Future Prospects for Plant-Made Biopharmaceuticals Against Rabies. *Mol. Biotechnol.* **57**, 869–879.

Roy, S., Tyagi, A., Tiwari, S., Singh, A., Sawant, S.V., Singh, P.K. and Tuli, R. (2010) Rabies glycoprotein fused with B subunit of cholera toxin expressed in tobacco plants folds into biologically active pentameric proteins. *Protein Expr. Purif.* **70**, 184–190.

Rustici, M., Santucci, A., Lozzi, L., Petreni, S., Spreafico, A., Neri, P., Bracci, L. *et al.* (1989) A monoclonal antibody to a synthetic fragment of rabies virus glycoprotein binds ligands of the nicotinic cholinergic receptor. *J. Mol. Recognit.* **2**, 51–55.

Sainsbury, F., Thuenemann, E.C. and Lomonossoff, G.P. (2009) pEAQ: versatile expression vectors for easy and quick transient expression of heterologous proteins in plants. *Plant Biotechnol. J.* **7**, 682–693.

Schwarze, S.R., Ho, A., Vocero-Akbani, A. and Dowdy, S.F. (1999) In vivo protein transduction: delivery of a biologically active protein into the mouse. *Science*, **285**, 1569–1572.

Shahid, N. and Daniell, H. (2016) Plant-based oral vaccines against zoonotic and non-zoonotic diseases. *Plant Biotechnol. J.* **14**, 2079–2099.

Shantavasinkul, P. and Wilde, H. (2011) Postexposure prophylaxis for rabies in resource-limited/poor countries. *Adv. Virus Res.* **79**, 291–307.

Siflinger-Birnboim, A., Del Vecchio, P.J., Cooper, J.A., Blumenstock, F.A., Shepard, J.M. and Malik, A.B. (1987) Molecular sieving characteristics of the cultured endothelial monolayer. *J. Cell. Physiol.* **132**, 111–117.

Thoulouze, M.I., Lafage, M., Montano-Hirose, J.A. and Lafon, M. (1997) Rabies virus infects mouse and human lymphocytes and induces apoptosis. *J. Virol.* **71**, 7372–7380.

Tzartos, S.J. and Changeux, J.P. (1983) High affinity binding of alpha-bungarotoxin to the purified alpha-subunit and to its 27,000-dalton proteolytic peptide from Torpedo marmorata acetylcholine receptor. Requirement for sodium dodecyl sulfate. *EMBO J.* **2**, 381–387.

Uwanyiligira, M., Landry, P., Genton, B. and de Valliere, S. (2012) Rabies postexposure prophylaxis in routine practice in view of the new Centers for Disease Control and Prevention and World Health Organization recommendations. *Clin. Infect. Dis.* **55**, 201–205.

Xiang, L., Zhou, R., Fu, A., Xu, X., Huang, Y. and Hu, C. (2011) Targeted delivery of large fusion protein into hippocampal neurons by systemic administration. *J. Drug Target.* **19**, 632–636.

Yamauchi, J.G., Nemecz, A., Nguyen, Q.T., Muller, A., Schroeder, L.F., Talley, T.T., Lindstrom, J. *et al.* (2011) Characterizing ligand-gated ion channel receptors with genetically encoded Ca2 + + sensors. *PLoS ONE*, **6**, e16519.

6

Stable production of cyanophycinase in *Nicotiana benthamiana* and its functionality to hydrolyse cyanophycin in the murine intestine

Daniel Ponndorf[1], Sven Ehmke[1,†], Benjamin Walliser[1], Kerstin Thoss[1], Christoph Unger[1], Solvig Görs[2], Gürbüz Daş[2], Cornelia C. Metges[2,*], Inge Broer[1] and Henrik Nausch[1]

[1]*Faculty of Agricultural and Environmental Sciences, Department of Agrobiotechnology and Risk Assessment for Bio- and Gene Technology, University of Rostock, Rostock, Germany*

[2]*Leibniz Institute for Farm Animal Biology (FBN), Institute of Nutritional Physiology 'Oskar Kellner', Dummerstorf, Germany*

*Correspondence
email metges@fbn-dummerstorf.de
†Present Address: Paraxel International GmbH, Klinikum am Westend, Haus 18, SpandauerDamm 130, 14050, Berlin, Germany.

Keywords: arginine, cyanophycin, cyanophycinase, dipeptide, digestion, *Nicotiana benthamiana*, protein stability.

Summary

Food supplementation with the conditionally essential amino acid arginine (Arg) has been shown to have nutritional benefits. Degradation of cyanophycin (CGP), a peptide polymer used for nitrogen storage by cyanobacteria, requires cyanophycinase (CGPase) and results in the release of β-aspartic acid (Asp)-Arg dipeptides. The simultaneous production of CGP and CGPase in plants could be a convenient source of Arg dipeptides. Different variants of the *cph*B coding region from *Thermosynechococcus elongatus* BP-1 were transiently expressed in *Nicotiana benthamiana* plants. Translation and enzyme stability were optimized to produce high amounts of active CGPase. Protein stability was increased by the translational fusion of CGPase to the green fluorescent protein (GFP) or to the transit peptide of the small subunit of RuBisCO for peptide production in the chloroplasts. Studies in mice showed that plant-expressed CGP fed in combination with plant-made CGPase was hydrolysed in the intestine, and high levels of ß-Asp-Arg dipeptides were found in plasma, demonstrating dipeptide absorption. However, the lack of an increase in Asp and Arg or its metabolite ornithine in plasma suggests that Arg from CGP was not bioavailable in this mouse group. Intestinal degradation of CGP by CGPase led to low intestinal CGP content 4 h after consumption, but after ingestion of CGP alone, high CGP concentrations remained in the large intestine; this indicated that intact CGP was transported from the small to the large intestine and that CGP was resistant to colonic microbes.

Introduction

Arginine (Arg) is an indispensable amino acid (AA) for young mammals and birds (Wu *et al.*, 2004, 2009). In addition to its function as a building block of proteins, Arg plays important roles in regulating gene expression, cell signalling, vascular development, reproduction and immunity (Bazer *et al.*, 2012; Wang *et al.*, 2012; Wu, 2014). Furthermore, Arg has nutritional benefits for athletes and the elderly or immune-compromised patients, but because its concentration is relatively low in food proteins, it has been used as a supplement for therapy and as an additive in food (Sallam and Steinbuchel, 2010). Supplemental free Arg is commonly produced by fermentation (Utagawa, 2004). The oral application of Arg-containing dipeptides may increase the uptake of Arg in the small intestine compared to Arg monomers (Matthews and Adibi, 1976; Wenzel *et al.*, 2001). Currently, dipeptides are synthesized by enzymatic, chemical and combined methods (Yagasaki and Hashimoto, 2008). Overexpression of the polypeptide cyanophycin (CGP) followed by cyanophycinase (CGPase)-mediated degradation results in β-aspartic acid (Asp)-Arg dipeptides, which could produce Arg (Sallam and Steinbuchel, 2009b, 2010; Sallam *et al.*, 2009). Cyanophycin is synthesized by cyanobacteria and several nonphotosynthetic bacteria via nonribosomal biosynthesis by the enzyme cyanophycin synthetase (CPHA) (Allen *et al.*, 1984; Simon, 1987; Simon and Weathers, 1976;

Ziegler *et al.*, 1998, 2002) and consists of an L-Asp backbone linked to L-Arg residues (Simon and Weathers, 1976). The expression of the CPHA-encoding gene from *Thermosynechococcus elongatus* BP-1 enables high and stable accumulation of CGP in tobacco and potato plastids (Hühns *et al.*, 2008, 2009; Neumann *et al.*, 2005). However, to the best of our knowledge, there are no reports on feeding CGP to animals and assessing its potential to produce Arg. CGP is highly stable and resistant to proteases (Simon and Weathers, 1976), except CGPase (Gupta and Carr, 1981). The cyanophycinase CPHB was described in cell extracts of *Anabaena 7120* (Gupta and Carr, 1981), *Aphanocapsa 6308* (Allen *et al.*, 1984) and *Synechocystis* sp. PCC 6803 (Richter *et al.*, 1999). Overexpression in *E.coli* and analysis of CPHB revealed that it is a serine (Ser)-type exopeptidase with a dimeric structure (Law *et al.*, 2009; Richter *et al.*, 1999), and its binding is highly specific for β-linked aspartyl peptides. If CGP and CGPase can be co-expressed in food plants, β-Asp-Arg dipeptides could become a source of dietary Arg. The degradation of CGP resulting in the release of β-Asp-Arg dipeptides might be achieved following two strategies: (i) accumulation of CGP in the plastid via separation of CGP (plastid) and CGPase (cytosol) leading to dipeptide formation after extraction when both components are joined and (ii) accumulation of the dipeptides in plastids by targeting cyanophycin synthetase (CPHA) and cyanophycinase (CPHB) to the chloroplast allowing dipeptide production during plant growth.

We used a transient expression system in *N. benthamiana* to determine whether plants can produce an active and stable form of CGPase that degrades CGP and whether the enzyme can be translocated to the chloroplast. In the second step, we formulated food pellets containing both, plant-made CGP and plant-made CGPase to investigate whether CGP is hydrolysed by CGPase and whether Arg from CGP is bioavailable in a mouse model.

Results

Cytosolic production and stabilization of CGPase in plants

A 5' truncated coding region of the *cphB*$_{tlr2169}$ gene (*cphB*-b) (Prof. Dr. Wolfgang Lockau (W.L.) Humbold Univesity Berlin) and two codon-optimized versions, *cphB*-s, designed to improve the efficiency of translation (Perlak *et al.*, 1991; Sharp and Li, 1987) and *cphB*-sA, where the sequence GCT TCCTCC encoding for Alanin (Ala)-Ser-Ser (Fig. 1), was added to improve efficiency of translation and protein stability (Sawant *et al.*, 2001) were transiently expressed in *N. benthamiana*. Total soluble protein (TSP) was assessed for the presence of CGPase using Western blot analysis (Fig. 2a). Infiltration with *cphB*-b did not result in

detection of CGPase in 50 μg TSP, but faint, not always reproducible, signals were visible at the expected size of approximately 29 kDa in 100 μg TSP. Expression of *cphB*-s and analysis of 50 μg TSP showed faint CGPase signals, while expression of *cphB*-sA resulted in more pronounced signals. Bands were detected not only at 29 kDa but also at 60, 130 and 200 kDa. These bands were not observed in the empty vector control and their size corresponds to potential di- and trimers and higher aggregates of CPHB-SA. Additionally, we were not able to detect CGPase without the addition of protease inhibitors, suggesting its instability in the crude plant extract. Greater stability of recombinant proteins can be achieved by fusion of the protein to a stable fusion partner at the *N*- or *C*-terminus. One promising fusion partner is GFP which was previously used successfully to improve protein stability (Piron *et al.*, 2014). Because our constructs carry a 6*Histag at the *C*-terminus for enzyme purification, we used *N*-terminal fusions. The fusion of *cphB*-s to the green fluorescent protein (GFP) coding region, resulting in *gfp*-*cphB*-s, led to an increase in protein yield as shown in Figure 2b. An additional band was visible at approximately 100 kDa, corresponding to the calculated size of a GFP:: CPHB-S dimer.

Because *N*-terminal fusions stabilized the protein, we analysed whether the complete CGPase protein (CPHB-S-C) described in the database is more stable compared to the truncated version. Therefore, we used a codon-optimized version of the complete sequence. After infiltration with *cphB*-s-c, three bands were detected in 50 μg TSP at approximately 27, 29 and 35 kDa. The 35-kDa protein corresponds to the calculated size of the complete protein. The smaller bands are potential degradation products, while the band at 29 kDa is the size of the truncated protein.

Subcellular targeting of CPHB to the chloroplast

To determine whether CPHB-S and CPHB-SA can be targeted to the chloroplast *cphB*-s was fused to the plastid leader peptide of the small subunit of RuBisCO (S) (Klimyuk *et al.*, 2004) resulting in S-*cphB*-s. Due to sequence incompatibility between S and the sequence GCT TCC TCC of Ala-Ser-Ser (A) in *cphB*-sA, it was necessary to adapt the sequence to GCC ATT GGA (A2) prior to the fusion to S, resulting in S-*cphB*-sA2. Western blot analysis of 25 μg TSP showed that S-*cphB*-s produced substantially more CGPase than S-*cphB*-sA2 (Figs 1 and 2b) and also showed a higher yield compared to *gpf*-*cphB*-s (Fig. 2b). To determine a possible effect of A2 on protein folding, we conducted *in silico* analysis of S-CPHB-S and S-CPHB-A2 and found different potential α-helices between AA 4-25, caused by the integration of Ala-Ser-Ser (Fig. S1). This might have an effect on protein stability or folding of the transit peptide. In addition to the expected band at 29 kDa, a band at approximately 35 kDa was detected, corresponding to the unprocessed proteins (Fig. 2b) for S-*cphB*-s and S-*cphB*-sA2. The same bands were present in isolated chloroplasts (data not shown).

Putative S-CPHB-S dimers of 70 kDa were also visible when analysing higher protein concentrations (data not shown). Expression of S-*chpB*-s led to the greatest amount of enzyme detected compared to all other constructs. To determine whether differences between the constructs might be caused by different RNA patterns, we conducted Northern blot analysis.

RNA analysis of different CGPase variants

Northern blot assays were performed to compare the RNA steady-state levels of the different constructs (Fig. 3a). For all

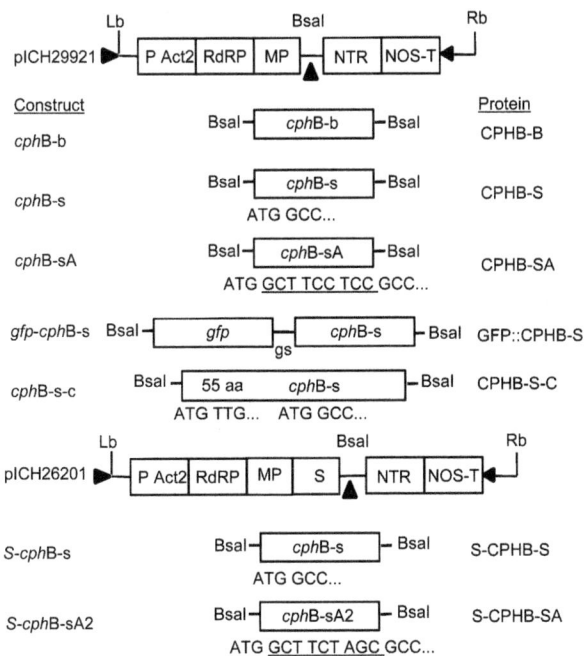

Figure 1 Plasmid constructs and corresponding labels of the respective protein variants: TMV-based viral vectors (Marillonnet *et al.*, 2005): pICH29912: empty vector, cphB-b: bacterial coding region of the cphB gene, cphB-s: cphB coding region adapted to the codon usage of *N. benthamiana*, cphB-sA: cphB-s with the addition of the amino acids Ala-Ser-Ser (A) (underlined sequence), gfp-cphB-s: fusion of gfp (sequence of the green fluorescent protein from pICH18711 (Marillonnet *et al.*, 2005)) and cphB-s, gs: linker, cphB-s-c: complete codon-optimized coding region of cphB as described in the database; pICH26201: containing a consensus sequence of the transit peptide of small subunit of RuBisCO (S) from dicotyl plants (Klimyuk *et al.*, 2004), cphB-sA2: cphB-s with the addition of the altered sequence of amino acids Ala-Ser-Ser (underlined sequence). LB and RB: left and right T-DNA borders; P Act2: Arabidopsis actin 2 promoter; RdRP: RNA-dependent RNA polymerase; MP: movement protein; NTR: 3' untranslated region of TMV; NOS-T: nos terminator.

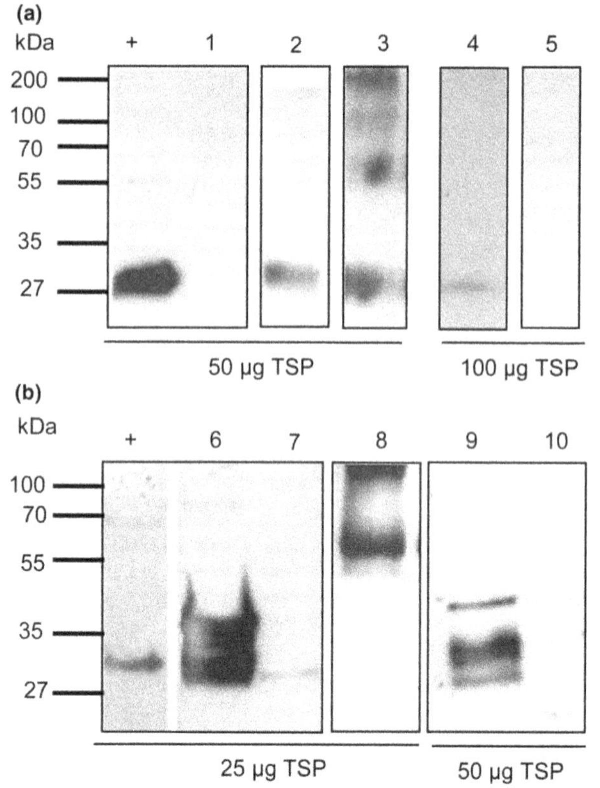

Figure 2 (a) Western blot analysis of 50 and 100 μg total soluble protein isolated from *N. benthamiana* leaves harvested 7 days post infiltration from one plant (dpi): + = Cyanophycinase-positive control isolated from *E. coli*; 1: CPHB-B; 2: CPHB-S; 3:CPHB-SA, 4: CPHB-B; 5: pICH29912 = empty vector control. b: 25 and 50 μg TSP harvested 7 dpi from one plant:+ = Cyanophycinase-positive control isolated from *E. coli* 6:S-CPHB-S, 7:S-CPHB-SA2; 8: GFP::CPHB-S; 9: CPHB-S-C; 10: pICH29912 = empty vector control. Plants were harvested 7 dpi.

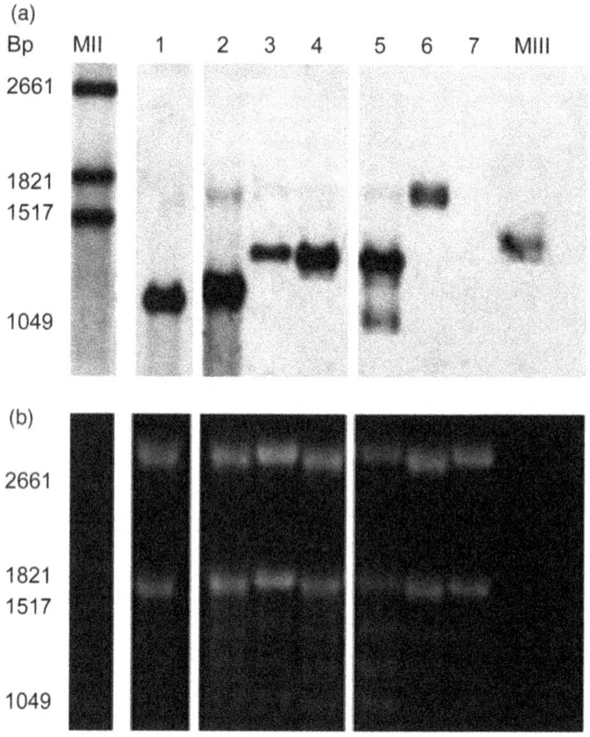

Lane	Construct	Approximate RNA size
1	*cph*B-s	1100
2	*cph*B-sA	1100
3	S-*cph*B-s	1300
4	S-*cph*B-sA2	1300
5	*cph*B-s-c	1300
6	*gfp*-*cph*B-s	1800

Figure 3 Northern blot (a) and RNA gel loading control (b) of 3 μg RNA isolated from *N. benthamiana* leaves. All samples showed the approximate expected size bands (c) but different signal strength. The loading control (b) shows that samples were loaded equally. MII: RNA marker II with a base pair (bp) range of 1516-6948 bp, MIII: RNA marker with a bp range of 310-1517 bp. 1: *cph*B-s, 2: *cph*B-sA, 3: S-*cph*B-s, 4: S-*cph*B-sA2, 5: *cph*B-s-c, 6: *gfp*-*cph*B-s, 7: empty vector control.

constructs, RNAs corresponding to the calculated size (Fig. 3c) were observed. For *cph*B-sA, an additional band at approximately 1700 bp was detected. This fragment was weakly detected for *cph*B-s-c, which also had a third band at approximately 1.000 bp. While the loading control showed equal amounts of total RNA (Fig. 3b), the strongest signal was observed with *cph*B-sA. CphB-s, S-*cph*B-sA2, *cph*B-s-c and *gpf*-*cph*B-s had similar signals. The weakest signals were found for S-*cph*B-s. This indicates a possible positive influence of A on transcript stability.

Activity of *N*-terminal-modified CGPase in crude plant extracts

The activity of the plant-produced, modified enzymes was determined by adding 100 or 200 μg purified CGP to 600 μg TSP isolated from *N. benthamiana* plants, which were infiltrated with the respective vectors. One reaction was stopped immediately (T0), while the other sample was incubated at room temperature (RT) overnight (T1). Both CGP samples were degraded with proteins from plants infiltrated with S-*cph*B-s (Fig. 4:1a) and *gfp*-*cph*B-s (Fig. 4:2a). Extracts from plants infiltrated with *cph*B-s-c (Fig. 4:1b) led to a nearly complete substrate reduction in the 100 μg CGP sample, while the 200 μg sample of CGP was only partially reduced. No degradation was found after incubation in plant material infiltrated with the GFP-expressing control vector pICH18711 (Marillonnet *et al.*, 2005)

(Fig. 4:2b) or with all other constructs used in this work (not shown).

Feeding CGP- and CGPase-containing pellets to mice results in the absorption of ß-Asp-Arg dipeptides

Feeding studies were performed using plant-made CGP and plant-made CGPase (S-CPHB-S) to investigate the activity of CGPase in the intestine and the bioavailability of the CGP constituent Arg and Asp. The mice were fed protein-free pellets supplemented with CGP, CGP+CGPase, Asp+Arg or none of these (CON). The pellet mass ingested was comparable between groups ($P > 0.6$) (Table S1). Intake of CGP was similar in mice fed pellets containing CGP or CGP+CGPase (Table S1). Intakes of Arg and Asp in the CGP, CGP+CGPase and Asp+Arg mouse groups were comparable, but intakes were zero in the CON group, as expected. Plasma Asp, ß-Asp-Arg, Arg and ornithine Orn

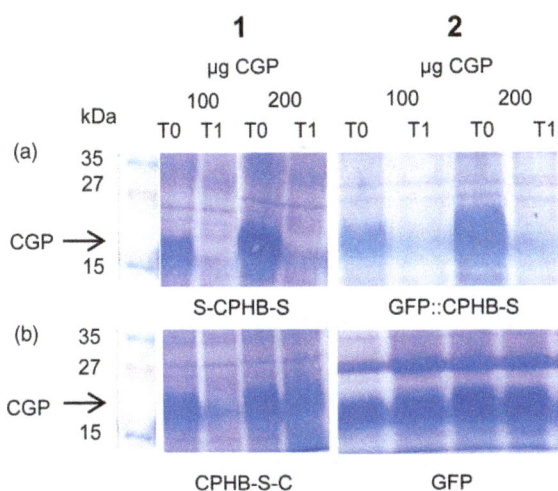

Figure 4 Enzyme activity test in 600 μg TSP crude plant extract. One hundred or 200 μg of cyanophycin was added, and the enzyme reaction was stopped by trichloroacetic acid (TCA) precipitation immediately (T0) or after 12h incubation at room temperature (T1). Total protein was precipitated with TCA and dissolved in 300 μL SDS sample buffer. Then, 30 μL of the sample was loaded on a 12% SDS gel and stained with Coomassie Brilliant Blue for 20 min; kDa: kilodalton; CGP: cyanophycin; 1a: S-CPHB-S, 2a: GFP:: CPHB-S, 1b: CPHB-S-C, 2b: GFP (green fluorescent protein expressed with vector pICH18711 (Marillonnet et al., 2005)) was used as a control.

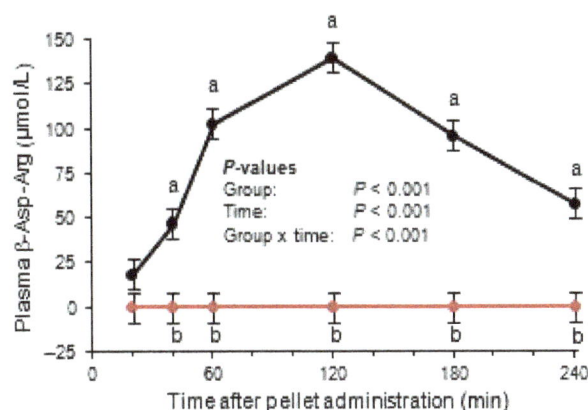

Figure 5 Course of ß-Asp-Arg concentrations in mouse plasma after administration of a pellet with cyanophycin only (red) and cyanophycin co-applied with cyanophycinase (black). LSMEANS ± SE (n = 6/group). Values with different letters (a, b) at the same time points differ between groups (Tukey, $P < 0.05$).

concentrations were affected by group, time after administration (with the exception of Orn), and group × time interaction (Table S1). In the plasma of the CGP+CGPase group, we found a relatively large peak, which was shown to be the ß-Asp-Arg dipeptide (Figs S2 and 5). This substance did not appear in the plasma of the CGP, Asp+Arg and CON mice, respectively (Fig. 5).

Arg from ß-Asp-Arg dipeptides is not bioavailable in mice

Free Arg concentrations in the plasma of Asp+Arg mice peaked 20–40 min after pellet intake and decreased thereafter to reach basal levels between 60 and 120 min ($P < 0.05$; Fig. 6a). In contrast, in the CON, CGP and CGP+CGPase groups, the courses of free plasma Arg, Asp (data not shown) and Orn, the product of Arg conversion, were similar and showed no increase (Fig. 6a,b). The course of plasma Orn concentrations in Asp+Arg mice followed the concentration curve of plasma Arg, although the maximal concentration of Orn was approximately twice that of Arg, and baseline was reached again at 180 min (Table S1; Fig. 6b). The pharmacokinetics of the plasma ß-Asp-Arg dipeptide in the CGP+CGPase group differed from those of the free plasma Arg in the group fed the pellets with free Asp and Arg (Figs 5, 6). The mean T_{max} and C_{max} of plasma Arg were 0.5 h and 172 μM, respectively, whereas plasma ß-Asp-Arg peaked with a T_{max} and C_{max} of 1.7 h and 224 μM, respectively ($P < 0.001$ and $P = 0.115$, for T_{max} and C_{max}, respectively). The plasma area under the curve (AUC) was greater for plasma ß-Asp-Arg than for free plasma Arg, 527 vs. 282 μM × h ($P = 0.003$), respectively, whereas plasma clearance (CL) for ß-Asp-Arg was lower than that for plasma Arg, 7.6 vs. 11.2 L/h ($P = 0.095$), respectively. Among the other proteinogenic AA, only plasma concentrations of glutamic acid (Glu), Ala, isoleucine (Ile) and lysine (Lys) showed a group effect ($P < 0.05$) with higher levels of Glu and Ala in the Asp+Arg group, and higher levels of Lys and Ile

in the CON group which received the protein-free pellet (data not shown).

Group, intestinal location and group × location interaction affected the residual intestinal CGP content 4 h after pellet intake ($P < 0.05$). Co-administration of CGP+CGPase resulted in a low residual CGP content in the small and large intestine (0.8 and 1.3 μg/mg of dry matter (DM); $P > 0.1$), while in mice fed the pellets with CGP only, relatively high contents of residual CGP were detected in the large intestine (33.4 μg/mg of DM), with smaller amounts of CGP (8 μg/mg) in the small intestine ($P < 0.05$).

Discussion

CGPase production in plants

In the present work, we showed for the first time the expression of cyanophycinase in plants and its cyanophycin degrading activity in the gastrointestinal tract (GIT) of mice. We used a high-yield MagnICON® transient expression system (Marillonnet et al., 2004) to determine whether plants can produce an active form of the enzyme cyanophycinase. In contrast to other researchers, who used similar vectors and described a high yield of recombinant protein of approximately 7% (Nausch et al., 2012a), 10% (Webster et al., 2009) and even 80% of the TSP (Gleba et al., 2005), we observed low levels of detectable protein. Because the expected RNA patterns were detected, possible reasons for the low protein yield may be instability of the RNA, low efficiency of translation and protein instability (reviewed by Egelkrout et al. (2012) and Ullrich et al. (2015)). Improvements in RNA stability and translation can be achieved by adapting the coding region to the codon usage of plants (Barahimipour et al., 2015; Perlak et al., 1991; Sharp and Li, 1987) and integrating the sequence GCT TCC TCC, which encodes Ala-Ser-Ser (A), downstream of the initial start codon (Sawant et al., 2001). Accordingly, both steps led to an increase in protein production, but the addition of protease inhibitors was necessary to detect the protein. As already assumed by Sawant et al. (2001) the results of this work indicate that the insertion of A led to an increased amount of RNA and also to an increase in protein prior to extraction, while the sensitivity to proteases was not changed.

In contrast to the aforementioned beneficial effect of the insertion of A, the combination of the RuBisCO transit peptide (S)

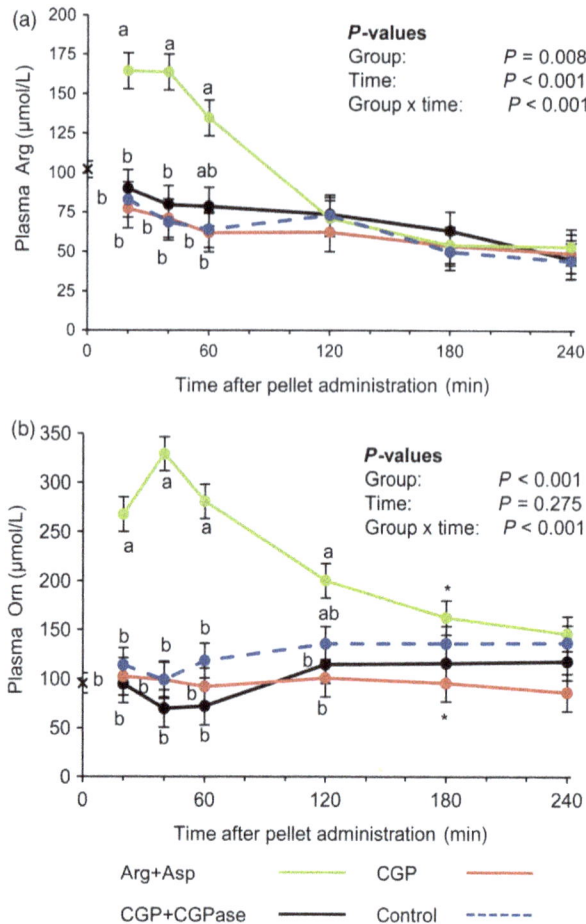

Figure 6 Course of Arg (a) and Orn (b) concentrations in mouse plasma after administration of a pellet with free Asp+Arg (green), cyanophycin only (red), cyanophycin co-ingested with cyanophycinase (black) and control pellet without supplement (dashed blue). LSMEANS ± SE (n = 6/ group). Values with different letters (a, b) at the same time points differ between groups (Tukey, $P < 0.05$). *Values sharing the sign at the same time point tend to differ (Tukey, $P < 0.10$). x: The sign on the y-axis indicates the overall average basal plasma concentration across the four groups.

and A2 in the variant S-CPHB-A2 substantially reduced protein accumulation compared to S-CPHB-S and RNA amounts were not increased, indicating that transcription was not responsible for the difference in protein accumulation. Hence, it appears that the difference is caused by different translational efficiencies or different protein stability *in planta*. *In silico* analysis of the protein indicates an influence in the secondary structure of the transit peptide which might result in aberrant folding and protein degradation. In general, protein targeting to the chloroplast led to high enzyme production. The presence of high amounts of processed protein indicates the successful import of the protein to the chloroplast. In addition, this may indicate that the native protein is protected in the chloroplast because the same protein is unstable in the cytosol. One reason may be the absence of cytosol-specific proteases in the chloroplast, as previously described for recombinant proteins (Benchabane *et al.*, 2008; Pillay *et al.*, 2014).

In addition to the processed protein, the unprocessed 35 – kDa protein was detected in the chloroplast fraction. Gils *et al.* (2005),

who used the same transit peptide with altered cleavage sites (valine-cysteine-Arg and proline-Ser-Arg instead of valine-gluta-mine-cysteine in our studies), made a similar observation and suggested that it was due to partially incorrect processing of the target protein. The unprocessed form may either be present in the cytosol or attached to the chloroplast membrane. Assuming that it is located in the cytosol, the *N*-terminal addition of the signal peptide may protect the protein from degradation as observed for the *N*-terminal addition of GFP or the *N*-terminal region of the complete protein.

While the fusion to GFP and S led to an increase in protein accumulation, it was less pronounced for the original *N*-terminus. The reduced stability of the complete protein was also indicated by the additional bands observed for CPHB-S-C, possibly repre-senting degradation products. The *N*-terminal modification may result in a decreased sensitivity to proteases, as suggested for GFP (Moreau *et al.*, 2010; Piron *et al.*, 2014), ubiquitin (Hondred *et al.*, 1999; Jang *et al.*, 2012) and elastin-like proteins (Floss *et al.*, 2008; Patel *et al.*, 2007). Nevertheless, we cannot exclude the possibility that the increased levels of GFP::CPHB-S and S-CPHB-S are due to altered translational efficiency.

The increase in protein yield was mirrored by the activity of the protein. S-CPHB-S and GFP::CPHB-S, which had the highest levels of CGPase, showed similar activities. Both enzyme variants decreased the CGP content for both amounts of substrate tested. CPHB-S-C, which was produced at lower levels, led to almost complete degradation of 100 µg CGP but only slightly reduced 200 µg CGP. No activity was detected for all other constructs. This indicates that the degradation was independent of the *N*-terminal modifications but depended on the amount of protein. This is consistent with the observations that the fusion to GFP did not impair biological function of an antigen (Piron *et al.*, 2014) and phytochrome B (Yamaguchi *et al.*, 1999).

Absorption of CGP-derived ß-Asp-Arg dipeptides in mice

We report in this study for the first time that co-application of CGP and CGPase, both isolated from plants, results in the enzymatic breakdown of CGP into dipeptides in the murine GIT, as shown by the increase in β-Asp-Arg concentrations in plasma. Upon luminal cleavage of CGP into β-Asp-Arg dipeptides by CGPase in the GIT, ß-Asp-Arg is apparently absorbed by peptide transporters (Klang *et al.*, 2005; Rubio-Aliaga and Daniel, 2008). Although the T_{max} of β-Asp-Arg dipeptide occurred at 1.7 h after intake, increased plasma concentration of the constituent Arg or Asp was not detected. ß-Asp-Arg may be partly degraded by peptidases in the intestinal epithelium, liberating Asp and Arg. However, because Arg is degraded by arginase to form Orn, urea and, to a lesser degree, nitric oxide and polyamine in the GIT, detectable amounts of Arg do not enter the systemic circulation (Wu *et al.*, 2009). As we did not observe an increase in plasma Orn concentrations in the CGP+CGPase group, we conclude that Arg cannot be liberated from β-Asp-Arg dipeptides. This is likely because β-Asp-Arg dipeptides contain an unusual bond between the C1 amino group of Arg and the C4 carboxy group of Asp. This rare phenomenon was also described for β-Ala-(Met)-His and Gly-Gly (Matthews and Adibi, 1976). Hence, the lack of an Arg and Orn increase in the plasma indicates that mice do not possess a suitable peptidase to degrade ß-Asp-Arg dipeptides. It has been shown that isoaspartyl dipeptidases have different activities in different species (Hejazi *et al.*, 2002). Consequently, β-Asp-Arg accumulates and is eventually transferred to the blood. The increased plasma ß-Asp-Arg concentrations from 40 to 240 min after pellet consumption in the CGP+CGPase group may be explained by two reasons: CGPase was stable at the pH in the small intestine and thus was active for 120 min after administration, after which plasma ß-Asp-Arg concentrations started to decline. Alternatively, upon entering the enterocytes, ß-Asp-Arg dipeptides accumulated but were only slowly released to the circulation and subsequently detoxified and excreted via urine, which we did not analyse.

Furthermore, the higher Glu and Ala plasma concentrations observed in the Asp+Arg group indicates the interconversion of Arg (via Orn) and Asp to Glu, whereas Asp is also biochemically related to Ala. The higher plasma levels of Lys and Ile in the CON group fed the protein-free pellet suggest cellular proteolysis and AA efflux to the plasma as a consequence of the lack of a suitable AA pattern necessary for protein synthesis.

The AUC in the Asp+Arg compared to the CGP+CGPase groups suggests that although equimolar amounts of Arg were consumed, a large portion of the free Arg was degraded in the small intestinal tissues to form Orn, as shown by the substantial increase in plasma Orn concentrations. Others have shown that supplemental Arg is rapidly catabolized to Orn by arginase present in the hepatocytes and also in plasma (Wu et al., 2009). Judged by the timing of the increase in plasma ß-Asp-Arg concentration compared to free Arg (1.7 vs. 0.5 h), we hypothesized that the major site of CGP degradation is the small intestine and not the stomach. This is further supported by the comparatively low residual CGP contents in the small intestine when mice were co-administered CGP and CGPase, which additionally indicates that CGP was not completely degraded by CGPase within 4 h. When mice were fed pellets with CGP only, CGP levels were high in the large intestine, suggesting that CGP was resistant to colonic fermentation, although bacteria expressing CGPase have been reported in the caecum microbiota of rabbits, sheep and carp (Sallam and Steinbuchel, 2009a).

In conclusion, we showed that plants are able to produce high amounts of active CGPase. Differences in enzyme activity were caused by different CGPase accumulation, indicating that the successful CGP degradation by CGPase depends on the enzyme amount. The greatest accumulation was observed for S-CPHB-S and GFP::CPHB-S; therefore, these two variants are suitable for further investigations related to the production of CGP in plants. The results obtained in the mouse study suggest that plant-derived CGPase, when co-ingested with CGP, is active in the mammalian intestine and hydrolyses CGP to form ß-Asp-Arg dipeptides, which can be absorbed. However, Arg from these dipeptides is not bioavailable owing to the lack of a suitable dipeptidase. This problem might be solved by the co-expression of a suitable dipeptidase in combination with CGP and CGPase in plants.

Experimental procedures

Construction of transient plant expression vectors

For transient expression, we used the MagnICON® vectors pICH29912 (cytosolic expression) (Marillonnet et al., 2005) and pICH26201 for chloroplast-targeted expression (Fig. 1), which were kindly provided by Nomad Bioscience (Halle/Saale, Germany). The cphB coding fragments were integrated using the BsaI cloning site. Vector pet22bcphB, carrying a 55 AA N-terminal truncated coding region of the cphB_{tlr2169} gene from Thermosynechococcus elongatus BP-1 (UniProt Accession No. P0C8P3) with a C-terminal 6xHistag, was provided by Prof. Dr. Wolfgang Lockau (Institute of Biology-Plant Biochemistry of the

Humboldt-University of Berlin, Germany). This sequence, called cphB-b, was adapted to the codon usage of N. tabacum (Eurofins MWG Operon, Ebersberg, Germany), resulting in cphB-s. The GCT TCC TCC sequence encoding Ala-Ser-Ser (A) was integrated downstream of the start codon to improve the efficiency of translation (Sawant et al., 2001), using primer BsaI-cphB-sA-fw (Table S2), resulting in cphB-sA. N-terminal modifications were added to the coding regions by cloning the corresponding PCR fragments. The sequence of green fluorescent protein (GFP) was amplified from vector pICH18711 (Marillonnet et al., 2005), flanked by BsaI (5') and BamHI (3') and subcloned into pJet (CloneJET PCR cloning kit, Thermo Scientific, Bonn, Germany). CphB-s was also flanked by BsaI (3') and BamHI (5') and integrated into pJet. Subsequently, gfp was fused to cphB-s using the BamHI sites. To create cphB-s-c, the unmodified coding region as described in the database, a synthetic sequence corresponding to the first 70 AA of the full-length cphB sequence (cphB-s-c-pI), was adapted to the codon usage of Nicotiana tabacum (Eurofins MWG Operon, Ebersberg, Germany) and cloned into vector pEXA2-cphB-s-pI. The sequence coding for AA 71-330 (cphB-s-c-pII) was amplified from cphB-s and flanked by BglII and SalI restriction sides. This fragment was integrated into pEXA2-cphB-s-c-pI, resulting in the vector pEXA2-cphB-s-c. S-cphB-s was constructed via integration of the coding region cphB-s into pICH26201. For vector pS-ccphB-sA2, the Ala-Ser-Ser AA sequence at position +4 to +12 was adapted from GCT TCC TCC to GCC ATT GGA using primer cphB-sA2-BsaI-fw. For propagation, all constructed vectors were transformed into E. coli TG1 and validated by sequencing (Eurofins MWG Operon, Ebersberg, Germany). For plant infiltration, the vectors were transformed into A. tumefaciens strain ICF320 (Bendandi et al., 2010).

Agroinfiltration of Nicotiana benthamiana plants

Transient expression in N. benthamiana plants was carried out as described by Nausch et al. (2012b) (cultivation and growth of bacteria and plants) and Leuzinger et al. (2013) (infiltration process) with modifications (for details see Data S1). Plants were infiltrated using vacuum infiltration with a freeze dryer (Alpha 1-4 Freeze Dryer, Christ, Germany) at 100 mBar for 2 min. Noninfiltrated leaves were removed, and plants were incubated in the dark overnight before they were returned to their regular growth conditions with a 16h day and 8h night cycle at 20–22 °C.

Sample preparation and analyses of protein content and RNA

Samples were taken 7, 8 and 9 days post infiltration (dpi), frozen immediately with liquid N_2 and stored at −80 °C. All leaves were harvested and pooled. Two to three plants per day and constructs were analysed as independent replications. TSP was measured according to Bradford (1976) using Pierce reagent and bovine serum albumin (BSA) (Thermo Scientific) as the standard. Isolation and preparation of total RNA were conducted as described previously (Nausch et al., 2012b), and Northern blot analysis was performed as described in Data S1 using the primers BsaI-cphB-b-fw, cphB-b-BsaI-rv, cphB-s-BsaI-rv and cphB-s-N-fw (Table S2).

Western blot analysis

Sample preparation and Western blot analysis were carried out as described by Nausch et al. (2012b) with modifications (for details see Data S1). CPHB-B isolated from E. coli was used as a positive control. The primary anti-CPHB antibody was produced in two Zika rabbits. Serum was obtained via centrifugation of the collected blood, and 0.04% sodium azide was added for storage. A commercial secondary antibody was used (goat anti-rabbit POD, Dianova, Hamburg Germany), and signals were detected using the ECL chemiluminescence system.

Analysis of enzyme activity in crude plant extracts

The pooled leaf sample was mixed with chilled phosphate-buffered saline (PBS) and homogenized using a Polytron (Pt-MR 2100, Kinematica AG, Switzerland) at maximum speed. TSP (600 µg) was incubated with 100 and 200 µg CGP. The reaction was neutralized (Law et al., 2009). Two samples per construct and CGP concentration were analysed. The reaction was stopped immediately or overnight using trichloroacetic acid precipitation. After centrifugation, the pellet was resolved in 300 µL 1× SDS sample buffer (Nausch et al., 2012b), and 30 µL of the sample was analysed by 12% SDS-PAGE. The gel was stained with Coomassie Brilliant Blue R250 (Carl Roth GmbH, Germany) for 20 min.

Isolation of CGP and CGPase and prediction of protein properties

CGP was isolated from Solanum tuberosum tubers (PsbY-cphA$_{TE}$-12) as described by Neubauer et al. (2012), and 10 mg was dissolved in 1 mL 0.1 M HCL pH < 2. CGPase was isolated from E. coli BL21 cells carrying pET22b-cphB-his and N. benthamiana leaves using Ni^{2+} NTA affinity chromatography as described in detail in the supplementary information (Data S1). The molecular weight of the proteins was predicted using the sequence manipulation suite home page (Stothard, 2000). Prediction of in silico protein folding was carried out using the Phyre2 server (Kelley and Sternberg, 2009; Kelley et al., 2015).

Mouse study and plasma amino acid analysis

The procedures performed in this study were in accordance with the German animal protection regulations and were approved by the relevant authorities (Landesamt für Landwirtschaft, Lebensmittelsicherheit und Fischerei, Mecklenburg-Vorpommern, Germany; permission No. 7221.3-1-017/14).

Male mice (age 49 days) of an unselected control strain (FZTDu; Dietl et al. (2004)) bred at the Leibniz Institute for Farm Animal Biology (FBN) in Dummerstorf were housed individually with sawdust bedding at 22 °C and a 12:12-h dark:light cycle. Mice were fed a standard rodent diet ad libitum (Altromin 1314, Altromin GmbH & Co. KG, Lage, Germany; 22.5% crude protein, 5% crude fat, 12.5 MJ ME/kg diet) and had free access to water. Balanced for litter and body weight (BW), the mice were randomly divided into four groups (n = 7/group) according to the type of test pellet fed: cyanophycin (CGP), cyanophycin + cyanophycinase (CGP+CGPase), free L-Arg and L-Asp (Asp+Arg) and control (CON).

Protein-free test pellets were based on 220 mg of a 1:1 mixture of corn starch (Backfee, OsnaNährmittel GmbH, Osnabrück, Germany) and powdered sucrose (Nordzucker AG, Braunschweig, Germany). The basal pellet mixture was supplemented with 30 mg CGP (from S. tuberosum tubers), 30 mg CGP + 10 mg CGPase (from N. benthamiana leaves), or 15 mg Arg (Degussa, Frankfurt/Main, Germany) plus 15 mg Asp (Reanal, Budapest, Hungary). This corresponded to 1 mg CGP, 0.33 mg CGPase, or 0.5 mg Arg or Asp, respectively, per g BW. The pellet for the control group (CON) contained none of these supplements. Dry matter was adjusted to 250 mg for each pellet with

the starch/sucrose mixture to achieve comparable energy contents. A volume of 40–45 µL of pH 3 water was added to the mixtures, the mixture was filled in 1-mL plastic syringes (Omnifix® 40 Solo, Braun, Melsungen, Germany), and pellets were pressed manually with a plunger. Pressed pellets were stored overnight at 4 °C before consumption.

At the age of 79–92 days, the mice were transferred to cages without sawdust after overnight food withdrawal. Test pellets were offered to the mice for 15 min, the remaining food was collected, and the animals were transferred back to their original cages. The tips of their tails were snipped to collect blood in sodium-heparinized microhematocrit capillary tubes (Marienfeld, Lauda-Königshofen, Germany) at 0, 20, 40, 60, 120, 180 and 240 min after pellet administration. The capillaries were immediately put on ice and were centrifuged for 3 min at 10 000 × g and 4 °C. The isolated plasma was diluted with ultrapure water and stored at −20 °C. Immediately after the 240 min, blood samples were taken, mice were killed by cervical dislocation. The abdomen was opened, the small and large intestine were isolated, and the total contents were rinsed with 3 mL cold PBS, weighed, and stored frozen at −20 °C The residual CGP concentration of the small and large intestine in the CGP, CGP+CGPase and free Asp+Arg mice were quantified as described in Data S1. We used only mice with >40% intake of their respective pellets. Thus, valid observations were obtained for 6, 6, 7 and 7 mice in the CGP, CGP+CGPase, Asp+Arg and CON groups, respectively. Plasma AA were analysed by HPLC separation with fluorescence detection of o-phthaldialdehyde derivatives on a 250 × 4 mm HyperClone ODS (C18) 120 Å (Phenomenex, Aschaffenburg, Germany) as described (Kuhla et al., 2010). Standard AA (A9906 Sigma, Munich, Germany) allowed assignment of retention times and quantification. The β-Asp-Arg dipeptide, eluting at a retention time of 3.8 min, was identified using isolated CGP degraded by CGPase (Fig. S2). Enzyme hydrolysis of CGP (200 µg) was performed as described (Law et al., 2009) with minor modifications. CGP (10 mg/mL in 0.1 M HCL) was diluted in PBS (pH 7.2), and 50 µg of S-CPHB-S (1 mg/mL) isolated from N. benthamiana was added. Samples were incubated at RT overnight. CGPase was removed from the hydrolysis mixture by centrifugation with Roti®-Spin MINI-3 cartridges (3 kDa, Carl Roth GmbH + Co. KG, Karlsruhe, Germany) for 20 min at 17 000 × g and 4 °C. The β-Asp-Arg dipeptide was quantified by the AUC using a four-point calibration. The detection limit was 0.5 µM. An aliquot of the β-Asp-Arg dipeptide filtrate (100 µL) was hydrolysed in 1 mL of 6 N HCl at 110 °C for 22 h with ascorbic acid as an antioxidant under a N$_2$ atmosphere to confirm the Asp and Arg constituents only. After HCl removal under a N$_2$ stream at 60 °C, the residue was diluted in 1 mL H$_2$O, centrifuged for 20 min at 17 000 × g and 4 °C and analysed for AA concentrations. Based on the concentrations of Arg and Asp, the amount of β-Asp-Arg was recalculated. Then, plasma concentrations of Asp, Arg, β-Asp-Arg dimer and Orn were quantified. For calculation of the pharmacokinetics of these metabolites, data were normalized for the pellet intake and BW after subtraction of the mean concentrations of the corresponding AA of the CON group. The AUC for the β-Asp-Arg dimer and Arg between 0 and 4 h after administration was calculated with TableCurve 2D V 5.01 software (SYSTAT Software Inc., Chicago, IL). The time (T_{max}) at which the maximal plasma concentration, C_{max}, was reached was computed using a best curve fit. Plasma clearance was calculated from the administered dose of CGP (converted to β-Asp-Arg equivalents) or Arg (µmol) divided by the AUC.

Statistical analysis of mouse study

Mouse data were evaluated with SAS 9.4 software (SAS Institute Inc. 2011, Cary, NC). A repeated-measures ANOVA implemented in a mixed model was used with the fixed effects of group (CGP+CGPase, CGP, Asp-Arg, CON) and time of blood sampling (20, 40, 60, 120, 180, and 240 min), as well as the interaction term. The effect of repeatedly sampled animals was considered random. The covariance structure was autoregressive (AR1). For singly measured variables, a one-way ANOVA was used with group as a fixed effect. Residual intestinal CGP contents were analysed by a two-factorial ANOVA with group and intestinal location (small and large intestine) as fixed factors and the interaction term. Pellet intake did not differ among groups and was not considered a factor. Effects were considered significant at $P \leq 0.05$, and group differences were tested using the Tukey–Kramer test. Significance levels of $P < 0.10$ were considered a statistical trend. Data are presented as least-square means (LSMEANS) and their standard errors (SE).

Acknowledgements

We are grateful for the support of Dr. Alain Steinmann and Bioserv GmbH Rostock in the production of anti-CPHB antibodies. Additionally, we thank Quentin L. Sciascia, PhD, for assistance during the mouse trial, particularly with collection of the blood samples. Care for the experimental animals by the team at the Tiertechnikum of the Leibniz Institute for Farm Animal Biology, Dummerstorf, is also greatly appreciated. The animal study was funded by the core budget to the Leibniz Institute for Farm Animal Biology, Dummerstorf. The plant work was funded by the scholarship of the 'Landesgraduiertenförderung' of the University of Rostock. The publication of this article was funded by the Open Access fund of the Leibniz Institute for Farm Animal Biology (FBN).

References

Allen, M.M., Morris, R. and Zimmerman, W. (1984) Cyanophycin granule polypeptide protease in a unicellular cyanobacterium. Arch. Microbiol. 138, 119–123.

Barahimipour, R., Strenkert, D., Neupert, J., Schroda, M., Merchant, S.S. and Bock, R. (2015) Dissecting the contributions of GC content and codon usage to gene expression in the model alga Chlamydomonas reinhardtii. Plant J., 84, 704–717.

Bazer, F.W., Kim, J., Ka, H., Johnson, G.A., Wu, G. and Song, G. (2012) Select nutrients in the uterine lumen of sheep and pigs affect conceptus development. J. Reprod. Develop. 58, 180–188.

Benchabane, M., Goulet, C., Rivard, D., Faye, L., Gomord, V. and Michaud, D. (2008) Preventing unintended proteolysis in plant protein biofactories. Plant Biotech. J. 6, 633–648.

Bendandi, M., Marillonnet, S., Kandzia, R., Thieme, F., Nickstadt, A., Herz, S. and Frode, R., et al. (2010) Rapid, high-yield production in plants of individualized idiotype vaccines for non-Hodgkin's lymphoma. Ann. Oncol. 21, 2420–2427.

Bradford, M.M. (1976) A rapid and sensitive method for the quantitation of microgram quantities of protein utilizing the principle of protein-dye binding. Anal. Biochem. 72, 248–254.

Dietl, G., Langhammer, M. and Renne, U. (2004) Model simulations for genetic random drift in the outbred strain Fzt: DU. Arch. Anim. Bree. 47, 595–604.

Egelkrout, E., Rajan, V. and Howard, J.A. (2012) Overproduction of recombinant proteins in plants. Plant Sci. 184, 83–101.

Floss, D.M., Sack, M., Stadlmann, J., Rademacher, T., Scheller, J., Stoger, E. and Fischer, R. et al. (2008) Biochemical and functional characterization of anti-HIV antibody-ELP fusion proteins from transgenic plants. Plant Biotech. J. 6, 379–391.

Gils, M., Kandzia, R., Marillonnet, S., Klimyuk, V. and Gleba, Y. (2005) High-yield production of authentic human growth hormone using a plant virus-based expression system. Plant Biotech. J. 3, 613–620.

Gleba, Y., Klimyuk, V. and Marillonnet, S. (2005) Magnifection—a new platform for expressing recombinant vaccines in plants. Vaccine 23, 2042–2048.

Gupta, M. and Carr, N.G. (1981) Enzyme-activities related to cyanophycin metabolism in heterocysts and vegetative cells of Anabaena Spp. J. Gen. Microbiol. 125, 17–23.

Hejazi, M., Piotukh, K., Mattow, J., Deutzmann, R., Volkmer-Engert, R. and Lockau, W. (2002) Isoaspartyl dipeptidase activity of plant-type asparaginases. Biochem. J. 364, 129–136.

Hondred, D., Walker, J.M., Mathews, D.E. and Vierstra, R.D. (1999) Use of ubiquitin fusions to augment protein expression in transgenic plants. Plant Physiol. 119, 713–723.

Hühns, M., Neumann, K., Hausmann, T., Ziegler, K., Klemke, F., Kahmann, U. and Staiger, D. et al. (2008) Plastid targeting strategies for cyanophycin synthetase to achieve high-level polymer accumulation in Nicotiana tabacum. Plant Biotech. J. 6, 321–336.

Hühns, M., Neumann, K., Hausmann, T., Klemke, F., Lockau, W., Kahmann, U. and Kopertekh, L. et al. (2009) Tuber-specific cphA expression to enhance cyanophycin production in potatoes. Plant Biotech. J. 7, 883–898.

Jang, I.C., Niu, Q.W., Deng, S.L., Zhao, P.Z. and Chua, N.H. (2012) Enhancing protein stability with retained biological function in transgenic plants. Plant J. 72, 345–354.

Kelley, L.A. and Sternberg, M.J.E. (2009) Protein structure prediction on the Web: a case study using the Phyre server. Nat. Protoc. 4, 363–371.

Kelley, L.A., Mezulis, S., Yates, C.M., Wass, M.N. and Sternberg, M.J.E. (2015) The Phyre2 web portal for protein modeling, prediction and analysis. Nat. Protoc. 10, 845–858.

Klang, J.E., Burnworth, L.A., Pan, Y.X., Webb, K.E. and Wong, E.A. (2005) Functional characterization of a cloned pig intestinal peptide transporter (pPepT1). J. Anim. Sci. 83, 172–181.

Klimyuk, V., Benning, G., Gils, M., Giritch, A. and Gleba, Y. (2004) Process of Producing a plastid-targeted protein in plant cells. AG, I.G. WO 2004/101797 A1

Kuhla, B., Kucia, M., Goers, S., Albrecht, D., Langhammer, M., Kuhla, S. and Metges Cornelia, C. (2010) Effect of a high-protein diet on food intake and liver metabolism during pregnancy, lactation and after weaning in mice. Proteomics 10, 2573–2588.

Law, A.M., Lai, S.W.S., Tavares, J. and Kimber, M.S. (2009) The structural basis of beta-Peptide-specific cleavage by the serine protease cyanophycinase. J. Mol. Biol. 392, 393–404.

Leuzinger, K., Dent, M., Hurtado, J., Stahnke, J., Lai, H., Zhou, X. and Chen, Q. (2013) Efficient agroinfiltration of plants for high-level transient expression of recombinant proteins. J. Vis. Exp., 77, 1–9.

Marillonnet, S., Giritch, A., Gils, M., Kandzia, R., Klimyuk, V. and Gleba, Y. (2004) In planta engineering of viral RNA replicons: efficient assembly by recombination of DNA modules delivered by Agrobacterium. Proc. Natl. Acad. Sci. USA 101, 6852–6857.

Marillonnet, S., Thoeringer, C., Kandzia, R., Klimyuk, V. and Gleba, Y. (2005) Systemic Agrobacterium tumefaciens-mediated transfection of viral replicons for efficient transient expression in plants. Nat. Biotechnol. 23, 718–723.

Matthews, D.M. and Adibi, S.A. (1976) Peptide absorption. Gastroenterology 71, 151–161.

Moreau, M.J.J., Morin, I. and Schaeffer, P.M. (2010) Quantitative determination of protein stability and ligand binding using a green fluorescent protein reporter system. Mol. BioSyst. 6, 1285–1292.

Nausch, H., Mikschofsky, H., Koslowski, R., Meyer, U., Broer, I. and Huckauf, J. (2012a) High-Level transient expression of ER-targeted human interleukin 6 in Nicotiana benthamiana. PLoS ONE, 7, 1–16.

Nausch, H., Mischofsky, H., Koslowski, R., Meyer, U., Broer, I. and Huckauf, J. (2012b) Expression and subcellular targeting of human complement factor C5a in Nicotiana species. PLoS ONE, 7, 1–13.

Neubauer, K., Huhns, M., Hausmann, T., Klemke, F., Lockau, W., Kahmann, U. and Pistorius, E.K. *et al.* (2012) Isolation of cyanophycin from tobacco and potato plants with constitutive plastidic cphA(Te) gene expression. *J. Biotech.* **158**, 50–58.

Neumann, K., Neumann, K., Stephan, D.P., Ziegler, K., Huhns, M., Broer, I., Lockau, W. and Pistorius, E.K. (2005) Production of cyanophycin, a suitable source for the biodegradable polymer polyaspartate, in transgenic plants. *Plant Biotech. J.* **3**, 249–258.

Patel, J., Zhu, H., Menassa, R., Gyenis, L., Richman, A. and Brandle, J. (2007) Elastin-like polypeptide fusions enhance the accumulation of recombinant proteins in tobacco leaves. *Transgenic Res.* **16**, 239–249.

Perlak, F.J., Fuchs, R.L., Dean, D.A., Mcpherson, S.L. and Fischhoff, D.A. (1991) Modification of the coding sequence enhances plant expression of insect control protein genes. *Proc. Natl. Acad Sci. USA* **88**, 3324–3328.

Pillay, P., Schluter, U., van Wyk, S., Kunert, K.J. and Vorster, B.J. (2014) Proteolysis of recombinant proteins in bioengineered plant cells. *Bioengineered* **5**, 15–20.

Piron, R., De Koker, S., De Paepe, A., Goossens, J., Grooten, J., Nauwynck, H. and Depicker, A. (2014) Boosting in planta production of antigens derived from the Porcine Reproductive and Respiratory Syndrome Virus (PRRSV) and subsequent evaluation of their immunogenicity. *PLoS ONE*, **9**, 1–16.

Richter, R., Hejazi, M., Kraft, R., Ziegler, K. and Lockau, W. (1999) Cyanophycinase, a peptidase degrading the cyanobacterial reserve material multi-L-arginyl-poly-L-aspartic acid (cyanophycin) - Molecular cloning of the gene of Synechocystis sp PCC 6803, expression in *Escherichia coli*, and biochemical characterization of the purified enzyme. *Eur. J. Biochem.* **263**, 163–169.

Rubio-Aliaga, I. and Daniel, H. (2008) Peptide transporters and their roles in physiological processes and drug disposition. *Xenobiotica* **38**, 1022–1042.

Sallam, A. and Steinbuchel, A. (2009a) Cyanophycin-degrading bacteria in digestive tracts of mammals, birds and fish and consequences for possible applications of cyanophycin and its dipeptides in nutrition and therapy. *J. Appl. Microbiol.* **107**, 474–484.

Sallam, A. and Steinbuchel, A. (2009b) *Process for the preparation of dipeptides from cyanophycin employing the isolated Pseudomonas alcaligenes DIP1 CGPase CphEal.* Münster, W.W.-U.

Sallam, A. and Steinbuchel, A. (2010) Dipeptides in nutrition and therapy: cyanophycin-derived dipeptides as natural alternatives and their biotechnological production. *Appl. Microbiol. Biotechnol.* **87**, 815–828.

Sallam, A., Kast, A., Przybilla, S., Meiswinkel, T. and Steinbuchel, A. (2009) Biotechnological process for production of beta-dipeptides from cyanophycin on a technical scale and its optimization. *Appl. Environ. Microbiol.* **75**, 29–38.

Sawant, S.V., Kiran, K., Singh, P.K. and Tuli, R. (2001) Sequence architecture downstream of the initiator codon enhances gene expression and protein stability in plants. *Plant Physiol.* **126**, 1630–1636.

Sharp, P.M. and Li, W.H. (1987) The codon adaptation index – a measure of directional synonymous codon usage bias, and its potential applications. *Nucleic Acids Res.* **15**, 1281–1295.

Simon, R.D. (1987) Inclusion bodies in the cyanobacteria: cyanophycin, polyphosphate, polyhedra bodies. In *The Cyanobacteria* (Fay, P. and Van Baalen, C., eds), pp. 192–222. Amsterdam, New York, Oxford: Elsevier.

Simon, R.D. and Weathers, P. (1976) Determination of structure of novel polypeptide containing aspartic-acid and arginine which is found in cyanobacteria. *Biochim. Biophys. Acta* **420**, 165–176.

Stothard, P. (2000) The sequence manipulation suite: javaScript programs for analyzing and formatting protein and DNA sequences. *Biotechniques* **28**, 1102-+.

Ullrich, K.K., Hiss, M. and Rensing, S.A. (2015) Means to optimize protein expression in transgenic plants. *Curr. Opin. Biotechnol.* **32**, 61–67.

Utagawa, T. (2004) Production of arginine by fermentation. *J. Nutr.* **134**, 2854s–2857s.

Wang, Y.X. , Zhang, L., Zhou, G., Liao, Z., Ahmad, H., Liu, W. and Wang, T. (2012) Dietary L-arginine supplementation improves the intestinal development through increasing mucosal Akt and mammalian target of rapamycin signals in intra-uterine growth retarded piglets. *Br. J. Nutr.* **108**, 1371–1381.

Webster, D.E., Wang, L., Mulcair, M., Ma, C., Santi, L., Mason, H.S. and Wesselingh, S.L. *et al.* (2009) Production and characterization of an orally immunogenic Plasmodium antigen in plants using a virus-based expression system. *Plant Biotech. J.* **7**, 846–855.

Wenzel, U., Meissner, B., Doring, F. and Daniel, H. (2001) PEPT1-mediated uptake of dipeptides enhances the intestinal absorption of amino acids via transport system b(0,+). *J. Cell. Physio.* **186**, 251–259.

Wu, G.Y. (2014) Dietary requirements of synthesizable amino acids by animals: a paradigm shift in protein nutrition. *J. Anim. Sci. Biotechnol.*, **5**, 34, 1–12.

Wu, G., Knabe, D.A. and Kim, S.W. (2004) Arginine nutrition in neonatal pigs. *J. Nutr.*, **134**, 2783S–2790S; discussion 2796S–2797S.

Wu, G.Y., Bazer, F.W., Davis, T.A., Kim, S.W., Li, P., Rhoads, J.M., Satterfield, M.C., Smith, S.B., Spencer, T.E. and Yin, Y. (2009) Arginine metabolism and nutrition in growth, health and disease. *Amino Acids* **37**, 153–168.

Yagasaki, M. and Hashimoto, S. (2008) Synthesis and application of dipeptides; current status and perspectives. *Appl. Microbiol. Biotechnol.* **81**, 13–22.

Yamaguchi, R., Nakamura, M., Mochizuki, N., Kay, S.A. and Nagatani, A. (1999) Light-dependent translocation of a phytochrome B-GFP fusion protein to the nucleus in transgenic Arabidopsis. *J. Cell Biol.* **145**, 437–445.

Ziegler, K., Diener, A., Herpin, C., Richter, R., Deutzmann, R. and Lockau, W. (1998) Molecular characterization of cyanophycin synthetase, the enzyme catalyzing the biosynthesis of the cyanobacterial reserve material multi-L-arginyl-poly-L-aspartate (cyanophycin). *Eur. J. Biochem.* **254**, 154–159.

Ziegler, K., Deutzmann, R. and Lockau, W. (2002) Cyanophycin synthetase-like enzymes of non-cyanobacterial eubacteria: characterization of the polymer produced by a recombinant synthetase of Desulfitobacterium hafniense. *J. Biosci.* **57**, 522–529.

7

A new location to split Cre recombinase for protein fragment complementation

Maryam Rajaee[1,2] and David W. Ow[1,*]

[1]Plant Gene Engineering Center, South China Botanical Garden, Chinese Academy of Sciences, Guangzhou, China
[2]University of Chinese Academy of Sciences, Beijing, China

*Correspondence
email dow@scbg.ac.cn

Keywords: site-specific recombination, gene stacking, protein fragment complementation, alpha-complementation.

Summary

We have previously described a recombinase-mediated gene stacking system in which the Cre recombinase is used to remove *lox*-site flanked DNA no longer needed after each round of Bxb1 integrase-mediated site-specific integration. The Cre recombinase can be conveniently introduced by hybridization with a *cre*-expressing plant. However, maintaining an efficient *cre*-expressing line over many generations can be a problem, as high production of this DNA-binding protein might interfere with normal chromosome activities. To counter this selection against high Cre activity, we considered a split-*cre* approach, in which Cre activity is reconstituted after separate parts of Cre are brought into the same genome by hybridization. To insure that the recombinase-mediated gene stacking system retains its freedom to operate, we tested for new locations to split Cre into complementing fragments. In this study, we describe testing four new locations for splitting the Cre recombinase for protein fragment complementation and show that the two fragments of Cre split between Lys244 and Asn245 can reconstitute activity that is comparable to that of wild-type Cre.

Introduction

A long-term aim of this laboratory has been to develop site-specific gene stacking to ease the introgression of transgenes from transformable lines to elite field cultivars (Hou *et al.*, 2014). Each step of the reiterative gene stacking scheme involves using the Bxb1 integrase to direct the insertion of new DNA into a genomic Bxb1 attachment site (*attP* or *attB* site), followed by introduction of the 343aa Cre recombinase to delete the 34 bp *lox*-site flanked DNA no longer needed after site-specific integration. While the Bxb1 integrase can be transiently introduced along with the integrating DNA, the Cre recombinase is most conveniently introduced from hybridization to a *cre*-expression line. Maintaining an efficient *cre*-expressing line over many generations, however, could potentially be a problem as high expression of *cre* in petunia and tomato has been associated with crinkled leaves and/or reduced fertility (Que *et al.*, 1998; Cappoolse *et al.*, 2003). As recombinases are DNA-binding proteins, it remains possible that high production of these proteins could interfere with normal chromosome activities. To insure against this possibility, we considered a split-*cre* approach as shown in Figure 1, in which the integrating vector brings along a portion of *cre*, and after integration, the rest of *cre* can be subsequently introduced from hybridization.

Figure 2a lists the various reports on splitting Cre into two peptides. The first approach is based on protein fragment complementation, where Cre activity is reconstituted from separate peptides. Casanova *et al.* (2003) first reported testing various pairs of nonfunctional N-terminal and C-terminal Cre fragments at break points between aa160 and aa203 and found one pair, aa1-196/aa182-343, that reconstituted activity in a transient assay in CV1-5B monkey cells, but at only 33%

efficiency compared to wild-type Cre. Seidi *et al.* (2007) also reported success with a split-Cre pair at the same region, aa1-194 and aa180-343, and with up to 68% excision efficiency in a transient assay in COS-7 monkey cells. More recently, Wen *et al.* (2014) reconstructed Cre from the pair aa1-59/aa60-343 with and without the SV40 NLS and in an excision assay of *A. rhizogene*-mediated transformation of tobacco root hair, up to 67% of wild-type Cre efficiency was reported.

Uses of heterodimerizing peptides to facilitate the reassembly of split-Cre complementating peptides have also been tested. Jullien *et al.* (2003) reported reconstituting the pairs aa19-59/aa60-343 and aa19-104/106-343, in which each N-terminal fragment (N-fragment) was fused to FKBP12 (FK506 binding protein) while each C-terminal fragment (C-fragment) was fused to FRB (binding domain of the FKBP12-rapamycin-associated protein). Upon addition of rapamycin that induces FKBP12/FRB interaction, Cre activity was reconstituted in both a transient and a stable excision assay in a Rat2/CALNLZ cell line. In the transient assay, the aa19-104/106-343 pair showed higher recombination activity, whereas in the stable excision assay, the aa19-59/aa60-343 pair was more efficient. In a study by Xu *et al.* (2007), each N- and C-fragment of the aa1-190/aa191-343 pair was fused to antiparallel leucine zippers. In transient and transgenic mouse cells, excision efficiencies were about 30% compared to wild-type Cre. Maruo *et al.* (2008) tested parallel binding modules and reported that a Zip(+)/(−) synthetic leucine zipper based on the vitellogenin-binding protein from a chicken b-ZIP family was most efficient among three heterodimerizing peptide pairs of aa19-59/aa60-343 tested in Cos-7 monkey cells and in immature mouse neurons, but its efficiency was not compared to wild-type Cre. Seidi *et al.* (2009) used the AP-1 transcription regulatory proteins bJun and bFos in a transient monkey COS7 cell assay that showed

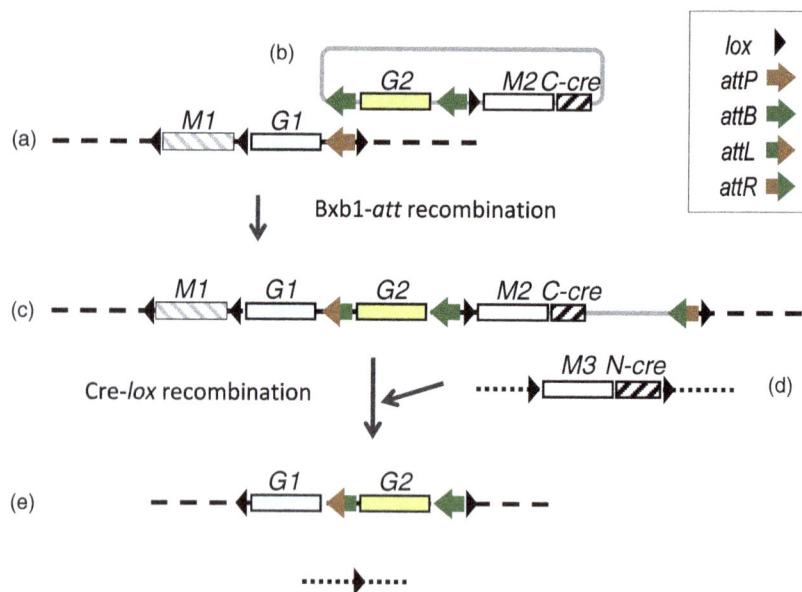

Figure 1 Reconstitution of Cre activity for removing DNA no longer needed after site-specific integration. Scheme for recombinase-mediated gene stacking shows the chromosomal target construct with an *attP* site (a) recombining with an integrating vector (b) that also includes a C-Cre peptide encoding gene. After site-specific integration from *attP* recombination with the *M2* distal *attB* site to yield the configuration shown in (c), the construct encoding N-Cre peptide (d) is introduced by genetic hybridization to yield the expected configuration shown in (e). *M1, M2 and M3* represent marker genes, and *G1 and G2* represent trait genes. Symbols of recombination sites defined in legend. Note that the *M3-N-cre* fragment is also excised in the F1 (e).

bJun-Cre (aa-1-194) and bFos-Cre (aa-184-343) yielded 23% deletion efficiency compared to wild-type Cre. Finally, Hirrlinger *et al.* (2009) reported using the coiled coil domain of the yeast transcriptional activator GCN4 on the split-Cre pair aa19-59/aa60-343 in transgenic mice and found 26% deletion efficiency.

In contrast to reconstituting activity from protein fragment complementation, split inteins can reconstruct a whole Cre protein by splicing together separate protein peptides. The first report of using split inteins to reconstruct Cre was described in a patent issued to Dupont (Yadav and Yang, 2007). The split inteins from *Cyanobacterium synechocystis DnaE* reconstructed Cre from the pair of aa1-155/aa156-343 fragments. Using this same approach, Wang *et al.* (2012) reconstructed Cre peptides aa19-59 and aa60-343 and reported high Cre activity in mouse brain tissue. Han *et al.* (2013) tested the same *Ssp DnaE* split intein on the Cre pair aa1-190/aa191-343 and obtained excision activity in tobacco. Based on the number of leaf explants resistant to the herbicide Basta, up to 77% of wild-type Cre efficiency was report. More recently, Ge *et al.* (2016) tested five Cre pairs (aa1-128/aa129-343, aa1-153/aa154-343, aa1-190/aa191-343, aa1-232/aa233-343 and aa1-281/aa282-343) in a transient tobacco explant assay and concluded that the best pair for split-intein reconstruction of Cre was aa1-232/aa233-343. In transgenic *Arabidopsis*, this split-intein reconstruction of Cre was reported to be near 100% efficiency in F1 hybrids.

In terms of location flexibility to split Cre, as well as restoration of Cre activity, the split-intein approach to reconstruct a whole Cre protein would seem preferable to complementation activity with separated peptides. However, as an important aim for developing our 'open-source' gene stacking system was to permit the freedom to operate for commercial crop improvement (Ow, 2016), we had to consider the intellectual property constraints of the Dupont intein-Cre patent that is in effect until 2023 (Yadav and Yang, 2007). With the Cre fragment complementation approach, there have been two patent applications filed based on the work of Jullien *et al.* (2003) and Xu *et al.* (2007) that claim the use of specific heterodimerizing peptides along with specified split-Cre locations, namely after aa 59, 104 and 190 (Gu and Xu, 2009; Herman and Jullien, 2004). As issued patents of these were

not found, it seems likely that these applications have been abandoned. Surprisingly, however, despite prior art on splitting Cre after aa 59, a Chinese patent was issued recently based on the work of Wen *et al.* (2014) that claimed splitting Cre after aa 59 (Gao *et al.*, 2015). Given the freedom-to-operate uncertainty, we thought it would be preferably to create our own version of Cre fragment complementation.

As indicated in Figure 2a, previous reports have led to relatively few break points that can yield Cre fragment complementation, and they are all within loops that connect alpha-helical segments or beta-sheets. Four Cre pairs were split at the region between aa180-196 within loops linking beta-sheets 1, 2 and 3; another four pairs split after aa59 between alpha-helixes B and C, and a single report of splitting at aa104-106 between alpha-helixes D and E (Figure 2b). Likewise, we thought it might be possible to split the protein within the loop connecting beta-sheets 4 and 5. Additionally, because it was convenient to generate C-fragments that begin with ATG start codon, we also included splitting Cre after Asn96, Val116 and Asp298 (Figure 2b), even though each of them disrupts an alpha-helical segment. Here, we report that the break point between Lys244 and Asn245 indeed yields two peptides that can undergo Cre fragment complementation activity in *E. coli* and in transgenic *Arabidopsis*. In *Arabidopsis*, the recombination efficiency by hybrid-reconstituted Cre was comparable to that of wild-type Cre and was obtained without the need of heterodimerizing peptides.

Results

Testing new split-Cre pairs for protein fragment complementation in *E. coli*

To examine whether the 343 amino acid Cre recombinase can be functionally reconstituted when split at regions other than the three previously reported, we chose to split Cre after Asn96 (Figure 2a, location 1), Val116 (location 2), Lys244 (location 3) and Asp298 (location 4). Locations 3 and 4 are within the C-terminal end of the protein that have not been previously tested (Figure 2a). Locations 1, 2 and 4 were also chosen because each could be split such that the C-terminal fragment begins with an ATG start codon, bypassing the need to add another amino acid to the C-terminal peptide, which was the case for splitting at location 3 (after Lys244). Based on the Cre X-ray crystal structure (Guo *et al.*, 1997) of five alpha-helical segments (A-E) connected by short loops in its amino-terminal domain and nine alpha-helical segments (F-N) along with five beta-sheets (1-5) in its carboxyl-terminal domain (Figure 2b), our location 1 cut is located in α-helix D between Asn96 and Met97 resulting in peptide 1N (1-96 aa) and 1C (97-343 aa) for N- and C-terminal moieties, respectively. Location 2 is in α-helix E between Val116 and Met 117 yielding 2N (1-116 aa) and 2C (117-343 aa) fragments. Location 3 cut is between the 4th and 5th beta-sheets after Lys244 resulting in 3N (1-244 aa) and 3C (245-343 aa) fragments, and location 4 is in the K alpha-helix between Asp298 and Met299 to yield 4N (1-298 aa) and 4C (299-343 aa) peptides (Figure 2b). By mixing N-terminal and C-terminal fragments of the different pairs, a total of 10 pairs, including six overlapping ones, were tested for Cre fragment complementation in *E. coli* (Figure 3b). For use as a positive control, we also generated a fifth split-Cre pair of aa1-192/aa181-343 because this region has been reported to work by four prior studies (ATG start codon added before aa181; referred to as 5N/5C).

Each N-terminal and C-terminal gene fragment was expressed in pETDuet and pRSFDuet vectors to produce the ampicillin-resistant pETDuet-N and kanamycin-resistant pRSFDuet-C series

of constructs, respectively (Figure 3a). A third plasmid that is chloramphenicol resistant, pACYCDuet-inv served as a reporter for site-specific recombination. This construct has a hygromycin coding region (*hpt*) flanked by oppositely oriented *lox* sites such that site-specific recombination inverts the *hpt* fragment to produce a predicted molecular structure detectable by PCR. For a positive control, the full-length *cre* gene was expressed from pETDuet-Cre. All three plasmid types (pETDuet, pRSFDuet and pACYCDuet) are compatible for co-propagation in the *recA1* homologous recombination impaired *E. coli* strain DH5α (Grant *et al.*, 1990).

In the absence of site-specific inversion, PCR products of pACYCDuet-inv can be detected using primer pairs 1 + 2 and 3 + 4, but not from primer pairs 1 + 3 and 2 + 4 (Figure 3a). If inversion takes place, which is a reversible event, PCR products can also be detected using primer pairs 1 + 3 and 2 + 4. As expected from the controls, 90% of the colonies derived from pACYCDuet-inv showed site-specific inversion when cotransformed by pETDuet-Cre, whereas none was found when cotransformed with the pETDuet empty vector. For the cotransformation of N and C pairs, which requires the co-introduction of three plasmids instead of two for the controls, the 5N/5C (aa1-192/aa181-343) pair indeed showed site-specific inversion in 67% of the colonies. This confirms earlier reports of reconstitution of Cre activity when split at the region between aa181-196.

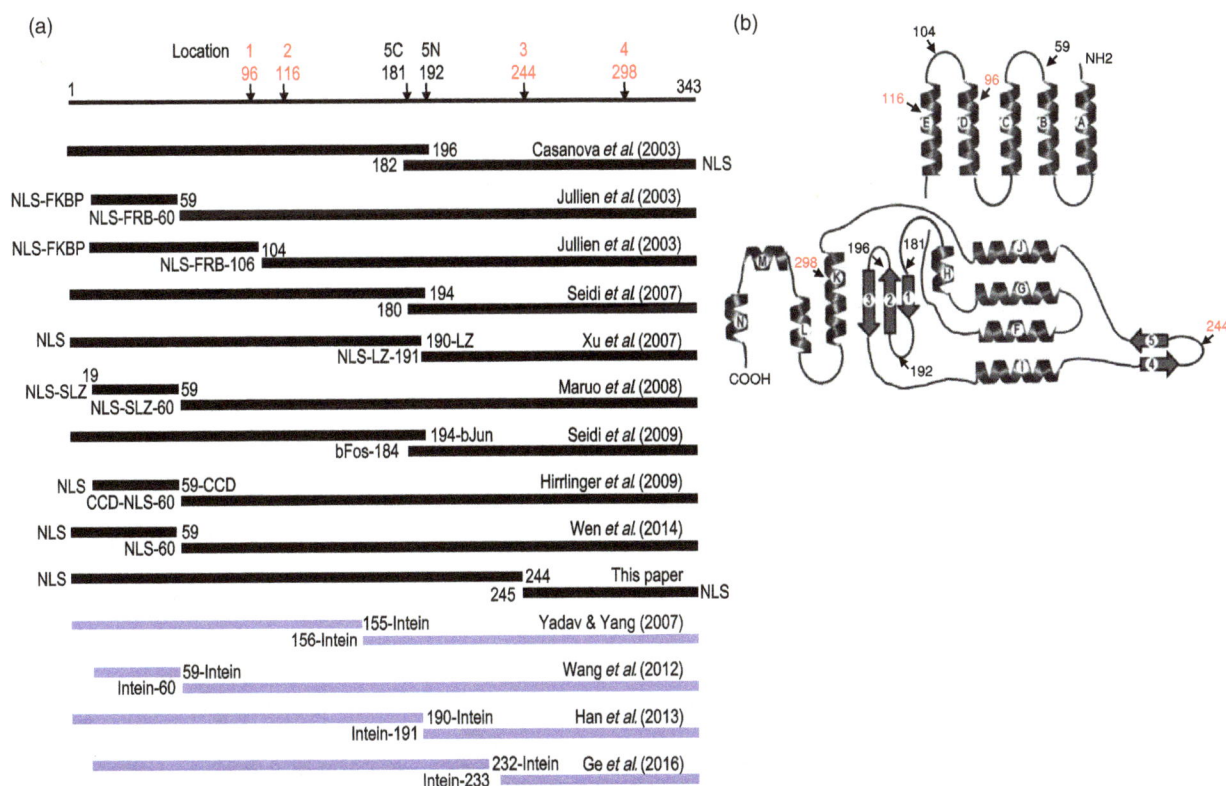

Figure 2 Linear depiction (a) and X-ray crystal structure (b) of Cre recombinase. (a) Split-Cre fragments that can reconstitute Cre activity through protein fragment complementation shown with grey bars. Split-Cre fragments for reconstruction of whole Cre protein through protein splicing shown with blue bars. Numbers refer to the aa residues. Fusion to the following: NLS = nuclear localization sequence; LZ = leucine zipper; SLZ = synthetic leucine zipper; bJun and bFos = AP-1 transcription regulatory proteins; CCD = coiled coil domain of the yeast transcriptional activator GCN4; FKBP12 = FK506 binding protein; FRB = binding domain of the FKBP12-rapamycin-associated protein. Intein= from *Cyanobacterium synechocystis* Ssp DnaE. ATG added to N-terminus of all C-Cre fragments shown. (b) X-ray crystal structure from Guo *et al.* (1997) showing the 14 alpha-helical segments (A-N) and five beta-sheets (1–5). Arrows point to relevant aa residues. Blue lettering, previously tested locations; red lettering, the four new locations tested in this study.

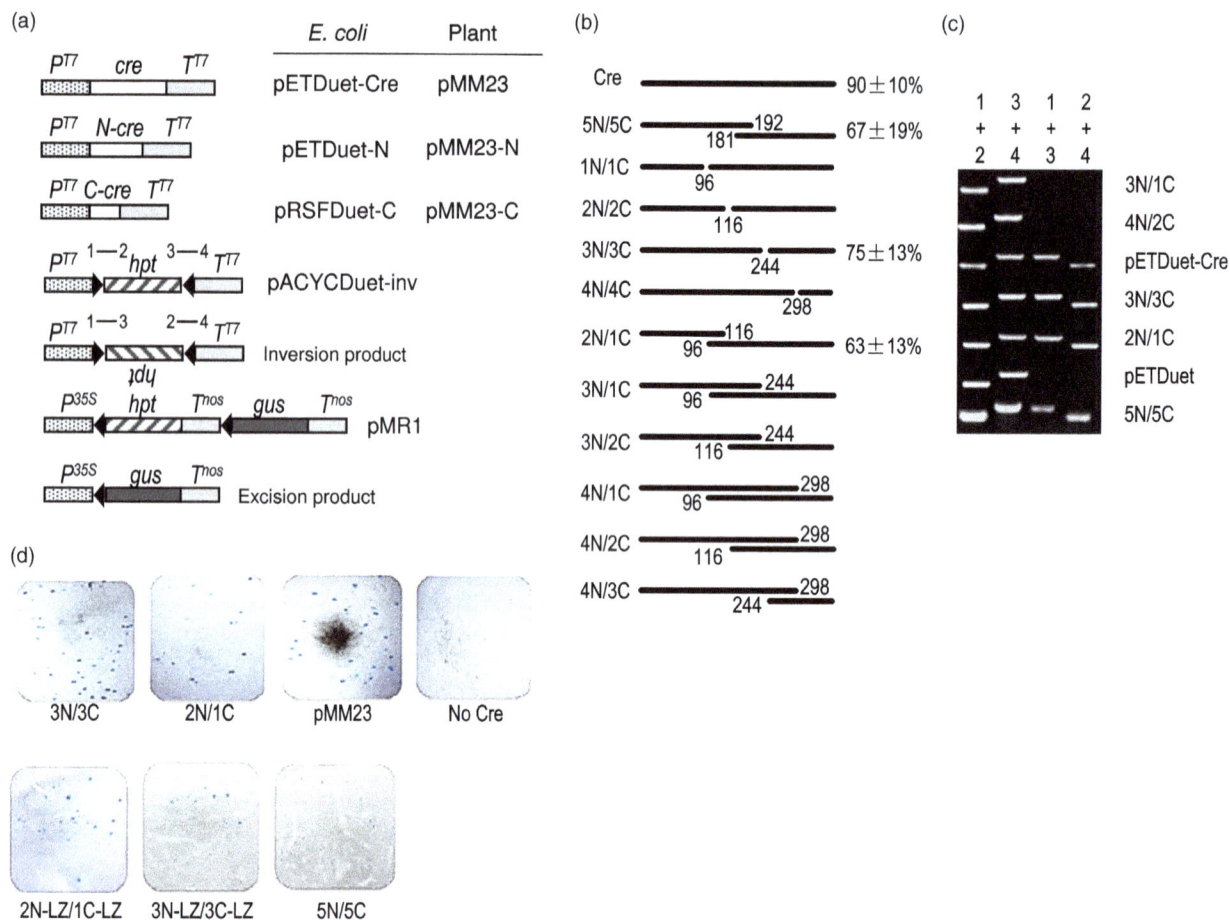

Figure 3 Reconstitution of Cre activity among split-Cre pairs in bacterial and plant cells. (a) Not to scale depiction of DNA constructs used for *E. coli* and plant cells, only relevant DNA segments shown. *N-cre* and *C-cre* encode N- and C-terminal fragments of Cre, respectively. P^{T7} = T7 phage promoter; P^{35S} = CaMV 35S RNA promoter; T^{T7} = T7 phage transcription terminator; T^{nos} = nos terminator; *hpt* = hygromycin phosphotransferase gene; *gus*=beta-glucuronidase gene; numbers 1–4 refer to PCR primers. Genes transcribed left to right except for *hpt* in the inversion product where upside-down lettering indicates transcription from right to left. (b) Depiction of 10 split-Cre pairs tested along with wild-type Cre and the 5N/5C positive control. Negative control not shown. % = % of *E. coli* colonies found to show Cre-mediated inversion (mean ± SD of three independent experiments; 16 colonies tested per experiments). 0% not shown. (c) Representative PCR analysis of the presence of the 1 + 3 and 2 + 4 PCR products indicative of site-specific inversion, 3N/1C and 4N/2C are representative pairs that failed to show Cre activity. (d) GUS histochemical staining of bombarded onion epidermis. Blue spots show GUS activity to indicate formation of excision product from site-specific excision of *hpt-T^{nos}* blocking DNA . LZ = leucine zipper of Max and Myc.

For the pairs corresponding to locations 1, 2, 3, 4, only 3N/3C (aa1-244/aa245-343) scored positive for reconstituted Cre activity. This may not be surprising as locations 1, 2 and 4 disrupt alpha-helical segments. However, among the six overlapping pairings of N- and C-fragments, reconstituted Cre activity was found in 2N/1C (aa1-116/aa96-343). It is interesting to note that 2N includes alpha-helix D, while 1C includes alpha-helix E, and the overlap provides a full set of alpha-helical segments. However, the same logic does not hold for the other overlapping pairs, 3N/1C, 3N/2C, 4N/1C, 4N/2C and 4N/3C. Given the positive results of 3N/3C and 2N/1C, we advanced them to the next step of testing for Cre fragment complementation in plant cells.

Reconstitution of Cre activity in plant transient assays

For scoring site-specific recombination in plant cells, a reporter construct was initially used in which the *gus* (β-glucuronidase) coding region is prevented from expression by an upstream blocking DNA that is itself flanked by a set of directly oriented

lox sites (pMR1, Figure 3a). The *hpt* coding region along with a *nos* terminator (*polyA*) region was used as the blocking DNA. Cre-mediated excision of the blocking DNA permits *gus* expression from the CaMV 35S RNA promoter (P^{35S}) in one of the excision products, and β-glucuronidase activity can be visually detected by blue staining. N- and C-fragments, and the full-length *cre* gene, were also expressed in separate constructs transcribed by P^{35S} (pMM23 series constructs, Figure 3a). Microparticle bombardment (biolistics) was used to deliver the N- and C-fragment-expressing constructs along with the reporter construct into onion epidermal cells. Blue spots were visible 12 hours after bombardment with both the 3N/3C and 2N/1C pairs, with 3N/3C nearly as efficient for recombination as the wild-type Cre control. Surprisingly, our positive split-Cre control pair 5N/5C was less effective. To examine whether heterodimerizing peptides could further enhance their reconstitution of activity, leucine zippers of transcription factors were added to both 3N/3C and 2N/1C, with *Myc* to the C-terminus of the N-Cre fragment and *Max* to the N-terminus of the C-

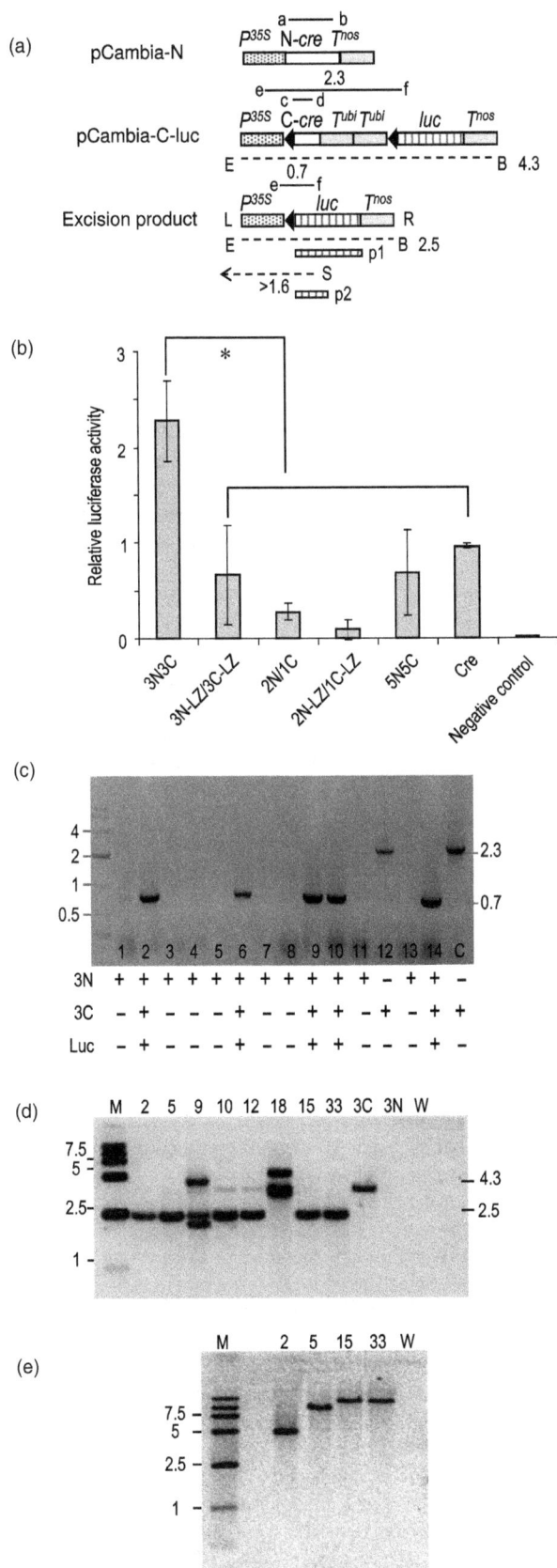

Figure 4 Reconstitution of Cre activity for recombination of plasmid and chromosomal DNA. (a) Not to scale depiction of constructs used, only relevant DNA segments shown. Genetic elements as in Figure 3(a). P^{35S} = CaMV 35S RNA promoter for plants; T^{nos} = nos terminator; T^{ubi} = ubiquitin gene terminator; luc = firefly luciferase gene; a to f = primers used. Dashed lines indicate extent of DNA hybridizing to probes p1 and p2 when cleaved with EcoRI (E) plus BstEII (B) or with SphI (S). Numbers refer to length of DNA size in kb. (b) Transient expression in Arabidopsis protoplasts. Values are mean ± SD; n = 9. *P < 0.05, Dunnett's t-test. (c) Representative PCR detection of deletion of chromosomal copy of lox-flanked DNA that indicates reconstitution of activity from 3N/3C split-Cre pair. Summary of detection of 3N- and 3C-specific PCR products and luciferase activity shown below gel. Primers a+b detects N-fragment, c + d detects C-fragment, e + f detects excision junction. Leftmost lane shows size markers in kb. (d) Southern blot of eight F1 plant lines. DNA cleaved with EcoRI (E) plus BstEII (B) and hybridized to probe p1. M = size markers in kb. 3C = pCambia-3C-luc line, 3N = pCambia-3N line, W = wild-type control. (e) Southern blot of F1 plant DNA cleaved with SphI (S) and hybridized to probe p2. M = size markers in kb. W = wild-type control.

Although the visual assay concluded that the 3N/3C and 2N/1C pairs work, a second test was conducted in Arabidopsis protoplasts to assess their relative recombination proficiency. Several modifications were made to the excision assay. First, for ease in quantitation, we switched to using luc (firefly luciferase gene) as a reporter. Second, in preparation of further testing in transgenic plants in which recombination activity in the nucleus is scored, an SV40 NLS was added to the N-terminus of the N-fragment and to the C-terminus of the C-fragment. Third, to reduce the cotransfer of constructs from 3 to 2, the corresponding C-cre fragment of each pair followed by two rice ubi1 terminators (T^{ubi1}) was used as the blocking DNA (pCambia-C-luc, Figure 4a). For the control experiment with the wild-type Cre, the luc reporter construct uses as the blocking DNA a bar (basta resistance) gene instead of C-cre (construct not shown).

Twenty hours after polyethylene glycol (PEG)-mediated uptake of DNA into Arabidopsis protoplasts, the cells were assayed for luc expression as an indication of site-specific excision of the blocking DNA. Compared to wild-type Cre, 2N/1C and 5N/5C showed 27% and 69% activity, respectively. However, 3N/3C was about 2.5-fold more active than wild-type Cre (Figure 4b). As in the β-glucuronidase assay in onion cells, the leucine zipper additions to 3N/3C and 2N/1C reduced rather than increased recombination.

Cre-fragment complementation in transgenic Arabidopsis

Having shown that the NLS-linked 3N/3C pair is the most active in the two transient assays, we sought to test for recombination of chromosomal DNA. The constructs pCambia-3N and pCambia-3C-luc were separately transformed into Arabidopsis. Putative single-copy lines were screened in T1 plants by qPCR, followed by testing for a 3:1 segregation of the transgene among T2 seedlings. These two assays narrowed the number of putative single-copy lines to 6 for pCambia-3N and 12 for pCambia-3C-luc. From these 18 lines, a total of 27 crosses were conducted between pCambia-3N and pCambia-3C-luc plants. The F1 plants from these crosses were PCR genotyped to determine the plants that harboured deletion of the C-cre-T^{ubi1}-T^{ubi1} fragment, which would fuse P^{35S} to luc. PCR primer pair e+f (Figure 4a) should

Cre fragment. However, neither leucine zipper-containing pairs were found to be as effective as the nonleucine zipper progenitors.

yield either a 2.3-kb product prior to excision or a 0.7-kb band after excision. As shown in Figure 4c, detection of the 0.7-kb PCR product was found in a representative sampling of F1 plants. Moreover, this correlated with expression of *luc*. All 27 independent crosses yielded progeny that showed reconstitution of Cre activity as defined by the PCR and luciferase assays. Both pCambia-3N and pCambia-3C-luc constructs, as determined by PCR pairs a + b and c + d, respectively, were detected in the F1 plants that showed excision. In contrast, plants that failed to show excision lacked either pCambia-3N or pCambia-3C-luc DNA in their genomes (Figure 4c).

Ten representative recombination-junction PCR products from the progenies of the 27 independent crosses were sequenced, and all showed precise site-specific recombination. Additionally, Southern blots were conducted on the F1 and F1 backcrossed (BC1) plants to rule out the possibility that the PCR data had arisen from PCR-mediated recombination of templates with a common *lox* site. *Eco*RI and *Bst*EII cleave just inside the border of the transgene and should yield a *luc*-hybridization band of ~4.3 kb before recombination and ~2.5 kb after recombination (Figure 4a). As shown in a Southern blot (Figure 4d) on plants previously found by PCR to have undergone excision, some lines (lines 9, 18) show multiple hybridizing bands to the *luc* probe p1. This suggests that their genomes harbour other imperfectly integrated T-DNA copies (Figure 4d). In some other lines (lines 10, 12), a faint before-excision band (~4.3 kb) was also detected in addition to the excision-specific product, suggesting that the excision was not complete. However, in the remaining four lines (lines 2, 5, 15 and 33), only a single ~2.5-kb excision-specific band was detected to indicate a homogenous recombination event. To rule out the possibility that the internal *Eco*RI-*Bst*EII fragment may represent the excision of more than a single pCambia-C-luc derived T-DNA, the genomic DNA of lines 2, 5, 15 and 33 were cleaved with *Sph*I and hybridized to *luc* probe p2. As *Sph*I cleaves only once within the T-DNA, p2 should detect the T-DNA left border band. As shown in Figure 4e, a single border band >1.6 kb was indeed detected in these four lines, confirming that each line harbours only a single T-DNA copy. However, it was surprising to find that lines 15 and 33 each showed a band of approximate the same size, thereby raising the possibility that they may be clonal due to a mix-up of the seeds. If that were the case, then we can only state that three of seven lines (rather than four of eight lines) examined showed efficient deletion of a single-copy T-DNA.

Testing transmission of the excision event

Given that the data show efficient recombination in somatic cells, we sought to test whether the recombination event could transmit to the next generation. For the F1 plants that showed excision, the N-fragment locus (*3N*) and C-fragment locus (*3C*) should be hemizygous (*NnCc*, lower case lettering indicating the absence of transgene) and four gamete types should be possible: *NC*, *Nc*, *nC* and *nc*. In an outcross to wild-type (*nc*), this would yield progeny with the following genotypes: *NnCc*, *Nncc*, *nnCc* and *nncc*. If recombination took place in germ-line cells, then we would expect to recover plants with the *nnCc* genotype but with an excision event at the *C* locus. Progenies from 16 of 27 F1 to wild-type backcrosses were randomly chosen for PCR detection of the excision event (~0.7 kb e + f PCR product) as well as for the *3N* and *3C* fragments (a + b and c + d PCR products, respectively). However, among 50 progeny from each backcross that were positive for the excision event, all

harboured the *3N* locus (data not shown). Hence, we conclude that the plants that harbour the excision configuration are genotypically *NnCc* and that the co-assortment of both the *3N* and *3C* loci permitted another round of reconstituted Cre-mediated site-specific recombination in the somatic cells of the backcross progenies.

Discussion

In our recombinase-mediated gene stacking scheme, Cre is used to delete DNA no longer needed after each site-specific integrations step, such as selectable markers and plasmid backbones. Cre-mediated excision of *lox*-flanked DNA is generally efficient, but depends on the *cre*-expression donor line used in the genetic cross. In recent work, we found that a *cre* line used previously was less efficient than we had expected, resulting in all F1 plants chimeric with a mixture of cells with or without recombination (Hou *et al.*, 2014). We had suspected that the *cre* line might have become less effective over time. Although gene silencing over the generations could have occurred with any transgene, there is some reason to suspect that high expression of a DNA-binding recombinase could interfere with normal chromosome activities. Hence, we sought to create *cre* lines with controllable *cre* DNA activity. Chemical induction (Joubes *et al.*, 2004; Zuo *et al.*, 2001) and heat induction of *cre* expression have been described (Cuellar *et al.*, 2006; Hoff *et al.*, 2001; Khattri *et al.*, 2011; Wang *et al.*, 2005; Zhang *et al.*, 2003). However, we favour using a split-*cre* system as it should be possible, as illustrated in Figure 1d, to flank the *M3-N-cre* DNA with directly oriented *lox* sites. With an inducible *cre* system, keeping a *cre* line over the generations might be difficult, as any leaky expression of *cre* would excise itself.

What prompted us to find a new split-Cre pair instead of using those described in the literature is due to our desire to keep the recombinase-mediated gene stacking system as an open-source system with freedom to operate (Chen and Ow, 2017; Ow, 2016). The split-intein patent (Yadav and Yang, 2007) has broad claims that are difficult to invent around. For the Cre fragment complementation approach, at least three applications were found that claim specific locations by which Cre can be split into two (Herman and Jullien, 2004; filed 2002; Gu and Xu, 2009; filed 2009; Gao *et al.*, 2015; filed 2013). At least one of these applications has since been issued (Gao *et al.*, 2015).

Under these circumstances, we undertook the task to a search for a new location for Cre fragment complementation. The positive outcome from this research is that we indeed found a new split-Cre pair that can reconstitute activity that is comparable to the wild-type Cre control. Unlike other studies that relied exclusively on scoring for recombination through reporter gene expression, or by PCR, both of which cannot detect the percentage of substrates that have not undergone recombination, we followed up our initial analysis with a Southern analysis which showed that from eight randomly selected progenies, the excision-specific hybridizing band was found in seven of them, and with only faint or undetectable hybridization for the band representing the lack of excision. This shows that the Cre pair brought together by hybridization was highly effective at the F1 generation.

Yet, despite the high efficiency of recombination in somatic cells, the one surprising outcome of this study was that backcrossed progenies that showed excision invariably harboured both the *N-cre* and *C-cre* loci. This suggests that the excision events were generated *de novo* from the reconstitution of *N-cre* and *C-cre* in the backcrossed progeny generation. Both *N-cre* and *C-cre* genes were transcribed by the CaMV 35S RNA promoter and this promoter has previously been used to express *cre* to cause excision that can be transmitted through the germ-line, although the efficiency varies depending on the particular *cre* donor line. It remains possible that a lower level of expression was caused by having both *N-cre* and *C-cre* transcribed by the same promoter. Whatever the reason, it remains a next engineering challenge to test different promoters, including germ-line-specific promoters for more effective transmission of the recombination event (Li *et al.*, 2007; Mlynarova *et al.*, 2006; Van Ex *et al.*, 2009; Verweire *et al.*, 2007). As we have now progressed to implementing the recombinase-mediated gene stacking system in rice (Li *et al.*, 2016), future testing of these germ-line-specific promoters will be conducted on this crop.

Experimental procedure

DNA constructs

Standard recombinant DNA methods were used throughout (Sambrook and Russell, 2001). Primer and DNA linker sequences shown in Table S1.

pETDuet-Cre: The *cre* gene was PCR amplified (Biolabs Phusion® High-Fidelity DNA Polymerase) from pMM23 (Qin *et al.*, 1994) using primers N-F and C-R and inserted between *Eco*RI and *Kpn*I restriction sites of pETDuet (pETDuet™-1, Novagen).

pETDuet-N series constructs: 1*N-cre* fragment was PCR amplified from pMM23 using primers N-F and 1N-R, 2*N-cre* used primers N-F and 2N-R, 3*N-cre* with primers N-F and 3N-R, 4*N-cre* with primers N-F and 4N-R, and 5*N-cre* with primers N-F and 5N-R. Each of the *N-cre* DNA was inserted between *Eco*RI and *Avr*II restriction sites of pETDuet (pETDuet™-1, Novagen).

pRSFDuet-C series constructs: 1*C-cre* fragment was PCR amplified from pMM23 using primers 1C-F and C-R, 2*C-cre* fragment amplified using primers 2C-F and C-R, 3*C-cre* fragment with primers 3C-F and C-R, 4*C-cre* fragment with primers 4C-F and C-R, and 5*C-cre* fragment with primers 5C-F and C-R. Each of the *C-cre* fragments was inserted between *Eco*RI and *Kpn*I restriction sites of pACYCD (pACYCDuet™-1, Novagen).

pACYCDuet-in: The *hpt* gene was PCR amplified from pZH36 (ZG Han, Ow lab, unpublished) using primers lhl-F and lhl-R with overhanging *lox* sites to create the *lox-hpt-*(inverted *lox*), which was then inserted between the *Eco*RI and *Avr*II restriction sites of pACYCD (pACYCDuet™-1, Novagen).

pMM23-N series and pMM23-C series constructs : 2*N-cre*, 3*N-cre*, 5*N-cre*, 1*C-cre*, 3*C-cre* and 5*C-cre* fragments were PCR amplified with same or analogous primers for the pETDuet-N and pETDuet-C series plasmids but with *Kpn*I and *Sph*I restriction sites (N-F1, 2N-R1, 3N-R1, 5N-R1 for *N-cre* fragments and 1C-F1, 3C-F1, 5C-F1 and C-R1 for *C-cre* fragments) for inserting between P^{35S} (*Kpn*I) and T^{nos} (*Sph*I) of pMM23.

pMM23-NLZ series constructs: 2N-LZ *cre* and 3N-LZ-*cre* fragments were PCR amplified from pMM23 using primers N-F2 and 2NLZ-R or 3NLZ-R for insertion into the intermediate vector pMD18-T (Takara) in between *Kpn*I and *Xho*I. Afterwards, a *Myc* (Ayer and Eisenman, 1993) linker (Beijing AUGCT Biotechnology Co., Ltd.) was inserted between the *Xho*I site and the vector-derived *Sph*I site of pMD-18T (C-terminus of the *N-Cre* fragment) to make 2NLZ and 3NLZ. Finally, 2NLZ and 3NLZ fragments were retrieved by cleavage with *Kpn*I and *Sph*I and inserted between P^{35S} (*Kpn*I) and T^{nos} (*Sph*I) of pMM23.

pMM23-CLZ series constructs: 1C-LZ-*cre* and 3C-LZ-*cre* fragments were PCR amplified from pMM23 using primers C-R2, 1CLZ-F or 3CLZ-F for insertion between the *Eco*RI and *Sph*I sites of pMD18-T (Takara). Afterwards, a *Max* (Ayer and Eisenman, 1993) linker (Beijing AUGCT Biotechnology Co., Ltd.) was inserted between the *Kpn*I and *Eco*RI sites of pMD-18T (N-terminus of the *C-cre* fragment) to make 1CLZ and 3CLZ. Finally, 1C-LZ-*cre* and 3C-LZ-*cre* fragments were retrieved by cleavage with *Kpn*I and *Sph*I and inserted between P^{35S} (*Kpn*I) and T^{nos} (*Sph*I) of pMM23.

pMR1: An *Xho*I-*gus*-*Sph*I fragment was PCR amplified from pZH36 (ZG Han, Ow lab, unpublished) using primers Gus-F and Gus-R for insertion into a T-vector (Takara) between the corresponding sites. An *hpt* fragment (with *nos* terminator) was also PCR amplified from pZH36, using primers lhl-F and lhl-R with overhanging *lox* sites to create a *Kpn*I-*lox*-*hpt*-*lox*-*Xho*I fragment for insertion upstream of the *gus* gene between . The *lox*-*hpt*-*lox*-*gus* fragment was then retrieved by cleavage with *Kpn*I and *Sph*I for insertion between P^{35S} (*Kpn*I) and T^{nos} (*Sph*I) of pMM23.

pCambia-N series constructs: 2*N-cre*, 3*N-cre*, 5*N-cre*, were PCR amplified from pMM23 with Nnls-F and 2N-R2, 3N-R2 or 5N-R2 primers (Nnls-F included a Kozak sequence, ATG and a SV40 NLS). 2N-LZ and 3N-LZ fragments were PCR amplified from pMM23-NLZ series constructs with Nnls-F and Nlz-R. Each fragment was inserted between P^{35} (*Bgl*II) and T^{nos} (*Bst*EII) of pCAMBIA1301 (http://www.cambia.org).

pCambia-C-luc series constructs: *C-cre* fragments were PCR amplified from the pMM23-C series constructs with primers C-R3 and either 1Cnls-F, 3Cnls-F or CLZnls-F that incorporated a *Kpn*I site, a Kozak sequence, *lox*, ATG and a SV40 NLS to form a *Kpn*I-*lox*-*C-cre*-*Nde*I fragment into pMD-18T (Takara). A first T^{ubi1} fragment was PCR amplified from pZH36 using primers T^{ubi1}-F with and T^{ubi1}-R and inserted downstream of *lox*-*C-cre* between *Nde*I and *Pst*I sites. A second T^{ubi1} fragment was created by PCR primers T^{ubi1}-F2 and T^{ubi1}-lox-R to create a *Pst*I-T^{ubi1}-*lox*-*Bgl*II fragment to insert behind the first T^{ubi1} to create *lox*-*C-cre*-$T^{ubi1}T^{ubi1}$*lox* linkage within pMD-18T, upon which the *Kpn*I-*Bgl*II fragment was transferred to pCambia1301. Finally, a *Bgl*II-*luc*-*Bst*EII fragment amplified by primers luc-F and luc-R from pYWP72 (Yau *et al.*, 2011) was inserted in place of the *gus* gene in pCambia1301.

pCambia-bar-luc: a *Kpn*I-*lox*-*bar*-*Nde*I fragment was PCR amplified from pZH210B (ZG Han, Ow lab, unpublished) using primers Bar-F and Bar-R to replace the *Kpn*I-*lox*-*C-cre*-*Nde*I fragment in pCambia-C-luc.

pCambia-Cre: A *Bgl*II-*cre*-*Bst*EII fragment was PCR amplified from pMM23 using Nnls-F and C-R4 to insert between P^{35S} (*Bgl*II) and T^{nos} (*Bst*EII) of pCAMBIA1301.

E. coli assay

E. coli DH5α (F- *endA1 glnV44 thi-1 recA1 relA1 gyrA96 deoR nupG Φ80dlacZΔM15 Δ(lacZYA-argF)U169 hsdR17(r_K^- m_K^+), λ–*) (Grant *et al.*, 1990) was used throughout for recovery of recombinant molecules. For the experiment shown in Figure 3b, 100 µL competent cells were transformed with 100 ng of each plasmid construct, and resistant colonies were scored on plates containing 34 mg/L chloramphenicol, 100 mg/L ampicillin and if needed 50 mg/L kanamycin.

Transient expression in onion epidermal cells

DNA was purified using Tiangene® High Pure Maxi Plasmid Kit. Onion inner epidermis was placed onto MS medium plates (supplemented by 200 μM of D-sorbitol and 200 μM D-mannitol) and subjected to microparticle bombardment as described (Altpeter *et al.*, 1996). Afterwards, the plates were incubated in the dark for 14–20 h and the onion epidermis stained for the Gus activity (Jefferson *et al.*, 1987).

Transient expression in *Arabidopsis* protoplasts

Protoplasts were isolated from 10 to 20 leaves of 3- to 4-week-old *A. thaliana* (cv. Columbia) plants before flowering and transformed by PEG as described (Yoo *et al.*, 2007), using for each sample ~1.5 × 10^5 protoplasts (200 μL volume) and 10 μg of each DNA construct (Tiangene® purified). Luciferase activity was assayed by the DualLuciferase Reporter Assay System (Promega) using a GloMax Multi JR detection Luminometer (Promega) on overnight resting protoplasts and normalized to total protein (Bradford assay kit, Thermo Scientific®).

Transgenic *Arabidopsis*

The constructs shown in Figure 4a were introduced into *Agrobacterium* strain *GV3101* through floral-dip transformation of *A. thaliana* (cv. Columbia) plants as described (Clough and Bent, 1998). T1 hygromycin-resistant transformants were selected on an MS medium with 40 μg/L hygromycin. For *N-cre* plants that lacked a cotransformed reporter gene, *N-cre* expression was checked by quantitative real-time PCR using F: 5′ ATTGGCAGAACGA AAACGCT 3′ and R: 5′-ATCAGCTACACCAGAGACGG-3′ primers. Total RNA was extracted using Hipure plant RNA mini kit (MAGEN). Reverse transcription was conducted using a PrimeScript™ RT reagent Kit with gDNA Eraser (TAKARA).

Copy number estimation by quantitative real-time PCR

Total DNA was extracted using HiPure Plant DNA Midi Kit (Magen, D3162) according to the manufacturer's recommendation. DNA was diluted to 100 μg/μL by ddH$_2$O, and 3 μL was used for PCR amplification. Quantitative RT-PCR reactions were performed in 384-well blocks using the Go Taq® qPCR Master Mix (Promega, A6001). The *hpt* gene was measured against *AtOXS1* used as an internal positive single-copy gene control. Primers use for *hpt*: 5′-TCGTCCATCACAGTTTGCC-3′ and 5′-TC GGTCAATACACTACATGGC-3′; for AtOXS1: 5′-ACTGTGTCAGA TAACCTGCCCGTTG-3′ and 5′-GGTTTCTCAGACTTGAGCCTT GGAA-3′. PCR reactions were 28 cycles, 95 °C 30 s, 60 °C 30 s and 72 °C 30 s. Relative quantification for copy number of T-DNA insertion was calculated by dividing the Cp values of *hpt* to Cp values of *AtOxs1*.

Southern hybridization

Southern blots were performed as described previously (Hou *et al.*, 2014). DNA was further purified through NucleoBond AX100 columns (Genopure plasmid MiDi kit, Roche, Germany) and concentration determined by a 2000c spectrophotometer (Thermo scientific, Wilmington). ^{32}P-probed membranes were exposed to a phosphor screen for 24 h and detected with Typhoon FLA 9500 (IP: 635 nm, PMT: 500 V, Pixel size 200 μm).

Acknowledgements

Funding was provided by Guangdong Province, China Talent Funds 2010 and Chinese Ministry of Agriculture Grant 2010ZX08010-001. The authors thank members of this laboratory for unpublished constructs (Z. Han) and technical advice (C. Wang, M. Liang). The authors declare no conflict of interests.

References

Altpeter, F., Vasil, V., Srivastava, V., Stoger, E. and Vasil, I.K. (1996) Accelerated production of transgenic wheat (Triticum aestivum L.) plants. *Plant Cell Rep.* **16**, 12–17.

Ayer, D.E. and Eisenman, R.N. (1993) A switch from Myc-Max to Mad-Max heterocomplexes accompanies monocyte/macrophage differentiation. *Genes Devel.* **7**, 2110–2119.

Casanova, E., Lemberger, T., Fehsenfeld, S., Mantamadiotis, T. and Schutz, G. (2003) Alpha complementation in the Cre recombinase enzyme. *Genesis*, **37**, 25–29.

Chen, W. and Ow, D.W. (2017) Precise, flexible and affordable gene stacking for crop improvement. *Bioengineered*, (in press) https://doi.org/10.1080/21655979.2016.1276679.

Clough, S.J. and Bent, A.F. (1998) Floral dip: a simplified method for Agrobacterium-mediated transformation of Arabidopsis thaliana. *Plant J.* **16**, 735–743.

Coppoolse, E.R., de Vroomen, M.J., Roelofs, D., Smit, J., van Gennip, F., Hersmus, B.J.M., John, H., Nijkamp, J. and van Haaren, M.J.J. (2003) Cre recombinase expression can result in phenotypic aberrations in plants. *Plant Mol. Biol.* **51**, 263–279.

Cuellar, W., Gaudin., A, Solorzano, D., Casas, A., Nopo, L., Chudalayandi, P., Medrano, G. et al. (2006) Self-excision of the antibiotic resistance gene nptII using a heat inducible Cre-loxP system from transgenic potato. *Plant Mol. Biol.* **62**, 71–82.

Gao, Y. K., Luo, K., Wen, M., Wang, L. and Li, C. (2015) *Recombinase Cre modification method, and application of modified recombinase Cre in plants.* China Patent Application 201310052609.9, filed Feb. 18, 2013, issued July 1, 2015.

Ge, J., Wang, L., Yang, C., Ran, L., Wen, M., Fu, X., Fan, D. et al. (2016) Intein-mediated Cre protein assembly for transgene excision in hybrid progeny of transgenic Arabidopsis. *Plant Cell Rep.* **35**, 2045–2053.

Grant, S.G.N., Jessee, J., Bloom, F.R. and Hanahan, D. (1990) Differential plasmid rescue from transgenic mouse DNAS into Escherichia-coli methylation-restriction mutants. *Proc. Natl Acad. Sci. USA*, **87**, 4645–4649.

Gu, G. and Xu, Y. (2009) *Cre complementation using leucine zipper.* U.S. Patent Application 20090291504. Filed May 20, 2008.

Guo, F., Gopaul, D.N. and van Duyne, G.D. (1997) Structure of Cre recombinase complexed with DNA in a site-specific recombination synapse. *Nature*, **389**, 40–46.

Han, X.Z., Han, F.Y., Ren, X.S., Si, J., Li, C.Q. and Song, H.Y. (2013) *Ssp* DnaE split-intein mediated split-Cre reconstitution in tobacco. *Plant Cell Tiss. Org.* **113**, 529–542.

Herman, J.-P. and Jullien, N.C.G. (2004) *Regulation of Cre recombinase using a dissociation/reassociation system of said recombinase.* EU Patent Application EP 1 413 586 A1. Filed Oct 21, 2002.

Hirrlinger, J., Scheller, A., Hirrlinger, P.G., Kellert, B., Tang, W.N., Wehr, M.C., Goebbels, S. et al. (2009) Split-Cre complementation indicates coincident activity of different genes In Vivo. *PLoS ONE*, **4**, e4286.

Hoff, T., Schnorr, K.M. and Mundy, J. (2001) A recombinase-mediated transcriptional induction system in transgenic plants. *Plant Mol. Biol.* **45**, 41–49.

Hou, L., Yau, Y.-Y., Wei, J., Han, Z., Dong, Z. and Ow, D.W. (2014) An open-source system for in planta gene stacking by bxb1 and cre recombinases. *Molec. Plant*, **7**, 1756–1765.

Jefferson, R.A., Kavanagh, T.A. and Bevan, M.W. (1987) GUS fusions: beta-glucuronidase as a sensitive and versatile gene fusion marker in higher plants. *EMBO J.* **6**, 3901–3907.

Joubes, J., De Schutter, K., Verkest, A., Inze, D. and De Veylder, L. (2004) Conditional, recombinase-mediated expression of genes in plant cell cultures. *Plant J.* **37**, 889–896.

Jullien, N., Sampieri, F., Enjalbert, A. and Herman, J.P. (2003) Regulation of Cre recombinase by ligand-induced complementation of inactive fragments. *Nucl. Acids Res.* **31**, e131.

Khattri, A., Nandy, S. and Srivastava, V. (2011) Heat-inducible Cre-*lox* system for marker excision in transgenic rice. *J. Biosci.* **36**, 37–42.

Li, Z., Xing, A., Moon, B.P., Burgoyne, S.A., Guida, A.D., Liang, H., Lee, C. *et al.* (2007) A Cre/*loxP* mediated self-activating gene excision system to produce marker gene free transgenic soybean plants. *Plant Mol. Biol.* **65**, 329–341.

Li, R., Han, Z., Hou, L., Kaur, G., Qian, Y. and Ow, D.W. (2016) Method for biolistic site-specific integration in rice and tobacco catalyzed by Bxb1 integrase. In: *Methods Mol Biol: Chromosome and Genomic Engineering in Plants*, vol. **1469** (Murata, M., ed), pp. 15–30. New York: Humana Press.

Maruo, T., Ebihara, T., Satou, E., Kondo, S. and Okabe, S. (2008) Cre complementation with variable dimerizers for inducible expression in neurons. *Neurosci. Res.* **61**, S279.

Mlynarova, L., Conner, A.J. and Nap, J.P. (2006) Directed microspore-specific recombination of transgenic alleles to prevent pollen-mediated transmission of transgenes. *Plant Biotechnol. J.* **4**, 445–452.

Ow, D.W. (2016) The long road to recombinase-mediated plant transformation. *Plant Biotechnol. J.* **14**, 441–117.

Que, Q., Wang, H.-Y. and Jorgensen, R. (1998) Distinct patterns of pigment suppression are produced by allelic sense and antisense chalcone synthase transgenes in petunia flowers. *Plant J.* **13**, 401–409.

Qin, M.M., Bayley, C., Stockton, T. and Ow, D.W. (1994) Cre recombinase-mediated site-specific recombination between plant chromosomes. *Proc. Natl Acad. Sci. USA*, **91**, 1706–1710.

Sambrook, J. and Russell, D.W. (2001) *Molecular Cloning. A Laboratory Manual*. Cold Spring Harbor: Cold Spring Harbor Laboratory Press.

Seidi, A., Mie, M. and Kobatake, E. (2007) Novel recombination system using Cre recombinase alpha complementation. *Biotechnol. Lett.* **29**, 1315–1322.

Seidi, A., Mie, M. and Kobatake, E. (2009) Recombination system based on Cre alpha complementation and leucine zipper fusions. *Appl. Biochem. Biotech.* **158**, 334–342.

Van Ex, F., Verweire, D., Claeys, M., Depicker, A. and Angenon, G. (2009) Evaluation of seven promoters to achieve germline directed Cre-*lox* recombination in *Arabidopsis thaliana. Plant Cell Rep.* **28**, 1509–1520.

Verweire, D., Verleyen, K., De Buck, S., Claeys, M. and Angenon, G. (2007) Marker-free transgenic plants through genetically programmed auto-excision. *Plant Physiol.* **145**, 1220–1231.

Wang, Y., Chen, B.J., Hu, Y.L., Li, J.F. and Lin, Z.P. (2005) Inducible excision of selectable marker gene from transgenic plants by the Cre/lox site-specific recombination system. *Transgenic Res.* **14**, 605–614.

Wang, P., Chen, T.R., Sakurai, K., Han, B.X., He, Z.G., Feng, G.P. and Wang, F. (2012) Intersectional Cre driver lines generated using split-intein mediated split-cre reconstitution. *Sci. Rep.* **2**, 497.

Wen, M.L., Gao, Y., Wang, L.J., Ran, L.Y., Li, J.H. and Luo, K.M. (2014) Split-Cre complementation restores combination activity on transgene excision in hair roots of transgenic tobacco. *PLoS ONE*, **9**, e110290.

Xu, Y.W., Xu, G., Liu, B.D. and Gu, G.Q. (2007) Cre reconstitution allows for DNA recombination selectively in dual-marker-expressing cells in transgenic mice. *Nucl. Acids Res.* **35**, e126.

Yadav, N. and Yang, J. (2007) *Method of controlling site-specific recombination*. U.S. Patent 7,238,854, filed Jan. 29, 2003, issued July 3, 2007.

Yau, Y.Y., Wang, Y., Thomson, J.G. and Ow, D.W. (2011) Method for Bxb1-mediated site-specific integration *in planta. Methods Mol. Biol.* **701**, 147–166.

Yoo, S.D., Cho, Y.H. and Sheen, J. (2007) *Arabidopsis* mesophyll protoplasts: a versatile cell system for transient gene expression analysis. *Nat. Protocols*, **2**, 1565–1572.

Zhang, W., Subbarao, S., Addae, P., Shen, A., Armstrong, C., Peschke, V. and Gilbertson, L. (2003) Cre/lox-mediated marker gene excision in transgenic maize (Zea mays L.) plants. *Theor. Appl. Genet.* **107**, 1157–1168.

Zuo, J.R., Niu, Q.W., Moller, S.G. and Chua, N.H. (2001) Chemical-regulated, site-specific DNA excision in transgenic plants. *Nat. Biotechnol.* **19**, 157–161.

8

High-yield secretion of recombinant proteins from the microalga *Chlamydomonas reinhardtii*

Erick Miguel Ramos-Martinez, Lorenzo Fimognari and Yumiko Sakuragi* (iD)

Department of Plant and Environmental Sciences, Copenhagen Plant Science Centre, University of Copenhagen, Frederiksberg C, Copenhagen, Denmark

*Correspondence
email ysa@plen.ku.dk

Summary

Microalga-based biomanufacturing of recombinant proteins is attracting growing attention due to its advantages in safety, metabolic diversity, scalability and sustainability. Secretion of recombinant proteins can accelerate the use of microalgal platforms by allowing post-translational modifications and easy recovery of products from the culture media. However, currently, the yields of secreted recombinant proteins are low, which hampers the commercial application of this strategy. This study aimed at expanding the genetic tools for enhancing secretion of recombinant proteins in *Chlamydomonas reinhardtii*, a widely used green microalga as a model organism and a potential industrial biotechnology platform. We demonstrated that the putative signal sequence from *C. reinhardtii* gametolysin can assist the secretion of the yellow fluorescent protein Venus into the culture media. To increase the secretion yields, Venus was C-terminally fused with synthetic glycomodules comprised of tandem serine (Ser) and proline (Pro) repeats of 10 and 20 units [hereafter $(SP)_n$, wherein n = 10 or 20]. The yields of the $(SP)_n$-fused Venus were higher than Venus without the glycomodule by up to 12-fold, with the maximum yield of 15 mg/L. Moreover, the presence of the glycomodules conferred an enhanced proteolytic protein stability. The Venus-$(SP)_n$ proteins were shown to be glycosylated, and a treatment of the cells with brefeldin A led to a suggestion that glycosylation of the $(SP)_n$ glycomodules starts in the endoplasmic reticulum (ER). Taken together, the results demonstrate the utility of the gametolysin signal sequence and $(SP)_n$ glycomodule to promote a more efficient biomanufacturing of microalgae-based recombinant proteins.

Keywords: *C. reinhardtii*, glycomodule, protein secretion, signal sequence, yellow fluorescent protein.

Introduction

The unicellular green microalga *Chlamydomonas reinhardtii* has a long history as a model organism and has helped in the understanding of fundamental biological processes such as metabolism, photosynthesis, phototaxis, chloroplast biology, circadian rhythmicity, cell cycle and mating (Harris *et al.*, 2009). As one of the best characterized algal species, *C. reinhardtii* has also been developed as a potential expression platform for the production of recombinant proteins with applications in different industries including biomaterials, bioenergy, therapeutics and nutraceuticals (Almaraz-Delgado *et al.*, 2014; Rasala and Mayfield, 2015; Specht *et al.*, 2010). Compared with prokaryotic hosts, recombinant proteins can be expressed from either chloroplast (Purton *et al.*, 2013) or nucleus (Jinkerson and Jonikas, 2015) of *C. reinhardtii*, making this microalga a versatile host. Proteins encoded by the nuclear genome can undergo post-translational modifications (PTMs) and can be targeted to different organelles or the culture media, whereas chloroplast-expressed proteins are retained inside the plastid. Significant efforts are being made to expand a molecular toolbox, allowing an efficient and robust expression of transgenes from the nuclear genomes of microalgae (Jinkerson and Jonikas, 2015; Mussgnug, 2015). This includes the generation of mutants with increased transgene expression (Neupert *et al.*, 2009), codon-optimized synthetic genes (Fuhrmann *et al.*, 2004; Shao and Bock, 2008), chimeric promoters and use of native introns (Eichler-Stahlberg *et al.*, 2009; Schroda *et al.*, 2000), enhanced transgene expression linked to a selection marker (Rasala *et al.*, 2012), insertion of promoterless genes fused to an antibiotic-resistant gene (Díaz-Santos *et al.*, 2016) and recently

the use of synthetic promoters for increasing nuclear gene expression (Scranton *et al.*, 2016). Moreover, a nuclear episomal vector that is capable of stable replication was recently developed for uses in diatoms (Karas *et al.*, 2015).

The secretion of expressed proteins into the medium is an attractive strategy and is widely employed in recombinant protein productions in heterotrophic microbial hosts (Demain and Vaishnav, 2009). In eukaryotes, secretion can ensure proper glycosylation of proteins, which plays important roles in determining the yield, biological activity, stability and half-life of a secreted recombinant protein (Lingg *et al.*, 2012; Mathieu-Rivet *et al.*, 2014). It can also simplify downstream processing and circumvent cost-ineffective and labour-intensive cell lysis (Hellwig *et al.*, 2004; Nikolov and Woodard, 2004). Moreover, the harvested algal biomass can be exploited as a co-product, adding more value to the process (Gangl *et al.*, 2015). Large-scale cultivation of transgenic *C. reinhardtii* in photobioreactors has been demonstrated for both wild-type and cell-wall-deficient strains, paving a path towards industrial exploitation of this microalga (Gimpel *et al.*, 2015; Zedler *et al.*, 2016).

Despite this progress, only a handful of investigations into the secretion of recombinant proteins have been made in microalgae. Thus far, over 30 recombinant proteins have been expressed in *C. reinhardtii*, of which only six have been secreted (Chavez *et al.*, 2016; Eichler-Stahlberg *et al.*, 2009; Lauersen *et al.*, 2013a,b, 2015b; Rasala *et al.*, 2012). Protein secretion is characterized by the presence of a signal sequence that targets a protein to the secretory pathway and ultimately into the culture media. Thus far, four signal sequences have been exploited in fusion with recombinant proteins in *C. reinhardtii*: a non-native signal sequence from *Gaussia princeps* luciferase (Ruecker *et al.*, 2008), native signal sequences from arylsulphatases (ARS1 and ARS2; Eichler-Stahlberg *et al.*, 2009; Rasala *et al.*, 2012) and carbonic anhydrase (CAH1; Lauersen *et al.*, 2013a). In these cases, the yields of recombinant proteins ranged from 100 μg/L to 10 mg/L (Eichler-Stahlberg *et al.*, 2009; Lauersen *et al.*, 2013a; Rasala *et al.*, 2012). Interestingly, other species like the diatom *P. tricornutum* are able to secrete up to 2.5 mg/L of a functional human IgG antibody without a signal peptide (Hempel and Maier, 2012), demonstrating how little we know about the secretion mechanism of photosynthetic microorganism. In general, secretion yields above 10 mg/L are considered the minimum for commercial process development (Hellwig *et al.*, 2004), and higher yields (>1 g/L) are typically obtained using heterotrophic host organisms. Although extensive optimization of growth parameters has been shown to increase secretion efficiency and productivity in *C. reinhardtii*, reaching up to 12 mg/L (Lauersen *et al.*, 2015a), further development is needed to compete with commonly used hosts.

In plant cell cultures, a successful strategy to enhance secretion yields and the stability of recombinant proteins has been demonstrated based on (SP)$_n$ glycomodules, which are *O*-glycosylated with arabinogalactan polysaccharides attached to hydroxylated Pro residues (Hyp) (Shpak *et al.*, 1999; Xu *et al.*, 2007). Briefly, in eukaryotes, many secretory and membrane proteins are glycosylated (Higgins, 2010). This occurs in the secretory pathway starting in ER and becomes elaborated in the Golgi apparatus. Glycosylation serves a variety of structural and functional roles and can be classified into two main categories: *N*-glycans are linked to the amide group of asparagine residues and *O*-glycans are linked to the hydroxyl group of Ser, threonine, hydroxy lysine or Hyp residues. It is well established that glycosylation increases the stability of proteins. The presence of *O*-linked glycans, for instance, strongly enhances the secretion yields and physicochemical properties of a protein by influencing protein folding, solubility, stability and resistance to heat or proteolysis (Gomord *et al.*, 2010; Walsh and Jefferis, 2006). In contrast to higher plants, little is known about the mechanisms of protein glycosylation and effects on protein secretion in microalgae (Mathieu-Rivet *et al.*, 2014).

To increase the yield of recombinant proteins in the culture media, we tested whether the putative signal sequence from the metalloprotease gametolysin can efficiently secrete recombinant proteins into the culture media. Using a yellow fluorescent protein, Venus, as a reporter, we confirmed that the gametolysin signal sequence can indeed secrete recombinant Venus. To enhance the secretion yields, we implemented the (SP)$_n$ glycomodules in *C. reinhardtii*. Venus was expressed as fusion glycoproteins, resulting in up to a 12-fold increase in the yield in culture media and a greater resistance to proteolytic degradation as compared to the untagged Venus. These results support the potential utilization of transgenic microalgae as a platform for secretion of recombinant proteins.

Results

Gametolysin signal sequence targets Venus into the medium

Bioinformatic analysis by SignalP 4.0 (Petersen *et al.*, 2011) revealed that the N-terminal sequence of the metalloprotease gametolysin (Kinoshita *et al.*, 1992; Matsuda *et al.*, 1987) is highly likely to be a signal sequence (D score of 0.88) and to be cleaved between alanine (position 28) and asparagine (position 29). Therefore, the ability of the N-terminal 28 residues to secrete a recombinant protein was tested by fusing it to the N-terminus of Venus. The construct was termed pERC-SSVenus and consisted of two separate expression cassettes, one for the expression of the gene of interest and another for the expression of an antibiotic resistance marker (paromomycin). As a control, a second construct lacking the signal sequence (pERC-Venus) was generated (Figure 1). To ensure high levels of expression, we included in our vector design the following genetic elements: the chimeric promoter HSP70A-RBSC2, RBSC2 intron 1, the RBSC2 intron 2 and the RBSC2 3' untranslated region (UTR) as a transcriptional terminator as previously used by Eichler-Stahlberg *et al.* (2009). The expression cassette was introduced into the nuclear genome of the cell-wall-deficient mutant UVM4, a strain known for enhanced transgene expression levels (Neupert *et al.*, 2009). Integration of the expression cassette was confirmed by colony PCR. Positive cells were screened for either intracellular or extracellular accumulations of Venus by dot blotting analysis using a monoclonal anti-GFP antibody.

To determine whether Venus was effectively secreted into the culture media, the transgenic lines were grown mixotrophically in tris-acetate phosphate (TAP) medium. Western blotting of the cell pellets and the cell-free media were performed using representative lines of the Venus and SSVenus strains generated with the pERC-Venus and pERC-SSVenus vectors, respectively. The parental UVM4 strain was used as a control. In the Venus line, Venus was exclusively found in the cell pellets and as expected had an apparent molecular weight of 27 kDa (Figure 2a). In contrast, only a very low amount of Venus was detected in the cell pellets of the SSVenus line, whereas most Venus was found in the media (Figure 2a). Notably, Venus that accumulated in the SSVenus line had a slightly higher molecular weight, by approximately ~2 kDa, both in the cell pellets and in the media. A similar increase in the molecular weight was previously observed for a fungal xylanase when secreted from *C. reinhardtii* (Rasala *et al.*, 2012). One possible explanation is that the gametolysin signal sequence is not properly cleaved. Another possible explanation is post-translational modification such as glycosylation and phosphorylation (Cui *et al.*, 2015; Mathieu-Rivet *et al.*, 2013).

Live-cell imaging was performed by confocal scanning laser microscopy to visualize the localization of the recombinant Venus proteins in the Venus and SSVenus lines (Figure 2b). Non-transformed cells had no fluorescent signal as expected. The Venus line showed a strong cellular yellow fluorescence signal from active Venus in the cytoplasm confined by the chloroplast, which is consistent with previous reports (Fuhrmann *et al.*, 1999; Plucinak *et al.*, 2015). In contrast, the SSVenus lines showed weak and punctate fluorescent signals in the cytoplasmic space, corroborating with the Western blot analysis that only a small amount of Venus was present inside the SSVenus cells and is likely to represent Venus in transit through the ER and Golgi before secretion into the culture medium. Taken together, these results demonstrate that the gametolysin signal sequence was able to target Venus to the culture medium.

Subsequently, secretion yields were determined from cell lines with highest relative abundances by dot blotting. To detect Venus, clarified media were concentrated by 10-fold prior to the analysis and compared with defined amounts of a recombinant Venus that had been expressed and purified from *Escherichia coli*. Supernatants from a Venus line and the parental strain UVM were

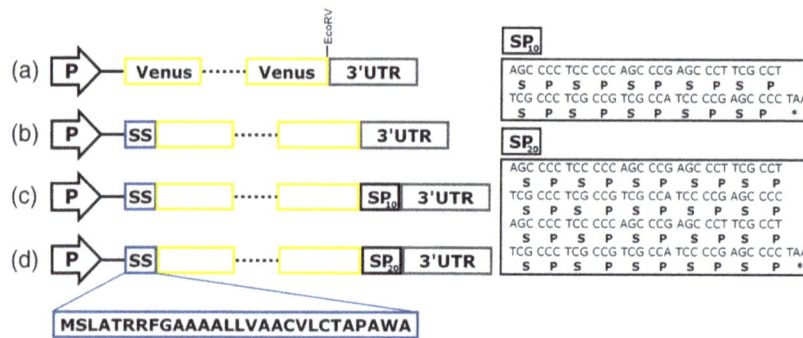

Figure 1 Schematic diagram of the expression cassettes. (a) pERC-Venus; (b) pERC-SSVenus; (c) pERC-SSVenus-(SP)$_{10}$; (d) pERC-SSVenus-(SP)$_{20}$. P, chimeric promoter RBCS2/HSP70A; solid line, RBCS2 intron 1; dotted line, RBSC2 intron 2; SS, gametolysin signal sequence. 3'UTR, BRSC2 terminator. The boxes on the right show the codon-optimized nucleotide sequences used in this study to encode (SP)$_n$ fusion tags. The peptide sequence of the gametolysin signal sequence is indicated.

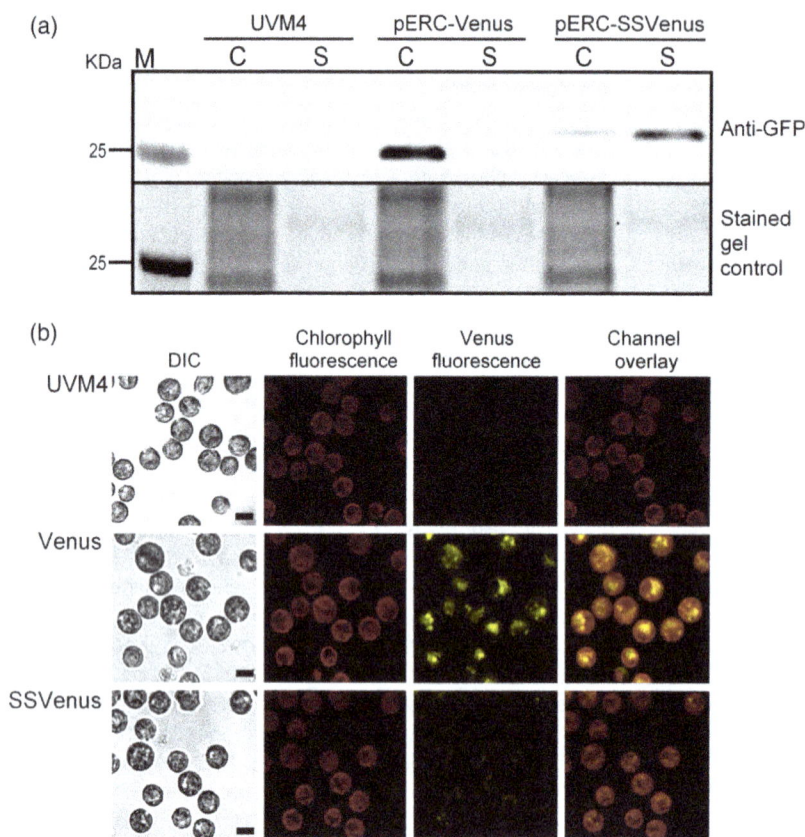

Figure 2 Subcellular localization and expression of fluorescent proteins from the Venus and SSVenus transgenic lines. (a) Western blot analysis of total proteins from cell lysates and supernatants (top panel). Total proteins were loaded in equal amounts and separated by SDS-PAGE and visualized upon the UV irradiation (lower panel, indicated as 'stained gel control'). Transgenic lines were grown until ~2 × 10^6 cells/mL in TAP media. Harvested cells (C) and supernatants (S) were resuspended in 4× Laemmli buffer. The untransformed UVM4 strain was used as a control. (b) Expression of Venus in *C. reinhardtii* cells visualized by confocal microscopy. Parental strain UVM4 was used as a control. Cells transformed with pERC-Venus (Venus line) and pERC-SSVenus (SSVenus line) were analysed. All images were acquired using the same settings for all strains. Scale bar indicates 5 μm.

used as controls. The five independent SSVenus lines, which bear the pERC-SSVenus construct, were able to secrete the recombinant protein into the media at levels between ~0.3 and 1.3 mg/L (Figure S1). This variation may be attributable to variable transcription activities due to different insertion positions and copy numbers (Jinkerson and Jonikas, 2015).

Secretion yields of (SP)$_n$-fused SSVenus

We aimed to increase the secretion yields by expressing Venus as fusion glycoproteins using synthetic (SP)$_n$ glycomodules. To this end, SSVenus was fused C-terminally with the (SP)$_{10}$ and (SP)$_{20}$ glycomodules, giving rise to the vectors pERC-SSVenus-(SP)$_{10}$ and pERC-SSVenus-(SP)$_{20}$ (Figure 1c, d). These constructs were used to transform the UVM4 strain and transgenic SSVenus-(SP)$_{10}$ and

SSVenus-(SP)$_{20}$ as described above. Cell-free culture media were screened by dot blotting using anti-GFP antibody, and five transgenic cell lines for each construct with the highest relative secretion yield were selected for further quantification. It should be noted that the (SP)$_n$-fused SSVenuses were readily detectable in the media of these strains without prior concentration. The highest yields of SSVenus-(SP)$_{10}$ and SSVenus-(SP)$_{20}$ fusion proteins in the culture medium were 7.5 mg/L in the line SP04 and 15 mg/L in the line SP14, respectively (Figure 3a). These values are sixfold and 12-fold higher, respectively, than the secreted SSVenus without the glycomodules. For further analysis, these two lines were selected. When grown on TAP agar plates, a halo of yellow fluorescence surrounding a colony of the SSVenus line was visible, while progressive increases in the fluorescence intensity were seen surrounding the colonies of the SSVenus-(SP)$_{10}$ and SSVenus-(SP)$_{20}$ lines (Figure 3b). To our best knowledge, this is the first time that the secretion of recombinant proteins has been enhanced in microalgae by using the (SP)$_n$ glycomodules.

(SP)$_n$-fused SSVenus are glycosylated

Molecular weights of the SSVenus, SSVenus-(SP)$_{10}$ and SSVenus-(SP)$_{20}$ proteins secreted from *C. reinhardii* into the media were analysed by Western blotting. Fusion proteins were detected as single bands at approximately 42 and 53 kDa for SSVenus-(SP)$_{10}$ and SSVenus-(SP)$_{20}$, respectively. These values were considerably larger than the sizes of the SSVenus control and the deduced molecular weights of the SSVenus-(SP)$_{10}$ (28.8 kDa) and SSVenus-(SP)$_{20}$ (30.7 kDa) (Figure 4a).

We investigated whether SSVenus-(SP)$_n$ proteins were glycosylated. The secreted proteins were subjected to chemical deglycosylation by treatment with trifluoromethanesulphonic acid (TFMS), a chemical reagent that efficiently cleaves *N*- and *O*-linked sugars from glycoproteins without affecting the integrity of the polypeptide (Edge *et al.*, 1981). The results presented in

Figure 4b showed that the TFMS treatment resulted in apparent mass shifts by ~14 and ~23 kDa for SSVenus-(SP)$_{10}$ and SSVenus-(SP)$_{20}$ to ~28 and ~30 kDa, respectively, which correspond to the unglycosylated forms of the respective proteins. Therefore, the 42- and 53-kDa forms of (SP)$_n$-fused SSVenuses represent glycosylated proteins.

Kinetics of protein secretion

The accumulation of SSVenus and (SP)$_n$-fused SSVenuses in the culture media was monitored for 7 days for the transgenic lines SS01, SP04 and SP14 representing SSVenus, SSVenus-(SP)$_{10}$ and SSVenus-(SP)$_{20}$, respectively. The untransformed cells were used as a control. All three transgenic cells grew similarly to the control cells (Table 1; Figure 5a). The average specific growth rates were calculated between 1.29 and 1.48 day^{-1} under the conditions tested (Table 1). Hence, secretion of the recombinant glycoproteins from transgenic lines does not impose an altered metabolic load that can decrease cell growth under the conditions tested. Under these conditions, the maximum yields of the secreted recombinant proteins in the culture media were 1.3, 7.5 and 15.1 mg/L for SSVenus, SSVenus-(SP)$_{10}$ and SSVenus-(SP)$_{20}$,

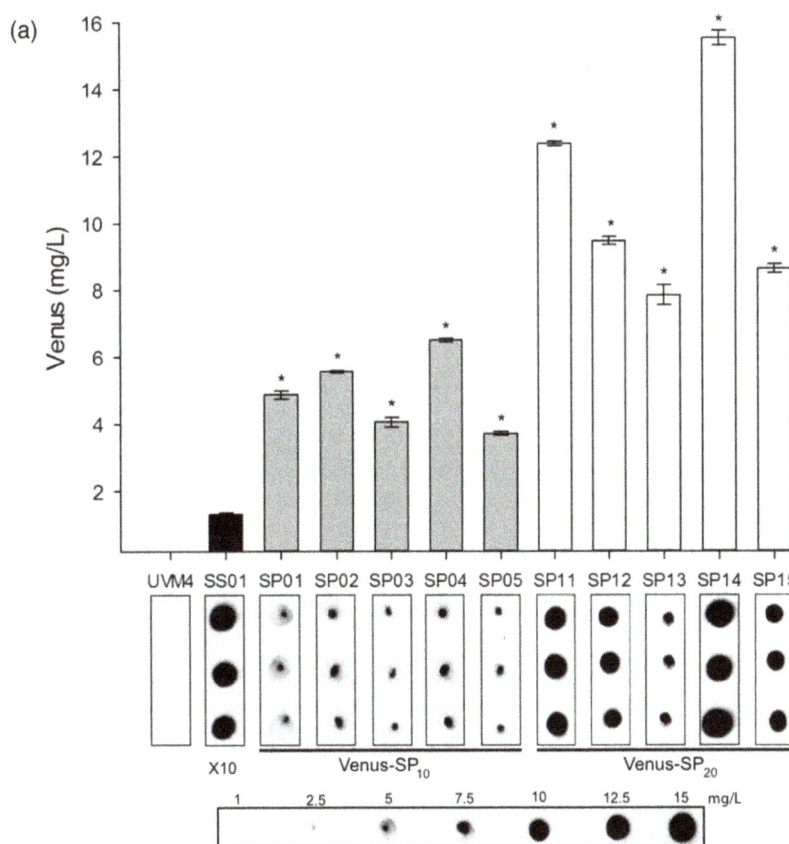

Figure 3 Introduction of (SP)$_n$ glycomodules enhances the yields of secreted Venus in *C. reinhardtii*. (a) Secretion yields of selected transgenic lines determined by dot blotting. Concentrations of Venus in the culture media from five lines for each construct were quantified using purified *E. coli*-derived Venus as a standard. Supernatants from SSVenus-(SP)$_{10}$ and SSVenus-(SP)$_{20}$ selected lines were collected after 7 days of cultivation and used directly for dot-blot analysis. Means of three technical replicates and standard errors are shown. Statistical analysis was performed using a one-way ANOVA ($P < 0.05$). The asterisk represents a significant difference from SS01. (b) Fluorescence emission from cells grown on agar plates detected with a stereo fluorescence microscope. Fluorescence was detected under the same settings for all colonies. i, untransformed UVM4; iii, Venus; iii, SSVenus; iv, SSVenus-(SP)$_{10}$; v, SSVenus-(SP)$_{20}$. Scale bars indicate 1 mm.

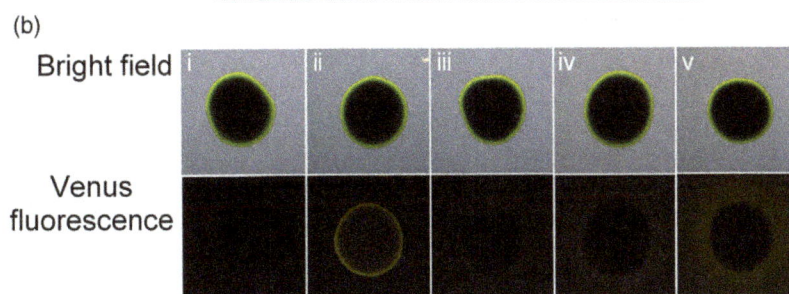

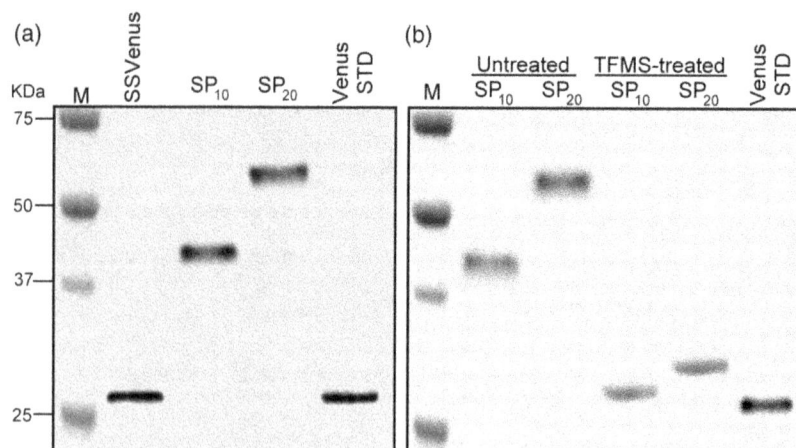

Figure 4 The secreted Venus-(SP)$_n$ proteins are glycosylated. (a) Western blot analysis of fusion glycoproteins secreted into the media by the transgenic
C. reinhardtii. M, molecular size markers. One hundred micrograms of total proteins in the media were loaded in each lane for SSVenus, SSVenus-(SP)$_{10}$
and SSVenus-(SP)$_{20}$, whereas 1 μg of purified *E. coli*-derived Venus was loaded as a control. (b) Effect of deglycosylation treatment with TFMS on the fusion
glycoproteins. SSVenus-(SP)$_{10}$ and -(SP)$_{20}$ proteins were chemically deglycosylated and analysed by Western blot. Untreated and TFMS-treated samples
were loaded on parallel lanes for comparison. *E. coli*-derived Venus, designated as Venus STD, was loaded as a control. M, molecular size markers. SP$_{10}$ and
SP$_{20}$ indicate SSVenus-(SP)$_{10}$ and SSVenus-(SP)$_{20}$, respectively.

respectively, which is consistent with the results described above (Figure 4a). The accumulation of Venus followed the cell growth, and the highest productivity of 2.13 mg/L day was observed for the transgenic cells expressing SSVenus-(SP)$_{20}$ (Table 1). Interestingly, non-glycosylated products were present at the end of the cultivation, probably as a result of the spontaneous lysis that occurs when the culture enters the stationary phase and the cells are under stress (Figure 5b).

Effect of Brefeldin A (BFA) on glycosylation of Venuses-(SP)$_n$ in the secretory pathway

Despite its importance, subcellular localization of the glycosylated recombinant proteins produced by *C. reinhardtii* has not been well understood at the molecular level. BFA is a fungal toxin that is widely used in studies of eukaryotic secretory pathways, and in *C. reinhardtii*, BFA inhibits the secretory pathway by destroying the Golgi apparatus (Hummel *et al.*, 2007). We took advantage of this effect of BFA and tested its impact on the glycosylation of the (SP)$_n$-fused Venuses. To do so, we first screened for the effective concentration of BFA on blocking the membrane trafficking. The SSVenus line was grown in TAP media to the mid-exponential phase (approximately 4×10^6 cells/mL), and the cells were treated with BFA for 4 h at final concentrations of 10,

Table 1 Growth parameters and yields of the secreted proteins from transgenic cell lines secreting Venus variants and the parent UVM4 strain

	Maximum biomass (cells/mL)	Specific growth rate (day^{-1})	Doubling time (days)	Maximum yields (mg/L)	Productivity (mg/L/day)
UVM4	2.20×10^7	1.38	0.50	–	–
SSVenus	2.26×10^7	1.29	0.54	1.3	0.19
Venus-(SP)$_{10}$	2.08×10^7	1.38	0.50	7.7	1.10
Venus-(SP)$_{20}$	2.25×10^7	1.43	0.48	15.1	2.13

25 and 50 μM or with dimethyl sulphoxide (DMSO) as a control. Untreated UVM4 strain and the Venus line were also included in the analysis as additional controls. Flow cytometry was used to monitor the changes in the fluorescence signal on the population basis. Distinctive distributions of fluorescence signals as a function of cell sizes were observed for the UVM4 strain, the DMSO-treated SSVenus line and the Venus line (Figure S2a–c). The BFA treatment of the SSVenus line progressively shifted the population distribution similar to the Venus line (Figure S2). Similar results have previously been reported for the secretion of the arylsulphatase from *C. reinhardtii* (Kagiwada *et al.*, 2004). At a concentration of 50 μM, BFA caused more than 95% of cells to accumulate Venus intracellularly (Figure S2d). Importantly, the effect of the BFA treatments was fully reversible. Based on these results, the following experiments were conducted with 50 μM BFA.

The SSVenus, SSVenus-(SP)$_{10}$ and SSVenus-(SP)$_{20}$ lines were treated with 50 μM BFA or DMSO for 4 h, and live-cell imaging was performed to analyse the localization of SSVenus and SSVenus-(SP)$_n$. DMSO-treated cells of the three transgenic lines displayed weak intracellular signals, consistent with Figures 2 and S2, whereas the cells treated with BFA considerably enhanced the intensities of the intracellular signals in all three lines in a reversible manner (Figure 6a). These results indicate that the BFA treatment successfully blocked the secretion of SSVenus and SSVenus-(SP)$_n$, likely by inhibiting the transport from ER to the Golgi apparatus as previously shown in tobacco plants (Boevink *et al.*, 1999).

Next, the molecular weights of SSVenus and SSVenus-(SP)$_n$ that accumulated intracellularly in BFA-treated cells were analysed by Western blotting. Firstly, in the absence of the BFA treatment, the SSVenus and SSVenus-(SP)$_n$ proteins migrated as single bands with apparent molecular weights that were consistent with the predicted polypeptide backbones, representing non-glycosylated forms (Figure 6b, Lane 2 and 3). The apparent absence of glycosylated Venus inside the cells suggests that glycosylated forms are more efficiently secreted than the corresponding non-glycosylated forms. Upon the BFA treatment,

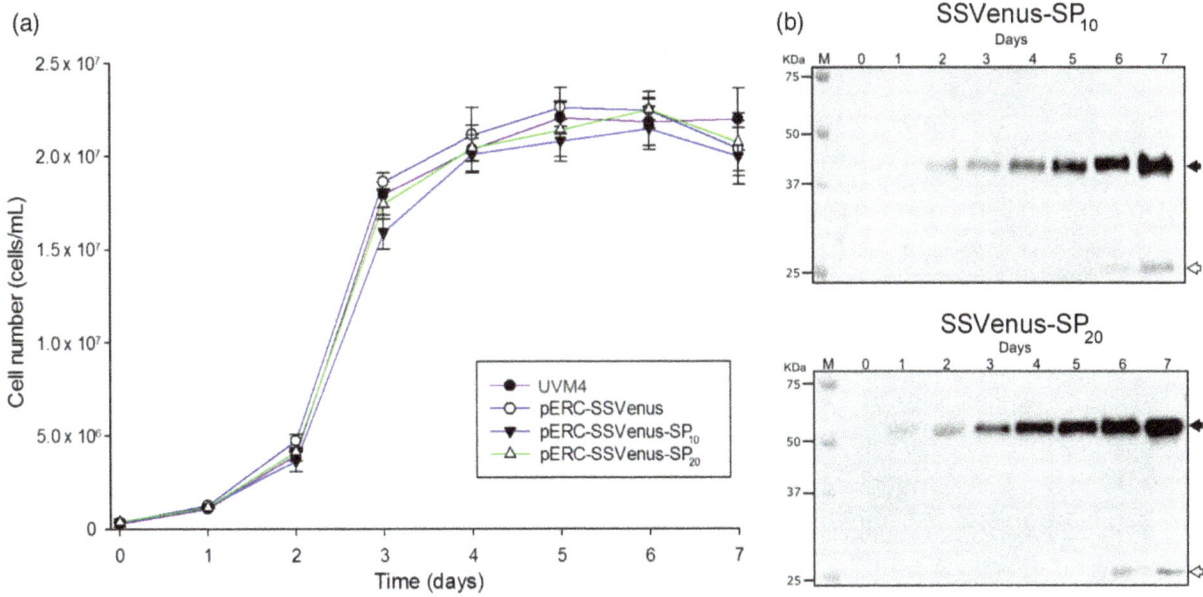

Figure 5 Growth of the transgenic lines of *C. reinhardtii* secreting Venus and fusion Venus is not affected by the production of the recombinant proteins. (a) Growth curves of the transgenic lines and the parental strain UVM4 under the mixotrophic conditions at 25 °C, under constant illumination at 120 μmol photons m^{-2} s^{-1}. Strains were inoculated in TAP media at the OD_{750nm} value of 0.05. Growth was monitored by cell counting for 7 days. Error bars represent standard errors of at least three independent cultures. There was no statistically significant difference between genotypes based on two-way ANOVA test ($P > 0.05$). (b) Time course of Venus-SP_{10} and Venus–SP_{20} secretions in the growth media. Cell line SP04 expressing SSVenus-$(SP)_{10}$ (upper panel); cell line SP14 expressing SSVenus-$(SP)_{20}$ (lower panel). Samples were taken from shake flasks every 24 h for 7 days. Twenty microlitres of clarified supernatants for each sample were separated on SDS-PAGE and analysed by Western blot. Lanes 1–7 correspond to samples from 0- to 7-day-old culture supernatants; molecular size markers are indicated as M. Solid arrow indicates the fusion proteins; open arrows represent non-glycosylated protein.

additional bands at higher molecular weights were detectable in the SSVenus-$(SP)_{10}$ and SSVenus-$(SP)_{20}$ lines (Figure 6b, lane 5 and 6). Their estimated molecular weights were ~32 and ~38 kDa, respectively, and are likely to represent partially glycosylated forms. These results may indicate that O-glycosylation in *C. reinhardtii* starts in the ER (Zhang *et al.*, 1989).

Determination of protein stability

Glycosylation is known to impact the stability of secreted proteins (Solá and Griebenow, 2009). Hence, proteolytic susceptibility of SSVenus-$(SP)_n$ was analysed *in vitro*, alongside SSVenus, by monitoring the kinetics of proteolysis with trypsin. SSVenus, SSVenus-$(SP)_{10}$ and SSVenus-$(SP)_{20}$ obtained from the respective transgenic culture media were incubated with trypsin for 4 h, and fluorescence was monitored for 240 min. SSVenus was completely degraded by 80 min, with the proteolytic half-life of 17 min. In contrast, SSVenus-$(SP)_{10}$ and SSVenus-$(SP)_{20}$ were found to be significantly more resistant to proteolysis than non-glycosylated Venus and their complete degradation required >240 min (Figure 7). The proteolytic half-lives of these proteins were 76 min and 130 min, respectively, which were fourfold and 17-fold longer as compared to SSVenus. These results demonstrate that Venuses tagged with the $(SP)_n$ glycomodules are more resistant to proteolytic degradation.

Discussion

The secretion of recombinant proteins is often desirable not only because it can simplify downstream processing (e.g. easier product recovery), but also because it enables post-translational

modifications, particularly glycosylation, for enhanced product yields and quality. Moreover, the secretion of recombinant proteins could allow direct application of the culture media without extensive purification, which could open up new opportunities in application of microalgae-derived products. In this study, we firstly showed that the gametolysin signal sequence successfully targeted Venus to the culture media giving rise to the maximum yield of 1.3 mg/L. Secondly, the introduction of synthetic $(SP)_n$ glycomodules at the C-terminus of Venus led to an increase in the molecular mass, indicating that the fusion proteins were glycosylated as demonstrated by chemical deglycosylation treatment. Thirdly, the glycomodules considerably increased the proteolytic stability of the fusion proteins. Finally, the yield of the fusion proteins secreted in culture media was enhanced by a 12-fold to 15 mg/L as compared to the untagged Venus, which is the highest secretion level achieved thus far for microalgae-based recombinant proteins.

Taken together, these results demonstrate that the gametolysin signal sequence and the synthetic glycomodules can be used to secrete recombinant proteins with higher yields than previously attained and can be added to the growing molecular toolbox in the engineering of *C. reinhardtii* and possibly other microalgae. To establish microalgae as alternative hosts for protein secretion, further enhancement in the yield of secreted proteins is needed. Increasing the length of the synthetic $(SP)_n$ glycomodule (Zhang *et al.*, 2016) has been shown to dramatically increase yields of secreted proteins in plants. A rigorous optimization of culturing conditions, for example by varying operating conditions of a photobioreactor, has been shown to increase the yield of a secreted recombinant protein in

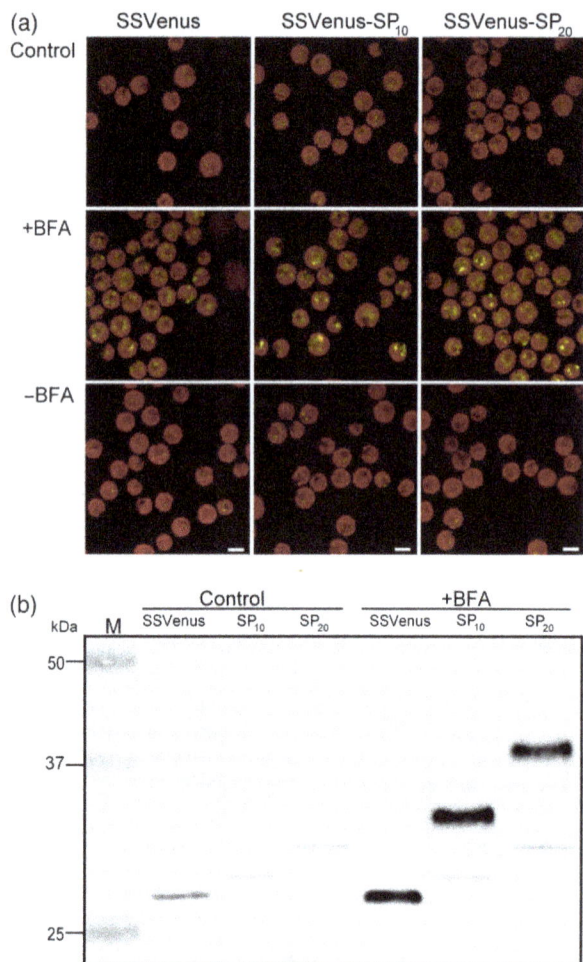

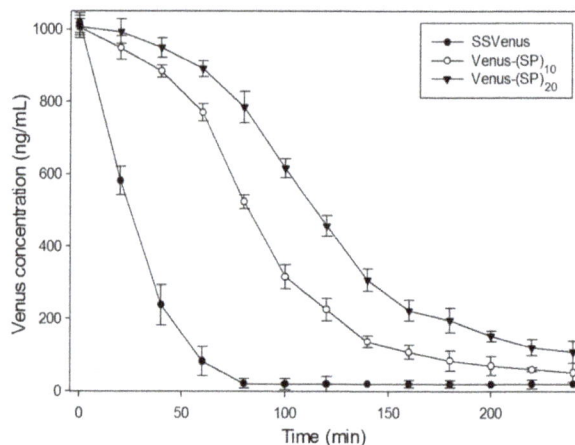

Figure 7 Profiles of proteolytic degradation of secreted fusion proteins. SSVenus, SSVenus-(SP)$_{10}$ and SSVenus-(SP)$_{20}$ fluorescence was monitored in the presence of trypsin for 4 h as described in Materials and methods. Quantification of the functional proteins was performed by using the standard curve prepared from purified *E. coli*-derived Venus. Error bars represent standard errors of at least three independent assays. SSVenus, SSVenus-(SP)$_{10}$ and SSVenus-(SP)$_{20}$ were statistically significantly different based on two-way ANOVA test ($P < 0.05$).

Figure 6 Effect of the BFA treatment on the subcellular localization and molecular masses of the Venus and fusion Venus. Transgenic cells in the exponential growth phase were treated with BFA at 50 μM or solvent control (DMSO) for 4 h. (a) Alteration of the fluorescence patterns in cells secreting Venus and fusion Venus upon treatment with BFA visualized by confocal microscopy. Images show that BFA treatment increases the fluorescence intensity in the transgenic cells as a result of the accumulation in the ER. All images were acquired using the same settings for all strains and treatments. Scale bars indicate 5 μm. (b) Western blot analysis of cell lysates from cells treated with BFA. Lanes 1 and 4, SSVenus; lanes 2 and 5, SSVenus-(SP)$_{10}$; lanes 3 and 6, SSVenus-(SP)$_{20}$. One hundred and fifty micrograms of TSP were loaded into each well.

C. reinhardtii (Lauersen *et al.*, 2015a). These optimization strategies may be applied in combination with the use of new mutants with improved transgene expression (Kong *et al.*, 2015; Kurniasih *et al.*, 2016) and genetic tools such as codon optimization (Barahimipour *et al.*, 2015) and novel synthetic algal promoters with increased performance (Scranton *et al.*, 2016).

The (SP)$_n$ glycomodules were originally developed for secretion of recombinant proteins from plant cell cultures, leading to extensive *O*-glycosylation and enhanced proteolytic stability of the expressed proteins (Shpak *et al.*, 1999; Xu *et al.*, 2007, 2008, 2010). From a biotechnological point of view, glycosylation of recombinant proteins is a desirable feature in recombinant protein secretion because it can strongly influence the physiochemical properties of a protein, such as folding, solubility, stability, biological activity and resistance to heat or proteolysis (Gomord *et al.*, 2010; Mathieu-Rivet *et al.*, 2014; Walsh and

Jefferis, 2006). Glycosylation of recombinant proteins (i.e. *xylanase 1* from *Trichoderma reesei*, a human erythropoietin, ice-binding protein) secreted by *C. reinhardtii* has been speculated to occur in previous studies (Eichler-Stahlberg *et al.*, 2009; Lauersen *et al.*, 2013b; Rasala *et al.*, 2012); however, detailed experimental evidence supporting the notion has been lacking. Our results show that the glycomodule can be readily applied as a modular unit to enhance the yield and proteolytic stability of recombinant proteins in *C. reinhardtii*. This strategy can be applied either alone or in combination with other strategies to further increase the secretion yield (Lauersen *et al.*, 2015a), thereby paving a path towards the development of cost-effective production of recombinant proteins.

The exact nature of glycoconjugates attached to Venus-(SP)$_n$ glycoproteins expressed in *C. reinhardtii* is yet to be elucidated. In plants, the Hyp-*O*-glycosylation code was established using synthetic glycomodules based on Hyp-rich glycoproteins (HGRPs), which are major components of the cell wall of plants and some microalgae including *C. reinhadtii* (Showalter, 1993; Woessner and Goodenough, 1994). Briefly, after hydroxylation of Pro residues, *O*-Hyp glycosylation is defined by a glycomodule. Branched and structurally heterogeneous arabinogalactan polysaccharides (AGPs) are added via a β-galactosyl glycosidic linkage to a glycomodule consisting of noncontiguous Hyp repeats [as in the (SP)$_n$ glycomodules], which is a common feature amongst arabinogalactan proteins (Shpak *et al.*, 1999; Tan *et al.*, 2003, 2004). On the other hand, a glycomodule consisting of contiguous repeats of Pro and Ser (e.g. Ser-Pro$_4$ repeats), widely conserved amongst extensin proteins, is decorated at the Hyp residues with simple arabinooligosaccharides of typically four monosaccharide units via a β-arabynosyl glycosidic linkage (Kieliszewski and Shpak, 2001; Shpak *et al.*, 2001). [Correction added on 28 June 2017, after first online publication: In the preceding sentences, "α-galactosyl" and "α-arabinosyl" were previously incorrect and this has been corrected in this current version.] The HRGPs present in members of Chlorocalles and Volvocales including *C. reinhardtii* contain contiguous and noncontiguous Hyp repeats (Voigt et al., 2014; Woessner and Goodenough, 1989, 1992). Some of the Hyp-bound glycoconjugates characterized to date in C. reinhardtii are relatively short, ranging in size and composition from single sugars (arabinose or galactose) to arabinooligosaccharides similar to extensins or arabinogalacto-oligosaccharides consisting of up to six monosaccharide units (Bollig et al., 2007; Ferris et al., 2001; Miller et al., 1972). Hence, it is plausible that the $(SP)_n$ glycomodule expressed in C. reinhardtii is also glycosylated with relatively short oligosaccharides.

Subcellular localization of O-glycan biosynthesis in higher plants is beginning to emerge, while it is still poorly understood in microalgae. Concerning the extensin-type O-glycosylation in C. reinhardtii, Pro residues are first hydroxylated by prolyl-4-hydroxylases localized in the ER (Keskiaho et al., 2007), and then, arabinosylation of the Hyp residues in the ER and Golgi apparatus by Hyp-O-arabinosyltransferases (HPATs) occurs (Zhang and Robinson, 1990; Zhang et al., 1989). In plants, these processes appear to take place in the Golgi apparatus (Ogawa-Ohnishi et al., 2013), although evidence suggests that it may also occur in the ER (Estevez et al., 2006). Concerning the AGP-type O-glycosylation, it has been shown in plants that Hyp-galactosyl-transferases (HPGT1 through HPGT3 and GALT3 through GALT6) localize to the Golgi apparatus (Basu et al., 2013; Ogawa-Ohnishi et al., 2013), while GALT2 was shown to localize to both the ER and the Golgi apparatus (Basu et al., 2013). In this study, the BFA treatment that blocks protein transport from ER to the Golgi apparatus led to the accumulation of partially glycosylated forms of $(SP)_n$-fused Venus in ER (Figure 6), which supports the notion that O-glycosylation starts in the ER (Zhang et al., 1989). In Arabidopsis, three HPATs involved in Hyp arabinosylation have been identified and shown to localize to the cis-Golgi (Ogawa-Ohnishi et al., 2013). Genome-wide search using the BLAST database for Arabidopsis HPATs homologues in C. reinhardtii identified three putative proteins (Cre01.g032600, Cre12.g531450, Cre16.g690000) that contain putative N-terminal transmembrane domains and share 20%–33% amino acid identities with Arabidopsis HPATs. In contrast, a genome-wide search for homologous proteins to plant Hyp-galactosyltrans-ferases (HPGTs, GALT2) failed to identify a significant hit (E value <0.001), suggesting that the mechanism of Hyp-galactosylation may be different between C. reinhardtii and plants.

In summary, microalgae offer great potential as a light-powered and low-cost platform for the secretion of recombinant proteins. We demonstrated that a plant-based system to enhance protein secretion can be readily transferred to C. reindhardtii. Therefore, it can potentially be applied to other algae species (Hempel and Maier, 2012). A further investigation into glycosylation processes would be beneficial for a better understanding of the underlining molecular mechanisms of O-glycosylation in microalgae. The set of synthetic glycomodules established in this study could facilitate this investigation. The molecular tools presented here allow a high-yield secretion of glycosylated proteins with a reduced susceptibility to proteases. These qualities are of biotechnological importance to minimize downstream processing costs.

Experimental procedures

Materials

All standard chemicals and reagents were purchased from Sigma-Aldrich, USA. Restriction enzymes were purchased from New England Biolabs, USA. All the oligonucleotides were purchased from Integrated DNA Technologies, Inc., USA.

Assembly of transformation vectors

Venus gene [a super enhanced YFP (Nagai et al., 2002)] was codon-optimized in silico based on the codon usage of the nuclear genome of C. reinhardtii using the IDT Codon Optimization tool (https://eu.idtdna.com/CodonOpt). The codon-optimized Venus gene was synthesized in fusion with the sequence of the gametolysin signal sequence consisting of 28 amino acids (Kinoshita et al., 1992) at the 5′ end, and an EcoRV restriction site was included at the 3′ end followed by a stop codon (Integrated DNA Technologies, Inc.). In the subsequent cloning, DNA parts were amplified by PCR using Phusion High-Fidelity DNA Polymerase (New England Biolabs) with a set of overlapping primers to generate pERC-Venus (Hcr123F, Hcr123R, Venus1F, Venus1R, RBSC2i2F, RBSC2i2, Venus2F, Venus2R) and pERC-SSVenus (Hcr123F, Hcr123R, SSVenus1F, SSVenus1R, RBSC2i2F, RBSC2i2, Venus2F, Venus2R; where SS denotes gametolysin signal sequence; Table S1). The PCR products were purified and prepared for Gibson Assembly using Gibson Assembly Master Mix (New England Biolabs). Vector pER123 was used as plasmid backbone; it contains the HSP70/RBCS2 chimeric constitutive promoter (Schroda et al., 2000), the 3′ UTR from RBSC2 as a terminator and the APHVIII resistance gene for selection on paromomycin, whose expression is controlled by the constitutively PSAD promoter. The final constructs are shown in Figure 1.

A synthetic gene encoding ten repeats of the dipeptide Ser-Pro was constructed from two codon-optimized and complementary oligonucleotides (SP10-F and SP10-R, Table S1) synthesized with 5′ phosphorylation. Three extra bases (ATC) at the 5′ end of the double-stranded fragment were added to restore the EcoRV site (Table S1). The oligonucleotides were resuspended in an annealing buffer (100 mM potassium acetate; 30 mM [4-(2-hydroxyethyl)-1-piperazineethanesulphonic acid (HEPES), pH 7.5] and mixed in equal molar amounts. The mixed oligonucleotides were incubated at 94 °C for 2 min in a water bath and allowed to cool down for 1 h at room temperature. The resulting double-stranded synthetic gene was inserted into EcoRV-digested and dephosphorylated pERC-SSVenus to generate pERC-SSVenus-$(SP)_{10}$. Vector pERC-SSVenus-$(SP)_{20}$ was constructed by ligation of the annealed oligonucleotides with EcoRV-digested and dephosphorylated pERC-SPVenus-$(SP)_{10}$. The extra nucleotides introduced in pERC-SSVenus-$(SP)_{20}$ for cloning purposes were removed via site-directed mutation using the Q5 Site-Directed Mutagenesis Kit (New England Biolabs) using primers mSP20-R and mSP20-F. Transformation of E. coli strain TOP10, plasmid isolation and confirmation of the DNA sequence were according to standard protocols.

Culture conditions

The cell-wall-deficient UVM4 strain of C. reinhardtii (Neupert et al., 2009) was cultivated in flasks containing TAP media (Gorman and Levine, 1965) at 25 °C under constant illumination (120 µmol photons m^{-2} s^{-1}) using cool fluorescent white light with constant agitation at 120 rpm in an orbital shaker. Cell concentrations in cultures were determined by counting the cell number in a Neubauer hemocytometer under a bright-field microscope. For each sample, three biological replicates were analysed. For quantification of Venus and Venus fused with the glycomodules, cells were grown to exponential or late exponential phase under the conditions indicated above and samples were taken every 24 h. When designated, cultures were grown on TAP agar plates, which consist of TAP media containing 1.5% (w/v) agar.

Transformation of C. reinhardtii

Chlamydomonas reinhardtii UVM4 was transformed according to the glass bead method (Kindle, 1990), using 1 μg of a designated vector linearized with ScaI. Cells were incubated overnight and were harvested at 1100 × g for 5 min and plated on TAP agar plates containing 10 μg/mL paromomycin. After 7 days of incubation, paromomycin-resistant colonies were picked and further cultivated to be screened for gene integration by colony PCR as described by Cao *et al.* (2009) using gene-specific primers GSP3 and GSP4 to amplify the full-length expression cassette (Table S1).

Immunoblotting

To detect the expression of Venus and Venus fused with the glycomodules, cultures of the transformed cell lines were subjected to centrifugation at 5000 × g for 5 min at 25 °C. Pellets were resuspended in 4× Laemmli buffer (BioRad, Copenhagen, Denmark) and denatured at 90 °C for 5 min. Supernatants were either used directly for analysis or concentrated 10-fold by freeze drying and denatured in the presence of 4× Laemmli buffer as described above. Either lysates and/or supernatants were subjected to protein quantification and detection by dot blotting or Western blotting. For Western blotting, samples were loaded on a 12% (w/v) Criterion™ TGX Stain-Free™ Protein Gel, which contains trihalo compounds that can react with tryptophan residues under the UV irradiation and gives rise to fluorescence (BioRad). Separated proteins were transferred to a polyvinylidene difluoride membrane (PVDF). For dot blotting, 1 μL of supernatant was spotted on a nitrocellulose membrane and left to dry at room temperature for 1 h; this was performed in technical triplicates for every positive strain. Purified *E. coli*-derived Venus was used as a standard for protein quantification or as a positive control. Immunodetection of Venus was carried out using anti-GFP mouse IgG (Roche Applied Science, Germany) in a 1 : 2000 dilution according to the manufacturer's instructions. No cross-reactivity with the proteins native to *C. reinhardtii* was detected by the anti-GFP antibody under the experimental conditions employed in this study. The secondary antibody (anti-mouse, IgG secondary antibody, horseradish peroxidase conjugate, Sigma-Aldrich) was used in a 1 : 2500 dilution. Total soluble protein concentration was measured using Micro BCA™ Protein Assay Kit according to the manufacturer's instructions (Thermo Fisher Scientific, Hvidovre, Denmark).

Fluorescence localization analysis

Live-cell imaging was performed using a Leica SP5-X confocal laser-scanning microscope with a 63× water-immersion objective. Localization of Venus was imaged using a 514 nm argon laser; fluorescence emission was detected between 520 and 550 nm. Chlorophyll fluorescence was observed independently upon excitation at 488 nm and emission 650–700 nm. All images were acquired using the same settings and were analysed by the Leica Application Suite Advanced Fluorescence software.

For direct fluorescence detection on TAP agar plates, 10 μL of cells was spotted on TAP agar plates containing 10 μg/mL paromomycin and incubated under standard growth conditions,

as indicated above, for 4 days. Algal colonies grown on the TAP agar plates were imaged by a fluorescence stereo microscope (Leica MZ FLII, Germany) using UV light source (Leica DN 5000B, Germany) using an excitation filter 470/40 nm and an emission filter of 525/50 nm.

Chemical deglycosylation

Chemical deglycosylation by TFMS was performed as previously described (Edge *et al.*, 1981) with a modification. Briefly, cells were removed from cultures by centrifugation at 5000 × g for 5 min, and the cell-free supernatants were desalted and lyophilized. One milligram of desalted and lyophilized cell-free supernatants containing the secreted Venus and $(SP)_n$ tagged Venus was mixed with 1 mL of a 9 : 1 (v/v) mixture of TFMS and anisole in a glass vial. The mixture was incubated at −80 °C for 5 min and then at −20 °C for 4 h; thereafter, it was neutralized with 1 mL of ice-cold 60% (v/v) aqueous pyridine solution followed by the addition of 100 μL of 100 mM NH_4HCO_3, pH 8.0. Deglycosylated proteins were recovered by centrifugation at 15 000 × g for 10 min at 4 °C; after centrifugation, the pellet was dissolved in 4× Laemmli buffer, as indicated above. Samples were analysed by Western blot as described above.

BFA treatment

BFA was dissolved in DMSO at a concentration of 10 mg/L and stored at −20 °C until use. For BFA treatment, cells cultures were grown to mid-exponential phase (4 × 10^6 cells/mL) in 24-well plated as indicated above. An aliquot of this stock solution was added to cell culture to obtain final concentrations of 10, 25 and 50 μM. The same volume of DMSO was added to make solvent control samples. The cultures with BFA were incubated for 4 h. For recovery experiments, the cells were further incubated for 4 h with BFA-free TAP media. After each treatment, 200 μL of cell cultures was transferred into a 96-wll plate and was analysed using a BD LSR Fortessa flow cytometer (DB, USA) fitted with multiwell plate sampling system using a YFP filter (excitation at 488 nm; emission at 545/35 nm). Fluorescence was recorded from ten thousand cells per sample. Data were collected, transformed to the logarithmic scale and analysed using BD FACSDiva™ v6.2 software. Cell imaging was performed on cells treated with 50 μM BFA and cells recovered as previously described. Western blotting was carried out on lysates of cells treated with 50 μM BFA as described above. DMSO-treated cells were used as control in all experiments.

Proteolytic assay for protein stability

Concentrations of the secreted Venus and $(SP)_n$ tagged Venus were adjusted using a standard curve prepared from *E. coli*-expressed Venus. Proteolysis assay was performed using trypsin at a ratio of 1 : 100 (trypsin : Venus) in a trypsin buffer (67 mM sodium phosphate buffer, pH 7.6). Each recombinant protein was analysed in triplicates in 96-well black flat-bottom plates. Proteolytic reactions were incubated at 25 °C, and fluorescence was monitored every 20 min for 4 h using a TriStar² LB 942 Multidetection Microplate Reader (BERTHOLD, Germany; excitation 510 nm, emission 550 nm). TAP media were used to determine the baseline.

Statistical analysis

In all experiments, three biological or technical replicates were performed for each sample and the mean value represents the average. Statistical analysis was carried out using the ANOVA test.

Acknowledgements

Vector pER123 was kindly provided by Hussam Hassan Nour-Eldin, University of Copenhagen. This work was supported by the People Programme (Marie Curie Actions) of the European Union's Seventh Framework Programme FP7/2007-2013/under REA grant agreement no 317184, the Danish National Advanced Technology Foundation (grant 'Biomass for the 21st Century', no. 001-2011-4) and Villumn Fonden (grant no. 13363).

References

Almaraz-Delgado, A.L., Flores-Uribe, J., Perez-Espana, V.H., Salgado-Manjarrez, E. and Badillo-Corona, J.A. (2014) Production of therapeutic proteins in the chloroplast of Chlamydomonas reinhardtii. AMB Express, 4, 57.

Barahimipour, R., Strenkert, D., Neupert, J., Schroda, M., Merchant, S.S. and Bock, R. (2015) Dissecting the contributions of GC content and codon usage to gene expression in the model alga Chlamydomonas reinhardtii. Plant J. 84, 704–717.

Basu, D., Liang, Y., Liu, X., Himmeldirk, K., Faik, A., Kieliszewski, M., Held, M. et al. (2013) Functional identification of a hydroxyproline-O-galactosyltransferase specific for arabinogalactan protein biosynthesis in Arabidopsis. J. Biol. Chem. 288, 10132–10143.

Boevink, P., Martin, B., Oparka, K., Santa Cruz, S. and Hawes, C. (1999) Transport of virally expressed green fluorescent protein through the secretory pathway in tobacco leaves is inhibited by cold shock and brefeldin A. Planta, 208, 392–400.

Bollig, K., Lamshöft, M., Schweimer, K., Marner, F.-J., Budzikiewicz, H. and Waffenschmidt, S. (2007) Structural analysis of linear hydroxyproline-bound O-glycans of Chlamydomonas reinhardtii—conservation of the inner core in Chlamydomonas and land plants. Carbohyd. Res. 342, 2557–2566.

Cao, M., Fu, Y., Guo, Y. and Pan, J. (2009) Chlamydomonas (Chlorophyceae) colony PCR. Protoplasma, 235, 107–110.

Chavez, M.N., Schenck, T.L., Hopfner, U., Centeno-Cerdas, C., Somlai-Schweiger, I., Schwarz, C., Machens, H.G. et al. (2016) Towards autotrophic tissue engineering: photosynthetic gene therapy for regeneration. Biomaterials, 75, 25–36.

Cui, J., Xiao, J., Tagliabracci, V.S., Wen, J., Rahdar, M. and Dixon, J.E. (2015) A secretory kinase complex regulates extracellular protein phosphorylation. eLife, 4, e06120.

Demain, A.L. and Vaishnav, P. (2009) Production of recombinant proteins by microbes and higher organisms. Biotechnol. Adv. 27, 297–306.

Díaz-Santos, E., Vila, M., Vigara, J. and León, R. (2016) A new approach to express transgenes in microalgae and its use to increase the flocculation ability of Chlamydomonas reinhardtii. J. Appl. Phycol. 28, 1611–1621.

Edge, A.S., Faltynek, C.R., Hof, L., Reichert, L.E. Jr. and Weber, P. (1981) Deglycosylation of glycoproteins by trifluoromethanesulfonic acid. Anal. Biochem. 118, 131–137.

Eichler-Stahlberg, A., Weisheit, W., Ruecker, O. and Heitzer, M. (2009) Strategies to facilitate transgene expression in Chlamydomonas reinhardtii. Planta, 229, 873–883.

Estevez, J.M., Kieliszewski, M.J., Khitrov, N. and Somerville, C. (2006) Characterization of synthetic hydroxyproline-rich proteoglycans with arabinogalactan protein and extensin motifs in Arabidopsis. Plant Physiol. 142, 458–470.

Ferris, P.J., Woessner, J.P., Waffenschmidt, S., Kilz, S., Drees, J. and Goodenough, U.W. (2001) Glycosylated polyproline II rods with kinks as a structural motif in plant hydroxyproline-rich glycoproteins. Biochemistry, 40, 2978–2987.

Fuhrmann, M., Hausherr, A., Ferbitz, L., Schödl, T., Heitzer, M., Hegemann, P. Monitoring dynamic expression of nuclear genes in Chlamydomonas reinhardtii by using a synthetic luciferase reporter gene. (2004) Plant Mol. Biol. 55, 869–881.

Fuhrmann, M., Oertel, W. and Hegemann, P. (1999) A synthetic gene coding for the green fluorescent protein (GFP) is a versatile reporter in Chlamydomonas reinhardtii. Plant J. 19, 353–361.

Gangl, D., Zedler, J.A., Rajakumar, P.D., Martinez, E.M., Riseley, A., Wlodarczyk, A., Purton, S. et al. (2015) Biotechnological exploitation of microalgae. J. Exp. Bot. 66, 6975–6990.

Gimpel, J.A., Hyun, J.S., Schoepp, N.G. and Mayfield, S.P. (2015) Production of recombinant proteins in microalgae at pilot greenhouse scale. Biotechnol. Bioeng. 112, 339–345.

Gomord, V., Fitchette, A.C., Menu-Bouaouiche, L., Saint-Jore-Dupas, C., Plasson, C., Michaud, D. and Faye, L. (2010) Plant-specific glycosylation patterns in the context of therapeutic protein production. Plant Biotechnol. J. 8, 564–587.

Gorman, D.S. and Levine, R.P. (1965) Cytochrome f and plastocyanin: their sequence in the photosynthetic electron transport chain of Chlamydomonas reinhardii. Proc. Natl. Acad. Sci. U. S. A. 54, 1665–1669.

Hellwig, S., Drossard, J., Twyman, R.M. and Fischer, R. (2004) Plant cell cultures for the production of recombinant proteins. Nat. Biotechnol. 22, 1415–1422.

Harris, E.H. (2009) The Chlamydomonas sourcebook: Vol I: Introduction to Chlamydomonas and its laboratory use. Academic Press, Cambridge, USA.

Hempel, F. and Maier, U.G. (2012) An engineered diatom acting like a plasma cell secreting human IgG antibodies with high efficiency. Micro. Cell Fact. 11, 126.

Higgins, E. (2010) Carbohydrate analysis throughout the development of a protein therapeutic. Glycoconj. J. 27, 211–225.

Hummel, E., Schmickl, R., Hinz, G., Hillmer, S. and Robinson, D.G. (2007) Brefeldin A action and recovery in Chlamydomonas are rapid and involve fusion and fission of Golgi cisternae. Plant Biol. 9, 489–501.

Jinkerson, R.E. and Jonikas, M.C. (2015) Molecular techniques to interrogate and edit the Chlamydomonas nuclear genome. Plant J. 82, 393–412.

Kagiwada, S., Nakamae, I., Kayukawa, M. and Kato, S. (2004) Cytoskeleton-dependent polarized secretion of arylsulfatase in the unicellular green alga, Chlamydomonas reinhardtii. Plant Sci. 166, 1515–1524.

Karas, B.J., Diner, R.E., Lefebvre, S., McQuaid, J., Philips, A.P.R., Noddings, C., Brunson, J.K. et al. (2015) Disigner diatom episomes delivered by bacterial conjugation. Nat. Commun. 6, 6925.

Keskiaho, K., Hieta, R., Sormunen, R. and Myllyharju, J. (2007) Chlamydomonas reinhardtii has multiple prolyl 4-hydroxylases, one of which is essential for proper cell wall assembly. Plant Cell, 19, 256–269.

Kieliszewski, M.J. and Shpak, E. (2001) Synthetic genes for the elucidation of glycosylation codes for arabinogalactan-proteins and other hydroxyproline-rich glycoproteins. Cell. Mol. Life Sci. 58, 1386–1398.

Kinoshita, T., Fukuzawa, H., Shimada, T., Saito, T. and Matsuda, Y. (1992) Primary structure and expression of a gamete lytic enzyme in Chlamydomonas reinhardtii: similarity of functional domains to matrix metalloproteases. Proc. Natl. Acad. Sci. U. S. A. 89, 4693–4697.

Kong, F., Yamasaki, T., Kurniasih, S.D., Hou, L., Li, X., Ivanova, N., Okada, S. et al. (2015) Robust expression of heterologous genes by selection marker fusion system in improved Chlamydomonas strains. J. Biosci. Bioeng. 120, 239–245.

Kurniasih, S.D., Yamasaki, T., Kong, F., Okada, S., Widyaningrum, D. and Ohama, T. (2016) UV-mediated Chlamydomonas mutants with enhanced nuclear transgene expression by disruption of DNA methylation-dependent and independent silencing systems. Plant Mol. Biol. 92, 629–641.

Lauersen, K.J., Berger, H., Mussgnug, J.H. and Kruse, O. (2013a) Efficient recombinant protein production and secretion from nuclear transgenes in Chlamydomonas reinhardtii. J. Biotechnol. 167, 101–110.

Lauersen, K.J., Vanderveer, T.L., Berger, H., Kaluza, I., Mussgnug, J.H., Walker, V.K. and Kruse, O. (2013b) Ice recrystallization inhibition mediated by a nuclear-expressed and -secreted recombinant ice-binding protein in the microalga Chlamydomonas reinhardtii. Appl. Microbiol. Biotechnol. 97, 9763–9772.

Lauersen, K.J., Huber, I., Wichmann, J., Baier, T., Leiter, A., Gaukel, V., Kartushin, V. et al. (2015a) Investigating the dynamics of recombinant protein secretion from a microalgal host. J. Biotechnol. 215, 62–71.

Lauersen, K.J., Kruse, O. and Mussgnug, J.H. (2015b) Targeted expression of nuclear transgenes in Chlamydomonas reinhardtii with a versatile, modular vector toolkit. Appl. Microbiol. Biotechnol. 99, 3491–3503.

Lingg, N., Zhang, P., Song, Z. and Bardor, M. (2012) The sweet tooth of biopharmaceuticals: importance of recombinant protein glycosylation analysis. Biotechnol. J. 7, 1462–1472.

Mathieu-Rivet, E., Kiefer-Meyer, M.C., Vanier, G., Ovide, C., Burel, C., Lerouge, P. and Bardor, M. (2014) Protein N-glycosylation in eukaryotic microalgae and its impact on the production of nuclear expressed biopharmaceuticals. *Front. Plant Sci.* **5**, 359.

Mathieu-Rivet, E., Scholz, M., Arias, C., Dardelle, F., Schulze, S., Le Mauff, F., Teo, G., *et al.* (2013) Exploring the N-glycosylation pathway in Chlamydomonas reinhardtii unravels novel complex structures. *Mol Cell Proteomics.* **12**, 3160–3183.

Matsuda, Y., Saito, T., Yamaguchi, T., Koseki, M. and Hayashi, K. (1987) Topography of cell wall lytic enzyme in *Chlamydomonas reinhardtii*: form and location of the stored enzyme in vegetative cell and gamete. *J. Cell Biol.* **104**, 321–329.

Miller, D.H., Lamport, D.T. and Miller, M. (1972) Hydroxyproline heterooligosaccharides in Chlamydomonas. *Science,* **176**, 918–920.

Mussgnug, J.H. (2015) Genetic tools and techniques for *Chlamydomonas reinhardtii. Appl. Microbiol. Biotechnol.* **99**, 5407–5418.

Nagai, T., Ibata, K., Park, E.S., Kubota, M., Mikoshiba, K. and Miyawaki, A. (2002) A variant of yellow fluorescent protein with fast and efficient maturation for cell-biological applications. *Nat. Biotechnol.* **20**, 87–90.

Neupert, J., Karcher, D. and Bock, R. (2009) Generation of Chlamydomonas strains that efficiently express nuclear transgenes. *Plant J.* **57**, 1140–1150.

Nikolov, Z.L. and Woodard, S.L. (2004) Downstream processing of recombinant proteins from transgenic feedstock. *Curr. Opin. Biotechnol.* **15**, 479–486.

Ogawa-Ohnishi, M., Matsushita, W. and Matsubayashi, Y. (2013) Identification of three hydroxyproline O-arabinosyltransferases in *Arabidopsis thaliana. Nat. Chem. Biol.* **9**, 726–730.

Petersen, T.N., Brunak, S., von Hejne, G. and Nielsen, H. (2011) SignalP 4.0: descriminating signal peptides from transmembrane regions. *Nat. Methods,* **8**, 785–786.

Plucinak, T.M., Horken, K.M., Jiang, W., Fostvedt, J., Nguyen, S.T. and Weeks, D.P. (2015) Improved and versatile viral 2A platforms for dependable and inducible high-level expression of dicistronic nuclear genes in *Chlamydomonas reinhardtii. Plant J.* **82**, 717–729.

Purton, S., Szaub, J.B., Wannathong, T., Young, R. and Economou, C.K. (2013) Genetic engineering of algal chloroplasts: progress and prospects. *Russian J. Plant Physiol.* **60**, 491–499.

Rasala, B.A. and Mayfield, S.P. (2015) Photosynthetic biomanufacturing in green algae; production of recombinant proteins for industrial, nutritional, and medical uses. *Photosynth. Res.* **123**, 227–239.

Rasala, B.A., Lee, P.A., Shen, Z., Briggs, S.P., Mendez, M. and Mayfield, S.P. (2012) Robust expression and secretion of Xylanase1 in *Chlamydomonas reinhardtii* by fusion to a selection gene and processing with the FMDV 2A peptide. *PLoS ONE,* **7**, e43349.

Ruecker, O., Zillner, K., Groebner-Ferreira, R. and Heitzer, M. (2008) Gaussia-luciferase as a sensitive reporter gene for monitoring promoter activity in the nucleus of the green alga *Chlamydomonas reinhardtii. Mol. Gen. Genom.* **280**, 153–162.

Schroda, M., Blöcker, D. and Beck, C.F. (2000) The HSP70A promoter as a tool for the improved expression of transgenes in Chlamydomonas. *Plant J.* **21**, 121–131.

Scranton, M.A., Ostrand, J.T., Georgianna, D.R., Lofgren, S.M., Li, D., Ellis, R.C., Carruthers, D.N. *et al.* (2016) Synthetic promoters capable of driving robust nuclear gene expression in the green alga *Chlamydomonas reinhardtii. Algal Res.* **15**, 135–142.

Shao, N., Bock, R. (2008) A codon-optimized luciferase from Gaussia princeps facilitates the in vivo monitoring of gene expression in the model alga Chlamydomonas reinhardtii. *Curr Genet.,* **53**, 381–388.

Showalter, A.M. (1993) Structure and function of plant cell wall proteins. *Plant Cell,* **5**, 9–23.

Shpak, E., Leykam, J.F. and Kieliszewski, M.J. (1999) Synthetic genes for glycoprotein design and the elucidation of hydroxyproline-O-glycosylation codes. *Proc. Natl Acad. Sci. U. S. A.* **96**, 14736–14741.

Shpak, E., Barbar, E., Leykam, J.F. and Kieliszewski, M.J. (2001) Contiguous hydroxyproline residues direct hydroxyproline arabinosylation in *Nicotiana tabacum. J. Biol. Chem.* **276**, 11272–11278.

Solá, R.J. and Griebenow, K.A.I. (2009) Effects of Glycosylation on the stability of protein pharmaceuticals. *J. Pharmaceut. Sci.* **98**, 1223–1245.

Specht, E., Miyake-Stoner, S. and Mayfield, S. (2010) Micro-algae come of age as a platform for recombinant protein production. *Biotechnol. Lett.* **32**, 1373–1383.

Tan, L., Leykam, J.F. and Kieliszewski, M.J. (2003) motifs that direct arabinogalactan addition to arabinogalactan-proteins. *Plant Physiol.* **132**, 1362–1369.

Tan, L., Qiu, F., Lamport, D.T. and Kieliszewski, M.J. (2004) Structure of a hydroxyproline (Hyp)-arabinogalactan polysaccharide from repetitive Ala-Hyp expressed in transgenic *Nicotiana tabacum. J. Biol. Chem.* **279**, 13156–13165.

Voigt, J., Stolarczyk, A., Zych, M., Malec, P. and Burczyk, J. (2014) The cell-wall glycoproteins of the green alga *Scenedesmus obliquus.* The predominant cell-wall polypeptide of *Scenedesmus obliquus* is related to the cell-wall glycoprotein gp3 of *Chlamydomonas reinhardtii. Plant Sci.* **215–216**, 39–47.

Walsh, G. and Jefferis, R. (2006) Post-translational modifications in the context of therapeutic proteins. *Nat. Biotechnol.* **24**, 1241–1252.

Woessner, J.P. and Goodenough, U.W. (1989) Molecular characterization of a zygote wall protein: an extensin-like molecule in *Chlamydomonas reinhardtii. Plant Cell,* **1**, 901–911.

Woessner, J.P. and Goodenough, U.W. (1992) Zygote and vegetative cell wall proteins in *Chlamydomonas reinhardtii* share a common epitope, (SerPro)X. *Plant Sci.* **83**, 65–76.

Woessner, J.P. and Goodenough, U.W. (1994) Volvocine cell walls and their constituent glycoproteins: an evolutionary perspective. *Protoplasma,* **181**, 245–258.

Xu, J., Tan, L., Goodrum, K.J. and Kieliszewski, M.J. (2007) High-yields and extended serum half-life of human interferon α2b expressed in tobacco cells as arabinogalactan-protein fusions. *Biotechnol. Bioeng.* **97**, 997–1008.

Xu, J., Tan, L., Lamport, D.T.A., Showalter, A.M. and Kieliszewski, M.J. (2008) The O-Hyp glycosylation code in tobacco and Arabidopsis and a proposed role of Hyp-glycans in secretion. *Phytochemistry,* **69**, 1631–1640.

Xu, J., Okada, S., Tan, L., Goodrum, K.J., Kopchick, J.J. and Kieliszewski, M.J. (2010) Human growth hormone expressed in tobacco cells as an arabinogalactan-protein fusion glycoprotein has a prolonged serum life. *Transgen. Res.* **19**, 849–867.

Zedler, J.A., Gangl, D., Guerra, T., Santos, E., Verdelho, V.V. and Robinson, C. (2016) Pilot-scale cultivation of wall-deficient transgenic *Chlamydomonas reinhardtii* strains expressing recombinant proteins in the chloroplast. *Appl. Microbiol. Biotechnol.* **100**, 7061–7070.

Zhang, Y.H. and Robinson, D.G. (1990) Cell-wall synthesis in *Chlamydomonas reinhardtii:* an immunological study on the wild type and wall-less mutants cw2 and cw15. *Planta,* **180**, 229–236.

Zhang, Y.-H., Lang, W.C. and Robinson, D.G. (1989) *In vitro* localization of hydroxyproline O-glycosyl transferases in *Chlamydomonas reinhardii. Plant Cell Physiol.* **30**, 617–622.

Zhang, N., Gonzalez, M., Savary, B. and Xu, J. (2016) High-yield secretion of recombinant proteins expressed in tobacco cell culture with a designer glycopeptide tag: process development. *Biotechnol. J.* **11**, 497–506.

9

Spearmint R2R3-MYB transcription factor MsMYB negatively regulates monoterpene production and suppresses the expression of geranyl diphosphate synthase large subunit (*MsGPPS.LSU*)

Vaishnavi Amarr Reddy[1,2] (ID), Qian Wang[1,a,#], Niha Dhar[1,#], Nadimuthu Kumar[1], Prasanna Nori Venkatesh[1],
Chakravarthy Rajan[1,b], Deepa Panicker[1,b], Vishweshwaran Sridhar[1], Hui-Zhu Mao[1] and Rajani Sarojam[1,*]

[1]*Temasek Life Sciences Laboratory, National University of Singapore, Singapore, Singapore*
[2]*Department of Biological Sciences, National University of Singapore, Singapore, Singapore*

*Correspondence
email rajanis@tll.org.sg
[a]Present address: College of Biological and Environmental Sciences, Zhejiang Wanli University, Ningbo, Zhejiang 315100, China.
[b]Present address: Singapore Centre on Environmental Life Sciences Engineering, Nanyang Technological University, Singapore 637551, Singapore
[#]These authors contributed equally.

Keywords: transcription factor, R2R3-MYB, secondary metabolism, spearmint, GPPS, terpene.

Summary

Many aromatic plants, such as spearmint, produce valuable essential oils in specialized structures called peltate glandular trichomes (PGTs). Understanding the regulatory mechanisms behind the production of these important secondary metabolites will help design new approaches to engineer them. Here, we identified a PGT-specific R2R3-MYB gene, *MsMYB*, from comparative RNA-Seq data of spearmint and functionally characterized it. Analysis of *MsMYB*-RNAi transgenic lines showed increased levels of monoterpenes, and *MsMYB*-overexpressing lines exhibited decreased levels of monoterpenes. These results suggest that MsMYB is a novel negative regulator of monoterpene biosynthesis. Ectopic expression of *MsMYB*, in sweet basil and tobacco, perturbed sesquiterpene- and diterpene-derived metabolite production. In addition, we found that MsMYB binds to *cis*-elements of *MsGPPS.LSU* and suppresses its expression. Phylogenetic analysis placed MsMYB in subgroup 7 of R2R3-MYBs whose members govern phenylpropanoid pathway and are regulated by miR858. Analysis of transgenic lines showed that *MsMYB* is more specific to terpene biosynthesis as it did not affect metabolites derived from phenylpropanoid pathway. Further, our results indicate that *MsMYB* is probably not regulated by miR858, like other members of subgroup 7.

Introduction

Plants produce an overwhelming variety of specialized metabolites essential for ecological interactions, among which terpenes form the largest and structurally diverse class of natural products. They are involved in mediating plant defence responses and plant pollinator attraction and facilitate plant–environment interactions (Bhargava et al., 2013; Pichersky and Gershenzon, 2002; Singh and Sharma, 2015). Apart from imparting ecological benefits to plants, terpenoids are of great economic importance to humans as well, as they are widely used in flavours, fragrances, cosmetics, pharmaceuticals, agricultural industries and chemical industries (Bouvier et al., 2005). Given the commercial and ecological importance of terpenoids, strategies to metabolically engineer them are of considerable interest. Monoterpenes form the C_{10} class of terpenoids and are generally colourless, lipophilic and volatile. They are responsible for the characteristic aromas and flavours of essential oils, floral scents and resin of aromatic plants (Loza-Tavera, 1999). The essential oil of spearmint (*Mentha spicata*) mainly consists of two monoterpenes, limonene and carvone, which are extensively exploited for their biological properties (Ringer et al., 2005). In spearmint, the essential oils are produced and stored in specialized structures called peltate glandular trichomes (Figure S1a). These so-called green biofactories are widely found on aerial surfaces of many aromatic plants, and they actively produce and store large quantity of volatile

metabolites (Champagne and Boutry, 2013; Lange and Turner, 2013; Turner et al., 2000).

Terpene biosynthesis pathway has been well studied in plants. The universal C_5 building blocks for all types of terpenes, isopentenyl diphosphate (IPP) and its isomer, dimethylallyl diphosphate (DMAPP), are synthesized either by the mevalonate (MVA) pathway in the cytosol or by the 2-C-methyl-D-erythritol 4-phosphate (MEP) pathway in plastids (Vranova et al., 2013). The plastidial MEP pathway is largely responsible for producing C_5 precursors for monoterpenes and diterpenes production, whereas the cytosolic MVA pathway generates C_5 precursors for sesquiterpene and triterpene production (Dubey et al., 2003). However, few studies indicate that, under certain conditions, an exchange of precursor metabolites can occur between the cytosolic MVA and plastidial MEP pathways (Hemmerlin et al., 2012; Vranova et al., 2013). IPP and DMAPP undergo successive condensation reactions, catalysed by a class of enzymes called prenyltransferases, to form intermediates geranyl diphosphate (GPP; C_{10}), farnesyl diphosphate (FPP; C_{15}) and geranylgeranyl diphosphate (GGPP; C_{20}). These terpene diphosphates form the immediate precursors of monoterpenes, sesquiterpenes and diterpenes, respectively (Kellogg and Poulter, 1997; Liang, 2009; Liang et al., 2002; Ogura and Koyama, 1998). Prenyltransferases are key to terpene production as they control the IPP flux into various branches of the terpene family (Liang et al., 2002; Oldfield and Lin, 2012).

Geranyl diphosphate synthase (GPPS) is the prenyltransferase enzyme that is mainly responsible for the production of monoterpene precursor GPP in the plastids. It catalyses a single condensation of terpene precursors DMAPP and IPP, to form GPP. Studies in few species of plants revealed that GPPS can function as a homodimer or heterodimer (Nagegowda, 2010). Heteromeric GPPSs have been characterized so far only in angiosperms *Mentha piperita, Antirrhinum majus, Clarkia breweri* and *Humulus lupulus*. All these plants produce large amounts of monoterpenes in specialized organs such as PGTs and flower petals (Burke *et al.*, 1999; Tholl *et al.*, 2004; Wang and Dixon, 2009). Heteromeric GPPS in mint consist of a large subunit (LSU) and a small subunit (SSU); both LSU and SSU are catalytically inactive alone (Burke and Croteau, 2002; Burke *et al.*, 1999; Croteau *et al.*, 2005). Interaction between the two subunits results in the formation of an active GPPS. Structural studies show that the LSU acts as the catalytic unit, whereas the SSU serves as the regulatory unit (Chang *et al.*, 2010).

In the case of spearmint, the monoterpene biosynthetic pathway is well characterized (Champagne and Boutry, 2013; Croteau *et al.*, 1991; Lange *et al.*, 2011; Munoz-Bertomeu *et al.*, 2008). Strategies to increase the yield of peppermint oil by manipulating genes that code for structural pathway enzymes have been reported (Diemer *et al.*, 2001; Mahmoud *et al.*, 2004). However, the developmental regulation of this secondary metabolite pathway still remains elusive. Transcription factors (TFs) can activate or repress multiple genes in a metabolic pathway; hence, they are ideal targets for pathway engineering (Grotewold, 2008; Iwase *et al.*, 2009). Only one TF *MsYABBY5* has been reported from spearmint that regulates monoterpene production (Wang *et al.*, 2016) but several TFs have been identified from other plants like *Artemisia*, cotton, *Taxus*, rubber and rice that control terpene biosynthesis (Chen *et al.*, 2012; Li *et al.*, 2013; Miyamoto *et al.*, 2014; Shen *et al.*, 2016; Xu *et al.*, 2004; Zhang *et al.*, 2012).

The R2R3-MYBs represent one of the largest families of plant TFs (Kranz *et al.*, 1998; Stracke *et al.*, 2001) and can function as activators or repressors of genes (Adato *et al.*, 2009; Legay *et al.*, 2007). Many R2R3-MYBs have been characterized as transcriptional activators of flavonoid, phenylpropanoid and glucosinolate pathway genes in plants (Dubos *et al.*, 2010; Zhao and Dixon, 2011; Zhong *et al.*, 2010). R2R3-MYBs belonging to subgroup 4 are known to negatively regulate monolignol and flavonoid biosynthetic pathways (Fornalé *et al.*, 2010; Jin *et al.*, 2000; Legay *et al.*, 2007; Preston *et al.*, 2004; Tamagnone *et al.*, 1998). Further, a single R2R3-MYB gene was also found to act as both repressor and activator of genes (Bhargava *et al.*, 2010). Redundancy, dual function and opposite action as activators and repressors on common target genes by R2R3-MYBs help fine tune the regulation of plant secondary metabolite pathways. Despite several studies on R2R3 MYB-mediated regulation of plant phenylpropanoid pathway, knowledge of its role in terpene secondary metabolism is lacking (Matus, 2016). Besides TFs, microRNAs also are known regulators of various processes in plant growth and development. However, knowledge of their role in controlling secondary metabolic pathways is limited. microRNAs mainly target TFs to regulate various plant developmental processes (Sunkar *et al.*, 2012; Wu *et al.*, 2013; Zhang, 2015; Zhang and Wang, 2015). In plants, MYB TFs are potential targets for miR858 (Addo-Quaye *et al.*, 2008; Guan *et al.*, 2014; Jeong *et al.*, 2013; Jia *et al.*, 2015; Xia *et al.*, 2012). Further, many studies have shown that miR858 can target R2R3-MYBs in

apple, grapes, salvia and Arabidopsis (Li and Lu, 2014; Rock, 2013; Sharma *et al.*, 2016; Xia *et al.*, 2012). Presently, miR858 is considered as a potential regulator of plant phenylpropanoid metabolite pathway through its action on R2R3-MYBs (Fahlgren *et al.*, 2007).

In this study, we identified a R2R3-MYB transcript which is preferentially expressed in spearmint PGTs from the comparative transcriptome data (Jin *et al.*, 2014) of spearmint leaf, leaf stripped of PGT (L-PGT) and PGT. Phylogenetic analysis assigned MsMYB to subgroup 7 of R2R3-MYBs of Arabidopsis. Members of subgroup 7 are closely related R2R3-MYBs which control flavonoid accumulation in different parts of the Arabidopsis seedling, and are regulated by miR858 (Fahlgren *et al.*, 2007; Sharma *et al.*, 2016; Yang *et al.*, 2012). Apart from Arabidopsis, several other R2R3-MYBs from different plants, similar to the members of subgroup 7, are also involved in flavonoid pathway (Adato *et al.*, 2009). We functionally characterized the role of this PGT-specific R2R3-MYB and found it to be a negative regulator of monoterpene production in spearmint. Suppression of *MsMYB* expression, by RNAi approach, led to increase in monoterpene levels and its overexpression led to decrease in monoterpene levels in transgenic lines. This is the first report of regulation of monoterpene production by a R2R3-MYB. Ectopic expression of *MsMYB* in sweet basil (*Ocimum basilicum*), an aromatic plant similar to spearmint, and in tobacco (*Nicotiana sylvestris*), affected the production of terpenes but not of flavonoids. We further found that MsMYB suppresses the expression of geranyl diphosphate synthase large subunit (*MsGPPS.LSU*) gene in spearmint. To investigate whether *MsMYB* is regulated by miR858, similar to other R2R3-MYBs of subgroup 7, we performed small RNA-Seq of spearmint PGT and L-PGT. We found low amounts of miR858 in spearmint PGT but high amounts in L-PGT. Although *in silico* analysis predicted *MsMYB* as a target for miR858, we were unable to detect its cleaved products. This suggests that the differential expression of miR858 and *MsMYB* in separate tissues prevents *MsMYB* regulation by miR858. Collectively, the above results provide insight into the production of essential oils in a valuable crop plant, such as spearmint, and expand our knowledge of R2R3-MYB-mediated regulation of plant secondary metabolism.

Results

MsMYB shows preferential expression in spearmint PGTs

From the transcriptome data of three tissues of spearmint, namely leaves, PGTs and leaves devoid of PGTs (Jin *et al.*, 2014), we annotated 45 R2R3-MYBs (Figure 1; Supporting Data S1). Of these, only two R2R3-MYBs, *MsMYB* and *MsMYB112*, showed high expression in PGTs. *MsMYB* was chosen for further characterization. RNA-Seq expression of *MsMYB* along various tissues was further validated by quantitative RT-PCR (qRT-PCR) (Figure 2a). Full-length open reading frame (ORF) of *MsMYB* was amplified from PGT cDNA and consisted of 813 base pairs (bp) encoding a polypeptide of 271 amino acids (Figure 3). It contained a R2, R3 repeat and five tryptophan residues within the R2 and R3 repeats which together forms a helix-turn-helix motif at the N-terminus. These features are generally conserved among all the R2R3-MYBs (Dubos *et al.*, 2010). MsMYB showed highest sequence similarity (~60%) to *Salvia miltiorrhiza* MYB-related transcription factor in BLAST analysis. A phylogenetic tree was constructed using the amino acid sequences of known *Arabidopsis thaliana* (At) R2R3-MYBs (Figure S2). MsMYB fell

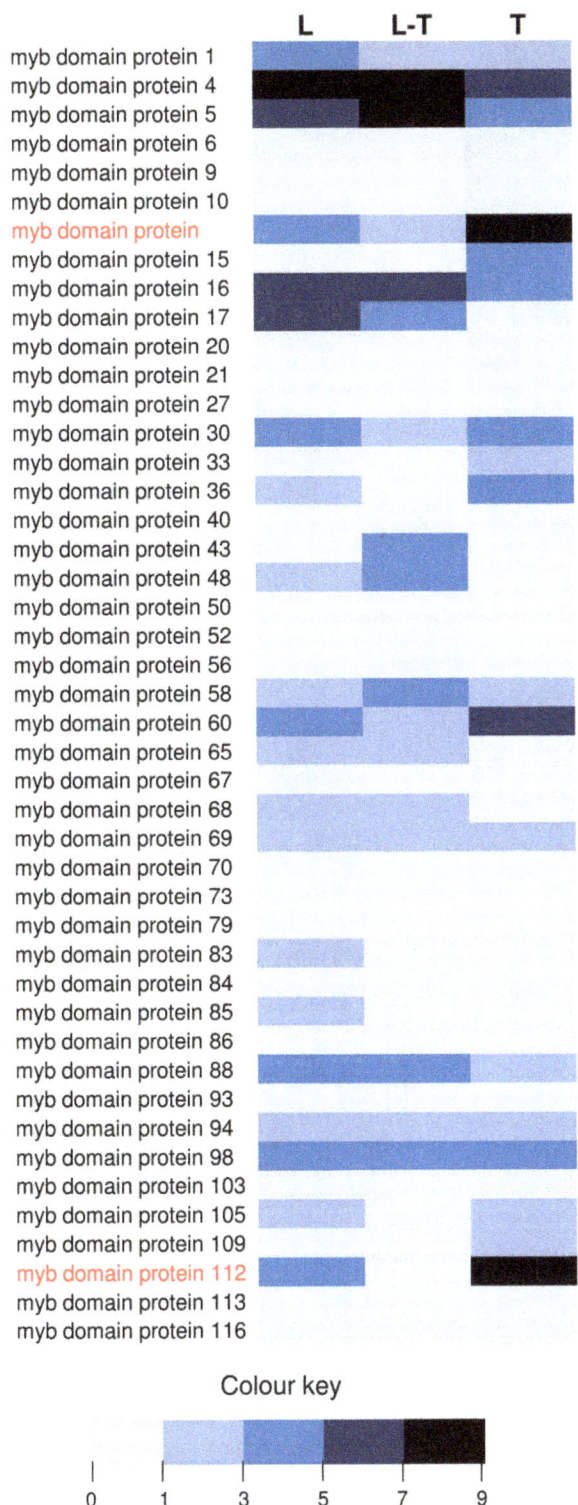

Figure 1 Heat map analysis of spearmint R2R3-MYBs. Differential expression pattern of 45 annotated R2R3-MYBs along various tissues [leaf (L), leaf stripped of PGTs (L-T) and PGTs (T)]. MYB and MYB112 are highlighted in red which show high expression in PGTs. Sequence data and expression values are available in Supplemental Data S1.

under subgroup 7 along with AtMYB111, AtMYB11 and AtMYB12 (Dubos et al., 2010). All these three Arabidopsis MYBs function as activators of flavonoid pathway (Stracke et al., 2007).

To examine subcellular localization of MsMYB, the ORF was fused with yellow fluorescent protein (YFP) and expressed under the control of CaMV 35S promoter in *Nicotiana benthamiana* leaves by agroinfiltration. As shown in Figure 2b, the recombinant protein specifically localized to the nucleus.

MsMYB promoter shows trichome specific expression

To identify the *cis*-acting regulatory elements in the promoter of *MsMYB*, we cloned a 615-bp genomic DNA fragment upstream of the translation start site by genome walking. Apart from the common TATA and CAAT box, several other *cis*-acting regulatory elements were identified (Figure 4a) using PlantCARE tool (http://bioinformatics.psb.ugent.be/webtools/plantcare/html/). Many of these elements were for hormones like ABRE-motif which is known for abscisic acid responsiveness (Basu et al., 2014), AuxRE-motif which is an auxin-responsive element (Ulmasov et al., 1995), CGTCA-motif and TGACG-motif which are known for MeJA responsiveness (Zhou et al., 2012). Interestingly an AC-II element was also found. In general, AC-rich regions are known to be bound by R2R3-MYBs (Koschmann et al., 2012; Prouse and Campbell, 2012). This suggests that *MsMYB* might be regulated by other R2R3-MYBs. To check for the expression pattern, the 615-bp promoter fragment was fused with a β-glucuronidase (GUS) reporter gene and transformed into *N. benthamiana* plants. Trichome-specific staining was observed in leaves and stems of tobacco transgenic plants (Figure 4b,c).

Manipulation of *MsMYB* expression levels affects monoterpene production in spearmint

To characterize the function of *MsMYB*, we generated transgenic spearmint lines harbouring RNAi construct and overexpression construct where *MsMYB* was under the control of CaMV 35S promoter. Four independent lines each confirmed by Southern blot were selected for further characterization (Figure S1b,c). qRT-PCR analysis showed significant reduction in levels of *MsMYB* transcripts in RNAi lines (Figure 5a) and higher levels of *MsMYB* transcripts in overexpression lines when compared to wild type (WT) (Figure 5b). Both overexpression and RNAi transgenic plants appeared phenotypically similar to WT plants. Scanning electron microscopy was performed on these plants to take a closer look at leaf cells and PGTs. No phenotypical changes were observed.

Further, gas chromatography–mass spectrometry (GC-MS) analysis was performed to evaluate any changes in volatiles production. Major secondary metabolites of spearmint are monoterpenes; limonene and carvone (Figure S3a). GC-MS analysis revealed significant changes in the amount of these secondary metabolites in transgenic lines. Compared to WT plants, total monoterpene production in *MsMYB*-RNAi lines was higher; about 2.3- to 4.5-fold increase was observed (Figure 5c, d), whereas in *MsMYB*-overexpressing lines a decrease of 0.5- to 0.7-fold was seen (Figure 5e,f). The above results indicated that *MsMYB* might be a negative regulator of secondary metabolism in spearmint.

MsMYB negatively regulates *MsGPPS.LSU*

To understand how MsMYB affects terpene metabolism, it is essential to know its downstream target genes. Towards this, we checked the expression levels of several transcripts encoding enzymes in various terpene precursor pathways, transporters and flavonoid pathway (Table S2), but none of them were significantly changed except for *MsGPPS.LSU*. It showed

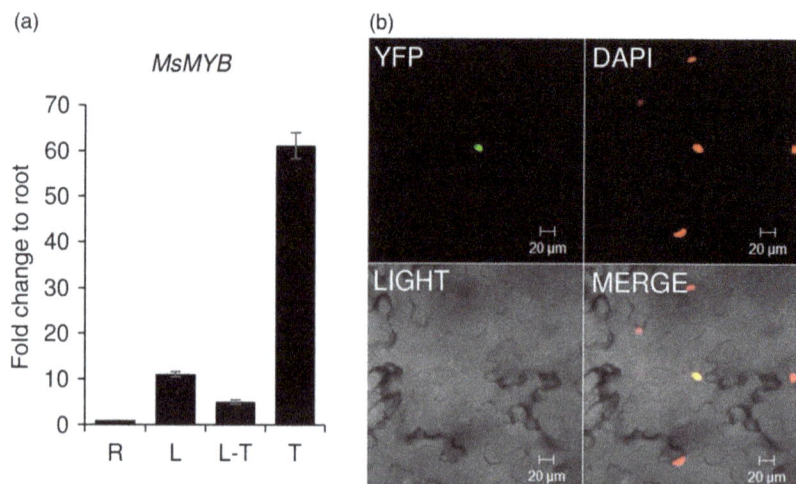

Figure 2 *MsMYB* expression and localization. qRT-PCR was performed to analyse the expression of *MsMYB* along the various tissues [leaf (L), leaf stripped of PGTs (L-T), root (R) and PGTs (T)]. (a) Expression levels of *MsMYB* showing preferential expression in PGTs. (b) Nucleus-specific localization of MsMYB in *N. benthamiana* leaf cells.

> *MsMYB* (KY081780)

```
ATGGGAAGAGCGCCGTGCTGTGAGAAAGTTGGGTTGAAGAGAGGGAGATGGACTGCAGAAGAA
GATGAAAAGCTCAGAAAATATATTCAGGAAAATGGTGAAGGCTGCTGGCGATCATTGCCCAAG
AATGCAGGTTTACTTAGATGTGGAAAGAGTTGCAGACTGAGATGGATTAATTATTTGAGATCA
GATGTGAAGAGAGGGAATATTTCTTCTCAAGAAGAAGAAATCATCATTAATCTCCATGCATCT
ATGGGCAACAGGTGGTCCCTGATCGCCGCGCACTTGCCGGGTAGAACAGACAATGAAATCAAA
AATTACTGGAACTCCCATTTGAGCAGAAAATTCCACGGTTTCCGCCCTAATCCACAGTTCATT
CCGCCGCCGCCGCCTCCGCCGCCATCCTCCAAGCCCAAGAAGACAAAGAACAGTAATAAGAAG
GCAAAGGCGGCAGCTAAAACCGCCACTGCCGCCGTCGTCATGCCGACTACCCCCACTCCGGAG
AAAGAGTCCTCGGTGGGCAGACCTGGGAAGGAGAGAGAAAGTGAAGCGAGAGAGAGCGGCAGC
TCCATGGTGGGAGAGTTGGACGATTTAAATATGACGGAGGACTTGAGCGGCCTCTGGGGCCCC
ACACTGGACTTCGGGACCGGAAGTGAAATTTCGGATCCGGGTCTGGGCCAGCTAGATAACTCG
TCGTTGCAAATTTACGAAGAAACGCTTTCGTGGATTTGGAACGACGAGGACGACAAGTGGAAT
TCCAACGTGGACAACGGAGAAATGGATGGTGCAATGCTTTCTTGGTTGTTGTCATGA
```

> MsMYB

MGRAPCQ[EKVGLKRGRWTAEEDEKLRKYIQE]NGEGCWRSLPKNAGLLRCGKSCRLRWINYL

RSDVKRGNISSQEEEIIINLHASMGNRWSLIAAHLPGRTDNEIKNYWNSHLSRKFHGFRPN

PQFIPPPPPPPPPSSKPKKTKNSNKKAKAAAKTATAAVVMPTTPTPEKESSVGRPGKERESE

ARESGSSMVGELDDLNMTEDLSGLWGPTLDFGTGSEISDPGLGQLDNSSLQIYEETLSWIW

NDEDDKWNSNVDNGEMDGAMLSWLLS

Figure 3 DNA and protein sequence of MsMYB. R2 and R3 repeats are highlighted in blue and red colours, respectively. Five conserved tryptophan residues are highlighted with grey boxes. Nuclear localization signal as predicted by cNLS mapper is shown by a black box.

increased levels in *MsMYB*-RNAi lines and decreased levels in *MsMYB*-overexpressing lines (Figure 6a,b). GPPSs are known to localize to both nongreen plastids and chloroplasts of photosynthetic cells where MEP pathway is active (Bouvier *et al.*, 2000). We checked the localization of MsGPPS.LSU and found it to be localized in the chloroplast (Figure 9b). To check whether the promoter of *MsGPPS.LSU* has putative MYB binding sites, we cloned a 549-bp genomic DNA fragment upstream of the translation start site by genome walking. Using PlantCARE tool we identified two MYB binding sites (MBS) along with several elements for hormones and light responsiveness (Figure 7a).

To verify MsMYB interaction with *GPPS.LSU* promoter, yeast one-hybrid assays were performed. The whole *GPPS.LSU* promoter along with two truncated regions, viz. site 1 and site 2, which had the conserved MYB binding domains and mutated versions of site 1 and site 2 was used as bait to generate five Y1HGold [Bait/AbAi] strains, that is pAbAi-GPPS-full, pAbAi-site1,

pAbAi-site2, pAbAi-mut-site1 and pAbAi-mut-site2. After cotransformation with functional *MsMYB* gene in pGADT7 vector (pGADT7-MYB), three Y1HGold strains, that is pAbAi-GPPS-full, pAbAi-site1 and pAbAi-site2, could grow on SD minus leucine with aureobasidin (SD/-Leu/AbA, 800 ng of AbA) auxotrophic medium but not the two mutant strains pAbAi-mut-site1 and pAbAi-mut-site2, confirming interaction of MsMYB with *GPPS.LSU* promoter *in vivo* (Figure 8). TFs can act as activators or repressors of transcription. Transgenic studies indicated that *MsMYB* is a negative regulator. To further determine the transcriptional repression activity of MsMYB in planta, transient expression assay in *N. benthamiana* was performed. We used $35S_{pro}$:GFP or $35S_{pro}$:MsMYB as effectors and MsGPPS.LSU$_{pro}$:GUS as reporter. Promoter activity of MsGPPS.LSU$_{pro}$:GUS was significantly reduced in $35S_{pro}$:MsMYB expressing leaves when compared to $35S_{pro}$:GFP expressing leaves (Figure 9a). These results show that *MsMYB* binds to the promoter of *MsGPPS.LSU* and suppresses its activity.

(a) CACTTATATATATTATTATAATTATATGAAATGAATTAGGGTTATGATTAG

GAGTGGGATTAAGATTTGAGAATGAGTGATTAGAGGGAAAAGGGAAAGG

ACE   **CGTCA- motif**
AGAAAGTGTCAACGTGGAAGCTCGACGCCCCGTCAGGAGGCGCTCGACA

G- box   **Box II ABRE**
GTGGAGGCGCCACGACGAGTCTACTCCTCTCTAGTCTACGTGGCGATTTC

box S
AGCCACCACTAATTCTTTAATTTCCATTTCTGCCCCCACATTTATTTTCCTC

                                              AC-II element
TCTCTCTCTCTTTCTCTTTCTCTCTCTCTCTATAAACCATAACCAACTTTCCT

CTCTGAGTCGCCCTTGTCTCTCTCTTTCTCCCACATCAACAATCAATTCAA

ACTTCAAACCCCATTACAAACAATAAAAGTCAAACCACGAGAATAATCCC

                    CGTCA- motif
CTTCTCTTTTTTTCTTGTATTTTTCTCACTCGTCACTTAAACAGTCGGGAAT

TATTTATATCGAAATTTAAGGTGGCATTGGCAACAACATACACTCCTATA

                AuxRE   **MRE**
ATAACTTGCCTCTTGTCTTGTCTCTATAAGTAGTACTCCATTAGGTTTTTG
                                              AATCCAA

TTGTCTAGTGAGAGAAGAGAGAGGTTTTTTGTTGTCTAGTGAGAGAAGAGAG

AGAGAG**ATG GGA AGA GCG**
      M   G   R   A

(b)

(c)

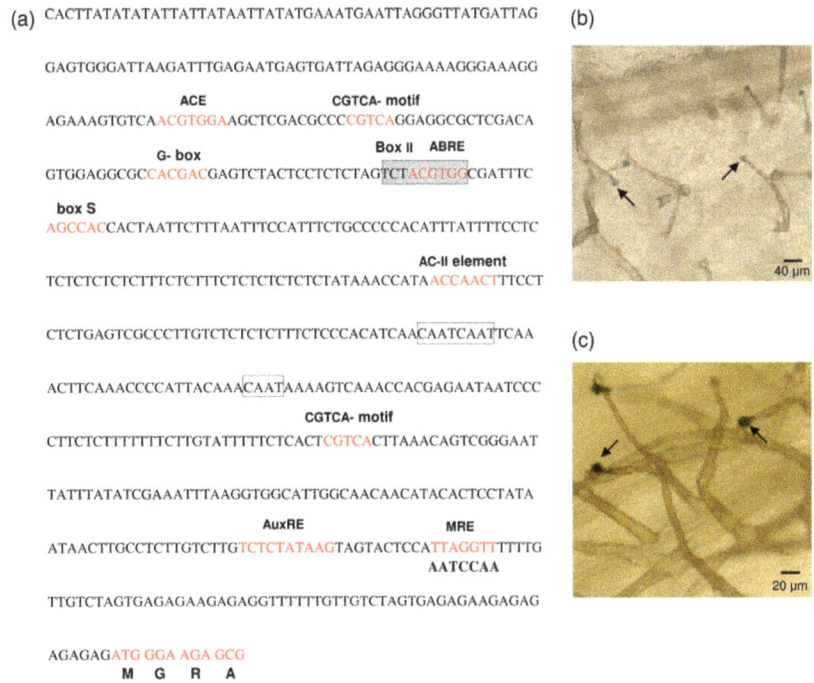

Figure 4 *MsMYB* promoter analysis and expression pattern in tobacco. (a) Regulatory elements present in the promoter region (−615 bp) of *MsMYB*. (b), (c) *N. benthamiana* plants transformed with *MsMYB_pro:GUS* showing trichome specific expression in leaves and stems. Black arrows point GUS stained trichome heads.

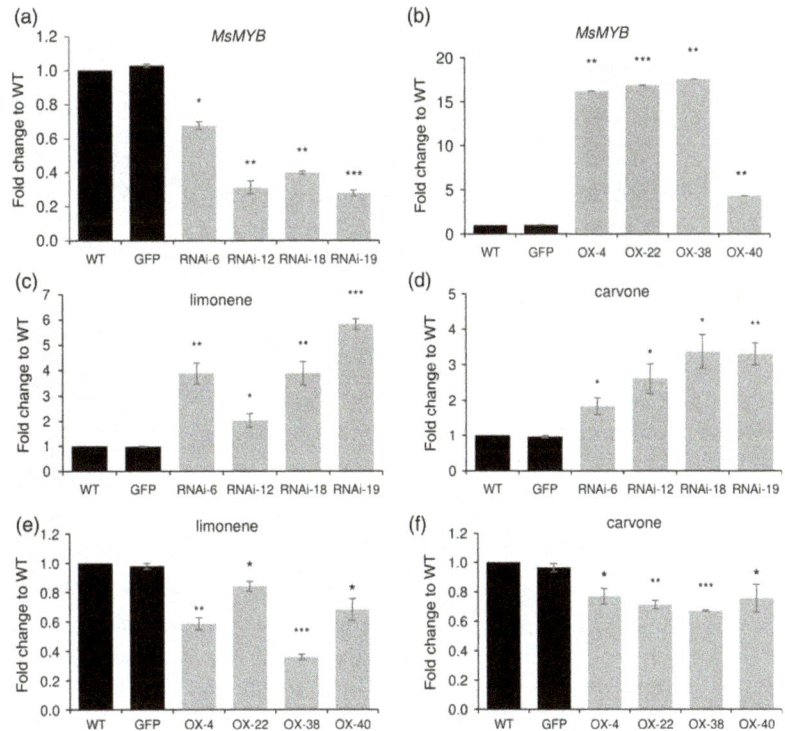

Figure 5 Transcript levels and GC analysis of transgenic plants. Reduced (a) and increased (b) levels of *MsMYB* in transgenic spearmint *MsMYB*-RNAi and *MsMYB*-overexpressing lines when compared to WT. (c), (d) Increased levels of limonene and carvone in *MsMYB*-RNAi lines when compared to WT. (e), (f) Reduced levels of limonene and carvone in *MsMYB*-overexpressing lines when compared to WT. GFP: GFP-overexpressing line. Data are indicated as mean ± SE. *$P < 0.05$; **$P < 0.01$; ***$P < 0.001$.

Terpene biosynthesis is known to be regulated by light in many plants (Cordoba *et al.*, 2009). As we found light-responsive elements in *MsGPPS.LSU* promoter, we checked whether *MsGPPS.LSU* is also regulated by light. WT spearmint plants were grown in dark and light conditions. Leaf samples were collected at 0-, 24-, 48- and 72-h interval to quantify *MsGPPS.LSU* transcript levels. Expression of *MsGPPS.LSU* was highly reduced in the plants grown under dark conditions suggesting that *MsGPPS.LSU* is regulated by light (Figure 7b). As GPPS is a key enzyme in the monoterpene pathway,

multiple layers of gene regulation to fine tune the pathway is plausible.

Ectopic expression of *MsMYB* affects secondary terpene metabolism in sweet basil and *Nicotiana sylvestris*

Similar to spearmint, sweet basil too produces its essential oil in PGTs. The constituents of sweet basil essential oil are derived from terpene and phenylpropanoid pathways (Figure S3b). It is a mixture of monoterpenes, sesquiterpenes and the main phenyl-propanoid derivative is eugenol. Ectopic expression of *MsMYB* in

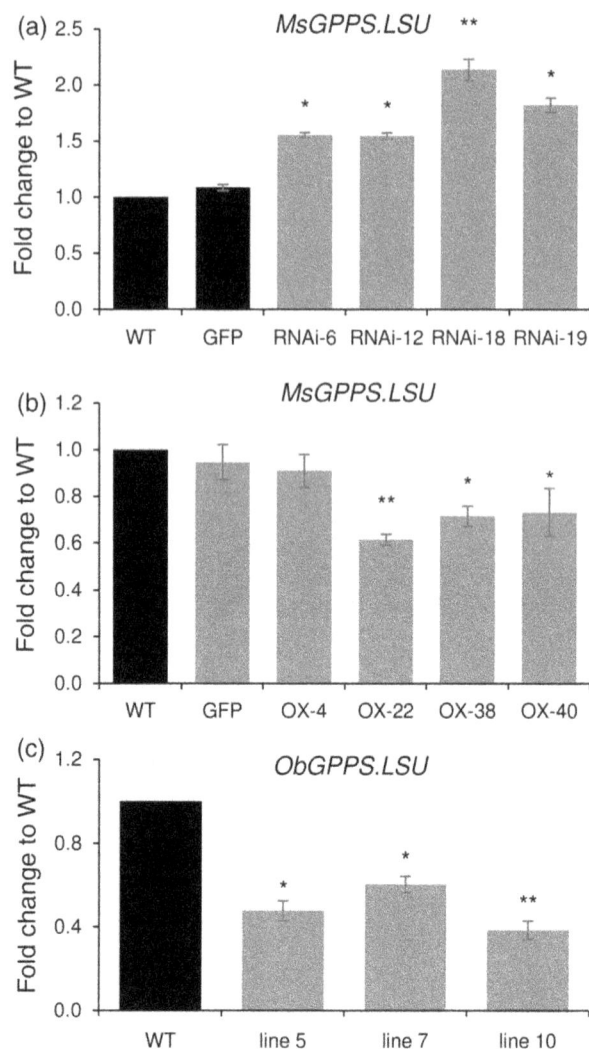

Figure 6 Transcript levels of *GPPS.LSU* in transgenic plants. (a) Increased levels of *MsGPPS.LSU* in *MsMYB*-RNAi lines. (b) Decreased levels of *MsGPPS.LSU* in *MsMYB*-OX lines. (c) Decreased levels of *ObGPPS.LSU* in sweet basil plants expressing *MsMYB*. Data are indicated as mean ± SE. *$P < 0.05$; **$P < 0.01$; ***$P < 0.001$.

sweet basil was pursued to evaluate whether *MsMYB* can affect sesquiterpene pathway and phenylpropanoid pathway as suggested by phylogenetic analysis. Three independent lines (line 5, line 7 and line 10) confirmed by Southern blot were further selected for characterization (Figure S1d). Ectopic expression of *MsMYB* in these lines was confirmed by qRT-PCR (Figure 10a). GC-MS analysis on T-2 plants revealed an overall 73%–85% decrease in total terpenes, both monoterpenes and sesquiterpenes (Figure 10b,c,d). However, the amount of eugenol derived from phenylpropanoid pathway remained unaltered (Figure 10e). Additionally, the leaves of transgenic lines were smaller in size when compared to WT (Figure S4a,b). From the NCBI database, we obtained the sequences of sweet basil geranyl diphosphate synthase large subunit (*ObGPPS.LSU*) and farnesyl diphosphate synthase (*ObFPPS*). To investigate whether similar to spearmint, the decrease in terpenes in sweet basil is due to the reduction in transcripts of *ObGPPS.LSU* or because of reduction in *ObFPPS* transcript, we measured their levels by qRT-PCR. When compared to WT, *ObGPPS.LSU* expression was significantly reduced in

transgenic plants (Figure 6c), but levels of *ObFPPS* were unaltered. Additionally, we measured the total flavonoid content in the leaves of sweet basil transgenic plants and found the total flavonoid content to be unaltered (Figure 10f). This suggests that *MsMYB* is a R2R3-MYB specific to terpene metabolism.

The sweet basil data showed that *MsMYB* can affect both monoterpene and sesquiterpene synthesis. To investigate whether it has a role in diterpene synthesis, we ectopically expressed *MsMYB* in *N. sylvestris*. Glandular trichomes of *N. sylvestris* mainly produce diterpenes; cembranoids (CBT-diol) are generally derived from the same MEP pathway as the monoterpenes. Three independent lines were chosen for further characterization and ectopic expression of *MsMYB* was confirmed by qRT-PCR (Figure S5a). Analysis of CBT-diols production in transgenic plants showed a 42%–50% reduction (Figure S5b). Similar to sweet basil plants, the leaves of tobacco transgenic lines were smaller in size when compared to WT (Figure S4c). The above results indicate that *MsMYB* can affect the flux in various branches of terpene metabolism.

MsMYB is probably not a target of microRNA858

microRNA858 is known to regulate several R2R3-MYBs, specifically the subgroup 7 R2R3-MYBs, *AtMYB11*, *AtMYB111* and *AtMYB12* (Fahlgren et al., 2007; Sharma et al., 2016). As MsMYB fell under subgroup 7, we were interested to know whether it is regulated by microRNA858. We generated small RNA sequencing data from PGTs and L-PGT and identified ms-miR858. The sequence similarity of ms-miR858 is shown in Figure 11a. Expression analysis from the sequencing data showed that ms-miR858 had higher expression in L-PGT tissue when compared with PGT. This was validated by TaqMan® qPCR assay using miR396 as internal control due to its stable expression in L-PGT and PGT tissues (Figure 11b). The Plant Small RNA Target Analysis Server (Dai and Zhao, 2011) predicted *MsMYB* as a probable target of ms-miR858 (Figure 11c). To test whether low levels of miR858 expression can target *MsMYB*, we looked for cleavage products through RNA ligase-mediated 5′ amplification of cDNA ends (RLM-RACE) and poly(A) polymerase-mediated (PPM)-RACE. However, no cleavage products of *MsMYB* were found in our conditions, thus reducing the possibility of it being regulated by miR858. *MsMYB* and miR858 appears to be differentially regulated at transcriptional level making their interaction less feasible.

Discussion

Extensive exploitation of aromatic plants for their commercially valued secondary metabolites calls for novel strategies for metabolic engineering to increase yield. A complete understanding of the secondary metabolism pathway by unravelling the enzymes and TFs controlling the pathway is indispensable for the same. PGTs are found on the aerial parts of many aromatic plants but very few studies have focused on identifying regulators of secondary metabolite production from these organs (Wang, 2014). In this study, we identified and characterized a R2R3-MYB gene, *MsMYB* from spearmint which is highly expressed in PGTs and acts as a negative regulator of monoterpene synthesis. GPPS enzyme is responsible for the formation of monoterpene precursor C_{10}-GPP in plastids and regulates flux into this pathway. Details about their function and regulation in plants are very scarce. The heteromeric GPPS in spearmint consists of a small and large subunit and interaction between them results in an active

(a)

GT1- motif   TGA- box
ATCATGGTTAAACATATGAAAAAATTGACGTAAACTTTGATTTTATACTAA

                                                        ABRE
TCCGTCTAGGCCGTTTAGGCACCGCCTAGACTCCCGCCTAGGCTGCCTAGG

CGCTAATCGCTCCCCCACCGTTCTACTAGCGTCTAGCGACTTTTAGAACACT

GATCAACCGTCGTTG[CAAT]AAAAGATGCAACACGGAAATG[CAAT]TGAGGA

TCCGATTCGAGAATCCATTGATGGTGGTGCATAAACTAGCGATTCAGATCA

                                   CGTCA- motif
TCTATTACATTTCCTTGTTACACCGTGGGATGATCGTCATGTATTTAATGTA

                                                       MBS (site2)
ATTGTGTTTATTGAATTATTTAAAGCAGTTTAGGGCTTTAAAAG**CAGTTAG**
                                                       GTCAAT

                          LRE
GCGTTAAAACTGCCAAACTCTCC[CAAT]CAAAATCTCCGCTCTTTTATTGCAAC

                                          TGACG- motif
ATCACATATATTCATTCACCTAATTGCCAAAA[CAATT]CTGACGTTGTTTCTT

MBS (site1)
CGCACAGATTTC**CAGTTG**CAGTGCTGGGTTGATATATATATAATCTCGTGT
GTCAAC

            GARE- motif
ATTTATATATATATATATATATTTCTGTTGGGTAAAAATG GGA AGA GCG
                                        M   G   R   A

(b)

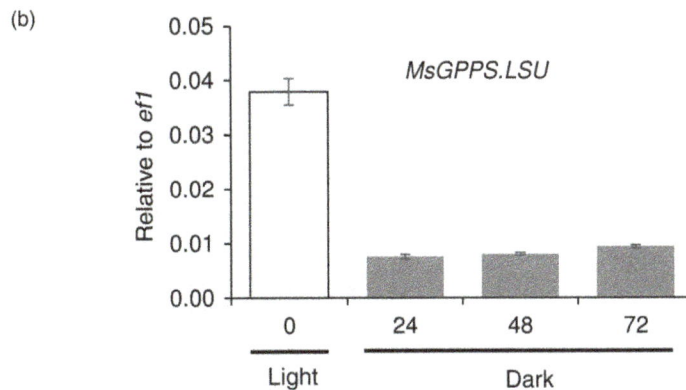

Figure 7 Promoter analysis of *MsGPPS.LSU*. (a) Regulatory elements present in the promoter region (−549 bp) of *MsGPPS.LSU*. (b) Decreased levels of *MsGPPS.LSU* in plants grown under dark when compared to plants grown under light. Data are indicated as mean ± SE.

GPPS. Structural and mutagenic studies of these two subunits have shown that the large subunit is catalytically active and presumably through protein–protein interactions, and the SSU shapes the active site cavity of LSU towards producing C_{10}-GPP and restricts the production of long-chain C_{15} or C_{20} prenylphosphates (Chang et al., 2010). In spearmint, both the subunits of GPPS are preferentially expressed in PGTs (Turner and Croteau, 2004). MsMYB was found to interact with *MsGPPS.LSU* promoter and repress its activity. The increased production of monoterpene in *MsMYB*-RNAi lines can be attributed to enhanced production of GPP through greater production of MsGPPS.LSU. Misexpression of either the regulatory or catalytic subunit are known to alter enzyme activity in plants and animals (Gergs et al., 2004; He et al., 2004; Hu et al., 2016; Tiessen et al., 2002). In snapdragon

and hop where heteromeric GPPS are found, only GPPS.SSU is closely correlated with monoterpene production, whereas the GPPS.LSU is more ubiquitous. But spearmint *GPPS.LSU* expression is more PGT-specific (Jin et al., 2014). Unlike the purified mint GPPS.LSU, which is inactive by itself, the potential GPPS.LSU from snapdragon and hops are active alone and functions as GGPPS (Tholl et al., 2004; Wang and Dixon, 2009). This suggests function of GPPS subunits from different plants might vary and also the mechanisms that regulate them.

Ectopic expression of *MsMYB* in sweet basil and tobacco revealed that it could also perturb metabolites derived from sesquiterpene and diterpene pathways. Sesquiterpenes and diterpenes are derived from C_{15}-FPP and C_{20}-GGPP which are produced in cytoplasm and plastids through MVA and MEP

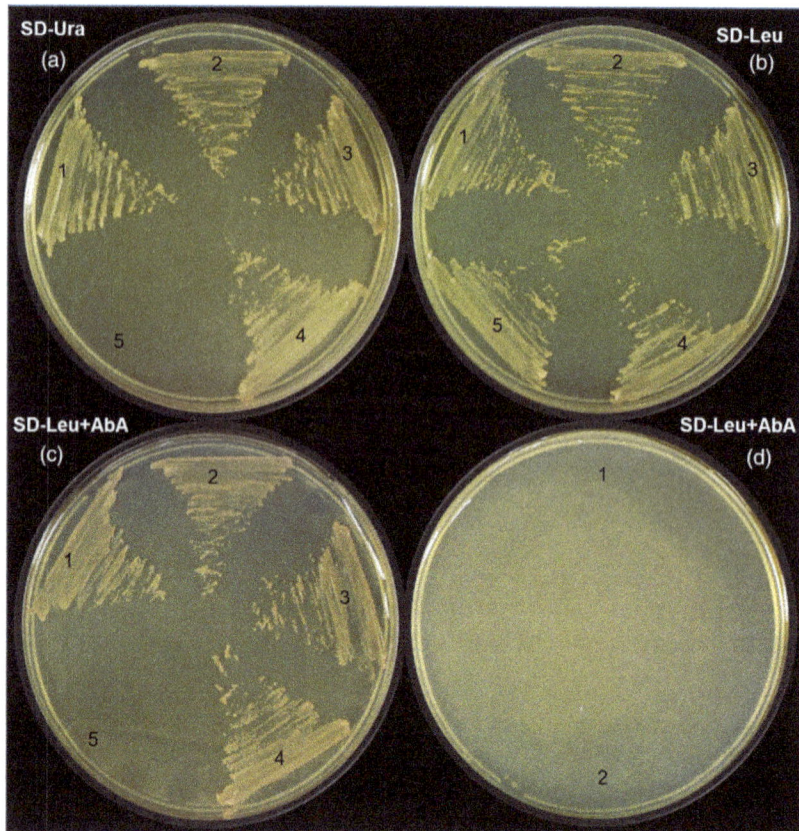

Figure 8 Yeast one-hybrid assay between MsMYB and *GPPS.LSU* promoter. (a) Bait strains of (1) pAbAi-GPPS-full, (2) pAbAi-site1, (3) pAbAi-site2, (4) pAbAi-p53 (positive control) and (5) auxotrophic Y1HGold (negative control) growing on SD/-Ura. (b) pGADT7-MYB prey expression in (1) pAbAi-GPPS-full, (2) pAbAi-site1 and (3) pAbAi-site2; pGADT7-p53 prey expression in (4) pAbAi-p53 and empty pGADT7 in (5) pAbAi-p53 on SD/-Leu. (c) DNA–protein interaction of (1) full *GPPS* promoter and MYB, (2) site 1 and MYB, (3) site 2 and MYB, (4) pAbAi-p53 and pGADT7-p53, and (5) no interaction of pAbAi-p53 and empty pGADT7 on SD/-Leu/AbA. (d) Absence of DNA–protein interaction of (1) pAbAi-mut-site1 and (2) pAbAi-mut-site2 with MYB. SD/-Ura, SD medium without Ura; SD/-Leu, SD medium without Leu; SD/-Leu/AbA, SD medium without Leu but containing aureobasidin A (800 ng).

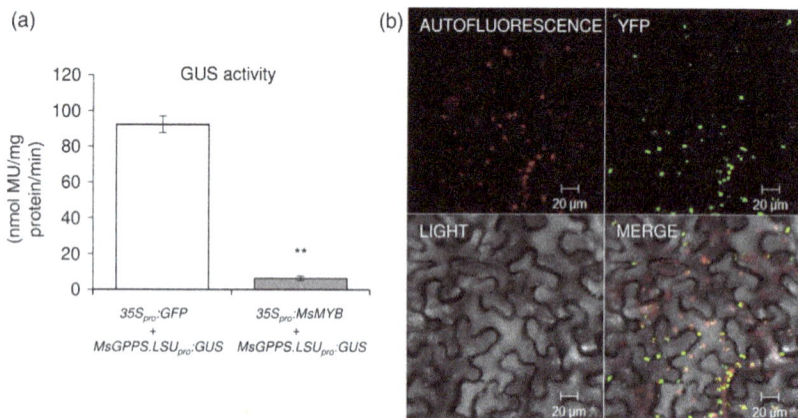

Figure 9 Transcriptional repression activity of MsMYB and localization of MsGPPS.LSU. (a) Reduced levels of GUS in *35Spro:MsMYB* expressing leaves (b) Chloroplast-specific localization of MsGPPS.LSU. Data are indicated as mean ± SE.

pathway, respectively. Studies have demonstrated that exchange of precursor metabolites can occur between the cytosolic MVA and plastidial MEP pathway. Metabolic intermediates like IPP, GPP and GGPP can be transported across plastidial membranes through transporters (Hemmerlin et al., 2012; Vranova et al., 2013). FPPS catalyses the sequential condensation of two molecules of IPP with DMAPP to form C_{15}-FPP, the immediate precursor for all sesquiterpenes (Cornish, 1993; Lange et al., 2000). But FPPS is also know to accept C_{10}-GPP as an allylic substrate to generate FPP (Guo et al., 2015; Hemmerlin et al., 2003). Further sesquiterpenes synthesized in cytoplasm can be derived from plastidial GPP too (Adam et al., 1999). Hence, an alteration in GPP produced in plastids can have an effect on sesquiterpene production. Ectopic expression of *MsMYB* in sweet basil resulted in reduced expression of *ObGPPS.LSU*. The decrease in sesquiterpene production observed in sweet basil indicates that

GPP produced might be used towards synthesizing sesquiterpene especially as no change was observed on sweet basil FPPS. The monoterpene and diterpene pathway are both largely localized to plastids. GGPPS is responsible for producing GGPP for diterpenes (C_{20}) and tetraterpenes (C_{40}) synthesis. Apart from IPP and DMAPP, *in vitro* data show that GPP or FPP can also be used by GGPPS as allylic substrates to generate GGPP (Hefner et al., 1998; Okada et al., 2000; Takaya et al., 2003). The decrease in diterpenes produced in *MsMYB*-overexpressing tobacco plants can be due to a reduction in endogenous pool of GPP. Tobacco generally produces low levels of monoterpenes and there is no information currently about endogenous GPPS from it. Although introduction of monoterpene synthases from other plants does result in the emission of volatile monoterpenes in tobacco, suggesting some amount of GPP availability (Lücker et al., 2004). Apart from alteration in terpene-derived secondary metabolites,

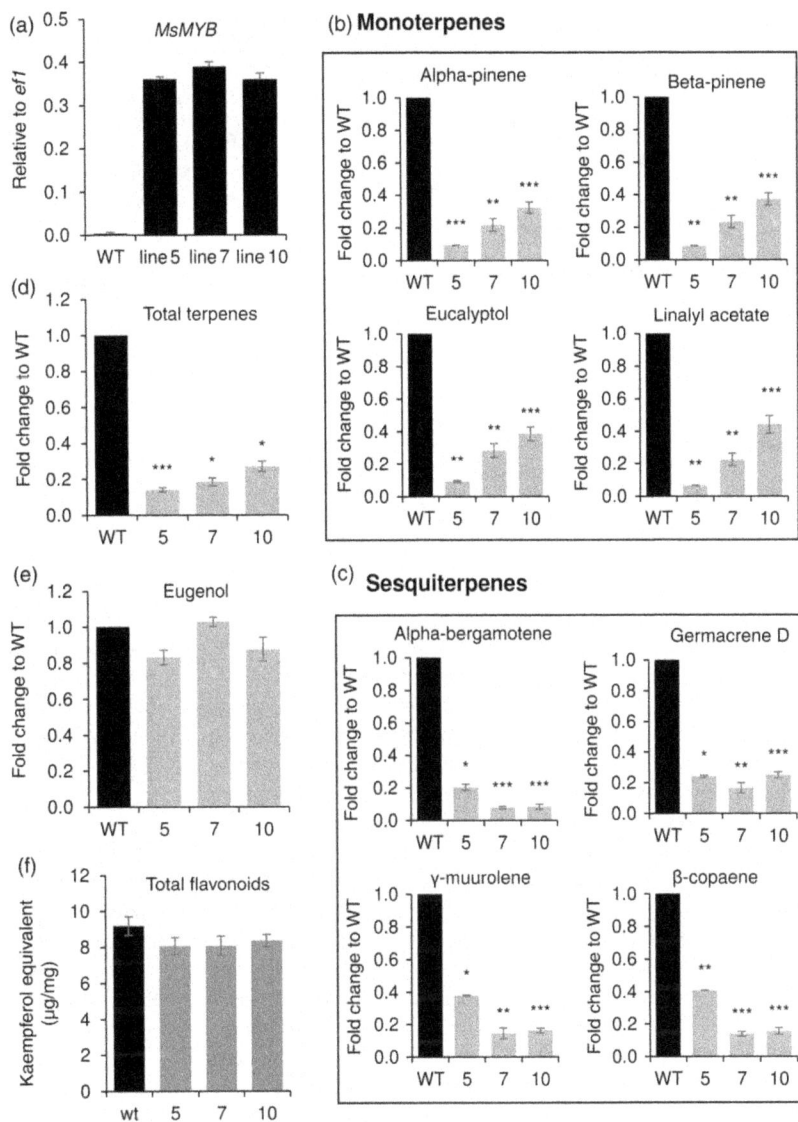

Figure 10 Ectopic expression of *MsMYB* in sweet basil alters the amount of secondary metabolites. (a) *MsMYB* expression in basil transgenic plants. (b), (c) Reduced levels of monoterpenes and sesquiterpenes. (d) Decreased levels of total terpenes. (e) Unaltered levels of phenylpropenes (eugenol). (f) Unaltered levels of total flavonoids. Data are indicated as mean ± SE. *$P < 0.05$; **$P < 0.01$; ***$P < 0.001$.

overexpression of *MsMYB* in tobacco and sweet basil generated smaller leaves. Several major plant hormones are produced through terpene pathway such as gibberellin, cytokinin, abscisic acid and strigolactone which control various aspects of plant development (Gomez-Roldan *et al.*, 2008; Umehara *et al.*, 2008). Further experiments need to be carried out to identify perturbation in which hormone/hormones leads to smaller leaf size. This phenotype was not observed in spearmint. TFs can have additional targets when ectopically expressed. For example, maize ZmMYB31 cannot bind to the promoter of maize *4CL* and *C3H* genes; however, it can repress the expression of Arabidopsis *4CL* and *C3H* when overexpressed in Arabidopsis (Fornalé *et al.*, 2010). There is a possibility that additional target genes related to primary terpene metabolism are regulated when MsMYB is overexpressed heterologously in these plants. Ectopic expression of *MsMYB* did not lead to any observable defect in chlorophyll formation which is a primary metabolite derived from diterpenes. An earlier study showed that GGPP used in chlorophyll biosynthesis is formed directly from DMAPP and IPP rather than from GPP and IPP. Thus, any alterations in the amount of GPP formed do not affect chlorophyll biosynthesis (Van Schie *et al.*, 2007).

TFs are known to regulate multiple genes. *MsGPPS.LSU* is probably one of the many downstream targets regulated by MsMYB. Hence, we cannot negate the fact that the effects seen on terpene metabolism in spearmint, sweet basil and tobacco can be due to other genes as well. Transcriptome analysis of *MsMYB*-RNAi lines can provide us with more candidate genes that are potentially regulated by *MsMYB*. We found that *MsGPPS.LSU* is further regulated by light. Several genes of MEP pathway are known to be regulated by light (Cordoba *et al.*, 2009). Secondary metabolite production is mainly for plants biotic and abiotic stress responses, so stringent multiple regulatory mechanisms involving activators and repressors must exist to determine the timing, amount and patterning of these compounds.

Majority of R2R3-MYBs characterized in plants are involved in the regulation of different classes of phenylpropanoid-derived compounds (Liu *et al.*, 2015). Phylogenetic analysis showed that MsMYB is very similar to subgroup 7 of Arabidopsis R2R3-MYBs which include MYB11, MYB12 and MYB111. These MYBs have been characterized as activators of biosynthetic enzymes of flavonoid pathway which is derived from the general phenylpropanoid pathway. Analysis of major

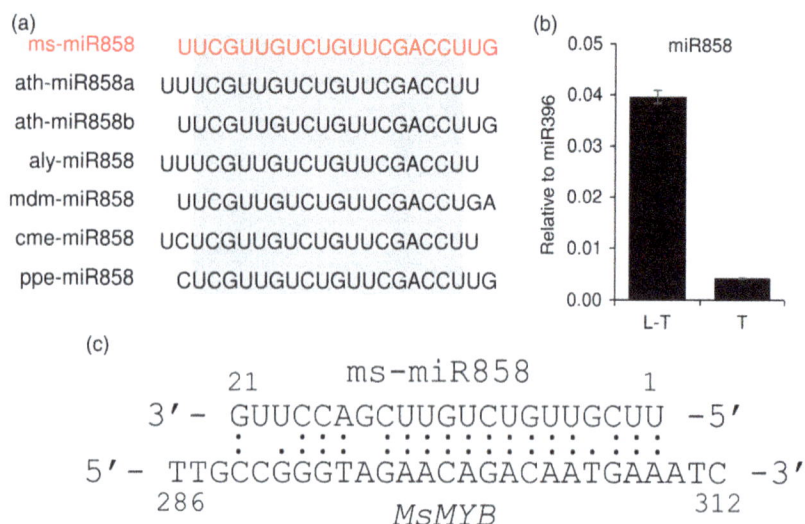

Figure 11 ms-miR858 analysis. (a) ms-miR858 sequence comparison with other known miR858 sequences from various species. miR858 sequences were obtained from miRBase. ms—*Mentha spicata*; ath—*Arabidopsis thaliana*; aly—*Arabidopsis lyrata*; mdm—*Malus domestica*; cme—*Cucumis melo*; ppe—*Prunus persica*. (b) Expression levels of ms-miR858 in leaf stripped of PGTs (L-T) and PGTs (T). (c) Probable binding of ms-miR858 to *MsMYB*.

flavonoid pathway genes from spearmint showed no change in *MsMYB* transgenic lines. Further no change was seen in either eugenol production which is derived from phenylpropanoid pathway or in total flavonoid content of leaves from sweet basil *MsMYB*-overexpressing lines. These results suggest *MsMYB* as a regulator of terpene metabolism and not affecting flavonoid production. This indicates towards diversification of function within the same subgroup. Although the N-terminus of R2R3-MYBs is conserved, C-terminus amino acids are very diverse and unique providing opportunity for divergence. Comparison of MYB proteins from different species will furnish new insights into the evolution of this important family of TF. In the past few years numerous studies have demonstrated miR858 can target R2R3-MYBs from various plants including the members of subgroup 7 of Arabidopsis R2R3-MYBs (Addo-Quaye *et al.*, 2008; Rock, 2013; Sharma *et al.*, 2016; Xia *et al.*, 2012). Small RNA-Seq data of spearmint PGTs and L-PGTs showed an inverse pattern of expression for miR858 and *MsMYB*. High expression was seen in L-PGTs and low in PGTs for miR858 which is opposite to *MsMYB* expression, indicating a differential transcriptional regulation. *MsMYB* promoter is able to drive trichome specific expression as seen in tobacco. As genome of spearmint is not available, we cannot verify the promoter activity of miR858. Although *in silico* analysis predicts *MsMYB* to be a potential target for miR858, no cleavage products could be observed. This suggests that the PGT specific expression of *MsMYB* is governed mainly by the *cis*-elements present in its promoter and may be by other *trans*-factors but not by miR858.

In conclusion, *MsMYB* is the first R2R3-MYB gene identified from spearmint that is related to terpene secondary metabolism. MsMYB represses *MsGPPS.LSU* and is able to modify terpene production. This opens up new avenues and targets to metabolically engineer plants for altering pathways.

Experimental procedures

Plant material and transformation

Secondary metabolites of commercial spearmint variety (*M. spicata*) and sweet basil (*O. basilicum*) were analysed by GC-MS and grown in green house under natural light conditions. *Agrobacterium* mediated transformation of spearmint was performed

according to previously published protocol (Niu *et al.*, 1998, 2000) with several modifications. The complete details of transformation are provided in the supplementary data. *Agrobacterium* mediated transformation of sweet basil was performed as previously described by Wang *et al.* (2016). Tobacco transformation was performed as described previously (Gallois and Marinho, 1995).

RNA isolation and quantitative real-time PCR (qRT-PCR)

For PGT RNA isolation, initially, PGTs were isolated from 2- to 3-cm leaves as described previously (Jin *et al.*, 2014). Later, total RNA was extracted from PGT using the Spectrum Plant total RNA kit from Sigma (UK). Total RNA from other tissues was extracted using an RNeasy® Plus Mini kit from Qiagen (Germany). 500 ng of RNA was reverse-transcribed to cDNA using iScript™ cDNA Synthesis kit form Bio-Rad (CA, USA). Expression levels of genes along various tissues were analysed using qRT-PCR that was carried out as described previously (Jin *et al.*, 2014). In current study, elongation factor 1 (*ef1*) was used as internal control for both spearmint and sweet basil, due to its stable expression in plant (Nicot *et al.*, 2005). For tobacco, elongation factor TuB (EF-TuB) was used as internal control. Error bars represent mean ± SE.

TaqMan® qPCR assay

7 ng of total RNA was reverse-transcribed using the TaqMan® microRNA reverse transcription kit (Applied Biosystems, USA) according to the manufacturer's protocol. Custom-made primers were used specifically for miR858 and miR396 cDNA synthesis. Custom TaqMan® small RNA assays (Applied Biosystems, USA) were carried out for microRNA qPCR validation using custom-made TaqMan® probe for miR858 and miR396. Expression values were calculated relative to the internal control miR396.

Promoter cloning and analysis

Genomic DNA was isolated from young leaves of spearmint using CTAB method. The flanking sequences of *MsMYB* and *MsGPPS.LSU* genes were amplified using a GenomeWalker™ Universal kit (Clontech, USA) and later ligated to pGEM®-T vector. The resulting product was transformed into *Escherichia coli* (*E. coli*) XL1-Blue cells and sequenced. The promoter was amplified with Phusion® High-Fidelity DNA Polymerase (NEB, USA) and subcloned into a gateway donor vector pENTR™/ D-TOPO® (Invitrogen, Germany). Further, the recombinant plasmid was introduced into destination vector pBGWFS7 by LR recombination. The destination plasmid was further transformed into *Agrobacterium* EHA105 by heat shock. The recombinant *Agrobacterium* EHA105 strain was used to generate transgenic tobacco lines.

Gene amplification and plasmid construction

Full-length ORF encoding MsMYB was obtained by performing 3' and 5' RACE using the SMARTer™ RACE cDNA amplification kit from Clontech (CA, USA). For sequencing of ORF, purified fragments were ligated with pGEM®-T vector and transformed into *E. coli* XL1-Blue cells. To overexpress *MsMYB*, firstly, the sequences were amplified with Phusion® High-Fidelity DNA Polymerase (NEB, USA). For *MsMYB*-RNAi, four primers with restriction enzymes located at flanking region were used to amplify the fragment showing low similarity to other *MYB* genes. The purified PCR products were then cloned into the donor vector and subsequently introduced into the destination vector pK7WG2D via LR recombination. The *MsMYB* gene was driven by 35S promoter in both overexpression and RNAi plants. All destination plasmids harbouring target gene were transformed into *Agrobacterium* EHA105 by heat shock. The recombinant *Agrobacterium* EHA105 strains were used for plant transformation. Sequences of primers used in this study are listed in Table S1.

Subcellular localization of TFs

The full-length cDNA of *MsMYB* without the stop codon was cloned into the gateway vector pENTR/D-TOPO (Invitrogen, Germany) and then subsequently transferred into the destination vector pBA-DC-YFP (Zhang *et al.*, 2005) which contains the CaMV 35S promoter and C-terminal in frame with YFP, to generate *MsMYB*-YFP. The construct was then introduced into *Agrobacterium* strain EHA105 by heat shock. 4',6-Diamidino-2-phenylindole (DAPI) was used as maker to stain nucleus. *N. benthamiana* seeds were germinated on MS plate and transferred into soil. Three weeks after growing in the glasshouse, the seedlings were used for agroinfiltration. Subcellular localization was performed as described by Wang *et al.* (2016).

Selection of transgenic lines

Initially, visual screening using GFP filter was pursued to isolate GFP-positive plants. Later, DNA was isolated from GFP-positive plants and insertion of gene of interest was checked using PCR. DNA-positive lines were then subjected to Southern blot using DIG wash and block buffer set from Roche (IN, USA). DNA probe against CaMV 35S promoter was generated using PCR DIG probe synthesis kit from Roche (IN, USA) (Hart and Basu, 2009). Concentration of probe was quantified by creating a dot plot using DIG nucleic acid detection kit from Roche (IN, USA) as described previously (Javelle *et al.*, 2011). A total of 15 µg genomic DNA was digested overnight with NdeI at 37 °C. The next day, digested product was electrophoresed on a 1% (w/v) agarose gel at 40 V for 5 h. After that, the gel was transferred to a nylon membrane and hybridized with CaMV 35S promoter probe followed by antibody binding and chemiluminescent reaction development by CDP-Star. Finally, the membrane was

exposed in a film and the number of T-DNA insertions were analysed. Spearmint plants were clonally propagated thrice before analysis. Basil plants were analysed in T2 generation and a minimum of six plants were sampled for each line. Tobacco plants were analysed in T1 generation. For each line, ~32 plants were used for phenotypical screening and six plants for GC-MS analysis.

GC-MS and total flavonoid content analysis

For GC-MS studies, spearmint plants were clonally propagated three times before collecting about 4–6 leaves of 2–3 cm which were ground to a fine powder using liquid nitrogen and homogenized using 500 µL ethyl acetate. Camphor was added as an internal control. For sweet basil, the analysis was performed in T-2 plants. From bottom, leaves of 3–4 cm at fourth node were used. For tobacco, 4- to 5-cm leaves were used from T1 plants. Diethyl sebacate was added as an internal standard in sweet basil and tobacco samples. Samples were incubated for 10 min at room temperature with vigorous shaking followed by a centrifugation for 10 min at 15871g. The top organic layer was transferred to a new tube and dehydrated using anhydrous Na_2SO_4. The samples were analysed using GC-MS (7890A with 5975C inert MSD with triple axis detector, Agilent Technologies, USA). 2 µL of samples was injected, and separation was achieved with a temperature programme of 50 °C for 1 min and increased at a rate of 8 °C/min to 300 °C and held for 5 min, on a 30 m HP-5 MS column (Agilent Technologies, USA). Estimation of total flavonoid content was performed using the aluminium chloride (AlCl₃) colorimetric method as described previously (Vaidya *et al.*, 2014).

Yeast one-hybrid assay

For ascertaining the interaction between MsMYB and *MsGPPS.LSU* promoter, a Matchmaker Gold yeast one-hybrid library screening system (Clontech, USA) was used. As *MsGPPS.LSU* promoter encompasses two different MYB binding sites, two promoter regions approximately 40–45 bp in size harbouring the two sites, namely site 1 and site 2, were cloned into the MCS of pAbAi vector independently in addition to two mutant versions of these two sites. Furthermore, complete promoter was also cloned in pAbAi vector and used as bait for further confirmation. The *MsMYB* gene was cloned into the pGADT7 vector to be used as prey in the form of functional fusion protein along with GAL4 transcription activation domain (GAL4 AD). As a positive control, p53 bait and prey provided in the kit was used. Mutated site 1 and mutated site 2 were used as negative control. Yeast transformation and validation of positive interactions were implemented as described in the Matchmaker Gold Yeast one-hybrid system user manual (www.clontech.com/ xxclt_ibcGetAttachment.jsp?cItemId=17599). The primers used are listed in Table S3.

Transactivation activity assay

The 549-bp promoter region of *MsGPPS.LSU* was amplified and inserted into pENTR™/D-TOPO®. The resulting plasmid was transformed into pBGWFS7 by LR recombination and further introduced into *Agrobacterium* EHA105. Leaves of *N. benthamiana* were agroinfiltrated with effector and reporter at a ratio of 1:1. Two days after infiltration, leaves were harvested to isolate crude protein. GUS quantitative assay was performed in triplicate as described in Li *et al.* (2014).

RLM-RACE and PPM-RACE

For RLM-RACE, 5′ RACE adapter was ligated to 1 μg of total RNA, and for PPM-RACE, poly(A) tailing of 1 μg of total RNA was performed using poly(A) polymerase (Invitrogen, Germany). RNA was then reverse-transcribed and amplified for 5′ and 3′ ends using FirstChiOce RLM-RACE kit (Thermo Fisher, USA) according to the manufacturer's protocol.

Small RNA library construction and high-throughput sequencing

The RNA libraries were prepared using the TruSeq RNA library prep Kit v2, set A (Illumina Inc., USA) according to manufacturer's instructions. The quality and size of DNA libraries for sequencing were checked using the Agilent 2200 TapeStation system. Three libraries were run on single lanes on HiSeq™ 2000 (Illumina Inc., USA), individually.

Bioinformatics analysis of sequencing data

After removing adapters by Cutadapt tool (Martin, 2011), modified sequences from 18 nt to 30 nt were used for further analysis. To begin with, sequences of rRNAs, tRNAs, snRNAs and snoRNAs available in Rfam11.0 database (ftp://ftp.ebi.ac.uk/pub/databases/Rfam/11.0/) were removed from the small RNA sequence reads. Deposited miR858 sequences from miRBase database (http://www.mirbase.org/search.shtml) were mapped to the remaining small RNA sequence reads by bowtie tool (Langmead and Salzberg, 2012) to identify ms-miRNA858. A maximum of one mismatch per sequence was set as a parameter.

Phylogenetic analysis

Phylogenetic tree was constructed using MEGA6 software by the neighbour-joining method with bootstrap values of 1000 replicates and edited in FigTree. *A. thaliana* R2R3-MYB sequences were obtained from TAIR website.

Statistical analysis

Data are indicated as 'mean ± SE' of three to six biological replicates each performed in triplicate. Statistical significance between transgenic plants and WT was analysed using a two-tailed Student's *t*-test and indicated by asterisks. * indicates $P < 0.05$; ** indicates $P < 0.01$; *** indicates $P < 0.001$.

Accession numbers

Sequence data of *A. thaliana* R2R3-MYBs used in Figure S2 can be found in TAIR library under accession numbers listed in Supplemental Data S2. Sequence data of *ObGPPS.LSU* and *ObFPPS* can be found in NCBI database under the sequence ID DY340136 and DY332783, respectively. Sequence data of *MsMYB* have been deposited in GenBank under the accession number KY081780.

Competing interests

The authors declare that they have no competing interests.

Acknowledgements

This research was funded by a grant from Singapore National Research Foundation (Competitive Research Programme Award No: NRF-CRP8-2011-02).

References

Adam, K.-P., Thiel, R. and Zapp, J. (1999) Incorporation of 1-[1-13C]Deoxy-d-xylulose in chamomile sesquiterpenes. *Arch. Biochem. Biophys.* **369**, 127–132.

Adato, A., Mandel, T., Mintz-Oron, S., Venger, I., Levy, D., Yativ, M., Dominguez, E. *et al.* (2009) Fruit-surface flavonoid accumulation in tomato is controlled by a SlMYB12-regulated transcriptional network. *PLoS Genet.* **5**, 23.

Addo-Quaye, C., Eshoo, T.W., Bartel, D.P. and Axtell, M.J. (2008) Endogenous siRNA and miRNA targets identified by sequencing of the Arabidopsis degradome. *Curr. Biol.* **18**, 758–762.

Basu, S., Roychoudhury, A. and Sengupta, D.N. (2014) Deciphering the Role of various cis-acting regulatory elements in controlling SamDC gene expression in Rice. *Plant Signal. Behavior*, **9**, e28391.

Bhargava, A., Mansfield, S.D., Hall, H.C., Douglas, C.J. and Ellis, B.E. (2010) MYB75 functions in regulation of secondary cell wall formation in the arabidopsis inflorescence stem. *Plant Physiol.* **154**, 1428–1438.

Bhargava, A., Ahad, A., Wang, S.C., Mansfield, S.D., Haughn, G.W., Douglas, C.J. and Ellis, B.E. (2013) The interacting MYB75 and KNAT7 transcription factors modulate secondary cell wall deposition both in stems and seed coat in Arabidopsis. *Planta*, **237**, 1199–1211.

Bouvier, F., Suire, C., d'Harlingue, A., Backhaus, R.A. and Camara, B. (2000) Molecular cloning of geranyl diphosphate synthase and compartmentation of monoterpene synthesis in plant cells. *Plant J.* **24**, 241–252.

Bouvier, F., Rahier, A. and Camara, B. (2005) Biogenesis, molecular regulation and function of plant isoprenoids. *Prog. Lipid Res.* **44**, 357–429.

Burke, C. and Croteau, R. (2002) Interaction with the small subunit of geranyl diphosphate synthase modifies the chain length specificity of geranylgeranyl diphosphate synthase to produce geranyl diphosphate. *J. Biol. Chem.* **277**, 3141–3149.

Burke, C.C., Wildung, M.R. and Croteau, R. (1999) Geranyl diphosphate synthase: Cloning, expression, and characterization of this prenyltransferase as a heterodimer. *Proc. Natl Acad. Sci. USA*, **96**, 13062–13067.

Champagne, A. and Boutry, M. (2013) Proteomic snapshot of spearmint (Mentha spicata L.) leaf trichomes: a genuine terpenoid factory. *Proteomics*, **13**, 3327–3332.

Chang, T.H., Hsieh, F.L., Ko, T.P., Teng, K.H., Liang, P.H. and Wang, A.H.J. (2010) Structure of a Heterotetrameric Geranyl Pyrophosphate Synthase from Mint (Mentha piperita) Reveals Intersubunit Regulation. *Plant Cell*, **22**, 454–467.

Chen, Y.Y., Wang, L.F., Dai, L.J., Yang, S.G. and Tian, W.M. (2012) Characterization of HbEREBP1, a wound-responsive transcription factor gene in laticifers of Hevea brasiliensis Muell. *Arg. Molecul. Biol. Rep.* **39**, 3713–3719.

Cordoba, E., Salmi, M. and Leon, P. (2009) Unravelling the regulatory mechanisms that modulate the MEP pathway in higher plants. *J. Exp. Bot.* **60**, 2933–2943.

Cornish, K. (1993) The separate roles of plant cis and trans prenyl transferases in cis-1,4-polyisoprene biosynthesis. *Eur. J. Biochem.* **218**, 267–271.

Croteau, R., Karp, F., Wagschal, K.C., Satterwhite, D.M., Hyatt, D.C. and Skotland, C.B. (1991) Biochemical-characterization of a spearmint mutant that resembles peppermint in monoterpene content. *Plant Physiol.* **96**, 744–752.

Croteau, R.B., Davis, E.M., Ringer, K.L. and Wildung, M.R. (2005) (−)-Menthol biosynthesis and molecular genetics. *Naturwissenschaften*, **92**, 562.

Dai, X.B. and Zhao, P.X. (2011) psRNATarget: a plant small RNA target analysis server. *Nucleic Acids Res.* **39**, W155–W159.

Diemer, F., Caissard, J.C., Moja, S., Chalchat, J.C. and Jullien, F. (2001) Altered monoterpene composition in transgenic mint following the introduction of 4S-limonene synthase. *Plant Physiol. Biochem.* **39**, 603–614.

Dubey, V.S., Bhalla, R. and Luthra, R. (2003) An overview of the non-mevalonate pathway for terpenoid biosynthesis in plants. *J. Biosci.* **28**, 637.

Dubos, C., Stracke, R., Grotewold, E., Weisshaar, B., Martin, C. and Lepiniec, L. (2010) MYB transcription factors in Arabidopsis. *Trends Plant Sci.* **15**, 573–581.

Fahlgren, N., Howell, M.D., Kasschau, K.D., Chapman, E.J., Sullivan, C.M., Cumbie, J.S., Givan, S.A. et al. (2007) High-throughput sequencing of Arabidopsis microRNAs: evidence for frequent birth and death of MIRNA genes. PLoS ONE, **2**, e219.

Fornalé, S., Shi, X., Chai, C., Encina, A., Irar, S., Capellades, M., Fuguet, E. et al. (2010) ZmMYB31 directly represses maize lignin genes and redirects the phenylpropanoid metabolic flux. Plant J. **64**, 633–644.

Gallois, P. and Marinho, P. (1995) Leaf Disk Transformation Using Agrobacterium tumefaciens-Expression of Heterologous Genes in Tobacco. In: Plant Gene Transfer and Expression Protocols (Jones, H. ed), pp. 39–48. New York: Springer.

Gergs, U., Boknik, P., Buchwalow, I., Fabritz, L., Matus, M., Justus, I., Hanske, G. et al. (2004) Overexpression of the catalytic subunit of protein phosphatase 2A impairs cardiac function. J. Biol. Chem. **279**, 40827–40834.

Gomez-Roldan, V., Fermas, S., Brewer, P.B., Puech-Pages, V., Dun, E.A., Pillot, J.-P., Letisse, F. et al. (2008) Strigolactone inhibition of shoot branching. Nature, **455**, 189–194.

Grotewold, E. (2008) Transcription factors for predictive plant metabolic engineering: are we there yet? Curr. Opin. Biotechnol. **19**, 138–144.

Guan, X.Y., Pang, M.X., Nah, G., Shi, X.L., Ye, W.X., Stelly, D.M. and Chen, Z.J. (2014) miR828 and miR858 regulate homoeologous MYB2 gene functions in Arabidopsis trichome and cotton fibre development. Nat. Commun. **5**, 3050.

Guo, D., Li, H.-L. and Peng, S.-Q. (2015) Structure conservation and differential expression of farnesyl diphosphate synthase genes in euphorbiaceous plants. Int. J. Mol. Sci. **16**, 22402–22414.

Hart, S.M. and Basu, C. (2009) Optimization of a digoxigenin-based immunoassay system for gene detection in Arabidopsis thaliana. J. Biomolecular Techniq.: JBT, **20**, 96–100.

He, X., Anderson, J.C., Pozo, O.D., Gu, Y.-Q., Tang, X. and Martin, G.B. (2004) Silencing of subfamily I of protein phosphatase 2A catalytic subunits results in activation of plant defense responses and localized cell death. Plant J. **38**, 563–577.

Hefner, J., Ketchum, R.E.B. and Croteau, R. (1998) Cloning and functional expression of a cDNA encoding geranylgeranyl diphosphate synthase fromtaxus canadensisand assessment of the role of this prenyltransferase in cells induced for taxol production. Arch. Biochem. Biophys. **360**, 62–74.

Hemmerlin, A., Rivera, S.B., Erickson, H.K. and Poulter, C.D. (2003) Enzymes encoded by the farnesyl diphosphate synthase gene family in the big sagebrush Artemisia tridentata ssp. spiciformis. J. Biol. Chem. **278**, 32132–32140.

Hemmerlin, A., Harwood, J.L. and Bach, T.J. (2012) A raison d'être for two distinct pathways in the early steps of plant isoprenoid biosynthesis? Prog. Lipid Res. **51**, 95–148.

Hu, R., Zhu, Y., Wei, J., Chen, J., Shi, H., Shen, G. and Zhang, H. (2016) Overexpression of PP2A-C5 that encodes the catalytic subunit 5 of protein phosphatase 2A in Arabidopsis confers better root and shoot development under salt conditions. Plant Cell Environ. **40**, 150–164.

Iwase, A., Matsui, K. and Ohme-Takagi, M. (2009) Manipulation of plant metabolic pathways by transcription factors. Plant Biotechnol. **26**, 29–38.

Javelle, M., Marco, C.F. and Timmermans, M. (2011) In situ hybridization for the precise localization of transcripts in plants. J. Vis. Exp. **57**, UNSP e3328.

Jeong, D.H., Schmidt, S.A., Rymarquis, L.A., Park, S., Ganssmann, M., German, M.A., Accerbi, M. et al. (2013) Parallel analysis of RNA ends enhances global investigation of microRNAs and target RNAs of Brachypodium distachyon. Genome Biol. **14**, R145.

Jia, X.Y., Shen, J., Liu, H., Li, F., Ding, N., Gao, C.Y., Pattanaik, S. et al. (2015) Small tandem target mimic-mediated blockage of microRNA858 induces anthocyanin accumulation in tomato. Planta, **242**, 283–293.

Jin, H.L., Cominelli, E., Bailey, P., Parr, A., Mehrtens, F., Jones, J., Tonelli, C. et al. (2000) Transcriptional repression by AtMYB4 controls production of UV-protecting sunscreens in Arabidopsis. EMBO J. **19**, 6150–6161.

Jin, J., Panicker, D., Wang, Q., Kim, M.J., Liu, J., Yin, J.-L., Wong, L. et al. (2014) Next generation sequencing unravels the biosynthetic ability of Spearmint (Mentha spicata) peltate glandular trichomes through comparative transcriptomics. BMC Plant Biol. **14**, 292.

Kellogg, B.A. and Poulter, C.D. (1997) Chain elongation in the isoprenoid biosynthetic pathway. Curr. Opin. Chem. Biol. **1**, 570–578.

Koschmann, J., Machens, F., Becker, M., Niemeyer, J., Schulze, J., Bülow, L., Stahl, D.J. et al. (2012) Integration of bioinformatics and synthetic promoters leads to the discovery of novel elicitor-responsive cis-regulatory sequences in Arabidopsis. Plant Physiol. **160**, 178–191.

Kranz, H.D., Denekamp, M., Greco, R., Jin, H., Leyva, A., Meissner, R.C., Petroni, K. et al. (1998) Towards functional characterisation of the members of the R2R3-MYB gene family from Arabidopsis thaliana. Plant J. **16**, 263–276.

Lange, B.M. and Turner, G.W. (2013) Terpenoid biosynthesis in trichomes-current status and future opportunities. Plant Biotechnol. J. **11**, 2–22.

Lange, B.M., Rujan, T., Martin, W. and Croteau, R. (2000) Isoprenoid biosynthesis: the evolution of two ancient and distinct pathways across genomes. Proc. Natl Acad. Sci. **97**, 13172–13177.

Lange, B.M., Mahmoud, S.S., Wildung, M.R., Turner, G.W., Davis, E.M., Lange, I., Baker, R.C. et al. (2011) Improving peppermint essential oil yield and composition by metabolic engineering. Proc. Natl Acad. Sci. USA, **108**, 16944–16949.

Langmead, B. and Salzberg, S.L. (2012) Fast gapped-read alignment with Bowtie 2. Nat. Methods, **9**, 357–U54.

Legay, S., Lacombe, E., Goicoechea, M., Briere, C., Seguin, A., Mackay, J. and Grima-Pettenati, J. (2007) Molecular characterization of EgMYB1, a putative transcriptional repressor of the lignin biosynthetic pathway. Plant Sci. **173**, 542–549.

Li, C.L. and Lu, S.F. (2014) Genome-wide characterization and comparative analysis of R2R3-MYB transcription factors shows the complexity of MYB-associated regulatory networks in Salvia miltiorrhiza. BMC Genom. **15**, 277.

Li, S., Zhang, P., Zhang, M., Fu, C. and Yu, L. (2013) Functional analysis of a WRKY transcription factor involved in transcriptional activation of the DBAT gene in Taxus chinensis. Plant Biol. **15**, 19–26.

Li, R., Weldegergis, B.T., Li, J., Jung, C., Qu, J., Sun, Y.W., Qian, H.M. et al. (2014) Virulence factors of geminivirus interact with MYC2 to subvert plant resistance and promote vector performance. Plant Cell, **26**, 4991–5008.

Liang, P.-H. (2009) Reaction kinetics, catalytic mechanisms, conformational changes, and inhibitor design for prenyltransferases. Biochemistry, **48**, 6562–6570.

Liang, P.-H., Ko, T.-P. and Wang, A.H.J. (2002) Structure, mechanism and function of prenyltransferases. Eur. J. Biochem. **269**, 3339–3354.

Liu, J., Osbourn, A. and Ma, P. (2015) MYB transcription factors as regulators of phenylpropanoid metabolism in plants. Molecul. Plant, **8**, 689–708.

Loza-Tavera, H. (1999) Monoterpenes in Essential Oils. In Chemicals via Higher Plant Bioengineering(Shahidi, F., Kolodziejczyk, P., Whitaker, J.R., Munguia, A.L. and Fuller, G., eds), pp. 49–62. Boston, MA: Springer, US.

Lücker, J., Schwab, W., van Hautum, B., Blaas, J., van der Plas, L.H.W., Bouwmeester, H.J. and Verhoeven, H.A. (2004) Increased and altered fragrance of tobacco plants after metabolic engineering using three monoterpene synthases from lemon. Plant Physiol. **134**, 510–519.

Mahmoud, S.S., Williams, M. and Croteau, R. (2004) Cosuppression of limonene-3-hydroxylase in peppermint promotes accumulation of limonene in the essential oil. Phytochemistry, **65**, 547–554.

Martin, M. (2011) Cutadapt removes adapter sequences from high-throughput sequencing reads. EMBnet.journal. **17**, 10–12.

Matus, J.T. (2016) Transcriptomic and metabolomic networks in the grape berry illustrate that it takes more than flavonoids to fight against ultraviolet radiation. Front. Plant Sci. **7**, 1337.

Miyamoto, K., Matsumoto, T., Okada, A., Komiyama, K., Chujo, T., Yoshikawa, H., Nojiri, H. et al. (2014) Identification of target genes of the bZIP transcription factor OsTGAP1, whose overexpression causes elicitor-induced hyperaccumulation of diterpenoid phytoalexins in rice cells. PLoS ONE, **9**, e105823.

Munoz-Bertomeu, J., Ros, R., Arrillaga, I. and Segura, J. (2008) Expression of spearmint limonene synthase in transgenic spike lavender results in an altered monoterpene composition in developing leaves. Metab. Eng. **10**, 166–177.

Nagegowda, D.A. (2010) Plant volatile terpenoid metabolism: biosynthetic genes, transcriptional regulation and subcellular compartmentation. FEBS Lett. **584**, 2965–2973.

Nicot, N., Hausman, J.F., Hoffmann, L. and Evers, D. (2005) Housekeeping gene selection for real-time RT-PCR normalization in potato during biotic and abiotic stress. *J. Exp. Bot.* **56**, 2907–2914.

Niu, X., Lin, K., Hasegawa, P.M., Bressan, R.A. and Weller, S.C. (1998) Transgenic peppermint (Mentha x piperita L.) plants obtained by cocultivation with Agrobacterium tumefaciens. *Plant Cell Rep.* **17**, 165–171.

Niu, X., Li, X., Veronese, P., Bressan, R.A., Weller, S.C. and Hasegawa, P.M. (2000) Factors affecting Agrobacterium tumefaciens-mediated transformation of peppermint. *Plant Cell Rep.* **19**, 304–310.

Ogura, K. and Koyama, T. (1998) Enzymatic aspects of isoprenoid chain elongation. *Chem. Rev.* **98**, 1263–1276.

Okada, K., Saito, T., Nakagawa, T., Kawamukai, M. and Kamiya, Y. (2000) Five geranylgeranyl diphosphate synthases expressed in different organs are localized into three subcellular compartments in Arabidopsis. *Plant Physiol.* **122**, 1045–1056.

Oldfield, E. and Lin, F.-Y. (2012) Terpene biosynthesis: modularity rules. *Angew. Chem. Int. Ed. Engl.* **51**, 1124–1137.

Pichersky, E. and Gershenzon, J. (2002) The formation and function of plant volatiles: perfumes for pollinator attraction and defense. *Curr. Opin. Plant Biol.* **5**, 237–243.

Preston, J., Wheeler, J., Heazlewood, J., Li, S.F. and Parish, R.W. (2004) AtMYB32 is required for normal pollen development in Arabidopsis thaliana. *Plant J.* **40**, 979–995.

Prouse, M.B. and Campbell, M.M. (2012) The interaction between MYB proteins and their target DNA binding sites. *Biochim. Biophys. Acta*, **1819**, 67–77.

Ringer, K.L., Davis, E.M. and Croteau, R. (2005) Monoterpene metabolism. Cloning, expression, and characterization of (-)-isopiperitenol/(-)-carveol dehydrogenase of peppermint and spearmint. *Plant Physiol.* **137**, 863–872.

Rock, C.D. (2013) Trans-acting small interfering RNA4: key to nutraceutical synthesis in 1 grape development? *Trends Plant Sci.* **18**, 601–610.

Sharma, D., Tiwari, M., Pandey, A., Bhatia, C., Sharma, A. and Trivedi, P.K. (2016) MicroRNA858 is a potential regulator of phenylpropanoid pathway and plant development. *Plant Physiol.* **171**, 944–959.

Shen, Q., Yan, T., Fu, X. and Tang, K. (2016) Transcriptional regulation of artemisinin biosynthesis in Artemisia annua L. *Sci. Bulletin*, **61**, 18–25.

Singh, B. and Sharma, R.A. (2015) Plant terpenes: defense responses, phylogenetic analysis, regulation and clinical applications. *3 Biotech*, **5**, 129–151.

Stracke, R., Werber, M. and Weisshaar, B. (2001) The R2R3-MYB gene family in Arabidopsis thaliana. *Curr. Opin. Plant Biol.* **4**, 447–456.

Stracke, R., Ishihara, H., Huep, G., Barsch, A., Mehrtens, F., Niehaus, K. and Weisshaar, B. (2007) Differential regulation of closely related R2R3-MYB transcription factors controls flavonol accumulation in different parts of the Arabidopsis thaliana seedling. *Plant J.* **50**, 660–677.

Sunkar, R., Li, Y.-F. and Jagadeeswaran, G. (2012) Functions of microRNAs in plant stress responses. *Trends Plant Sci.* **17**, 196–203.

Takaya, A., Zhang, Y.-W., Asawatreratanakul, K., Wititsuwannakul, D., Wititsuwannakul, R., Takahashi, S. and Koyama, T. (2003) Cloning, expression and characterization of a functional cDNA clone encoding geranylgeranyl diphosphate synthase of Hevea brasiliensis. *Biochimica et Biophysica Acta (BBA) – Gene Struct. Expression*, **1625**, 214–220.

Tamagnone, L., Merida, A., Parr, A., Mackay, S., Culianez-Macia, F.A., Roberts, K. and Martin, C. (1998) The AmMYB308 and AmMYB330 transcription factors from antirrhinum regulate phenylpropanoid and lignin biosynthesis in transgenic tobacco. *Plant Cell*, **10**, 135–154.

Tholl, D., Kish, C.M., Orlova, I., Sherman, D., Gershenzon, J., Pichersky, E. and Dudareva, N. (2004) Formation of monoterpenes in Antirrhinum majus and Clarkia breweri flowers involves heterodimeric geranyl diphosphate synthases. *Plant Cell*, **16**, 977–992.

Tiessen, A., Hendriks, J.H.M., Stitt, M., Branscheid, A., Gibon, Y., Farré, E.M. and Geigenberger, P. (2002) Starch synthesis in potato tubers is regulated by post-translational redox modification of ADP-glucose pyrophosphorylase: a novel regulatory mechanism linking starch synthesis to the sucrose supply. *Plant Cell*, **14**, 2191–2213.

Turner, G.W. and Croteau, R. (2004) Organization of monoterpene biosynthesis in mentha. immunocytochemical localizations of geranyl diphosphate synthase, limonene-6-hydroxylase, isopiperitenol dehydrogenase, and pulegone reductase. *Plant Physiol.* **136**, 4215–4227.

Turner, G.W., Gershenzon, J. and Croteau, R.B. (2000) Distribution of peltate glandular trichomes on developing leaves of peppermint. *Plant Physiol.* **124**, 655–664.

Ulmasov, T., Liu, Z.B., Hagen, G. and Guilfoyle, T.J. (1995) Composite structure of auxin response elements. *Plant Cell*, **7**, 1611–1623.

Umehara, M., Hanada, A., Yoshida, S., Akiyama, K., Arite, T., Takeda-Kamiya, N., Magome, H. *et al.* (2008) Inhibition of shoot branching by new terpenoid plant hormones. *Nature*, **455**, 195–200.

Vaidya, N.B., Brearley, T.A. and Joshee, N. (2014) Antioxidant capacity of fresh and dry leaf extracts of sixteen Scutellaria species. *J. Medicinally Active Plants*, **2**, 42–49.

Van Schie, C.C.N., Ament, K., Schmidt, A., Lange, T., Haring, M.A. and Schuurink, R.C. (2007) Geranyl diphosphate synthase is required for biosynthesis of gibberellins. *Plant J.* **52**, 752–762.

Vranova, E., Coman, D. and Gruissem, W. (2013) Network Analysis of the MVA and MEP Pathways for Isoprenoid Synthesis. In: *Annual Review of Plant Biology*, Vol 64 (Merchant, S.S. ed), pp. 665–700. ANNUAL REVIEWS: USA.

Wang, G. (2014) Recent progress in secondary metabolism of plant glandular trichomes. *Plant Biotechnol.* **31**, 353–361.

Wang, G.D. and Dixon, R.A. (2009) Heterodimeric geranyl(geranyl) diphosphate synthase from hop (Humulus lupulus) and the evolution of monoterpene biosynthesis. *Proc. Natl Acad. Sci. USA*, **106**, 9914–9919.

Wang, Q., Reddy, V.A., Panicker, D., Mao, H.Z., Kumar, N., Rajan, C., Venkatesh, P.N. *et al.* (2016) Metabolic engineering of terpene biosynthesis in plants using a trichome-specific transcription factor MsYABBY5 from spearmint (Mentha spicata). *Plant Biotechnol. J.* **14**, 1619–1632.

Wu, L., Liu, D.F., Wu, J.J., Zhang, R.Z., Qin, Z.R., Liu, D.M., Li, A.L. *et al.* (2013) Regulation of FLOWERING LOCUS T by a MicroRNA in Brachypodium distachyon. *Plant Cell*, **25**, 4363–4377.

Xia, R., Zhu, H., An, Y.Q., Beers, E.P. and Liu, Z.R. (2012) Apple miRNAs and tasiRNAs with novel regulatory networks. *Genome Biol.* **13**, R47.

Xu, Y.H., Wang, J.W., Wang, S., Wang, J.Y. and Chen, X.Y. (2004) Characterization of GaWRKY1, a cotton transcription factor that regulates the sesquiterpene synthase gene (+)-delta-cadinene synthase-A. *Plant Physiol.* **135**, 507–515.

Yang, C.Q., Fang, X., Wu, X.M., Mao, Y.B., Wang, L.J. and Chen, X.Y. (2012) Transcriptional regulation of plant secondary metabolism. *J. Int. Plant Biol.* **54**, 703–712.

Zhang, B.H. (2015) MicroRNA: a new target for improving plant tolerance to abiotic stress. *J. Exp. Bot.* **66**, 1749–1761.

Zhang, B.H. and Wang, Q.L. (2015) MicroRNA-based biotechnology for plant improvement. *J. Cell. Physiol.* **230**, 1–15.

Zhang, X., Garreton, V. and Chua, N.-H. (2005) The AIP2 E3 ligase acts as a novel negative regulator of ABA signaling by promoting ABI3 degradation. *Genes Dev.* **19**, 1532–1543.

Zhang, Q., Zhu, J., Ni, Y., Cai, Y. and Zhang, Z. (2012) Expression profiling of HbWRKY1, an ethephon-induced WRKY gene in latex from Hevea brasiliensis in responding to wounding and drought. *Trees*, **26**, 587–595.

Zhao, Q. and Dixon, R.A. (2011) Transcriptional networks for lignin biosynthesis: more complex than we thought? *Trends Plant Sci.* **16**, 227–233.

Zhong, R.Q., Lee, C.H. and Ye, Z.H. (2010) Functional characterization of poplar wood-associated NAC domain transcription factors. *Plant Physiol.* **152**, 1044–1055.

Zhou, M., Wu, L., Liang, J., Shen, C. and Lin, J. (2012) Expression analysis and functional characterization of a novel cold-responsive gene CbCOR15a from Capsella bursa-pastoris. *Mol. Biol. Rep.* **39**, 5169–5179.

10

Tailoring the composition of novel wax esters in the seeds of transgenic *Camelina sativa* through systematic metabolic engineering

Noemi Ruiz-Lopez[1,2], Richard Broughton[2], Sarah Usher[2], Joaquin J. Salas[3], Richard P. Haslam[2], Johnathan A. Napier[2,*] and Frédéric Beaudoin[2]

[1]*IHSM-UMA-CSIC, Universidad de Málaga, Málaga, Spain*
[2]*Department of Biological Chemistry, Rothamsted Research, Harpenden, Herts, UK*
[3]*Instituto de la Grasa, Universitario Pablo de Olavide, Seville, Spain*

*Correspondence
email johnathan.napier@rothamsted.ac.uk

Summary

The functional characterization of wax biosynthetic enzymes in transgenic plants has opened the possibility of producing tailored wax esters (WEs) in the seeds of a suitable host crop. In this study, in addition to systematically evaluating a panel of WE biosynthetic activities, we have also modulated the acyl-CoA substrate pool, through the co-expression of acyl-ACP thioesterases, to direct the accumulation of medium-chain fatty acids. Using this combinatorial approach, we determined the additive contribution of both the varied acyl-CoA pool and biosynthetic enzyme substrate specificity to the accumulation of non-native WEs in the seeds of transgenic Camelina plants. A total of fourteen constructs were prepared containing selected FAR and WS genes in combination with an acyl-ACP thioesterase. All enzyme combinations led to the successful production of wax esters, of differing compositions. The impact of acyl-CoA thioesterase expression on wax ester accumulation varied depending on the substrate specificity of the WS. Hence, co-expression of acyl-ACP thioesterases with *Marinobacter hydrocarbonoclasticus* WS and *Marinobacter aquaeolei* FAR resulted in the production of WEs with reduced chain lengths, whereas the co-expression of the same acyl-ACP thioesterases in combination with *Mus musculus* WS and *M. aquaeolei* FAR had little impact on the overall final wax composition. This was despite substantial remodelling of the acyl-CoA pool, suggesting that these substrates were not efficiently incorporated into WEs. These results indicate that modification of the substrate pool requires careful selection of the WS and FAR activities for the successful high accumulation of these novel wax ester species in Camelina seeds.

Keywords: metabolic engineering, seed lipids, Camelina, wax esters.

Introduction

There is a growing recognition of the potential of so-called green factories to help address some of the major societal challenges that now face the human race (Yuan and Grotewold, 2015). In particular, the ultimate need (irrespective of timescale) to transition from dependence on fossil fuels such as petroleum to more renewable and sustainable sources remains a very high priority (Beaudoin *et al.*, 2014; Carlsson *et al.*, 2011; Vanhercke *et al.*, 2013). Although it is now generally accepted, if only because of the volumes required, that bio-renewable forms could not hope to replace the current utilization levels of fossil fuels for combustible energy, it is equally appreciated that bio-based compounds could substitute for other, less voluminous petrochemically derived forms (Carlsson *et al.*, 2011). These could include acting as replacements for the chemical feedstocks used to create plastics and polymers such as nylon, and also to substitute for fossil fuel-derived lubricants (Bates *et al.*, 2013). In particular, a renewable, bio-based lubricant would be highly desirable, combining sustainability with improved biodegradability (Carlsson *et al.*, 2011). However, the demanding physicochemical properties which define a high-performance lubricant such as might be found in vehicle transmission or hydraulic systems preclude the use of most currently available forms of

biologically derived compounds (Jaworski and Cahoon, 2003). For example, oils (usually in the form of triacylglycerols—TAGs) derived from plant seeds lack both the thermal stability and melting point normally associated with petrochemically derived compounds. However, one naturally occurring compound which has been proven to have utility as a biolubricant is spermaceti oil, used extensively in the first half of the 20th century in the automotive industry (Iven *et al.*, 2016). Remarkably, it was even used in the NASA Apollo 'moon-shots' of the 1960s, reportedly on the basis of its superior ability to withstand extremes of temperature and pressure (Hoare, 2009). However, as sourcing this compound required the killing one of the largest sentient mammals on the planet, good sense prevailed and the commercial hunting of whales for the oil has been globally banned since the 1980s.

Yet interest has remained in understanding better the particular properties of spermaceti oil and the identification of alternative, less destructive, sources. In fact, spermaceti oil is predominantly comprised of wax esters (WE, 76%) and significantly lower amounts of TAGs (23%), and it is the presence of the WE (and their associated ester-bonded acyl-alcohols) that confers its desirable properties (Nevenzel, 1970). Unfortunately, although higher plants make a wide variety of different oils and fatty acids (Napier, 2007), there are hardly any known examples of plant species which accumulate WE in their seeds, the best characterized example being jojoba (*Simmondsia chinensis*) (Dyer *et al.*, 2008). Jojoba is a desert shrub which is poorly adapted for modern agriculture, and the specific form of WE present in the plant is distinct in its chemical properties to the WEs found in spermaceti oil (Miwa, 1971). Thus, in terms of both composition and production capacity, jojoba is not a viable substitute for spermaceti oil. However, the biosynthesis of jojoba WEs has provided insights into how these useful compounds are made, and the identification of the key enzymes and genes has permitted a biotechnological approach, in which the capacity to synthesize WEs is transferred to a suitable agronomically viable host crop (Lardizabal *et al.*, 2000; Metz *et al.*, 2000).

Based on studies in jojoba and other WE-accumulating organisms (such as some bacteria and protists), it is known that synthesis of WEs proceeds via a two-step reaction—firstly the conversion of an acyl chain to a fatty alcohol via a fatty acid reductase (FAR), followed by the condensation of the fatty alcohol with a fatty acid by a wax synthase (WS) to yield the final product, the WE (Figure 1a; Ishige *et al.*, 2003; Jaworski and Cahoon, 2003). In both cases, the enzymes utilize acyl-CoAs as their substrates and the reactions are predicted to occur in the endoplasmic reticulum. Polypeptides associated with the FAR and WS activities were purified from jojoba microsomal membrane fractions, knowledge which allowed the further identification of cDNA clones encoding these enzymes (Lardizabal *et al.*, 2000;

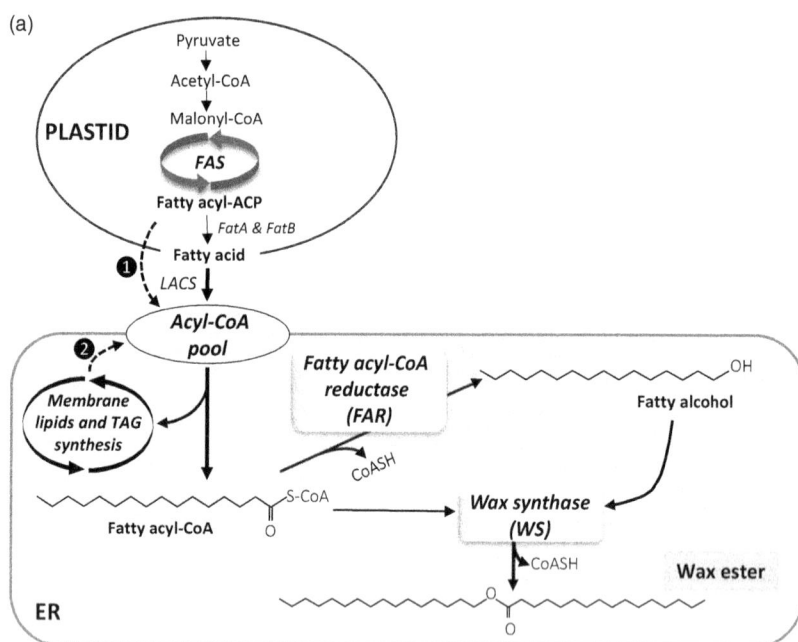

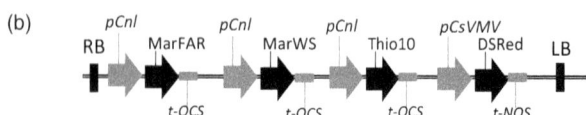

Figure 1 Metabolic engineering strategy for tailoring the composition of Wax Ester (WE) produced in transgenic Camelina seeds. (a) Schematic illustration of the approach used for investigating the contribution of heterologous enzymes activity and the acyl-CoA substrate pool to wax ester content and composition. ❶ Modulation of the acyl-CoA substrate pool by the co-expressing acyl-ACP thioesterases which prematurely terminate fatty acid synthesis in the plastid resulting in the accumulation of medium-chain fatty acids. ❷ Acyl exchange pathway(s) required for the transfer of unusual fatty acids (e.g. hydroxylated fatty acids) from membrane glycerolipids to the acyl-CoA substrate pool. The enzymes required for WE synthesis in the endoplasmic reticulum (ER) are a fatty acyl-CoA reductase (FAR) and a wax synthase (WS). (b) Diagram of the constructs used for seed-specific expression of different combinations of wax biosynthetic and lipid/fatty acid modifying enzymes. (c) Combinations of wax biosynthetic and lipid modifying genes tested for manipulating the composition of WEs produced in Camelina seeds. A simple abbreviated nomenclature was adopted for the definition of each gene set. The first two letters of each gene combination represent the wax synthase, the next two letters the fatty alcohol reductase and the last two the presence or the absence (Wt) of an additional gene. For example, the abbreviation MoMa10 indicates the combination of Mouse WS, *Marinobacter* FAR and *Cuphea hookeriana* acyl-ACP C10 thioesterase (Thio10). MoMaWt represents the same WS and FAR combination in a wild-type fatty acid background.

Abbreviation	WS gene	FAR gene	Additional gene	
MaMa10	*M. hydrocarbonoclasticus*	*M. aquaeolei*	*C. hookeriana*	(Thio10)
MaMa12	*M. hydrocarbonoclasticus*	*M. aquaeolei*	*U. Californica*	(Thio12)
MaMa14	*M. hydrocarbonoclasticus*	*M. aquaeolei*	*C. Palustris*	(Thio14)
MaMaWt	*M. hydrocarbonoclasticus*	*M. aquaeolei*	-	
MaMaOH	*M. hydrocarbonoclasticus*	*M. aquaeolei*	*C. purpurea*	(FAH12)
MoMa10	*M. musculus*	*M. aquaeolei*	*C. hookeriana*	(Thio10)
MoMa12	*M. musculus*	*M. aquaeolei*	*U. californica*	(Thio12)
MoMa14	*M. musculus*	*M. aquaeolei*	*C. palustris*	(Thio14)
MoMaWt	*M. musculus*	*M. aquaeolei*	-	
MoMaOH	*M. musculus*	*M. aquaeolei*	*C. purpurea*	(FAH12)
AcMa14	*A. calcoaceticus*	*M. aquaeolei*	*C. palustris*	(Thio14)
TeMa14	*T. thermophila*	*M. aquaeolei*	*C. palustris*	(Thio14)
MaTe14	*M. Hydrocarbonoclasticus*	*T. thermophila*	*C. palustris*	(Thio14)
MaTeWt	*M. hydrocarbonoclasticus*	*T. thermophila*	-	

Metz *et al.*, 2000). Functional characterization in transgenic plants demonstrated the predicted activities, opening the possibility of producing 'tailored wax esters' in the seeds of a suitable host plant (Lardizabal *et al.*, 2000; Metz *et al.*, 2000). However, despite these early successes, there has been only limited progress in increasing and optimizing the accumulation of WEs in seeds, and it is equally clear that the host plant plays an important role in determining the levels achieved (Dyer *et al.*, 2008; Iven *et al.*, 2016; Zhu *et al.*, 2016; Horn *et al.*, 2013). Moreover, a significant challenge remains in terms of directing the flux of substrate acyl-CoAs towards FAR and WS whilst minimizing their incorporation into TAG (which represents a metabolic dead end as far as WE synthesis is concerned). This must be balanced with a requirement to synthesize sufficient TAG to allow seed germination (Bates *et al.*, 2013), and also to avoid the overaccumulation of deleterious fatty alcohols which might arise from nonstoichiometric activities of FAR and WS. Finally, the composition of the fatty acids in the acyl-CoA pool may not be well matched with the substrate preferences of the FAR and WS enzymes, resulting in low overall synthesis of WEs. Through the adoption of iterative metabolic engineering approaches (Haslam *et al.*, 2016), incremental but important gains in our capacity to direct the accumulation of WEs in the seeds of GM plants have been achieved, although these have mainly been focused on producing a jojoba-like WE in which the majority of acyl chains are longer than 20 carbons (C40 + WEs; Zhu *et al.*, 2016). Attempts to generate shorter WEs, which would have physiochemical properties closer to a potential biolubricant, have been primarily focused on evaluating enzyme activities from different source organisms and enhancing substrate flux through cotargeting of the FAR and WS to specific organelles (Iven *et al.*, 2013, 2016). In this study, in addition to systematically evaluating a wide panel of WE biosynthetic activities, we have also modulated the acyl-CoA substrate pool through the co-expression of acyl-ACP thioesterases which prematurely terminate plastidial fatty acid synthesis and result in the accumulation of medium-chain fatty acids and their acyl-CoA derivatives (Figure 1c; Dehesh *et al.*, 1996a,b). By this approach, we determined the combinatorial contribution of both the acyl-CoA pool and FAR/WS substrate specificity to the accumulation of non-native WEs in the seeds of transgenic Camelina plants.

Results and discussion

Camelina was chosen as an experimental platform for seed-specific engineering as it is an oilseed crop that is amenable to rapid gene testing: it is easily transformed using *Agrobacterium*-based methods and seed-specific selection markers (e.g. DsRed) and it has short life cycle (Haslam *et al.*, 2016). The general strategy for producing wax esters in seeds of Camelina is shown in Figure 1a. Acyl-ACP products, primarily 16:0-and 18:1-ACP, synthesized by the plastidial fatty acid synthase (FAS) are hydrolysed by the FAtB and FATA thioesterases, respectively, and activated to CoA esters by long-chain acyl-CoA synthetases (LACS) during their export from the plastid. At that point, they enter the endoplasmic reticulum (ER)-associated acyl-CoA pool where they become available as the primary substrates for wax ester synthesis. As outlined above, this can be achieved by introduction and seed-specific expression of a fatty acyl reductase (FAR) and a wax synthase (WS) leading the production of novel wax esters. In this study, FAR genes from the bacteria

Marinobacter aquaeolei (Iven *et al.*, 2013) and *Tetrahymena thermophile* (Dittrich-Domergue *et al.*, 2014) were expressed in combinations with WS genes from *Marinobacter hydrocarbonoclasticus*, Mouse (*Mus musculus*) (Cheng and Russell, 2004), *Acinetobacter calcoaceticus* (Kalscheuer *et al.* 2003) and *T. thermophile* (Biester *et al.*, 2012). These FAR and WS combinations were also co-expressed with a number of additional genes (i.e. acyl-ACP thioesterases or a fatty acid hydroxylase) intended to further modify seed fatty acid composition (Figure 1c). For ease of reference, a simple abbreviated nomenclature was adopted for the definition of each gene set—for example, the abbreviation MoMa10 indicates the combination of <u>Mo</u>use WS, <u>Mari</u>nobacter FAR *and Cuphea hookeriana 10:0*-ACP thioesterase (Thio10).

Modulation of the substrate pool for wax ester biosynthesis

Earlier studies (Zhu *et al.*, 2016) have indicated how the endogenous seed oil fatty acid composition influences the outcome of any metabolic strategy for the production of industrial oils. Furthermore, it can be assumed that the availability of specific acyl-CoA substrates has a major effect on the composition of the wax ester molecular species produced. Typically, *de novo* fatty acid synthesis in plants generates C16 and C18 fatty acids; however, previous work has shown that the expression of functionally divergent FatB acyl-ACP thioesterases from Cuphea species in Camelina seeds results in the accumulation of medium-chain fatty acids with chain lengths ranging from C8 to C14 (Kim *et al.*, 2015). To explore how the availability of specific acyl-CoA substrate might impact wax ester synthesis and composition (Figure 1a; pathway 1), selected acyl-ACP thioesterases were evaluated using a combinatorial approach. Candidate FatB acyl-ACP thioesterases were selected from published reports of their substrate specificities: a 10:0-ACP thioesterases from *Cuphea hookeriana* (Thio 10; Dehesh *et al.*, 1996a), a 12:0-ACP thioesterase from *Umbellularia californica* (Thio12; Voelker *et al.*, 1992) and a 14:0-ACP thioesterase gene from *Cuphea palustris* (Thio14; Dehesh *et al.*, 1996b). Analysis of the acyl-CoA pool in developing seeds of the transgenic Camelina (Table S1) confirmed the suitability of this strategy for manipulating the substrate pool. The cumulative amount of 10:0-CoA, 12:0-CoA and 14:0-CoA was <0.5% in Wt but up to 19.7% of total acyl-CoAs in MoMa and MaMa lines. Additionally, in an attempt to engineer the accumulation of more unusual (and industrially valuable) acyl moieties in wax esters, the seed-specific synthesis of hydroxylated fatty acids substrates was investigated. Functional expression of a Δ^{12}-oleate hydroxylase (CpFAH) from *Claviceps purpurea* (Meesapyodsuk and Qiu, 2008) under the control of a seed-specific promoter was previously shown to result in high levels of ricinoleic acid (12-hydroxyoctadec-cis-9-enoic acid) accumulation in Arabidopsis. CpFAH was therefore chosen to provide a source of such hydroxylated fatty acids in Camelina. CpFAH is active on substrates esterified to membrane phospholipids, and therefore, hydroxylated fatty acids would only be made available for both wax ester and TAG synthesis via endogenous acyl exchange, primarily from PC into the acyl-CoA pool (Figure 1a; pathway 2). In this study, the level of 12-OH-18:1Δ^9 accumulated in TAG ranged from 0.4% to 3% of total fatty acids (Table 1). To investigate our approach to modulating substrate supply and tailoring wax ester synthesis, a total of fourteen constructs were prepared containing selected FAR and WS genes in combination with either an acyl-ACP thioesterase or

Table 1 Camelina seed fatty acid composition. Fatty acid methyl esters were analysed by GC-FID and expressed as mol %. The transgenes present in the different lines are detailed in Figure 1

	10:0	12:0	14:0	16:0	18:0	18:1n9	18:2n6	18:3n3	20:0	20:1n11	20:2n6	22:0	22:1n9	24:1n9	18:1OH	Others
Wt	0.0 ± 0.0	0.0 ± 0.0	0.0 ± 0.0	6.0 ± 0.1	3.0 ± 0.1	13.2 ± 0.5	17.9 ± 0.0	33.1 ± 0.5	2.0 ± 0.0	14.9 ± 0.3	2.2 ± 0.1	1.5 ± 0.1	3.3 ± 0.1	0.8 ± 0.0	0.0 ± 0.0	2.2 ± 0.3
MaMaWt	0.0 ± 0.0	0.0 ± 0.0	0.0 ± 0.0	6.2 ± 0.2	4.1 ± 0.1	14.2 ± 0.4	14.4 ± 0.1	33.1 ± 0.5	2.9 ± 0.0	14.8 ± 0.4	1.5 ± 0.0	1.5 ± 0.0	3.6 ± 0.1	0.8 ± 0.1	0.0 ± 0.0	2.8 ± 0.0
MaMa10	0.7 ± 0.2	0.0 ± 0.0	0.1 ± 0.1	8.1 ± 0.4	4.4 ± 0.4	14.3 ± 1.1	26.7 ± 0.9	24.6 ± 1.8	2.2 ± 0.0	10.5 ± 0.7	1.7 ± 0.1	0.7 ± 0.1	2.5 ± 0.2	0.8 ± 0.0	0.0 ± 0.0	2.8 ± 0.1
MaMa12	0.0 ± 0.0	2.3 ± 0.0	0.8 ± 0.0	7.8 ± 0.1	3.7 ± 0.1	12.7 ± 0.4	26.1 ± 0.8	24.4 ± 0.4	2.0 ± 0.1	11.8 ± 0.6	1.9 ± 0.1	0.8 ± 0.1	2.5 ± 0.2	0.8 ± 0.1	0.0 ± 0.0	2.4 ± 0.1
MaMa14	0.0 ± 0.0	0.0 ± 0.0	14.0 ± 1.1	15.1 ± 1.1	2.8 ± 0.0	8.5 ± 0.1	11.6 ± 0.3	27.1 ± 0.8	2.3 ± 0.1	10.2 ± 0.9	1.2 ± 0.1	1.1 ± 0.1	3.2 ± 0.4	0.6 ± 0.1	0.0 ± 0.0	2.3 ± 0.1
MaMaOH	0.1 ± 0.1	0.0 ± 0.0	0.0 ± 0.1	7.2 ± 0.1	3.9 ± 0.1	12.2 ± 1.3	22.4 ± 0.5	26.7 ± 2.4	2.7 ± 0.1	13.8 ± 1.0	2.1 ± 0.2	1.0 ± 0.1	3.6 ± 0.6	0.8 ± 0.2	0.4 ± 0.5	3.1 ± 0.2
MoMaWt	0.0 ± 0.0	0.0 ± 0.0	0.0 ± 0.0	5.6 ± 0.1	3.3 ± 0.1	14.4 ± 0.5	17.5 ± 0.2	31.2 ± 0.7	2.0 ± 0.1	16.6 ± 1.0	2.0 ± 0.0	1.3 ± 0.0	2.9 ± 0.2	0.8 ± 0.1	0.0 ± 0.0	2.3 ± 0.1
MoMa10	1.8 ± 0.3	0.3 ± 0.1	0.3 ± 0.1	6.4 ± 0.6	4.0 ± 0.1	12.6 ± 1.1	19.7 ± 1.3	26.8 ± 1.7	3.4 ± 0.1	12.7 ± 1.0	2.0 ± 0.0	1.1 ± 0.1	4.6 ± 0.4	1.0 ± 0.2	0.0 ± 0.0	3.3 ± 0.1
MoMa12	0.0 ± 0.1	1.0 ± 0.2	0.4 ± 0.0	6.5 ± 0.2	3.5 ± 0.1	11.0 ± 0.3	18.9 ± 0.0	28.9 ± 0.4	2.1 ± 0.0	17.7 ± 0.9	2.3 ± 0.1	1.2 ± 0.1	3.5 ± 0.2	1.0 ± 0.1	0.0 ± 0.0	2.2 ± 0.1
MoMa14	0.0 ± 0.0	0.0 ± 0.0	4.9 ± 0.6	10.8 ± 0.5	3.6 ± 0.1	10.3 ± 0.1	16.7 ± 0.2	27.5 ± 0.2	1.9 ± 0.0	15.8 ± 0.8	1.8 ± 0.1	1.1 ± 0.1	2.9 ± 0.2	0.8 ± 0.1	0.0 ± 0.0	2.0 ± 0.1
MoMaOH	0.0 ± 0.0	0.0 ± 0.0	0.0 ± 0.1	8.1 ± 0.1	7.8 ± 0.4	22.3 ± 0.4	18.2 ± 0.3	17.2 ± 0.5	2.7 ± 0.1	15.5 ± 0.6	0.8 ± 0.1	0.4 ± 0.0	1.4 ± 0.1	0.1 ± 0.1	3.0 ± 0.3	2.5 ± 0.1
AcMa14	0.0 ± 0.0	0.0 ± 0.0	2.1 ± 0.0	9.0 ± 0.1	3.7 ± 0.1	10.7 ± 0.2	18.2 ± 0.6	26.3 ± 0.5	2.1 ± 0.1	18.7 ± 0.7	2.1 ± 0.1	1.1 ± 0.1	3.2 ± 0.2	0.7 ± 0.1	0.0 ± 0.0	2.1 ± 0.1
TeMa14	0.0 ± 0.0	0.0 ± 0.0	10.1 ± 0.5	13.8 ± 0.5	3.6 ± 0.1	8.2 ± 0.1	20.7 ± 0.3	20.1 ± 0.6	2.5 ± 0.1	12.8 ± 0.1	1.8 ± 0.0	0.7 ± 0.0	3.0 ± 0.2	0.7 ± 0.0	0.0 ± 0.0	1.8 ± 0.1
MaTeWt	0.0 ± 0.0	0.0 ± 0.0	0.0 ± 0.0	6.6 ± 0.2	2.9 ± 0.1	9.4 ± 0.1	16.0 ± 0.4	36.6 ± 0.6	1.8 ± 0.0	15.3 ± 0.4	2.4 ± 0.1	2.0 ± 0.1	3.4 ± 0.2	0.9 ± 0.1	0.0 ± 0.0	2.8 ± 0.1
MaTe14	0.0 ± 0.0	0.0 ± 0.0	17.0 ± 0.1	17.8 ± 0.2	2.7 ± 0.0	5.8 ± 0.2	13.0 ± 0.3	25.7 ± 0.5	2.2 ± 0.1	8.1 ± 0.0	1.5 ± 0.0	1.1 ± 0.0	2.7 ± 0.2	0.6 ± 0.1	0.0 ± 0.0	1.7 ± 0.1

CpFAH under the control of seed-specific promoters (Figure 1b) and identified by the abbreviations listed in Figure 1c.

Production of wax esters in seeds of Camelina

Analysis of lipid extracts by HPLC-ELSD from ten to fifteen independent T2 transgenic (thirty seeds per analysis) events for each gene combination indicated the successful production of wax esters (Figure 2). A very small amount of ester lipids (3.6 nmol seed^{-1}) was detected in wild-type (Wt) seeds, and this level was established as the baseline for successful wax ester biosynthesis. The transgenic Camelina lines accumulated between 4.6 and 87.6 nmol seed^{-1} of wax esters—this represented the maximum and minimum levels observed across all events (see also mean accumulation levels in Table 2); with the lowest levels of accumulation measured in lines MaTe (WS and FAR) combinations (~6 nmol seed^{-1}). Wax ester synthesis in combination with the production of hydroxy fatty acids yielded modest but greater levels than Wt (7.3 and 7.8 nmol seed^{-1}; MaMa and MoMa, respectively), demonstrating the feasibility of this approach. Maximal amounts of wax ester synthesis occurred in combinations of MaMa and MoMa with acyl-ACP thioesterases, namely MaMa12 (12:0-ACP thioesterase) 44.8 nmol seed^{-1} and MoMa10 (10:0-ACP thioesterase) 74.6 nmol seed^{-1}. Further analysis of MaMa14 (67.4 nmol seed^{-1}) and MoMa14 (77.6 nmol seed^{-1}) T3 lines revealed very significant increases in the accumulation of wax esters (98% and 130%, respectively) compared to T2 seeds (Figure 2). Previous work with constructs containing FAR and WS genes from *M. aquaeolei* and *M. musculus* (Heilmann et al., 2012 and Iven et al., 2016; respectively) has demonstrated the efficacy of these enzymes for in planta wax ester synthesis, with a reported range in Camelina of 7–47 mg/g seed for combinations of *M. musculus*, *M. aquaeolei* and *Simmondsia chinensis* FAR and WS. Here, total wax ester yields in seeds of Camelina ranged between 9.8 and 34.3 mg/g seed for MoMa(Wt) and MaMa(Wt) combinations, respectively (Fig. S1a). Because of significant variation in seed weight observed in this study (as discussed below), the quantity of WE produced per seed illustrates heterologous enzyme activity in each transgenic line more faithfully than WE quantity per g of seeds which is a function of seed weight and displayed significantly different patterns (e.g. for MoMa10, MoMa12 and T3 MoMa14 lines; Figures 2 and S1a). The addition of acyl-ACP thioesterase activity generated maximal levels of 55.9 nmol seed^{-1} in MaMa12 and 87.6 nmol seed^{-1} in MoMa10. The impact of acyl-ACP thioesterase expression on wax ester accumulation, modifying substrate availability, varies depending on the substrate specificity of the WS. A comparison of MaMaWt (*M. hydrocarbonoclasticus* WS, *M. aquaeolei* FAR and endogenous Camelina thioesterase activity) with lines expressing either acyl 10:0-ACP thioesterase, 12 or 14, shows no beneficial impact on total wax ester production. However, the same comparison of endogenous versus introduced acyl-ACP thioesterase combinations in lines expressing a mouse WS gene demonstrated a substantial increase in average levels of wax ester accumulation, for example MoMaWt (27 nmol seed^{-1}) and MoMa10 (74.6 nmol seed^{-1}). The mouse wax synthase has been suggested to prefer shorter-chain (10:0, 12:0 and 14:0) acyl-CoA species (Cheng and Russell, 2004), and this may contribute to the outcomes observed in this study. Unlike many FARs, the *M. aquaeolei* FAR has substantial activity with diverse substrates of different carbon chain length (C10 to C20), unsaturation and other substitutions (Hofvander et al., 2011) although this wide range was not fully reflected in our data. The

M. aquaeolei FAR has reportedly the highest activity with 18:1-CoA, and although capable of using ricinoleoyl-CoA, fatty alcohol production is reduced by ~75%, which might explain the low wax ester yield in CpFAH-expressing lines (MaMaOH and MoMaOH). The expression of a multifunctional *WS from A. calcoaceticus* ADP1 (a broad specificity bifunctional WS that also functions as an acyl-CoA:diacylglycerol acyltransferase (DGAT)) and *Tetrahymena thermophila* WS in Camelina resulted in moderate levels of wax ester accumulation (Figure 2; AcMa14 and TeMa14). whereas expression of the bifunctional FAR gene (TtFARAT) from *Tetrahymena thermophila* generated only low levels of wax ester accumulation in combination with the *M. hydrocarbonoclasticus* WS (Figure 2; MaTeWt and MaTe14).

Enzyme combinations influence wax ester composition

Seed-specific expression in Camelina of WS, FAR and thioesterase combinations was examined as a strategy to effectively tailor wax ester composition and thereby produce industrial oils with properties mimicking those of spermaceti oil. In this first iteration, it is necessary to understand how, in a boutique crop such as Camelina, different enzyme combinations effectively manipulate wax ester composition. Extraction and GC analysis of wax esters from the different transgenic lines showed specific compositional changes. The expression of *M. hydrocarbonoclasticus* WS and *M. aquaeolei* FAR (MaMa) in combination with endogenous or additional (C10:0, C12:0 and C14:0) acyl-ACP thioesterases generated a broad distribution (mostly from C30 to C42) of wax esters (Figure 3a). Wax esters consisting of C36 and C38 (27% and 36%, respectively) molecular species predominated in

MaMaWt lines. Additionally, the expression of heterologous acyl-ACP thioesterases produced a shift in the distribution, reducing the accumulation of C36 and C38 species and, in particular with Thio14, increasing the contribution of C30–C34 and decreasing C36–C42. The expression of CpFAH generated a very specific (C34–C38; C36 in excess of 40%) wax ester composition. Then, purified WEs were transformed into fatty acid methyl esters and fatty alcohols by acid-catalysed methanolysis, and analysis of the resulting acyl and alcohol moieties revealed how the use of specific thioesterases changed the composition. In the wax esters of MaMaWt C18/C20 (47% and 28%, respectively), saturated and monoenoic acyl and alcohol moieties predominate (Figure 4), with only small amounts (<10%) of C16/C22/C24. The expression of thioesterases 10, 12 and 14 resulted in a reduction in most fatty acyl moieties C18–C24, specifically with C18 declining from 47% to 34% in MaMa14 lines (Figure 4a). Conversely, in the same lines the contribution of shorter acyl moieties increased with C16 and species ≤C14 shifting from 6.1% and 0.6% to 18.6% and 13.3%, respectively. Remodelling of the alcohol moieties involved a reduction (61%–50%) in C20 species and an increase in C16 (5%–16%). The contribution of the predominant saturated or monoenoic alcohol moieties remained unchanged. Overall, the expression of *M. hydrocarbonoclasticus* WS and *M. aquaeolei* FAR generated wax ester species likely containing 18:1/18:1 (C36) and 18:1/20:1 (C38), and the expression of 14:0-ACP thioesterase changed this to 14:0/18:1 (C32) and 14:0/20:1 (C34).

In an alternative iteration, the expression in Camelina of a mouse *M. musculus* WS and *M. aquaeolei* FAR (MoMa) generated a different and very specific wax ester composition of C36 (30%) and C38 (55%) species (Figure 3b). Despite earlier reported wide substrate specificity (Cheng and Russell, 2004), the mouse WS in Camelina preferentially incorporated C18/C20 alcohol moieties (27% and 65%, respectively; Figure 4) and monoenoic C18 acyl moieties (~80%; Figure 4) as reported by Heilmann *et al.* (2012) and Iven *et al.* (2016). The co-expression of the (C10:0, C12:0 and C14:0) acyl-ACP thioesterases in

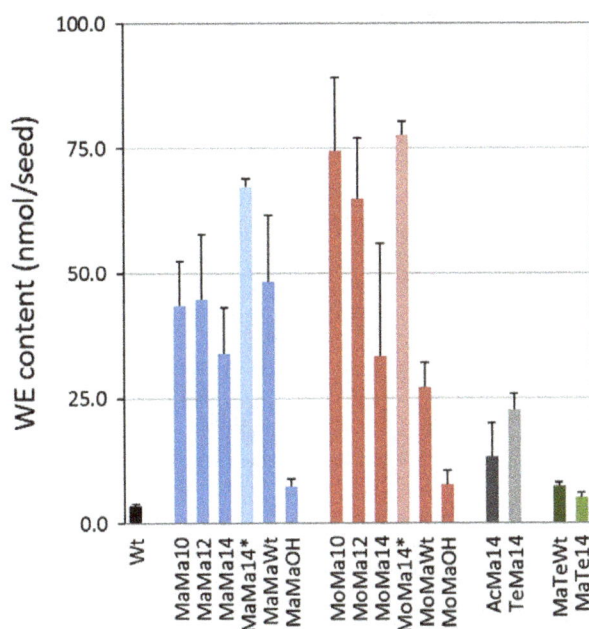

Figure 2 Wax ester content in the seeds of wild-type and transgenic *Camelina sativa* lines expressing different combinations of wax biosynthetic and fatty acid modifying enzymes. This data shows the absolute WE content (in nmol/seed) in T2 seeds for all gene combinations and in T3 Seeds for MaMa14 and MoMa14 (*). Total lipid was extracted from 30 seeds and WE was separated from TAG and quantified by HPLC-ELSD as described in the experimental procedures. For each construct the average WE content (±SE) was calculated from three independent transgenic lines.

Table 2 The mean seed accumulation of WE for each different construct. Wax esters are expressed as nmol/seed, and SD indicated.

Line	WE (nmol seed^{-1})
Wt	3.6 ± 0.3
MaMa10	43.5 ± 9.1
MaMa12	44.8 ± 13.0
MaMa14	34.0 ± 9.2
MaMaWt	48.3 ± 13.3
MaMaOH	7.3 ± 1.6
MoMa10	74.6 ± 14.7
MoMa12	65.1 ± 12.0
MoMa14	33.5 ± 22.5
MoMaWt	27.1 ± 5.1
MoMaOH	7.8 ± 2.8
AcMa14	13.2 ± 6.8
TeMa14	22.7 ± 3.2
MaTeWt	7.3 ± 0.7
MaTe14	5.2 ± 0.9
MaMa14_T3	67.4 ± 1.7
MoMa14_T3	77.6 ± 2.8

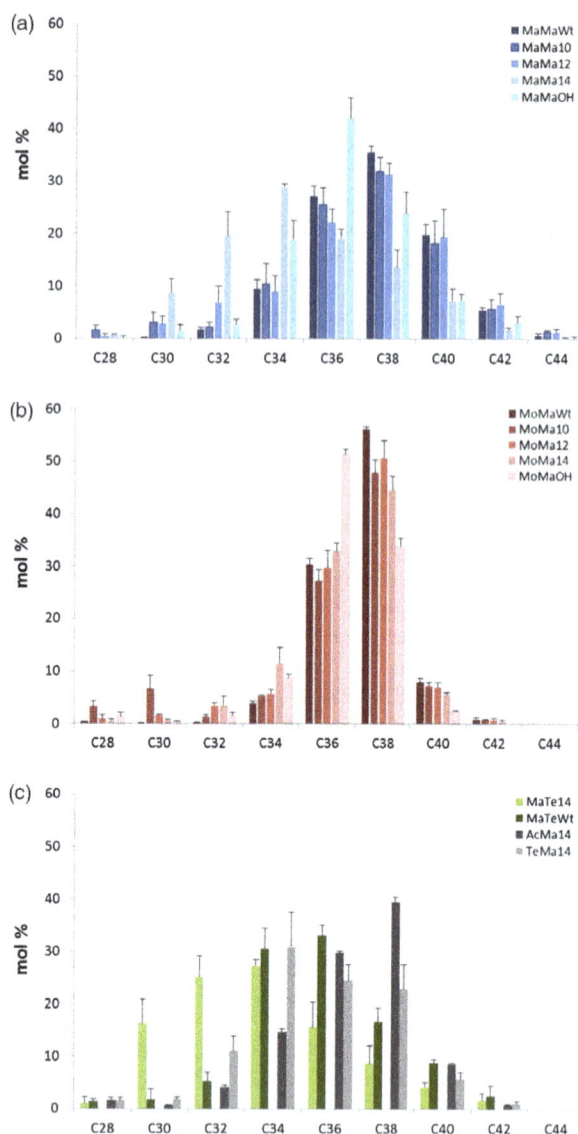

Figure 3 Wax ester composition in the seeds of T2 transgenic *Camelina sativa* lines expressing different combinations of wax biosynthetic and fatty acid modifying enzymes. Intact WEs were analysed by GLC as described in the experimental procedures and combined according to their total carbon number. The relative content (in mol %) in molecular species with chain lengths ranging from 28 to 44 carbons is shown for each gene combination. Transgenic lines expressing combinations of (a) *M. hydrocarbonoclasticus* WS and *M. aquaeolei* FAR or (b) *M. musculus* WS and *M. aquaeolei* FAR in different fatty acid backgrounds. (c) Lines expressing combinations of *M. hydrocarbonoclasticus* WS and *T. thermophila* FAR with *C. palustris* Thio14 or in Wt background, or combination of *A. calcoaceticus* or *T. thermophile* WSs with *M. aquaeolei* FAR and *C. palustris* Thio14. The average content (±SE) of three independent transgenic lines is shown for each gene combination.

combination with mouse WS and *M. aquaeolei* FAR made little impact on the overall final wax composition. Analysis of the acyl-CoA pool in the developing seeds of the transgenic MoMa Camelina lines indicated that thioesterase expression had resulted in substantial remodelling of the acyl-CoA pool (Table S1); however, these substrates were not efficiently incorporated into WEs. Conversely, in the MaMa14 lines, the acyl-CoA pool showed

only a small increase in 14:0-CoA, possibly reflecting the more efficient incorporation of that substrate into WEs by the *M. hydrocarbonoclasticus* WS. Only the expression of Thio14 (MoMa14) significantly reduced the accumulation of C38 (55%–45%) and concomitantly increased C34 content (5%–12%). Although accumulating much lower total levels of wax esters, the co-expression of CpFAH with this construct combination (MoMaOH) produced a change in the ratio of C36 : C38 more pronounced than that observed in MaMaOH lines (Figure 3b; C36 52% and C38 33%). The incorporation of novel hydroxylated species in the wax ester pool is possible, as the two iterations have demonstrated (Figure 2); however, modification of the substrate pool requires careful selection of the WS and FAR activities for the successful high accumulations of these novel wax ester species. This is illustrated by the differential accumulation of Δ12-OH-C18:1 in TAGs and WEs in MaMaOH lines (0.4% and 4.7% of total FA) and MoMaOH lines (3% and <0.5% of total FAs, respectively; Table 1; data not shown). Overall, the wax ester composition resulting from the MoMa combination was very specific and most likely composed primarily of 18:1/20:1 (C38) and 18:1/18:1 (C36). However, it should be noted that in the absence of analysis of intact WEs, the precise acyl and alcohol composition of individual WEs remains to be proven. This is particularly the case for lines expressing the CpFAH hydroxylase.

Changes in wax ester composition (and potential industrial utility) between the MaMa and MoMa WS/FAR enzyme combinations prompted examination of further combinations, for example MaTe14, AcMa14 and TeMa14 (Figure 1c). A comparison of the overall wax ester composition (Figure 5) clearly illustrates how different enzyme combinations can be used to generate very specific wax ester compositions. For example, in MaTe14 the combination of *M. hydrocarbonoclasticus* WS, *T. thermophile* FAR and 14:0-ACP thioesterase generated wax esters predominantly composed of C30–C36 (C34 27%; Figure 3c) whilst *A. calcoaceticus* WS, *M. aquaeolei* FAR and 14:0-ACP thioesterase (AcMa14) has a distribution of C34–C38 (C38 39%; Figure 3c) and *T. thermophile* WS, *M. aquaeolei* FAR and 14:0-ACP thioesterase (TeMa14) of C32 to C38 (C34 30%; Figure 3c). For metabolic engineering, it is important to recognize how the different WS/FAR combinations generate wax esters with different compositions from the same substrate pool. Moreover, using different WS enzymes in combination with *M. aquaeolei* FAR illustrated how compositional changes can occur only in the acyl moiety of the wax esters; that is, TeMa14 incorporated greater amounts of shorter (<C16:0) saturated acyl moieties than AcMa14 (Figure 4a) with a similar alcohol profile (Figure 4b). Inversely, in MaTe14 lines the *T. thermophile* FAR more efficiently converted shorter-chain acyl-CoA than *M. aquaeolei* FAR resulting in a greater content in C12 and C14 fatty alcohols (8.8% C12 + C14; data not shown) and shorter WEs being produced (Figures 3c and 5). Finally, in MaMa and MoMa, combinations substrate supply and enzyme specificity combined to generate wax esters with changes in both the acyl and alcohol moieties.

Properties of Camelina seed engineered for wax ester synthesis

Camelina seed TAGs typically contains approximately 20%–25% very long-chain fatty acids (~15% of C20:1), low levels (<3.5%) of erucic acid (C22:1) and high levels of 18 carbon desaturated fatty acids such as α-linolenic acid (30%–35% of 18 : 3). This was also reflected in the wax esters produced by the engineered Camelina lines, in which the acyl moieties predominantly

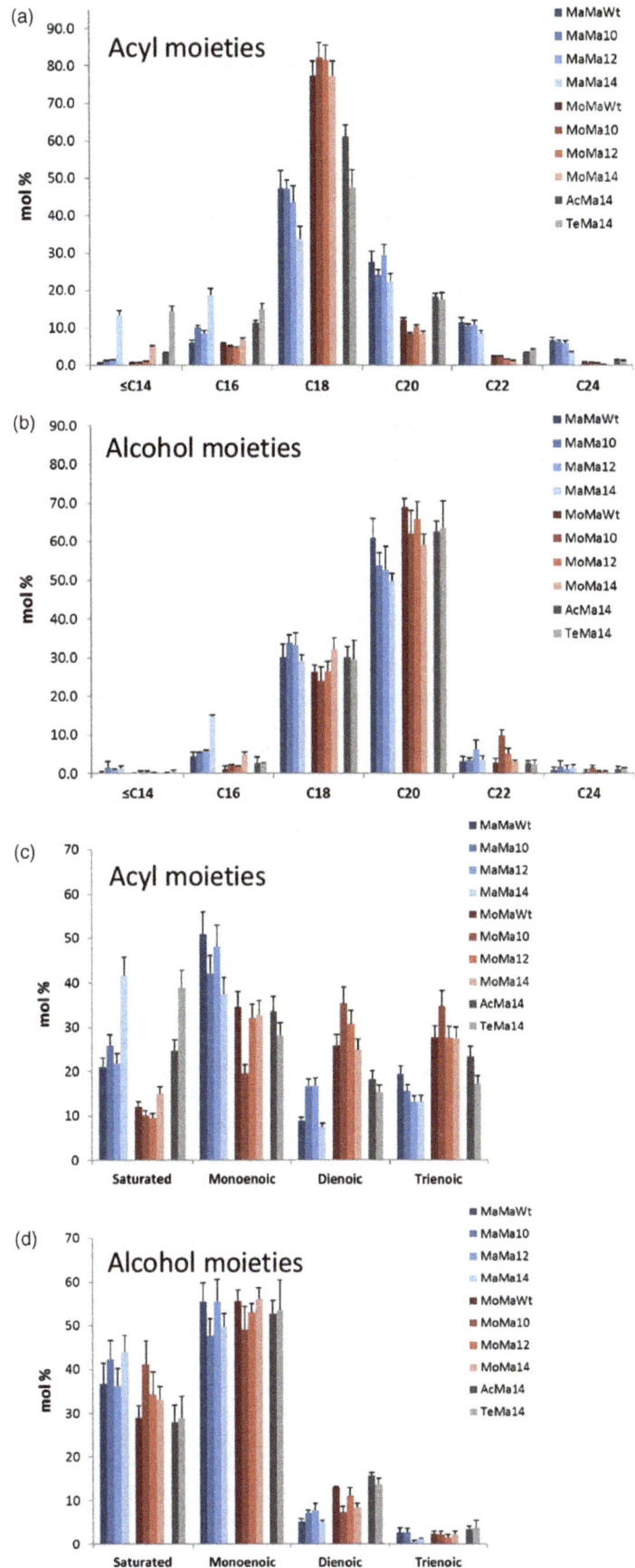

Figure 4 Composition (fatty acid and fatty alcohol) of wax esters produced in T2 seeds of transgenic *Camelina sativa* lines expressing different combinations of wax biosynthetic and fatty acid modifying enzymes. WE were hydrolysed and the resulting fatty acid and fatty alcohol moieties were derivatized, separated and quantified by GLC as described in the experimental procedure. This data shows the relative abundance of the wax ester moieties (in mol %) combining acyl chains according to the total carbon number (a, b) or the saturation degree (c, d) of the fatty acyl (a, c) or the fatty alcohol (b, d) moieties. Shown is the average content (±SE) of three independent transgenic lines for each gene combination.

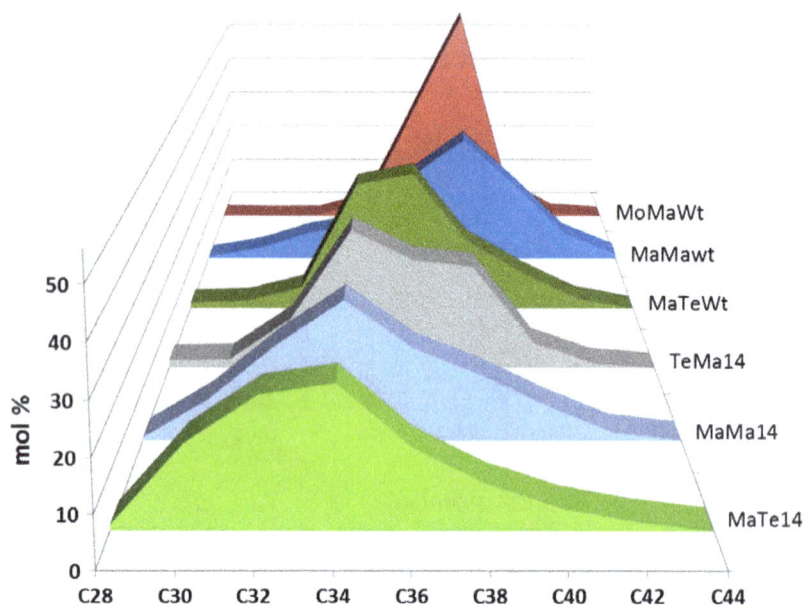

Figure 5 Comparison of wax ester profiles in T2 seeds of transgenic *Camelina sativa* lines expressing different wax biosynthetic and fatty acid modifying enzyme combinations. This figure illustrates the relative contribution (mol %) and distribution of WE molecular species combined according to their carbon chain length. The average of three independent transgenic lines is shown for each gene combination.

contained C18 mono-, di- and trienoic species, whereas the alcohol moieties typically contained C20 monoenoic species (Figure 4). A further aim of this study was to examine the influence seed wax ester synthesis has on TAG assembly, as recent research has indicated an associated reduction in oil content in transgenic lines (Zhu *et al.*, 2016). Analysis of TAG accumulation in the seed of Camelina engineered for wax synthesis was determined by quantification of the fatty acid methyl esters using GC-FID following separation of the TAG fraction by HPLC-ELSD. The seed TAG content varied from 16% to 38% (36% in Wt; Figure 6) in transgenic lines. T2 lines accumulating the highest amounts of wax esters, for example MoMa10 and MaMa10 (Figure 2), had a correspondingly low TAG content (16% and 17%, respectively; Figure 6). However, certain iterations, for example MoMaWt (37% TAG) and MaMa14 (30%), retained wild-type levels of seed oil accumulation. Furthermore, it should be noted that because of seed weight variation the relative TAG content in % (w/w) is not always representative of total seed oil content. For example, MoMaWt seeds contain 22% more oil than WT (0.544 and 0.447 mg/seed, respectively) in addition to ~15 μg (27 nmol) WE, hence bigger seeds. Those lines expressing CpFAH for the production of hydroxylated fatty acids were largely compromised in oil synthesis (24%–26% TAG), reflecting the difficulty of incorporating these novel species into wax ester and oil biosynthetic pathways and also the indirect impact of hydroxylated fatty acids on glycerolipid metabolism (van Erp *et al.*, 2011). The composition of TAG in seed oil was also modified, typically as a result of acyl-ACP thioesterase expression. Medium-chain fatty acids could be identified in the seed oil (Table 1); for example, MaMa10, MaMa12 and MaMa14 contained 0.8%, 3.1% and 14% medium-chain fatty acids (C10:0 to C14:0), whereas MoMa10, MoMa12 and MoMa14 accumulated 2.4%, 1.4% and 4.9%, respectively. The early termination of FAS by acyl-ACP thioesterase expression also had an impact on the C18 content of TAG, typically 67% in Wt; this was reduced to 50% in the most effective MaMa14 iterations. As a general trend, seed oil accumulation declined with wax ester production and underwent compositional changes in response to the expression of different WS, FAR and acyl-ACP thioesterase combinations. However, as

discussed above this was not systematic and we have also identified lines in which both the TAG and the WE contents and were significantly increased compared to Wt Camelina seeds. Particularly, T3 MoMa14 lines displayed a 130% increase in seed WE content compared to T2 seeds (77.6 and 33.5 nmol/seed, respectively; Figure 2) but also contained 16.7% more TAG than WT seeds (0.522 and 0.447 mg/seed, respectively; Figure 6). This suggest that with the right enzyme combinations and tailored substrate pool compositions it is possible to engineer a wax biosynthetic path in the seed of Camelina that does not compete directly with endogenous lipid biosynthetic pathways. On average, there was a decline in seed weight with the expression of different gene combinations (Figure 6); however, a consistent pattern did not emerge. For example, of the high accumulating wax ester lines MaMa10 had a reduction in seed weight, whilst MoMa10 increased in seed weight. However, any difference in seed weight or oil content did not translate into any germination phenotype or detectable plant developmental defect. All Camelina T2 and T3 seeds germinated at similar rates and grew into plants indistinguishable from wild type.

Towards the development of a new method for pilot-scale extraction and separation of wax esters from seed oil

Seed oil WEs and TAGs can easily be extracted and separated in the laboratory using simple solvent extraction and separation by chromatography either on TLC or by HPLC. However, one aspect which will determine the economic viability of WEs produced in seeds as an alternative to nonrenewable sources is the ability to simply and cheaply extract them and separate them from TAGs at an industrial scale. There has been only one report suggesting that very long-chain wax esters (C42–C48) produced in Camelina seeds could be at least partially separated from the TAGs by chilling the oil to 4 °C. This procedure called winterization leads to the formation of a solid fraction enriched in WEs that can be partially separated from the liquid oil (Zhu *et al.*, 2016). However, this technique relies on WE species with melting points higher than that of seed oil TAGs and is therefore likely to be more challenging with shorter-chain WE as the ones present in spermaceti (typically C28–C40) and the lines generated in this

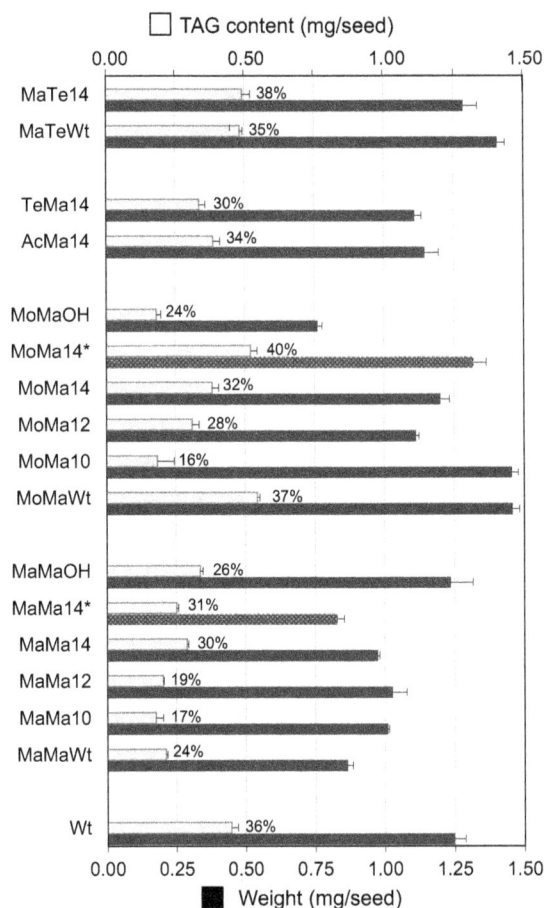

Figure 6 Seed weight and oil content in the seeds of wild-type and transgenic *Camelina sativa* lines expressing different combinations of wax biosynthetic and fatty acid modifying enzymes. This data shows the absolute TAG content and the average seed weight (both in mg/seed) in T2 seeds for all gene combinations and in T3 Seeds for MaMa14 and MoMa14 (*). Average seed weight (±SD) was estimated by measuring the weight of 30 seed batches with 4 technical repeats for each transgenic line analysed. Relative seed oil content is shown as a percentage of seed weight (% w/w) at the top of the bars illustrating absolute TAG content.

study. To test the possibility to efficiently extract and purify a sufficient amount of spermaceti-like WEs from Camelina seeds for performance testing, a pilot-scale experiment was conducted using the MaMa14 line described above. Two T3 MaMa14 lines were selected for their relatively high WE content (47 and 30 nmol/seed or 30 and 20 mg WE/g seed, respectively). Approximately 1200 plants were cultivated in containment glass houses, yielding 6.9 kg of T4 seeds. After pressing and hexane extraction 2.04 kg of oil was recovered (29.6%, w/w) and the WE concentration was estimated by HPLC at 25 mg/g oil corresponding to over 50 g of WE that could be potentially extracted (Table S2). This amount of WE was considered sufficient for performance testing as a lubricant therefore purification was undertaken.

WEs were initially purified using molecular distillation (MD) (Figure 7a), a procedure that can be used at both laboratory and industrial scale for purification of heat sensitive compounds. MD has been used to purify omega-3 fatty acids (EPA and DHA) from fish oil (Bergeron *et al.* 2012) but also tocotrienols and tocopherols from palm and rapeseed oil deodorizer distillates,

respectively (Jiang *et al.*, 2006; Posada *et al.*, 2007). MD presents many advantages such as the absence of a requirement for an organic solvent and reduced or no degradation of compounds due to high vacuum. Under high-vacuum conditions, the boiling point of the components to be separated is 200–300 °C lower than at atmospheric pressure (see Table S4a, b), allowing separation of high molecular weight compounds with minimal degradation. Therefore, it was decided to test the efficiency of this method for separating WEs from TAGs using different experimental conditions (Figure 7a). As described in the Supp. Methods and Figs S1–S6, this method can be used for efficient extraction of nondegraded WEs from refined oil (over 80% recovery; Table S2). However, due to codistillation with short-chain saturated TAG molecular species, only a moderate two fold enrichment was possible.

Winterization is a process commonly used industrially to remove components with high melting point (e.g. waxes) from vegetable oils such as sunflower, rice bran and cotton seed oil. Winterizing mainly consists of cooling the oil gradually and filtering it at a low temperature using special filters. As previous work has shown that long-chain wax ester could be solidified and partially separated from Camelina seed oil (Zhu *et al.*, 2016), the methods was tested for the purification of spermaceti-like medium-chain WEs. This method was less successful than molecular distillation as we were only able to recover less than 20% of WE present in the refined oil fraction (Table S3). However, the WE recovered in the crystallized fraction was heavily enriched in saturated species (80% of total WE content) suggesting that recovery could be greatly improved by saturating the oil by hydrogenation, a procedure easily and cheaply used industrially.

Conclusion

In this study, the seed-specific expression in Camelina of WS, FAR and thioesterase combinations was examined as a strategy to effectively tailor wax ester composition and thereby produce industrial oils with properties mimicking those of spermaceti oil. A comparison of the overall WE compositions generated illustrated how different enzyme combinations can yield very specific wax ester compositions. Depending on the enzymes used, compositional changes occurred in either or both the fatty acyl and the fatty alcohol moiety of the wax esters, although it is still (frustratingly) not possible to accurately predict the performance of a heterologous enzyme *in planta* (Napier *et al.*, 2014). As a general trend, seed weight and oil content declined with wax ester production and TAG underwent compositional changes in response to the expression of different WS, FAR and fatty acid modifying enzyme combinations. However, a consistent pattern did not emerge and differences in seed weight or oil content did not translate into any germination phenotype or obvious plant developmental defect. Interestingly, we also identified lines producing bigger seeds in which both the TAG and the WE contents were significantly increased suggesting that, given the right enzyme combinations and tailored substrate pool compositions, it is possible to engineer wax biosynthetic pathways in Camelina seeds that do not compete directly with TAG synthesis. Finally, preliminary purification experiments suggest that it should be possible to develop a method for the efficient extraction and enrichment of WE from Camelina seed oil, and this is currently being investigated. This method will likely involve a sophisticated combination of biorefining techniques including both MD and winterizations steps.

(a)

(b)

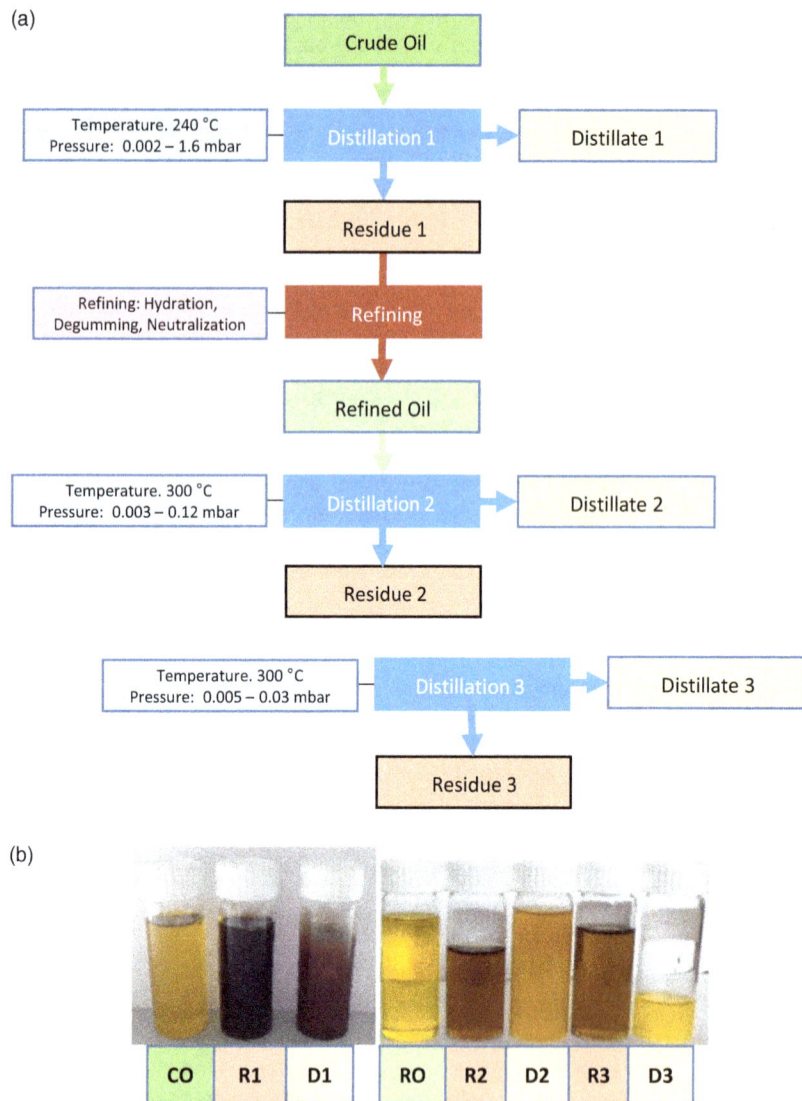

Figure 7 (a) Schematic overview of the Molecular Distillation (MD) and refining steps tested for purification of wax ester form Camelina seed oil. (b) Camelina seed oil fractions before and after refining and MD steps. CO, crude oil after cold pressing and solvent extraction; RO, refined oil after hydration, degumming and neutralization; D1-3, distillates recovered; R1-3, oil residues after MD steps.

Experimental procedures

Plant material and growth conditions

Camelina sativa (Camelina) were grown in a controlled-environment chamber at 23 °C day/18 °C night, 50%–60% humidity, and kept on a 16-h, 250 µmol m^{-2} s^{-1}, photoperiod (long day).

Generation of transgenic plants

Transgenic Camelina lines were generated as previously described with minor modifications (Lu and Kang, 2008; Ruiz-Lopez *et al.*, 2014). The designed vectors were transferred into *Agrobacterium tumefaciens* strain AGL1. Camelina inflorescences were immersed into the Agrobacterium suspension for 30 s without applying any vacuum. Transgenic seeds were identified by visual screening for DsRed activity. Seeds harvested from transformed plants were illuminated by a green LED flashlight. Fluorescent seeds were visualized using red-lens filter.

Vector construction

Fourteen constructs (Figure 1 C) were built using the Gateway recombination system (Invitrogen, Paisley, UK) as previously described in Ruiz-Lopez *et al.* (2015). Respective genes were inserted as NcoI/PacI or AscI/PacI fragments into the promoter/terminator cassettes and then moved into pENTRY vectors. All open reading frames for fatty acid acyl-CoA reductases, wax synthases and thioesterases or hydroxylases were resynthesized (GenScript Corporation, Piscataway, NJ) and codon-optimized for expression for Camelina. The destination vector contained a DsRed gene as the selection marker, driven by the constitutive CsVMV promoter. All constructs contained two or three expression cassettes where each gene was under control of the conlinin1 promoter (pCnl) (Truksa *et al.*, 2003) and linked to a terminator region of OCS, octopin synthase gene of *Agrobacterium tumefaciens*. In particular, MaMaWt construct contained a *M. aquaeolei fatty acyl reductase gene* (Hofvander *et al.*, 2011) *and a M. hydrocarbonoclasticus wax synthase gene (GenBank accession ABO21021)*. MaMa10, MaMa12, MaMa14 and MaMaOH constructs shared the same FAR and WS genes than MaMaWt, but they also included a *Cuphea hookeriana 8:0- and 10:0-ACP specific thioesterase gene* (Dehesh *et al.*, 1996a) *or a Umbellularia californica* 12:0-ACP specific thioesterase gene (Voelker *et al.*, 1992) or a 14:0-ACP specific thioesterase gene from *C. palustris* (Dehesh *et al.*, 1996b), or a *Claviceps purpurea* fatty acid hydroxylase gene (Meesapyodsuk and Qiu, 2008), respectively. Additionally, MoMaWt, MoMa10, MoMa12, MoMa14 and MoMaOH constructs included the same genes as the constructs detailed above, but they contained a wax synthase gene from *M. musculus* (Cheng and Russell, 2004) replacing the *M. hydrocarbonoclasticus* WS. In addition, AcMa14 and TeMa14 constructs also comprised of three expression cassettes including (i) a FAR gene from *M. aquaeolei*, (ii) a 14:0-ACP specific thioesterase gene from *C. palustris* and (iii) AcMa14 included a WS gene from *A. calcoaceticus* (Kalscheuer and Steinbuchel, 2003), but TeMa14 included a WS gene from *T. thermophile* (Biester *et al.*, 2012). Finally, MaTe14 and MaTeWt constructs were built with (i) a WS gene from *M. hydrocarbonoclasticus* and (ii) a FAR gene from *T. thermophile* (Dittrich-Domergue *et al.*, 2014), both flanked by conlinin1 promoters and OCS terminators as in the above constructs. Additionally, MaTe14 construct also included a 14:0-ACP specific thioesterase gene from *C. palustris*.

Acyl-CoA profiling

Camelina seeds (ten seeds, 28 DAF) were harvested and frozen in liquid nitrogen and extracted after Larson and Graham (2001) for reverse-phase LC with electrospray ionization tandem mass spectrometry (multireaction monitoring) in positive ion mode. LC-MS/MS MRM analysis followed the methods described by Haynes *et al.* (2008). (Agilent 1200 LC system; Gemini C18 column, 2 mm inner diameter, 150 mm with 5-mm particles). For the purpose of identification and calibration, standard acyl-CoA esters with acyl chain lengths from C14 to C20 were purchased from Sigma as free acids or lithium salts.

Lipid extraction, separation and quantification by HPLC-ELSD

Wax esters and TAG were extracted following the Hara and Radin (1978) method with some modifications. Thirty Camelina seeds (approximately 20 mg) were heated in 0.8 mL of isopropanol at 85 °C for 10 min. Then, 1.2 mL of hexane was added and seeds were homogenized using a mortar and pestle. Anhydrous sodium sulphate was added to remove any residual moisture and to facilitate the separation of lipid phase. The homogenate was centrifuged, supernatant collected and the pellet re-extracted with hexane : isopropanol (7 : 2, by vol). Both extracts were pooled, evaporated and diluted in 200 μL of hexane. An evaporative light scattering detection (ELSD, Agilent 1200 Infinity Evaporative Light Scattering Detector)-based high-performance liquid chromatography (HPLC, Agilent 1100 LC system) method was used based on the Aragón *et al.* (2011) method with some modifications. The separation was carried out on a 250 mm × 4 mm i.d. column packed with modified silica (Lichrospher 5 μm Si 60; Phenomenex) maintained at 25 °C. The initial composition of the eluent (95 : 5 hexane : ethyl acetate; v/v) at a flow rate of 0.5 mL/min was maintained for 8.5 min. Then, the flow was increased to 0.7 mL/min and the gradient was varied to reach 50% ethyl acetate within 0.5 min and maintained for 5 min. To assure the complete elimination of impurities, the gradient was set to 80% ethyl acetate and it was maintained for 2 min. For the analysis of the samples, 100 μL of the oil sample prepared as described above was injected. The ELSD was maintained at 40 °C throughout. The nebulizer (nitrogen) gas pressure was set at 3.5 bar and the detector gain was set at 2. WE and TAG peaks were identified by comparison with standards (Sigma-Aldrich, UK), and they were collected at their specific elution times using a fraction collector. WE

quantification was performed by area comparison of C28:0, C30:0, C32:0, C34:0, C36:0, C38:0 and C40:0 standards from Sigma-Aldrich. To adjust for the effects of the nonlinear response curve of the ELSD detector, calibration curves for WEs were generated using C28 to C44 commercially available standards. The calibration was made with lipid standard mixtures as close as possible in composition to those observed in the experimental samples.

Wax ester analysis

Intact wax esters were separated and quantified by GLC with an Agilent 6890 gas chromatograph (Palo Alto, CA), using a HP-1 ms capillary column from Agilent J&W (30 m, 0.25 mm, 0.25 μm). Helium was used as carrier gas, and injector and detector temperatures were 280 and 325 °C, respectively. Splitless injection was used, with a flow rate set at 1.5 mL/min. After injection at 70 °C, the oven temperature was raised to 200 °C at a rate 50 °C min^{-1}, then to 325 °C at a rate of 3 °C min^{-1}, and held constant for 10 min. The detector used was a flame ionization detector. Wax esters peaks were identified by comparing their retention times with those of authentic standards (C28–C46; Nu-Chek Prep Inc, Elysian, MN, USA.) and quantified using calibration curves essentially as described in Razeq *et al.* (2014).

Analysis of fatty acid composition in WE and TAG

Total fatty acids from wax esters and TAG were transmethylated by heating the samples at 80 °C for 2 h with 2 mL of a solution containing methanol/toluene/dimethoxypropane/H_2SO_4 (66 : 28 : 2 : 1 by volume) as previously described (Ruiz-Lopez *et al.*, 2015). Methyl ester derivatives of total fatty acids were extracted using hexane, and they were analysed by GC-FID and confirmed by GC-MS. Quantification of fatty acids from TAG was performed using tri-heptadecanoyl glycerol (Sigma-Aldrich) as an internal standard. Data are presented as representative numbers derived from three replicated analyses.

Fatty alcohol separation and analysis

Fatty alcohols were separated of FAMES using SPE silica gel 60 columns (Merck). The mixture of FAMEs and fatty alcohols was dissolved in hexane and poured on the column prepared in hexane. FAMEs were eluted after adding 6 mL of hexane : diethyl ether (95 : 5, v/v). Fatty alcohols were eluted by adding 6 mL of hexane:ethyl acetate (6/1, v/v). Fatty alcohols were silylated by adding excess of BSTFA and 1% TMCS and heating the samples at 80 °C for 1 h prior to analysing them by GC, as below.

Acknowledgements

This work was supported by the EU grant 'Industrial crops producing value-added oils for novel chemicals and energy (ICON)' from the 7th Framework Programme (FP&-KBBE-3007-1) and a BBSRC IBTI grant (BB/H00449/1) 'Engineering oilseeds to synthesise designer wax esters'. We thank Pilot Pflanzenöltechnologie Magdeburg e. V. (PPM) for their excellent technical assistance and service in the purification of wax esters. Rothamsted Research receives grant-aided support from the BBSRC.

References

Aragón, A., Cortés, J.M., Toledano, R.M., Villén, J. and Vázquez, A. (2011) Analysis of wax esters in edible oils by automated on-line coupling liquid chromatography-gas chromatography using the through oven transfer adsorption desorption (TOTAD) interface. *J. Chromatogr. A*, **1218**, 4960–4965.

Bates, P.D., Stymne, S. and Ohlrogge, J. (2013) Biochemical pathways in seed oil synthesis. *Curr. Opin. Plant Biol.* **16**, 358–364.

Beaudoin, F., Sayanova, O., Haslam, R.P., Bancroft, I. and Napier, J.A. (2014) Oleaginous crops as integrated production platforms for food, feed, fuel and renewable industrial feedstock-Manipulation of plant lipid composition via metabolic engineering and new opportunities from association genetics for crop improvement and valorisation of co-products. *OCL* **21**, D606.

Biester, E.-M., Hellenbrand, J. and Frentzen, M. (2012) Multifunctional acyltransferases from *Tetrahymena thermophile*. *Lipids*, **47**, 371–381.

Bergeron, C., Carrier, D. J. and Ramaswamy, S. (2012) Separation and Purification of Phytochemicals as Co-Products in Biorefineries. In Biorefinery Co-Products: Phytochemicals, Primary Metabolites and Value-Added Biomass Processing., Chichester, UK: John Wiley & Sons, Ltd. doi: 10.1002/9780470976692.ch3.

Carlsson, A.S., Yilmaz, J.L., Green, A.G., Stymne, S. and Hofvander, P. (2011) Replacing fossil oil with fresh oil – with what and for what? *Eur. J. Lipid Sci. Technol.* **113**, 812–831.

Cheng, J.B. and Russell, D.W. (2004) Mammalian wax biosynthesis: II. Expression cloning of wax synthase cDNAs encoding a member of the acyltransferase enzyme family. *J. Biol. Chem.* **279**, 37798–37807.

Dehesh, K., Jones, A., Knutzon, D.S. and Voelker, T.A. (1996a) Production of high levels of 8:0 and 10:0 fatty acids in transgenic canola by overexpression of ChFatB2, a thioesterase cDNA from *Cuphea hookeriana*. *Plant J.* **9**, 167–172.

Dehesh, K., Edwards, P., Hayes, T., Cranmer, A.M. and Fillatti, J. (1996b) Two novel thioesterases are key determinants of the bimodal distribution of acyl chain length of *Cuphea palustris* seed oil. *Plant Physiol.* **110**, 203–210.

Dittrich-Domergue, F., Joubès, J., Moreau, P., Lessire, R., Stymne, S. and Domergue, F. (2014) The bifunctional protein TtFARAT from Tetrahymena thermophila catalyzes the formation of both precursors required to initiate ether lipid biosynthesis. *J. Biol. Chem.* **289**, 21984–21994.

Dyer, J.M., Stymne, S., Green, A.G. and Carlsson, A.S. (2008) High-value oils from plants. *Plant J.* **54**, 640–655.

van Erp, H., Bates, P.D., Burgal, J., Shockey, J. and Browse, J. (2011) Castor phospholipid:diacylglycerol acyltransferase facilitates efficient metabolism of hydroxy fatty acids in transgenic Arabidopsis. *Plant Physiol.* **15**, 683–693.

Hara, A. and Radin, N. (1978) Lipid extraction of tissues with a low-toxicity solvent. *Anal. Biochem.* **90**, 420–426.

Haslam, R.P., Sayanova, O., Kim, H.J., Cahoon, E.B. and Napier, J.A. (2016) Synthetic redesign of plant lipid metabolism. *Plant J.* **87**, 76–86.

Haynes, C.A., Allegood, J.C., Sims, K., Wang, E.W., Sullards, M.C. and Merril, A.H. (2008) Quantitation of fatty acyl-coenzyme as in mammalian cells by liquid chromatography-electrospray ionisation tandem mass spectrometry. *J. Lipid Res.* **49**, 1113–1125.

Heilmann, M., Iven, T., Ahmann, K., Hornung, E., Stymne, S. and Feussner, I. (2012) Production of wax esters in plant seed oils by oleosomal co-targeting of biosynthetic enzymes. *J. Lipids Res* **53**, 2153–2161.

Hoare, P. (2009) *Leviathan or, the Whale*. Pp. 352 Fourth Estate

Hofvander, P., Doan, T.T.P. and Hamberg, M. (2011) A prokaryotic acyl-CoA reductase performing reduction of fatty acyl-CoA to fatty alcohol. *FEBS Lett.* **585**, 3538–3543.

Horn, P.J., Silva, J.E., Anderson, D., Fuchs, J., Borisjuk, L., Nazarenus, T.J., Shulaev, V. *et al.* (2013) Imaging heterogeneity of membrane and storage lipids in transgenic *Camelina sativa* seeds with altered fatty acid profiles. *Plant J.* **76**, 138–150.

Ishige, T., Tani, A., Sakai, Y. and Kato, N. (2003) Wax ester production by bacteria. *Curr. Opin. Microbiol.* **6**, 244–250.

Iven, T., Herrfurth, C., Hornung, E., Heilmann, M., Hofvander, P., Stymne, S., Zhu, L.H., *et al.* (2013) Wax ester profiling of seed oil by nano-electrospray ionization tandem mass spectrometry. *Plant Methods* **9**, 1.

Iven, T., Hornung, E., Heilmann, M. and Feussner, I. (2016) Synthesis of oleyl oleate wax esters in *Arabidopsis thaliana* and *Camelina sativa* seed oil. *Plant Biotechnol. J.* **14**, 252–259.

Jaworski, J. and Cahoon, E.B. (2003) Industrial oils from transgenic plants. *Curr. Opin. Plant Biol.* **6**, 178–184.

Jiang, S.T., Shao, P., Pan, L.J. and Zhao, Y.Y. (2006) Molecular distillation for recovering tocopherol and fatty acid methyl esters from rapeseed oil deodoriser distillate. *Biosyst. Eng.* **93**, 383–391.

Kalscheuer, R. and Steinbuchel, A. (2003) A novel bifunctional wax ester synthase/acyl-CoA: diacylglycerol acyltransferase mediates wax ester and triacylglycerol biosynthesis in *Acinetobacter calcoaceticus* ADP1. *J. Biol. Chem.* **278**, 8075–8082.

Kim, H.J., Silva, J.E., Vu, H.S., Mockaitis, K., Nam, J.-W. and Cahoon, E.B. (2015) Toward production of jet fuel functionality in oilseeds: identification of FatB acyl-acyl carrier protein thioesterases and evaluation of combinatorial expression strategies in Camelina seeds. *J. Exp. Bot.* **66**, 4251–4265.

Lardizabal, K.D., Metz, J.G., Sakamoto, T., Hutton, W.C., Pollard, M.R. and Lassner, M.W. (2000) Purification of a jojoba embryo wax synthase, cloning of its cDNA, and production of high levels of wax in seeds of transgenic Arabidopsis. *Plant Physiol.* **122**, 645–655.

Larson, T.R. and Graham, I.A. (2001) A novel technique for the sensitive quantification of acyl CoA esters from plant tissues. *Plant J.* **25**, 115–125.

Lu, C. and Kang, J. (2008) Generation of transgenic plants of a potential oilseed crop *Camelina sativa* by Agrobacterium-mediated transformation. *Plant Cell Rep.* **27**, 273–278.

Meesapyodsuk, D. and Qiu, X. (2008) An oleate hydroxylase from the fungus *Claviceps purpurea*: cloning, functional analysis, and expression in Arabidopsis. *Plant Physiol.* **147**, 1325–1333.

Metz, J.G., Pollard, M.R., Anderson, L., Hayes, T.R. and Lassner, M.W. (2000) Purification of a jojoba embryo fatty acyl-coenzyme A reductase and expression of its cDNA in high erucic acid rapeseed. *Plant Physiol.* **122**, 635–644.

Miwa, T.K. (1971) Jojoba oil wax esters and derived fatty acids and alcohols: gas chromatographic analyses. *J. Am. Oil Chem. Soc.* **48**, 259–264.

Napier, J.A. (2007) The production of unusual fatty acids in transgenic plants. *Ann. Rev. Plant Biol.* **58**, 295–319.

Napier, J.A., Haslam, R.P., Beaudoin, F. and Cahoon, E.B. (2014) Understanding and manipulating plant lipid composition: metabolic engineering leads the way. *Curr. Opin. Plant Biol.* **19**, 68–75.

Nevenzel, J.C. (1970) Occurrence, function and biosynthesis of wax esters in marine organisms. *Lipids*, **5**, 308–319.

Posada, L.R., Shi, J., Kakuda, Y. and Xue, S.J. (2007) Extraction of tocotrienols from palm fatty acid distillates using molecular distillation. *J. Sep. Purif. Technol.* **57**, 220–229.

Razeq, F.M., Kosma, D.K., Rowland, O. and Molina, I. (2014) Extracellular lipids of *Camelina sativa*: characterization of chloroform-extractable waxes from aerial and subterranean surfaces. *Phytochemistry*, **106**, 188–196.

Ruiz-Lopez, N., Haslam, R.P., Napier, J.A. and Sayanova, O. (2014) Successful high-level accumulation of fish oil omega-3 long-chain polyunsaturated fatty acids in a transgenic oilseed crop. *Plant J.* **77**(2), 198–208.

Ruiz-Lopez, N., Haslam, R.P., Usher, S., Napier, J.A. and Sayanova, O. (2015) An alternative pathway for the effective production of the omega-3 long-chain polyunsaturated EPA and ETA in transgenic oilseeds. *Plant Biotechnol. J.* **13**, 1264–1275.

Truksa, M., MacKenzie, S.L. and Qiu, X. (2003) Molecular analysis of flax 2S storage protein conlinin and seed specific activity of its promoter. *Plant Physiol. Biochem.* **41**, 141–147.

Vanhercke, T., Wood, C.C., Stymne, S., Singh, S.P. and Green, A.G. (2013) Metabolic engineering of plant oils and waxes for use as industrial feedstocks. *Plant Biotechnol. J.* **11**, 197–210.

Voelker, T.A., Worrell, A.C., Anderson, L., Bleibaum, J., Fan, C., Hawkins, D.J., Radke, S.E. *et al.* (1992) Fatty acid biosynthesis redirected to medium chains in transgenic oilseed plants. *Science*, **257**, 72–74.

Yuan, L. and Grotewold, E. (2015) Metabolic engineering to enhance the value of plants as green factories. *Metab. Eng.* **27**, 83–91.

Zhu, L.-H., Krens, F., Smith, M.A., Li, X., Qi, W., van Loo, E.B., Iven, T. *et al.* (2016) Dedicated industrial oilseed crops as metabolic engineering platforms for sustainable industrial feedstock production. *Sci. Rep.* **6**, 22181.

11

The wheat Lr34 multipathogen resistance gene confers resistance to anthracnose and rust in sorghum

Wendelin Schnippenkoetter[1], Clive Lo[2], Guoquan Liu[3], Katherine Dibley[1], Wai Lung Chan[2], Jodie White[4], Ricky Milne[1], Alexander Zwart[5], Eunjung Kwong[1], Beat Keller[6], Ian Godwin[3], Simon G. Krattinger[6] and Evans Lagudah[1,7,]*

[1]CSIRO Agriculture and Food, Canberra, ACT, Australia

[2]School of Biological Sciences, The University of Hong Kong, Hong Kong, China

[3]School of Agriculture and Food Sciences, The University of Queensland, St Lucia, QLD, Australia

[4]Centre for Crop Health, University of Southern Queensland, Toowoomba, QLD, Australia

[5]CSIRO Data61, Canberra, ACT, Australia

[6]Department of Plant and Microbial Biology, University of Zurich, Zurich, Switzerland

[7]School of Life and Environmental Sciences, University of Sydney, Sydney, NSW, Australia

*Correspondence
email evans.lagudah@csiro.au

Keywords: multiple disease resistance, Lr34, rust, anthracnose, flavonoid phytoalexin.

Summary

The ability of the wheat Lr34 multipathogen resistance gene (*Lr34res*) to function across a wide taxonomic boundary was investigated in transgenic *Sorghum bicolor*. Increased resistance to sorghum rust and anthracnose disease symptoms following infection with the biotrophic pathogen *Puccinia purpurea* and the hemibiotroph *Colletotrichum sublineolum*, respectively, occurred in transgenic plants expressing the *Lr34res* ABC transporter. Transgenic sorghum lines that highly expressed the wheat *Lr34res* gene exhibited immunity to sorghum rust compared to the low-expressing single copy *Lr34res* genotype that conferred partial resistance. Pathogen-induced pigmentation mediated by flavonoid phytoalexins was evident on transgenic sorghum leaves following *P. purpurea* infection within 24–72 h, which paralleled *Lr34res* gene expression. Elevated expression of *flavone synthase II*, *flavanone 4-reductase* and *dihydroflavonol reductase* genes which control the biosynthesis of flavonoid phytoalexins characterized the highly expressing *Lr34res* transgenic lines 24-h post-inoculation with *P. purpurea*. Metabolite analysis of mesocotyls infected with *C. sublineolum* showed increased levels of 3-deoxyanthocyanidin metabolites were associated with *Lr34res* expression, concomitant with reduced symptoms of anthracnose.

Introduction

Sorghum (*Sorghum bicolor*) is ranked as the fifth most commonly cultivated cereal in the world (FAOSTAT, 2016). Some of its useful attributes are tolerance to dry environments, high sugar content, high yields of forage biomass per unit of cultivated area and as a rich source of distinct phytochemicals such as dhurrin, sorgoleone and 3-deoxyanthocyanidins. While sorghum provides a useful resource for industrial purposes, for example the generation of ethanol, fibre and paper, its primary use is still for feed and food especially in the semi-arid tropics. Protecting yield losses from diseases such as anthracnose (*Colletotrichum sublineolum*) and rust (*Puccinia purpurea*), which can be variable in different agro-ecological regions continues to be a goal of sorghum improvement. Grain size and yield losses over 50% have been reported with anthracnose epidemics (Thakur and Mathur, 2000). Rust is particularly problematic in late-sown crops (White *et al.*, 2012) with yield losses up to 65% (Bandyopadhyay, 2000).

Genetic solutions to protect crop plants against pathogens are often preferable to the use of agrochemicals. Numerous studies on plant-microbe interaction have led to an increased understanding of the molecular basis of the plant defense system, depicted by multiple layers of the plants ability to resist pathogen proliferation (Dangl *et al.*, 2013). While most resistance genes tend to be short-lived, certain forms of plant defense genes provide more durable resistance. Studies in wheat with defined races of *Puccinia* (rust) and *Blumeria* (mildew) pathogen species have resulted in over 220 catalogued resistance genes, most of which individually provide resistance to a few races of a specific pathogen (McIntosh *et al.*, 2013). However, a small number (e.g. *Lr34*, *Lr46* and *Lr67*) have been identified that confer adult plant, broad spectrum partial resistance to multiple pathogen species. Most notable among the latter class of resistance genes is the *Lr34* multipathogen resistance gene (Dyck and Samborski, 1979; McIntosh, 1992; Singh, 1992; Spielmeyer *et al.*, 2005), which has been successfully deployed in wheat cultivation and provided durable field resistance to rust pathogens for over 100 years (Kolmer *et al.*, 2008). Significantly, the multipathogen resistance conferred by *Lr34* was not due to a cluster of resistance genes, but rather by a single gene encoding an ABC transporter (Krattinger *et al.*, 2009; Risk *et al.*, 2012). *Lr34* also differs from the other cloned multipathogen resistance gene *Lr67*, which encodes a sugar transporter from the STP13 lineage of monosaccharide transporters (Moore *et al.*, 2015). The Lr34 resistance allele (*Lr34res*) differs from the susceptible or wild-type allele,

Lr34sus, by changes to two amino acids; loss of a phenylalanine residue (F546) and a tyrosine to histidine substitution (Y634H) located in two separate, predicted transmembrane helices. The mechanism of disease resistance conferred by *Lr34res* remains unknown as does the substrate(s) it may translocate. Nonetheless, transfer of *Lr34res* as a transgene to other crops such as barley and rice has demonstrated its capability in conferring resistance against other pathogens that are unadapted to wheat (Krattinger *et al.*, 2016; Risk *et al.*, 2013).

While *Lr34res* in wheat confers partial adult plant resistance to predominantly biotrophic pathogens, the observation of *Lr34res* efficacy in transgenic rice against the hemibiotroph, *Magnaporthe grisea* (causal agent of rice blast; Krattinger *et al.*, 2016), led us in this study to examine the effect of *Lr34res* against the hemibiotroph pathosystem of anthracnose in sorghum. Secondly, we also note that while *Lr34res* effectiveness has been demonstrated against *Puccinia* species (*P. hordeii*, *P. striiformis*, *P. graminis*, and *P. triticina*) adapted to cool season crops, it remains unknown as to its function against *Puccinia* species that are virulent on warm season crops such as *P. purpurea* in sorghum. Thirdly, we take advantage of the pathogen-inducible visible pigmentation changes resulting from the synthesis of a unique class of flavonoid phytoalexins in sorghum, in furthering our understanding of potential signalling pathways triggered by *Lr34res*.

Results

Transgenic *Lr34res* expression confers resistance to sorghum rust (*Puccinia purpurea*) infection

We introduced the complete wheat genomic sequence of *Lr34res*, encompassing 2.4 kb of native promoter and 1.5 kb native terminator sequence, by stable transformation in the genetic background of sorghum cultivar (cv.) Tx430. Four independent T0 transformants with the full-length *Lr34res* were obtained, of which three independent genotypes were fertile (Lr34-2, Lr34-5 and Lr34-6). Subsequent genomic and phenotypic analyses at T1–T3 generations were carried out on these three independent transgenic lines. Genomic blot analysis showed that line Lr34-2 carried a single copy of the *Lr34res* gene, whereas multiple copies were detected in lines Lr34-5 (three copies) and Lr34-6 (seven copies; Figure S1). Analysis of individual T3 plants from the multicopy *Lr34res* events showed identical genomic hybridization patterns to the Lr34 probe that were unique to either Lr34-5 or Lr34-6 T3 progeny. This suggests that multicopy events in Lr34-5 and Lr34-6 were inserted at single sites or in close proximity and consequently the absence of segregation of the transgenes.

Given that Lr34 transgenic barley and rice plants exhibited the leaf tip necrosis (Ltn) phenotype at an earlier developmental stage than wheat (Krattinger *et al.*, 2016; Risk *et al.*, 2013), we monitored the transgenic sorghum plants for similar morphological changes. Phenotypically, the control plants (sib lines without the transgene) and the *Lr34res* single copy line, Lr34-2, were very similar at the seedling stage until the onset of booting. Yellowing leaf margins and leaf tips, typical of the Ltn phenotype, occurred earlier in Lr34-2 compared with adult plants lacking the transgene. In contrast, Lr34-5 and Lr34-6 lines showed a progressive development of a blotchy bronze/purple leaf coloration in adult plants from about the penultimate leaf development stage onwards (Figure S2). The penultimate leaves of adult plants had high *Lr34res* transcript levels in lines Lr34-5 and Lr34-

6, which was 8–13 fold higher than that detected in the single copy Lr34-2 genotype (Figure 1). Thus, the strong leaf coloration phenotypes correlated with *Lr34res* expression.

Sorghum rust pathogenesis on plants infected by *P. purpurea* urediospores at the 5-leaf stage was analysed microscopically at 7 days post-inoculation (dpi) and for sporulation at 12–14 dpi. Microscopic analysis of wheat germ agglutinin-fluorescein isothiocyanate (WGA-FITC) binding to fungal cell walls revealed extensive hyphal development in control plants and sib lines without the *Lr34res* transgene (Figure 2a–c, e). In contrast, hyphal growth from infection sites in all transgenic *Lr34res* lines was restricted (Figure 2d, f). Macroscopically, spores from uredinia developed on all the non*Lr34res* plants, whereas no sporulation was detected on *Lr34res* transgenic genotypes (Figures 3a and S3). Further quantification of the sorghum rust fungal biomass on transgenic plants showed the presence of the *Lr34res* transgene reduced fungal colonization by 75%–80% (Figure 3b). Interestingly, by 28–30 dpi, uredinia had developed on the Lr34-2 transgenic line, albeit at low frequency compared to sib lines lacking *Lr34res* (Figure S4). Estimation of fungal biomass at this late period showed approximately a 25% reduction in fungal colonization, which is indicative of the slow rusting phenotype that typifies the partial resistance often seen with *Lr34res* in wheat. In contrast, no sporulation was detected on Lr34-5 and Lr34-6 genotypes, even at this late stage.

From previous analysis of the *Sorghum bicolor* genome, two adjacent Lr34 orthologs, Sb01g016770 and Sb01g016775, were considered to have arisen by gene duplication, of which Sb01g016770 was deduced to be a pseudogene (Krattinger *et al.*, 2013). Sb01g016770 and Sb01g016775 share 71% and 75% identity respectively with the protein sequence of LR34. Of the two critical amino acids that distinguish LR34RES from LR34SUS, Sb01g016775 shared the same phenylalanine and tyrosine residues found in the wild-type variant of LR34SUS. We investigated by site directed mutagenesis whether changes to Sb01g016775 involving the two critical amino acids to a modified Sb01g016775 with a deleted phenylalanine (ΔF525) and tyrosine to histidine (Y613H) was capable of conferring resistance to sorghum rust as observed with *Lr34res*. Five independent stable transgenic lines of Sb01g016775$^{-\Delta F525,\ Y613H}$ were generated, all of which expressed transcripts carrying the modified gene. Plants infected with *P. purpurea* developed similar levels of sporulation as control plants or sib lines by 14 dpi and failed to exhibit the

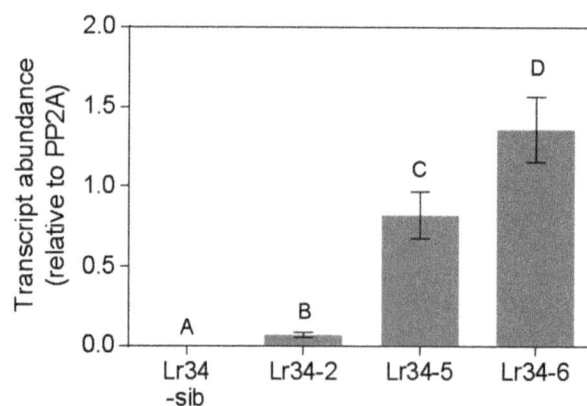

Figure 1 *Lr34res* expression levels in adult plants of transgenic sorghum. Lr34-sib negative line, Lr34-2 single copy line, Lr34-5 3 copy line, Lr34-6 7 copy line. Data shown as mean ± SE from three biological replicates.

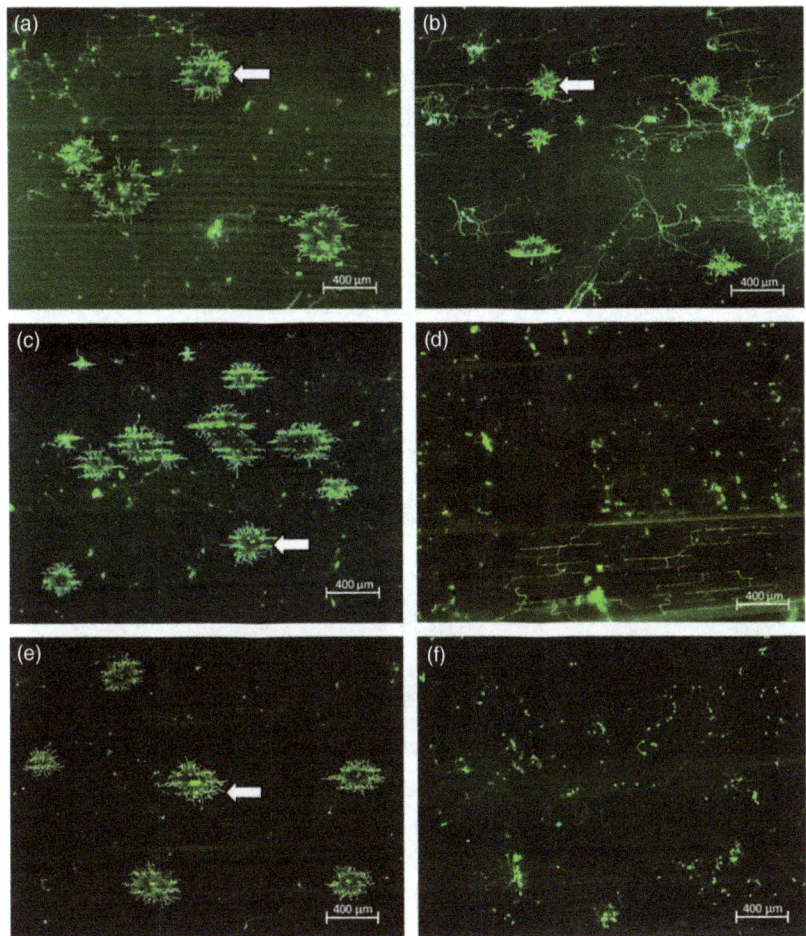

Figure 2 Micrographs of sorghum rust development following WGA-FITC staining at 7 days post-inoculation of fifth leaves. (a) Wild-type Sorghum cultivar Tx430. (b) Sorghum landrace. (c) and (e) Segregate Sib lines Lr34-2 and Lr34-5 respectively not harbouring *Lr34res* gene – Infection sites (arrows) developing from germinated rust spores on leaf surfaces. (d) and (f) Lr34-2 and Lr34-5 transgenic sorghum respectively–germinated spores and hyphae present on leaf surface but with no infection sites. Micrographs of Lr34-6 negative sib and transgenic lines yielded similar results to that of the negative sib and transgenic lines of Lr34-2 and Lr34-5, respectively.

resistance phenotype that accompanied the introduction of the wheat *Lr34res* in sorghum (Figure S5).

Pathogen-induced leaf pigmentation, expression of genes involved in the flavonoid phytoalexin synthesis pathway and metabolite analysis

Within the first 2 days following *P. purpurea* inoculation, reddish brown pigmented spots were observed on leaves of control and transgenic plants. Leaf area coverage and size of pigmented spots were larger in transgenic plants when compared to non-transgenic sibs and the control genotype (Figure S6). Furthermore, the *Lr34res* multicopy genotypes Lr34-5 and Lr34-6 consistently exhibited more pigmented areas than the single copy Lr34-2 transgenic line. To test whether the magnitude of the pathogen-induced pigmentation was associated with *Lr34* expression, transcript levels of *Lr34res* were quantified over a 48-h period post-inoculation. An increase in the *Lr34res* transcript occurred within 24-h post-inoculation (hpi) and declined by 48 hpi (Figure 4). More than threefold increased expression occurred in Lr34-5 at 24 hpi compared to Lr34-2, which parallels the extent of pigmentation noted on the leaves. We also examined the expression levels of the *S. bicolor* orthologous *Lr34* gene, Sb01g016775 under mock and rust inoculation in comparison with the introduced *Lr34res* transgene. Interestingly, Sb01g016775 expression was negatively responsive to *P. purpurea* inoculation in contrast to the pathogen responsiveness of the wheat *Lr34res* demonstrated through increased expression (Figure S7).

Pathogen-inducible synthesis of flavanone derived metabolites, some of which have been implicated in plant defense, has previously been described in sorghum (Lo *et al.*, 1996, 1999; Nicholson *et al.*, 1987). Analysis of expression levels of key enzymes involved in 3-deoxyanthocyanidin and flavone biosynthesis (Figures 4b–d, 5) revealed similar trends to effects of the *Lr34res* transgene. Enzymatic steps encoded by *SbFNSII* (flavone synthase II, a cytochrome P450 pathogen-inducible gene), *SbFNR* (flavanone 4-reductase) and *SbDFR3* (dihydroflavonol reductase) were elevated in gene expression at 24 hpi and declined at 48 hpi (Figure 4b–d). The high expressing *Lr34res* lines, typified by Lr34-5 genotype, exhibited over 15-, 75- and 140-fold increases in expression of *SbDFR3*, *SbFNSII* and *SbFNR*, respectively, at the peak period of 24 hpi. By contrast the control sib line showed 5-, 15- and 20-fold increases for *SbDFR3*, *SbFNSII* and *SbFNR*, respectively. Approximately an eightfold elevation of *SbDFR3* was detected in the low *Lr34res* expressing genotype, Lr34-2, over the same period, whereas *SbFNSII* and *SbFNR* showed similar quantitative changes in the control line and Lr34-2 (Figure 4b–d). Taken together, the early induction of this group of genes which form part of the pathway in converting naringenin flavanones to 3-deoxyanthocyanidin and flavone biosynthesis (Figure 5) is enhanced by the introduction of the wheat *Lr34res* gene upon pathogen infection.

We further investigated the production of metabolites that belong to the 3-deoxyanthocyanidin class upon infection using the well-studied *C. sublineolum*-sorghum pathogen-host interaction. Pathogen-induced formation of purple pigments has

Figure 3 (a) *P. purpurea* pustule development on control and transgenic sorghum leaves at 14 dpi. (b) Quantification of fungal biomass on corresponding plants in (a). Data shown as mean ± SD.

attributed this colour change to structurally related compounds, 3-deoxyanthocyanidins (luteolinidin and apigeninidin; Dykes and Rooney, 2006). These compounds accumulate within inclusions in the epidermal cells as a defense response to pathogen attack (Snyder and Nicholson, 1990; Snyder *et al.*, 1991). As part of the metabolite analysis we also included the flavone aglycones, luteolin and apigenin, that have also been implicated to differentially accumulate as sorghum phytoalexins in response to pathogen infection (Du *et al.*, 2010). Metabolite analysis was conducted on elongated mesocotyls inoculated with *C. sublineolum* at 48, 72 and 96 hpi. Significantly enhanced levels of luteolinidin were detected in genotypes carrying *Lr34res* at 72 hpi (Figure 6). Differences in metabolite accumulation were not as significant at the other two time points (Figure S8). At 48 hpi, metabolite amounts were still rather low and by 96 hpi, the 3-deoxyanthocyanidins started to be degraded. Methoxyluteolinidin and methoxyapigenidin levels were also significantly higher in most transgenic lines at 72 hpi, but their levels were not as high as luteolinidin. As expected, the flavones (luteolin and apigenin) accumulated at considerably lower levels than luteolinidin, while

some elevation of luteolin levels was detected in the high expressing *Lr34res* genotypes, L34-5 and L34-6 (Figure 6).

Transgenic *Lr34res* expression confers resistance to sorghum anthracnose (*C. sublineolum*)

In addition to the metabolite analysis, we investigated the effect of the *Lr34res* transgene on disease symptoms caused by infection with *C. sublineolum*. Necrotic lesion phenotypes on elongated mesocotyls were examined at 7 dpi. Mild symptoms were characterized by single localized lesions, whereas strong symptoms were associated with multiple or complete lesions along the entire length of the mesocotyl (Figure S9). Strong anthracnose symptoms developed on 65% of the control lines compared with 30% in genotypes carrying the *Lr34res* transgene (Figure 7). Analysis of the total symptoms showed approximately 33% and 26% reduction in disease severity associated with the high and low expressing Lr34res lines, respectively (Figure 7). However, mild symptoms occurred twice as much in the single copy transgenic line compared with the higher *Lr34res* expressing genotype.

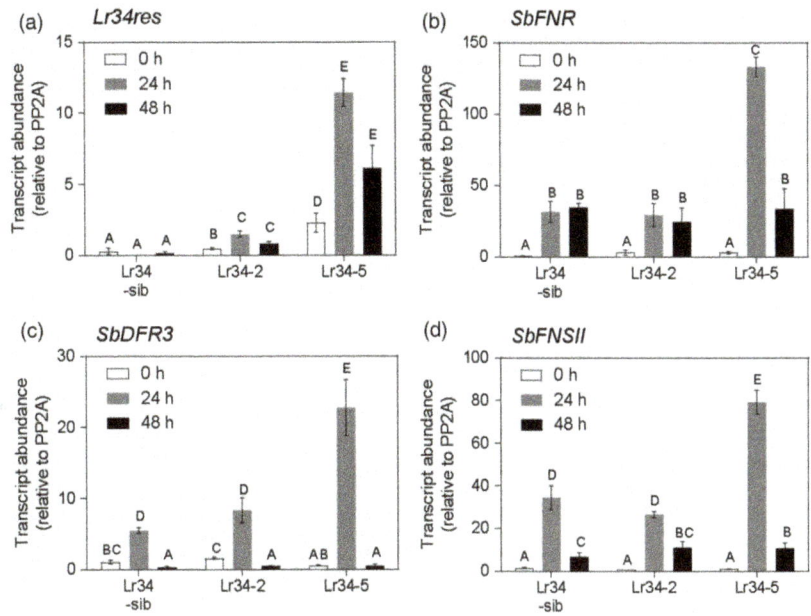

Figure 4 Comparative pathogen-induced gene expression pre-and post-inoculation with *P. purpurea*. (a) *Lr34res*. (b) *SbFNR*. (c) *SbDFR3*. (d) *SbFNSII*. Data shown as mean ± SE from three biological replicates.

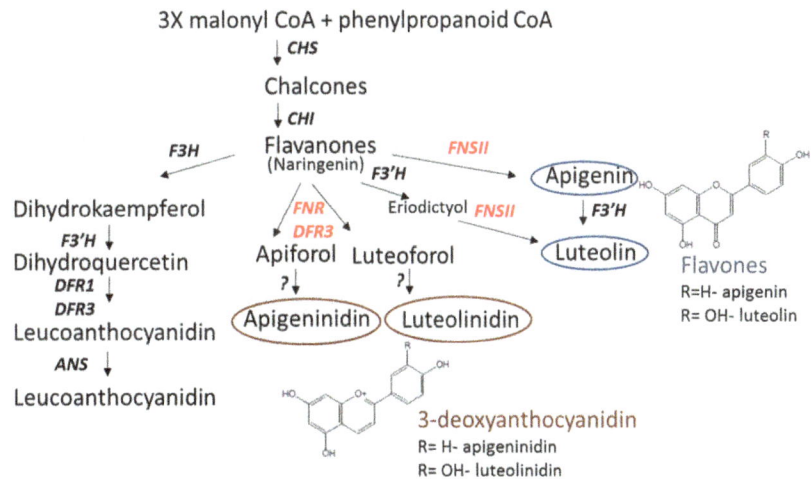

Figure 5 Flavonoid phytoalexin and anthocyanidin biosynthetic pathway (Kawahigashi *et al.*, 2016; Liu *et al.*, 2010). Genes highlighted in red and products circled were quantified in this study.

Effect of the *Lr34res* transgene on plant vigour

As a general observation, no differences in plant growth vigour were noticed among the control sib lines and *Lr34res* transgenic lines during the seedling stage and even after the 5-leaf stage when rust inoculations were conducted. However, as the plants approached booting, it was evident that the high expressing *Lr34res* genotypes (Lr34-5 and Lr34-6) were less vigorous in growth compared to the single copy line (Lr34-2) and sib lines lacking the transgene. To quantify the growth effects and subsequent effect on reproductive development and yield, aspects of panicle morphology and yield were measured. Panicle size tended to be smaller in genotypes with increased *Lr34res* gene copy number and expression (Figure 8). The mean panicle weight declined by 33% and 67%, respectively, in the single copy and multicopy Lr34res lines, respectively, as compared with the negative sib lacking *Lr34res* (Figure 9a). The mean peduncle diameter in comparison with the negative sib lacking *Lr34res* (measured immediately below the node of the basal rachis), was reduced by 1.0 mm in the single copy *Lr34res* line and 3.2 mm in the multicopy Lr34res genotypes (Figure S10). The grain yield component of 100-seed weight remained unchanged between the control sib and the Lr34-2 line, whereas a reduction of 0.5–1.2 g occurred in the multicopy *Lr34res* lines (Figure 9).

Discussion

We demonstrate in this study that the ABC transporter encoded by the wheat *Lr34res* gene functions in sorghum and confers resistance to sorghum-adapted rust and anthracnose causing pathogens, while *Lr34res*-mediated resistance to rust caused by *Puccinia* species has previously been confined to species in the Triticeae (Dyck and Samborski, 1982; Rinaldo *et al.*, 2016; Risk *et al.*, 2013), our findings together with the recently reported observations in maize (Sucher *et al.*, 2016) extends the efficacy of *Lr34res* to the warm season adapted *Puccinia* species with pathogenesis on Andropogoneae taxa. The successful incorporation of *Lr34res*-mediated resistance into sorghum suggests that the necessary components required for biosynthesis of the Lr34 putative substrate, and proteins involved in signalling and defense response, are also present in sorghum. This finding is of importance as it opens alternate avenues to explore the genetic dissection of *Lr34res*-mediated resistance. Indeed, the well characterized features of pathogen-induced pigmentation in

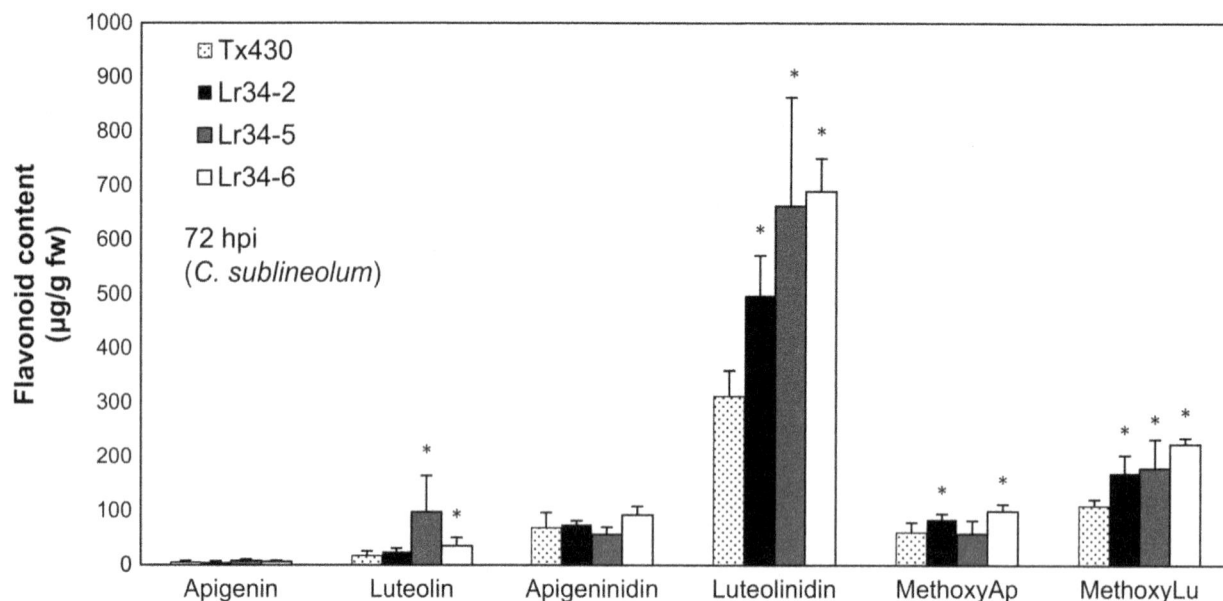

Figure 6 Metabolite analysis of 3-deoxyanthocyanidn and flavone phytoalexins measured in sorghum mesocotyls 72-h post-inoculation with *C. sublineolum*. Data shown as mean ± SD from three biological replicates. *$P < 0.05$ (*t*-test).

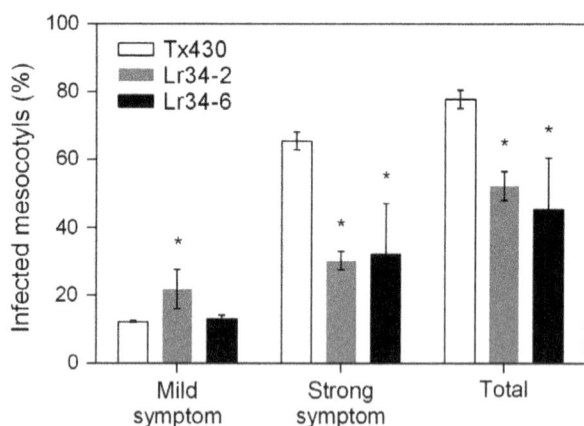

Figure 7 Spectrum of infection following *C. sublinoleum* inoculation of mesocotyls of control and transgenic sorghum lines. Data shown as mean ± SD from three biological replicates. *: Significantly different from the corresponding values in Tx430 (*t*-test, *P*-value <0.05); Total = mild + strong

sorghum, and associated flavonoid phytoalexin biosynthesis defense response, are avenues that were further investigated in this study.

While most of the pathogen-inducible pigments formed in sorghum have been reported with the *C. sublineolum* and *Cochliobolus heterostrophus* pathosystems, we show in this study that *P. purpurea* infection triggers similar phenotypes as part of the early host response in the sorghum cv. Tx430. Such visible phenotypes in wheat plants with *Lr34res* or other rust resistance genes at 24–72 hpi are yet to be reported. The correlation of *Lr34res* expression and strength of pathogen-induced pigmentation suggests that the *Lr34* transgene interacts with the signalling response, triggering pigmentation. Because the pigments responsible for the pathogen-induced colour changes formed in sorghum are derived predominantly from the 3-deoxyanthocyanidin flavonoids luteolinidin and

apigeninidin, which accumulate as a site-specific response to fungal infection (Nicholson *et al.*, 1987; Snyder and Nicholson, 1990), it is conceivable that the presence of the *Lr34res* transgene may contribute to their elevated accumulation in infected plants. Accumulation of the 3-deoxyanthocyanidins, in particular luteolinidin, occurs much faster in infected cells of resistant genotypes than susceptible genotypes, implicating early phytoalexin accumulation in preventing disease spread by restricting proliferation of fungal hyphae (Basavaraju *et al.*, 2009; Poloni and Schirawski, 2014; Wharton and Julian, 1996). In infected cells, the 3-deoxyanthocyanidins migrate to the site of attempted penetration dependent on nuclear migration, cytoplasmic streaming and intracellular pH to provide an environment for inclusion trafficking and release of the phytoalexins (Nielsen *et al.*, 2004). Exactly how *Lr34res* fits into this transport processes remains to be defined. Notwithstanding, it is noteworthy that at 24 hpi by *P. purpurea*, the strong expressing *Lr34res* transgenic lines exhibited higher expression levels for *FNR*, *DFR3* and *FNSII* genes that form part of the flavonoid phytoalexin biosynthesis pathway. Interestingly, the introduction of *Lr34res* into barley also resulted in constitutive up-regulation of genes involved in the flavonoid pathway and in the biosynthesis of barley defense compounds, such as *anthranilate synthase*, *anthranilate N-benzoyltransferase*, *agmatine coumaroyl transferase* and *flavonoid 7-O-methyl transferase* (Chauhan *et al.*, 2015).

Lr34res in hexaploid wheat typically provides partial resistance to rusts and mildew in adult plants, although under low-temperature conditions (>10°C) seedling resistance can be detected. In the current study, we show that *Lr34res* functions in seedlings and obviates the need for low-temperature induction in transgenic sorghum to provide resistance to sorghum rust and anthracnose. This observation corroborates seedling resistance by *Lr34res* against other pathogens reported in barley, rice, maize and durum wheat. The obvious difference in sorghum to other species being the highly expressed pathogen-inducible purple coloration due to phytoalexin production. Expression levels of *Lr34res* in hexaploid wheat seedlings are elevated under low

Figure 8 Effects of the *Lr34res* transgene on panicle morphology. T2 generation primary panicles grouped from respective sorghum Lr34res transgenic lines.

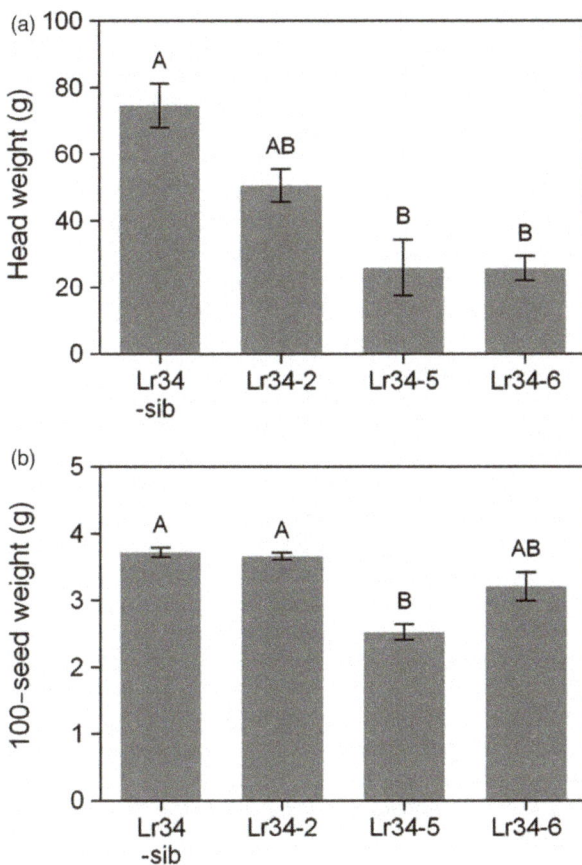

Figure 9 Effects of the *Lr34res* transgene on panicle yield. (a) Panicle weights and (b) 100-seed weight of transgenic sorghum lines. Data shown as mean ± SE from 4 to 6 biological replicates.

previous studies with the rice orthologue (Krattinger *et al.*, 2016). Given the amino acid sequence identity of 75% between the wheat and sorghum orthologues of the LR34 ABC transporter, it is possible that other regions of the LR34RES absent in sorghum are required for resistance function. Additionally, the *Lr34res* in transgenic sorghum is up-regulated upon pathogen infection, but the corresponding sorghum *Lr34* orthologue showed a weak negative response to pathogen challenge, which may also account for the lack of resistance phenotype associated with the modified sorghum LR34 orthologue. Thus any attempts at further modifications to the sorghum LR34 orthologue towards engineering resistance will likely require changes beyond the coding region to include pathogen responsive regulatory sequences.

Overexpression of *Lr34res* typified by the multicopy transgenic Lr34-5 and Lr34-6 genotypes in this study results in plants with reduced yield components, despite exhibiting immunity against *P. purpurea*. Conversely, the single copy low-expressing line (Lr34-2) had similar growth vigour as the non-transgenic or sib line control plants and less detrimental effects on reproductive yield. As Lr34-2 showed no rust symptoms 2 weeks post-infection and reduced rust sporulation after 4 weeks, indicative of the characteristic slow rusting response of the *Lr34res*, and it holds promise for the use of *Lr34* as a transgene for sorghum improvement. It may also be useful to explore high expression of *Lr34res* for plant immunity to various pathogens using pathogen-inducible promoters in an attempt to overcome detrimental reproductive yield effects associated with constitutive overexpression in adult plants. Our findings in sorghum that *Lr34res* confers resistance to sorghum rust and anthracnose demonstrates that the multipathogen resistance of the wheat *Lr34* gene extends to biotrophic and hemibiotrophic adapted pathogens across the Triticeae, Oryzeae and Andropogoneae taxa.

Experimental procedures

Production of transgenic Lr34 sorghum

The genomic construct of *Lr34res* under the native promoter and terminator sequences was cloned into plasmid pWGEM-NZf as previously described (Risk *et al.*, 2012) and subsequently transformed into the sorghum inbred line Tx430 via microprojectile-mediated transformation (Liu and Godwin 2012). The presence of the transgene in T0 plants was initially assessed by PCR with *Lr34res*-specific primers (Lagudah *et al.*, 2009) and subsequently by genomic blots probed with the Lr34 3'UTR DNA fragment. A genomic construct containing the sorghum Lr34 ortholog

temperatures in rust-infected plants (Rinaldo *et al.*, 2016; Risk *et al.*, 2013) and may account for the low temperature-induced resistance. In transgenic sorghum, barley, rice, maize and durum wheat seedlings, it is likely that the expression level of the *Lr34res* transgene upon pathogen infection reaches a threshold level that is sufficient to trigger resistance by curbing pathogen proliferation. The leaf tip necrosis/early senescence phenotype of lines carrying *Lr34res* suggests a common pathway confers resistance.

Orthologues of the wheat *Lr34* gene are present in the sorghum and rice genomes. Targeted changes to the two amino acids in the sorghum LR34 orthologue to mimic the wheat LR34RES failed to generate a resistance phenotype similar to

(Sb01g016775) was generated from an EagI 16.3 kb DNA fragment from a sorghum BAC clone (CUGI BAC#156N20) subcloned into pWGEM-NZf. Site directed mutagenesis using primers Sb2Quickchange 1F/1R (1F-TGGGAGCATTATATTTTTC CATCATCATTATGCTAAATGGCATACC/1R- GGTATGCCATTTAG CATAATGATGATGGAAAAATATAATGCTCCCA) and Sb2Quick-change 2F/R (2F- CATCAATCAGTAATGGCGTTCCATCGATTTG TCGCTTCTTATG/2R- CATAAGAAGCGACAAATCGATGGAACGC CATTACTGATTGATG) was used to generate the derived subclone Sb01g016775$^{-\Delta F525, Y613H}$ as per protocols in Krattinger et al. (2016). Transgenic plants with the genomic construct of Sb01g016775$^{-\Delta F525, Y613H}$ with its native promoter and terminator sequences were also generated by microprojectile bombardment.

Identification of transformants and Lr34 copy number

Leaf samples (2–3 g) from T1 and T3 plants were ground in liquid nitrogen using pestle and mortars and sand. Frozen leaf material was transferred into 3 mL CTAB extraction buffer (100 mM Tris-HCl pH 8.0, 20 mM EDTA, 1.4M NaCl, 0.5% $Na_2S_2S_5$, 2% CTAB, and 1% β-mercaptoethanol) and processed for DNA isolation in accordance with Collins et al. (1998). About 12 µg of each gDNA sample was subjected to NotI and EcoRV restriction endonuclease digestions to ascertain the presence of full-length Lr34 gene constructs and the copy number of Lr34 in transgenic sorghum lines, respectively. Digested gDNA samples were loaded on 1% agarose gels and run at 60V (at approx. 50 mA) for 18–20 h, capillary transferred onto Hybond-N+® filter using 20xSSC buffer and UV cross-linked. Filter was subjected to 5-h prehybridization in 30 mL prehybridization solution containing salmon sperm DNA at 65°C in a hybridization tube. Subsequent hybridization incorporating 50 ng of probe DNA (^{32}P-labelled Lr34-3′UTR probe from the amplicon generated using ABCTEX1314F- CAGAACACCTA CAGAAGAATATC and ABCR9- GGCAAGTAGCTATATCTGTAAC) was performed according to Lagudah et al. (1991).

Sorghum rust inoculation

Sorghum seed was germinated and grown in pots at 20°C in a glasshouse or in a growth chamber. Plants at the 5–7 leaf stage were placed in closed inoculation chambers and allowed to acclimatize for 18–24 h. Puccinia purpurea urediniospores collected from Hermitage Research Station, Warwick, Queensland (28.2102°S, 152.1041°E) in Queensland, Australia (White et al., 2015) were suspended in 100–150 mL distilled water with 1–2 drops of Tween 20, and sprayed onto plants using a Preval® compressed air atomizer (Preval Sprayer Division). A complementary set of plants were treated identically but without urediniospores as controls. Inoculation chambers were closed to maintain high (80%+) humidity and placed in darkness at 20°C for 24 h before being transferred to diurnal conditions (16 h light, 8 h dark) to allow rust infection to develop.

Anthracnose infection

Sorghum seeds were sown in rolls of germination paper and kept in darkness for 4 days at 28°C (Lo et al., 1996). Etiolated seedlings with elongated mesocotyls were inoculated with spore suspensions of C. sublineolum, at the concentration of ~3.0 × 10^6 conidia/mL with gelatin as a wetting agent (0.25%). Inoculated plants were incubated at 100% relative humidity at room temperature for 24 h. Three independent infections were performed on elongated mesocotyls. The phenotypes were examined after 7 days following inoculation on an average of 35 plants per genotype. For each

genotype the number of plants with mild symptom, strong symptom or no symptoms were recorded.

Microscopy

At 6–8 dpi, fourth or fifth leaf samples were collected and submerged in 1M KOH and incubated for 48 h at 37°C with gentle agitation. The KOH solution was replaced with fresh 1M KOH solution after 18–24 h. The KOH solution was discarded, and the leaf material was washed gently 2–3 times with 50 mM Tris-HCl, allowing material to incubate in the Tris-HCl solution for 10–20 min per wash. 1–2 mL 50 mM Tris-HCl and 10–20 µL 1 mg/mL wheat germ agglutinin conjugated to fluorescein isothiocyanate (WGA-FITC, Sigma-Aldrich, Castle Hill, NSW, Australia) were added, and samples incubated at ambient temperature for an hour or kept at 4°C before mounting on microscope slides. Stained leaf samples were mounted on slides using a few drops of 40% glycerol before covering with cover slips. GFP3 fluorescence filters were used on a Leica MZFLIII fluorescence dissecting microscope or a Zeiss Axioimager upright fluorescence microscope to score the presence of P. purpurea infection sites in the sampled leaves.

Rust biomass assays

Chitin assays were carried out as described by Ayliffe et al. (2014). Three biological replicates of the 6th leaves were sampled 14 and 28 dpi and weighed. Leaves were cut into 1.5–2.0 cm fragments and submerged in 1M KOH with 0.15 Silwet L-77 in Falcon tubes. Leaf samples are autoclaved at 121°C and 15 psi for 20 min, then washed gently three times in 50 mM Tris-HCl pH 7.5. Plant samples were suspended in 50 mM Tris-HCl pH 7.5 at the rate of 200 mg fresh weight per mL and homogenized by sonication for 1–2 min to form a fine uniform suspension. About 4 × 100 µL of each homogenate was transferred to 4 × 200 µL PCR tubes. About 10 µL of 1 mg/mL WGA-FITC (Sigma-Aldrich) was added to each homogenate in PCR tubes and left at ambient temperature for an hour. Homogenates were washed three times by centrifuging at 250 **g** for 3–5 min and carefully replacing the supernatants with 50 mM Tris-HCl pH 7.5 using a micropipette. The final washed suspensions were transferred to a 96-well fluorometer microtiter plate. Fluorescence values of each sample were measured in a Wallac Victor 1420 multilabel counter at 485 nm excitation and 535 nm emission wavelengths with a 1.0 s measurement time. Means of the technical replicate fluorescence values were calculated, and the standard errors were ascertained for biological replicates.

Metabolite analysis

Samples of mesocotyl tissue (~200 mg each) from uninfected and C. sublineolum inoculated plants at 48, 72 and 96 hpi were collected, cut into segments, weighed and placed in acidified (0.1%) HPLC-grade methanol. Metabolites were allowed to leach from the tissue at 4°C overnight. The composition of plant extracts was then determined by liquid chromatography-tandem mass spectrometry (LC-MS) in accordance with the protocols as described (Du et al., 2010; Lo et al., 1999). Authentic standards of luteolinidin, apigeninidin, luteolin and apigenin (Sigma) were used for metabolite identification and quantification.

qRT-PCR

The sixth leaf of plantlets were sampled at 0, 24 and 48 hpi, snap-frozen in liquid nitrogen and stored at −80°C. RNA was isolated with the RNeasy® Plant Mini Kit (QIAGEN, Chadstone Centre, VIC,

Table 1 Primers used in QPCR gene expression analyses

Gene	Primer	Primer sequence (5′–3′)	Amplicon size (bp)	Reference
SbActin	Forward	CTAGCAGCATGAAGATCAAGGTG	134	Pavli *et al.* (2011)
	Reverse	GCCAGACTCGTCGTACTCAG		
SbPP2A	Forward	AACCCGCAAAACCCCAGACTA	138	Reddy *et al.* (2016)
	Reverse	TACAGGTCGGGCTCATGGAAC		
Lr34res	Forward	GGGAGCATTATTTTTTTCCATCA	242	This paper
	Reverse	ACTGGCAGAAGAACCTTGAAACA		
SbL34 (Sb01g016775)	Forward	GGGAGCATTATATTTTTCCATCT	247	This paper
	Reverse	TAACTGGCAGAAGAACCTGGAAG		
Flavone Synthase II (*SbFNSII*, Sb02g000220)	Forward	CGCAAGACCACCGTCTTCTT	209	Du *et al.* (2010)
	Reverse	GCCGGCACGGCCTGCATGGC		This paper
Dihydroflavonol 4-reductase 3 (*SbDFR3*; Sb04g004290)	Forward	CGGATGTGACGATTGTTTGA	123	Liu *et al.* (2010)
	Reverse	GGGCATATTGGTTTGGAACTT		
Flavanone 4-reductase (*SbFNR*; Sb06g029550)	Forward	GGGTAACAAGAAGACGATGAAGA	287	Kawahigashi *et al.* (2016)
	Reverse	CTGGATCCTGTGCCTCGAAGT		

Australia) according to manufacturer's instructions. About 1–2 µg RNA samples were subjected to first-strand DNA synthesis in 20 µL reactions using Superscript® III reverse transcriptase Life Technologies (Mulgrave, VIC, Australia). About 3 µL of 1-in-10 dilutions of first-strand synthesis products were subjected to qPCR reactions using the C1000 Touch™ thermocycler with the CFX96™ Real-Time System (Bio-Rad, Gladesville, NSW, Australia). Reaction conditions included an initial denaturization at 95°C for 3 min; 40 cycles of denaturization at 95°C for 10 s and annealing/elongation at 60°C for 30 s, followed by a melt step range of 65–95°C with increments of 0.5°C. The sorghum *actin* gene (Pavli *et al.*, 2011) was used as a reference gene for each qRT-PCR experiment, and each qRT-PCR experiment was repeated using the more stable sorghum reference gene *PP2A* (Reddy *et al.*, 2016). qPCR primers specific for Lr34*res*, *SbL34* (Lr34 orthologue), *SbFNR*, *SbFNSII* and *SbDFR3* were used to measure the relative gene expressions at the different post-inoculation time points are listed in Table 1. Experiments included three technical replicates of each of three respective biological replicates. Means of the ΔCq values were calculated, and Standard Errors were determined for the data. Gene expression values were log(base 2)-transformed, and a repeated measures analysis was performed via the linear mixed model software asreml (Butler, 2009) in R (R Core Team, 2015). Means and SE bars in Figure 4 are back-transformed to the scale of the raw expression levels.

Reproductive yield components

Sorghum plants at the T2 generation and four replications were grown to physiological maturity, and intact panicles were harvested and dried at 37°C for 48 h. Individual panicles were weighed for each of the negative and positive Lr34 lines. Peduncle diameters were measured 1 mm above the last node using Vernier calipers. Kernels were separated from the panicles and 100 kernel quantities from transgenic and non-transgenic lines were weighed.

Acknowledgements

The authors wish to acknowledge the Bill and Melinda Gates foundation for financial support through grant # OPP1060218 and OPP1131636. This work was supported by the Research Grants Council of Hong Kong, China (grant no. 17 123 315).

References

Ayliffe, M., Periyannan, S.K., Feechan, A., Dry, I., Schumann, U., Lagudah, E. and Pryor, A. (2014) Simple quantification of in planta fungal biomass. In *Plant–Pathogen Interactions: Methods and Protocols* (Birch, P., Jones, J.T. and Bos, J.I.B., eds), pp. 159–172. Totowa, NJ: Humana Press.

Bandyopadhyay, R. (2000) Rust. In *Compendium of Sorghum Diseases*, 2nd ed. (Frederiksen, R.A. and Odvody, G.N., eds), pp. 23–24. St Paul: APS.

Basavaraju, P., Shetty, N.P., Shetty, H.S., de Neergaard, E. and Jørgensen, H.J.L. (2009) Infection biology and defence responses in sorghum against *Colletotrichum sublineolum*. *J. Appl. Microbiol.* **107**, 404–415.

Butler, D. (2009) *asreml: asreml() fits the linear mixed model*. R Package, version 3.0-1.

Chauhan, H., Boni, R., Bucher, R., Kuhn, B., Buchmann, G., Sucher, J., Selter, L.L. *et al.* (2015) The wheat resistance gene *Lr34* results in the constitutive induction of multiple defense pathways in transgenic barley. *Plant J.* **84**, 202–215.

Collins, N.C., Webb, C.A., Seah, S., Ellis, J.G., Hulbert, S.H. and Pryor, A.J. (1998) The isolation and mapping of disease resistance gene analogs in maize. *Mol. Plant-Microbe Interact.* **11**, 968–978.

Dangl, J.L., Horvath, D.M. and Staskawicz, B.J. (2013) Pivoting the plant immune system from dissection to deployment. *Science*, **341**, 746–751.

Du, Y., Chu, H., Wang, M., Chu, I.K. and Lo, C. (2010) Identification of flavone phytoalexins and a pathogen-inducible flavone synthase II gene (SbFNSII) in sorghum. *J. Exp. Bot.* **61**, 983–994.

Dyck, P.L. and Samborski, D.J. (1979) Adult plant resistance in PI250413, an introduction of common wheat. *Can. J. Plant Sci.* **59**, 329–332.

Dyck, P.L. and Samborski, D.J. (1982) The inheritance of resistance to *Puccinia recondita* in a group of common wheat cultivars. *Can. J. Genet. Cytol.* **24**, 273–283.

Dykes, L. and Rooney, L.W. (2006) Sorghum and millet phenols and antioxidants. *J. Cereal Sci.* **44**, 236–251.

FAOSTAT. (2016). *Crops: FAO Global Statistical Yearbook*. http://www.fao.org/faostat/en/#data/QC

Kawahigashi, H., Kasuga, S., Sawada, Y., Yonemaru, J., Ando, T., Kanamori, H., Wu, J. *et al.* (2016) The sorghum gene for leaf color changes upon wounding (P) encodes a flavanone 4-reductase in the 3-deoxyanthocyanidin biosynthesis pathway. *G3 (Bethesda, Md.)*, **6**, 1439–1447.

Kolmer, J.A., Singh, R.P., Garvin, D.F., Viccars, L., William, H.M., Huerta-Espino, J., Ogbonnaya, F.C. *et al.* (2008) Analysis of the Lr34/Yr18 rust resistance region in wheat germplasm. *Crop Sci.* **48**, 1841–1852.

Krattinger, S.G., Lagudah, E.S., Spielmeyer, W., Singh, R.P., Huerta-Espino, J., McFadden, H., Bossolini, E. *et al.* (2009) A putative ABC transporter confers durable resistance to multiple fungal pathogens in wheat. *Science*, **323**, 1360–1363.

Krattinger, S.G., Jordan, D.R., Mace, E.S., Raghavan, C., Luo, M.C., Keller, B. and Lagudah, E.S. (2013) Recent emergence of the wheat Lr34 multi-pathogen resistance: insights from haplotype analysis in wheat, rice, sorghum and *Aegilops tauschii*. *Theor. Appl. Genet.* **126**, 663–672.

Krattinger, S.G., Sucher, J., Selter, L.L., Chauhan, H., Zhou, B., Tang, M., Upadhyaya, N.M. *et al.* (2016) The wheat durable, multipathogen resistance gene Lr34 confers partial blast resistance in rice. *Plant Biotechnol. J.* **14**, 1261–1268.

Lagudah, E.S., Appels, R. and McNeil, D. (1991) The Nor-D3 locus of *Triticum tauschii*: natural variation and linkage to chromosome 5 markers. *Genome*, **34**, 387–395.

Lagudah, E.S., Krattinger, S.G., Herrera-Foessel, S., Singh, R.P., Huerta-Espino, J., Spielmeyer, W., Brown-Guedira, G. *et al.* (2009) Gene-specific markers for the wheat gene *Lr34/Yr18/Pm38* which confers resistance to multiple fungal pathogens. *Theor. Appl. Genet.* **119**, 889–898.

Liu, H., Du, Y., Chu, H., Shih, C.H., Wong, Y.W., Wang, M., Chu, I.K. *et al.* (2010) Molecular dissection of the pathogen-inducible 3-deoxyanthocyanidin biosynthesis pathway in sorghum. *Plant Cell Physiol.* **51**, 1173–1185.

Liu, G. and Godwin, I.D. (2012) Highly efficient sorghum transformation. *Plant Cell Rep.* **31**, 999–1007. https://doi.org/10.1007/s00299-011-1218-4

Lo, S.C., Weiergang, I., Bonham, C., Hipskind, J., Wood, K. and Nicholson, R.L. (1996) Phytoalexin accumulation in sorghum: identification of a methyl ether of luteolinidin. *Physiol. Mol. Plant Pathol.* **61**, 179–188.

Lo, S.C., de Verdier, K. and Nicholson, R.L. (1999) Accumulation of 3-deoxyanthocyanidin phytoalexins and resistance to *Colletotrichum sublineolum* in sorghum. *Physiol. Mol. Plant Pathol.* **55**, 263–273.

McIntosh, R.A. (1992) Close genetic linkage of genes conferring adult-plant resistance to leaf rust and stripe rust in wheat. *Plant. Pathol.* **41**, 523–527.

McIntosh, R.A., Dubcovsky, J., Rogers, W.J., Morris, C., Appels, R. and Xia, X.C. (2013). *Catalogue of Gene Symbols for Wheat.* http://wheat.pw.usda.gov/GG2/Triticum/wgc/2013/

Moore, J.W., Herrera-Foessel, S., Lan, C., Schnippenkoetter, W., Ayliffe, M., Huerta-Espino, J., Lillemo, M., *et al.* (2015) Recent evolution of a hexose transporter variant confers resistance to multiple pathogens in wheat. *Nat. Genet.* **47**, 1494–1498.

Nicholson, R.L., Kollipara, S.S., Vincent, J.R., Lyons, P.C. and Cadena-Gomez, G. (1987) Phytoalexin synthesis by the sorghum mesocotyl in response to infection by pathogenic and nonpathogenic fungi. *Proc. Natl Acad. Sci. USA*, **84**, 5520–5524.

Nielsen, K.A., Gotfredsen, C.H., Buch-Pedersen, M.J., Ammitzbøll, H., Mattsson, O., Duus, J.Ø. and Nicholson, R.L. (2004) Inclusions of flavonoid 3-deoxyanthocyanidins in *Sorghum bicolor* self-organize into spherical structures. *Physiol. Mol. Plant Pathol.* **65**, 187–196.

Pavli, O.I., Ghikas, D.V., Katsiotis, A. and Skaracis, G.N. (2011) Differential expression of heat shock protein genes in sorghum (*Sorghum bicolor* L.) genotypes under heat stress. *Aust. J. Crop Sci.* **5**, 511–515.

Poloni, A. and Schirawski, J. (2014) Red card for pathogens: phytoalexins in sorghum and maize. *Molecules (Basel, Switzerland)*, **19**, 9114–9133.

R Core Team. (2015) *A Language and Environment for Statistical Computing.* Vienna: R Foundation for Statistical Computing. www.R-project.org.

Reddy, P.S., Reddy, D.S., Sivasakthi, K., Bhatnagar-Mathur, P., Vadez, V. and Sharma, K.K. (2016) Evaluation of Sorghum [*Sorghum bicolor* (L.)] reference genes in various tissues and under abiotic stress conditions for quantitative real-time PCR data normalization. *Front. Plant Sci.* **7**, 529.

Rinaldo, A., Gilbert, B., Boni, R., Krattinger, S.G., Singh, D., Park, R.F., Lagudah, E. *et al.* (2016) The *Lr34* adult plant rust resistance gene provides seedling resistance in durum wheat without senescence. *Plant Biotechnol. J.* doi:10.1111/pbi.12684.

Risk, J.M., Selter, L.L., Krattinger, S.G., Viccars, L.A., Richardson, T.M., Buesing, G., Herren, G. *et al.* (2012) Functional variability of the *Lr34* durable resistance gene in transgenic wheat. *Plant Biotechnol. J.* **10**, 477–487.

Risk, J.M., Selter, L.L., Chauhan, H., Krattinger, S.G., Kumlehn, J., Hensel, G., Viccars, L.A. *et al.* (2013) The wheat Lr34 gene provides resistance against multiple fungal pathogens in barley. *Plant Biotechnol. J.* **11**, 847–854.

Singh, R.P. (1992) Genetic association of leaf rust resistance gene *Lr34* with adult plant resistance to stripe rust in bread wheat. *Phytopathology*, **82**, 835–838.

Snyder, B.A. and Nicholson, R.L. (1990) Synthesis of phytoalexins in sorghum as a site-specific response to fungal ingress. *Science*, **248**, 1637–1639.

Snyder, B.A., Leite, B., Hipskind, J., Butler, L.G. and Nicholson, R.L. (1991) Accumulation of sorghum phytoalexins induced by *Colletotrichum graminicola* at the infection site. *Physiol. Mol. Plant Pathol.* **39**, 463–470.

Spielmeyer, W., McIntosh, R.A., Kolmer, J. and Lagudah, E.S. (2005) Powdery mildew resistance and *Lr34/Yr18* genes for durable resistance to leaf and stripe rust cosegregate at a locus on the short arm of chromosome 7D of wheat. *Theor. Appl. Genet.* **111**, 731–735.

Sucher, J., Boni, R., Yang, P., Rogowsky, P., Büchner, H., Kastner, C., Kumlehn, J. *et al.* (2016) The durable wheat disease resistance gene Lr34 confers common rust and northern corn leaf blight resistance in maize. *Plant Biotechnol. J.* **15**, 489–496.

Thakur, R.P. and Mathur, K. (2000) Anthracnose. In *Compendium of Sorghum Diseases* (Frederiksen, R.A. and Odvody, G.N., eds), 0890542406, pp. 10–12. St. Paul, MN: APS Press.

Wharton, P. and Julian, A. (1996) A cytological study of compatible and incompatible interactions between *Sorghum bicolor* and *Colletotrichum sublineolum*. *New Phytol.* **134**, 25–34.

White, J.A., Ryley, M.J., George, D.L., Kong, G.A. and White, S.C. (2012) Yield losses in grain sorghum due to rust infection. *Australas. Plant Pathol.* **41**, 85–91.

White, J.A., Ryley, M.J., George, D.L. and Kong, G.A. (2015) Identification of pathotypes of the sorghum rust pathogen, *Puccinia purpurea*, in Australia. *Australas. Plant Pathol.* **44**, 1–4.

12

Simultaneous stimulation of sedoheptulose 1,7-bisphosphatase, fructose 1,6-bisphophate aldolase and the photorespiratory glycine decarboxylase-H protein increases CO₂ assimilation, vegetative biomass and seed yield in Arabidopsis

Andrew J. Simkin[1], Patricia E. Lopez-Calcagno[1], Philip A. Davey[1], Lauren R. Headland[1], Tracy Lawson[1], Stefan Timm[2], Hermann Bauwe[2] and Christine A. Raines[1]*

[1]School of Biological Sciences, University of Essex, Colchester, UK
[2]Plant Physiology Department, University of Rostock, Rostock, Germany

*Correspondence
email rainc@essex.ac.uk

Keywords: chlorophyll fluorescence imaging, FBP aldolase, glycine decarboxylase-H protein, photosynthesis, SBPase, transgenic.

Summary

In this article, we have altered the levels of three different enzymes involved in the Calvin–Benson cycle and photorespiratory pathway. We have generated transgenic Arabidopsis plants with altered combinations of sedoheptulose 1,7-bisphosphatase (SBPase), fructose 1,6-bisphophate aldolase (FBPA) and the glycine decarboxylase-H protein (GDC-H) gene identified as targets to improve photosynthesis based on previous studies. Here, we show that increasing the levels of the three corresponding proteins, either independently or in combination, significantly increases the quantum efficiency of PSII. Furthermore, photosynthetic measurements demonstrated an increase in the maximum efficiency of CO₂ fixation in lines over-expressing SBPase and FBPA. Moreover, the co-expression of GDC-H with SBPase and FBPA resulted in a cumulative positive impact on leaf area and biomass. Finally, further analysis of transgenic lines revealed a cumulative increase of seed yield in SFH lines grown in high light. These results demonstrate the potential of multigene stacking for improving the productivity of food and energy crops.

Introduction

The accumulated photosynthate produced over the season determines the yield of a crop, but improvements in photosynthesis have not been used in traditional breeding approaches to identify high-yielding varieties. The reasons for this are twofold, (i) methodologies to make accurate field measurements have only been available in the last 10–20 years, and also, (ii) there is a lack of evidence to determine whether there is a correlation between the rate of photosynthesis on a leaf area basis and final yield of the crop (Evans, 2013; Fischer et al., 1998; Gifford and Evans, 1981). There is now an urgent need to increase crop productivity and yields to meet the nutritional demands of a growing world population, and there is growing evidence that this may be achieved through improvement of photosynthetic energy conversion to biomass (von Caemmerer and Evans, 2010; Ding et al., 2016; Lefebvre et al., 2005; Long et al., 2006, 2015; Simkin et al., 2015). Evidence from a combination of theoretical studies and transgenic approaches has provided compelling evidence that manipulation of the Calvin–Benson (CB) cycle can improve energy conversion efficiency and lead to an increase in yield potential (Long et al., 2006; Poolman et al., 2000; Raines, 2003, 2006, 2011; Zhu et al., 2007, 2010).

Previous studies have demonstrated that even small reductions in individual CB cycle enzymes such as sedoheptulose 1,7-bisphosphatase (SBPase) and fructose 1,6-bisphosphate aldolase (FBPA) negatively impact on carbon assimilation and growth,

indicating that these enzymes exercise significant control over photosynthetic efficiency (Ding et al., 2016; Haake et al., 1998, 1999; Harrison et al., 1998, 2001; Lawson et al., 2006; Raines, 2003; Raines and Paul, 2006; Raines et al., 1999). Furthermore, the disruption of the chloroplastic fructose-1,6-bisphosphatase (FBPase) gene was also shown to impact negatively on carbon fixation (Kossmann et al., 1994; Rojas-González et al., 2015; Sahrawy et al., 2004). These results strongly suggested that improvements in photosynthetic carbon fixation may be achieved by increasing the activity of individual CB cycle enzymes. Evidence supporting this hypothesis came from transgenic tobacco plants over-expressing SBPase (Lefebvre et al., 2005), the cyanobacterial bifunctional SBPase/FBPase (Miyagawa et al., 2001) or FBPA (Uematsu et al., 2012). These single manipulations resulted in increase in photosynthetic carbon assimilation, enhanced growth and an increase in biomass. More recently, Simkin et al. (2015) demonstrated that the combined over-expression of SBPase and FBPA in tobacco resulted in a cumulative increase in biomass and that these increases could be further en'hanced by the over-expression of the cyanobacterial inorganic carbon transporter B (ictB), proposed to be involved in CO₂ transport, although its function was not established in these plants (Simkin et al., 2015). These results demonstrate the potential for the manipulation of photosynthesis, using multigene stacking approaches, to increase biomass yield (Simkin et al., 2015).

The efficiency of CO₂ fixation by the CB cycle is compromised by the oxygenase activity of ribulose-1,5-bisphosphate carboxylase/oxygenase (Rubisco) which directly competes with CO₂ fixation at the active site, resulting in the formation of 2-phosphoglycolate (2PG) and subsequently significant energy costs and CO_2 losses in the photorespiratory pathway, resulting in significant losses in yield (Bowes et al., 1971; Tolbert, 1997; Walker et al., 2015, 2016). Therefore, a major target to improve photosynthesis has been to reduce photorespiration, either through protein engineering to improve Rubisco catalysis or by limiting the flux through this pathway, none of which have as yet yielded positive results due to both the complexity of the Rubisco catalytic and assembly processes (Cai et al., 2014; Carmo-Silva et al., 2015; Lin et al., 2014a; Orr et al., 2016; Sharwood et al., 2016; Whitney et al., 2011). More ambitious approaches to this problem are now being taken, including the introduction of cyanobacterial or algal CO_2-concentrating mechanisms, novel synthetic metabolic pathways and the introduction of the C4 pathway into C3 crops (Betti et al., 2016; Lin et al., 2014b; McGrath and Long, 2014; Meyer et al., 2016; Montgomery et al., 2016). However, to date the only successful approach to limiting photorespiration which has resulted in an improvement in photosynthesis has been through the introduction of alternative routes to metabolize 2PG and return CO_2 for use in the CB cycle (Dalal et al., 2015; Kebeish et al., 2007; Maier et al., 2012; Nolke et al., 2014; Peterhänsel et al., 2013; Xin et al., 2015). Reductions in the flux through the photorespiratory cycle by targeted knock-down of GDC-P in potato and GDC-H in rice have been shown to lead to reductions in photosynthesis and growth rates (Engel et al., 2007; Heineke et al., 2001; Lin et al., 2016). The opposite approach, namely over-expression of the glycine decarboxylase (GDC)-H protein (GDC-H) and glycine decarboxylase (GDC)-L protein (GDC-L) in Arabidopsis thaliana (Arabidopsis), resulted in an improvement of photosynthesis and increased vegetative biomass when compared to wild-type plants (Timm et al., 2012, 2015, 2016). Although the underlying mechanism responsible for this effect has not been fully elucidated, these authors proposed that stimulation of the CB cycle is brought about by the increase in GDC activity, resulting in a reduction in the steady-state levels of photorespiratory metabolites that may negatively impact on the function of the CB cycle (e.g. 2PG, glycolate, glyoxylate or glycine (Anderson, 1971; Kelly and Latzko, 1976; Eisenhut et al., 2007; Lu et al., 2014; Timm et al., 2015, 2016)).

In the light of the results from Timm et al. (2012, 2015), the aim of this study was to explore the possibility that the simultaneous increase in the activity of enzymes of both the CB cycle and the photorespiratory pathway could lead to a cumulative positive impact on photosynthetic carbon assimilation and yield. To test this, we have taken a proof-of-concept approach using the model plant Arabidopsis in which we have over expressed SBPase, FBPA and GDC-H either alone or in combination. We have shown that the simultaneous manipulation of multiple targets can lead to a cumulative impact on biomass yield under both low- and high-light growing conditions. Interestingly, we have also shown that manipulation of the photorespiratory pathway alone resulted in an increase in vegetation biomass but not seed yield. In contrast, manipulation of both the CB cycle and photorespiratory pathway increased both biomass and seed yield.

Results

Production and selection of arabidopsis transformants

The full-length Arabidopsis SBPase (At3 g55800) and FBPA cDNA (At4 g38970) were used to generate three over-expression constructs PTS1-SB, PTS1-FB and PTS1-SBFB in vector pGWPTS1 (Figure S1). These transgenes were under the control of the photosynthetic tissue-specific (PTS) rbcS2B (1150 bp; At5 g38420) promoter. These constructs were transformed into Arabidopsis using the floral dip method (Clough and Bent, 1998), and the resulting transgenic plants were selected on kanamycin-/hygromycin-containing medium. T2 plants expressing the integrated transgenes were screened by immunoblotting and allowed to self-fertilize to generate seeds for T3 plants. Following initial characterisation of primary independent lines generated, 3–4 independent lines overexpressing either SBPase (S3, S8, S12) or FBPA (F6, F9, F11) and SBPase and FBPA together (SF4, SF6, SF7, SF12) were selected for further study.

Further analysis was carried out on T3 plants grown at 130 µmol/m²/s in an 8-h/16-h light/dark cycle and total extractable SBPase and FBPA activity determined in extracts from newly fully expanded leaves. The results are represented as a percentage (%) of total activity for SBPase (6.7 µmol/m²/s) and FBP aldolase (22 µmol/m²/s) determined in wild type (WT). This analysis showed that these plants had increased levels of SBPase (137%–185%) and FBPA (146%–180%) activity (Figure 1) compared to WT and nontransformed azygous (A) controls (azygous-control plants used in this study were recovered from the segregating population and verified by PCR). Interestingly, a small increase in endogenous FBPA activity (125%–136%) was also observed in SBPase over-expressing lines (Figure 1a), but no significant increase in SBPase activity was observed in lines over-expressing FBPA.

Plants over-expressing SBPase (S), FBPA (F) and the GDC-H protein (H) were generated by crossing two SBPase + FBPA (SF) lines (SF6 and SF12) with two Flaveria pringlei GDC-H protein (Kopriva and Bauwe, 1995) over-expressing lines (FpHL17 and FpHL18) originally generated by Timm et al. (2012) under the control of the leaf-specific and light-regulated Solanum tuberosum ST-LS1 promoter (Stockhaus et al., 1989). Four independent lines (SFH4, SFH20, SFH23 and SFH31) over-expressing SBPase, FBPA and GDC-H (SFH) were identified by PCR and SBPase and FBPA enzyme activities. SBPase and FBPA protein levels were found to be similar to those observed in SF lines (Figure 1b). No significant difference in SBPase or FBPA activities was observed in lines over-expressing GDC-H alone compared to WT/A controls (C). The full set of assays showing the variation between plants for both SBPase and FBPA activities can be seen in Figure S2.

In addition to total extractable enzyme activity, immunoblot analysis of the T3 progenies of S, F, SF, H and SFH lines was carried out using WT/A as controls (C). This analysis identified a number of plants over-expressing SBPase or FBPA and others with increased levels of both SBPase and FBPA (Figures 1a,b and S3). Interestingly, the over-expression of SBPase in Arabidopsis led to an increase in endogenous FBPA protein levels (Figure 1a) in agreement with the observed increase in enzyme activity. The original H lines and the newly generated SFH plants were shown to accumulate GDC-H when compared to both nontransformed control plants and other transgenic lines (Figure 1a,b). Given the change in FBPA protein levels in the SBPase over-expressing line, we used immunoblot analysis to determine whether there were any changes in other CB cycle enzymes. No detectable changes in the levels of transketolase (TK), phosphoribulokinase (PRK), fructose-1,6-bisphosphatase (FBPase), Rubisco or the ADP glucose pyrophosphorylase (ssAGPase) small protein were observed when compared to levels in C plants (Figure 2).

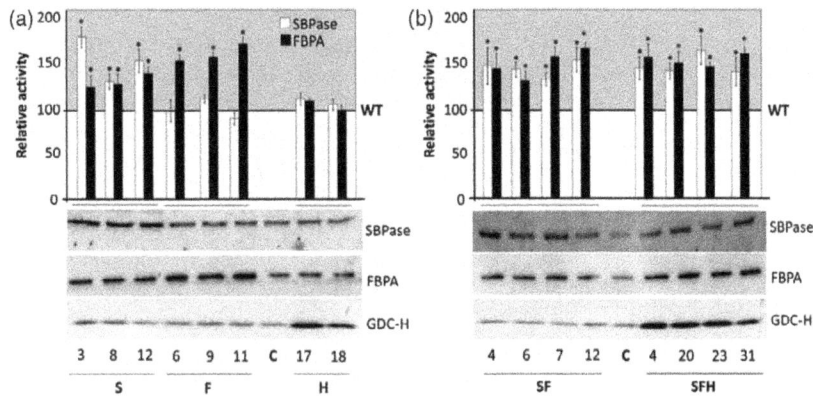

Figure 1 Molecular and biochemical analysis of the transgenic plants over-expressing SBPase (S), FBPA (F) and GDC-H (H). SBPase and FBPA enzyme activity (SBPase, FBPA) and immunoblot blot analysis (SBPase, FBPA, GDC-H) of protein extracts from two independent leaves of (a) S, F and H lines and (b) SF and SFH lines used in this study compared to non-transformed control (C). Enzyme assays represent data from 12 to 24 independent plants per group compared to 12–16 C plants. The results are represented as a percentage (%) of total activity for SBPase (6.7 $\mu mol/m^2/s$) and FBP aldolase (22 $\mu mol/m^2/s$) determined in wild type (WT). Enzyme activities per plant can be seen in Figure S2. Columns represent mean values, and standard errors are displayed. Lines that are significantly different to C are indicated (*$P < 0.05$).

Chlorophyll fluorescence imaging reveals increased photosynthetic efficiency in young over-expressing seedlings

To explore the impact of increased levels of SBPase, FBPA and the GDC-H protein on photosynthesis, plants were grown at 130 $\mu mol/m^2/s$ in an 8-h/16-h light/dark cycle and the quantum efficiency of PSII photochemistry (F_q'/F_m') analysed using chlorophyll a fluorescence imaging (Baker, 2008; Murchie and Lawson, 2013). Plants over-expressing SBPase and FBPA, either independently or in combination (including with GDC-H), had a significantly higher F_q'/F_m' at an irradiance of 200 $\mu mol/m^2/s$ when compared to C plants (Figure 3a,b). Plants over-expressing GDC-H alone showed a small increase in the average levels of F_q'/F_m' compared to C ($P = 0.061$). When measurements were made at a higher light level (600 $\mu mol/m^2/s$), all lines analysed, with the exception of SFH, showed a significant increase in F_q'/F_m' compared to C plants (Figure S4a). From images taken as part of the chlorophyll fluorescence analysis, leaf area was determined and shown to be significantly larger for all transgenic lines compared with WT and azygous (A) controls (Figure 3c). Interestingly, SFH plants showed the greatest leaf area in all experiments. No significant differences in leaf area were observed

between WT and A. From this point on, C plants used were the combined data from WT and A plants.

Photosynthetic CO_2 assimilation rates are increased in mature plants grown in low light

To explore the impact of changes in the levels of enzymes in both the CB cycle and photorespiratory pathway, CO_2 assimilation rates were determined as a function of light intensity (Figure 4a, b). From these light response curves, the maximum light-saturated rate of photosynthesis (A_{sat}) was shown to be significantly higher in all transgenic plants when compared to C plants (Figure 4c). Small differences in CO_2 assimilation rates (A) were also observed in the S, F, SF and SFH plants even at light intensities as low as 150 $\mu mol/m^2/s$, which is close to that of the growth conditions (Figure S5).

We also determined A as a function of internal CO_2 concentration (C_i) in the same plants (Figure 4d,e). In all transgenic plants, except those over-expressing GDC-H alone, A was significantly greater at C_i concentrations above 400 $\mu mol/mol$ than in C plants (Figure 4d,e). Although A in SFH plants was higher than in the control plants at 400 $\mu mol/mol$, it was lower than that observed in the S, F or SF plants. The maximum rate of CO_2 assimilation (A_{max}) was significantly higher in lines S, F, SF

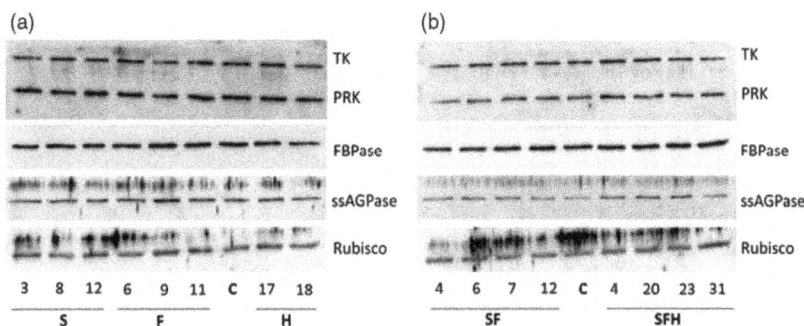

Figure 2 Molecular and biochemical analysis of the transgenic plants over-expressing SBPase (S), FBPA (F) and GDC-H (H). Immunoblot blot analysis of protein extracts from two independent leaves of (a) S, F and H lines and (b) SF and SFH lines used in this study compared to C. Transketolase (TK), phosphoribulokinase (PRK), fructose-1,6-bisphosphase (FBPase) the small subunit of ADP glucose pyrophosphoryalse (ssAGPase) and Rubisco.

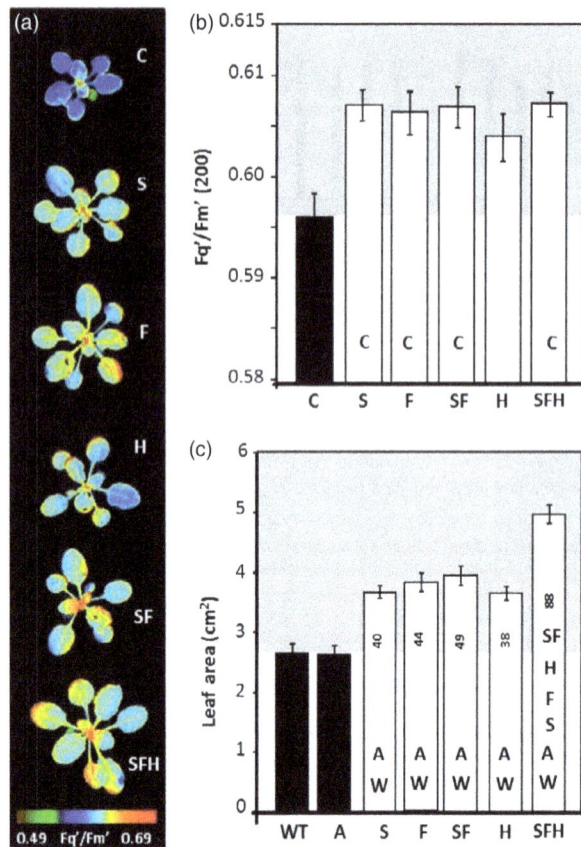

Figure 3 Photosynthetic capacity and leaf area in transgenic seedlings determined using chlorophyll fluorescence imaging. C and transgenic plants were grown in controlled environment conditions with a light intensity 130 µmol/m²/s, 8-h light/16-h dark cycle for 15 days and chlorophyll fluorescence imaging used to determine F_q'/F_m' (maximum PSII operating efficiency) values of the whole plant at (a, b) 200 µmol/m²/s and (c) leaf area at time of analysis. Azygous controls (A) recovered from a segregating population. Lines over-expressing SBPase (S), FBPA (F), GDC-H protein (H), SBPase and FBPA (SF) and SBPase, FBPA and GDC-H (SFH) are represented. The data were obtained using six individual plants from two (H), three (S, F, SF) or four (SFH) independent transgenic lines (18–24 plants total) compared to 12 C. Columns represent mean values, and standard errors are displayed. Significant differences between lines ($P < 0.05$) are represented as capital letters indicating whether each specific line is significantly different from another (i.e. SBPase lines (S) are significantly bigger than wild type (WT) and azygous lines (A)). Numbers indicate % increases over WT.

and SFH compared to C; however, no significant differences were observed between these lines (Figure 4f). Interestingly, the H plants show no increase in A_{max} when compared to C plants. Further analysis of the A/C_i curves using the equations published by von Caemmerer and Farquhar (1981) illustrated that the maximum rate of carboxylation by Rubisco (Vc_{max}: Figure S4b) in lines S, SF and SFH was significantly greater than in C, and Vc_{max} in these lines was also significantly greater than in lines expressing GDC-H alone. Maximum electron transport rates (J_{max}: Figure S4c) were also elevated in lines S, F, SF and SFH compared to C and were also shown to be significantly elevated compared to H.

To further assess the effect of the manipulation of the CB cycle and/or the GDC-H protein, the rates of photosynthetic carbon

assimilation and electron transport were determined in mature plants as a function of light intensity at 2% [O₂] to eliminate photorespiration (Figure 5a). Electron transport rates through PSII in H and SFH over-expression plants were significantly greater than in the C and SF plants at light levels above 300 µmol/m²/s (Figure 5b). A_{sat} was also significantly higher, 12%–19%, in all lines compared to C although no significant differences were observed between the different transgenic lines (Figure 5c).

Increased SBPase and FBPA activity and over-expression of the glycine decarboxylase-H protein stimulates growth in low light

The growth of the different transgenic and C plants was determined using image analysis of total leaf area over a period of 38 days from planting (Figure 6a), which showed all transgenic lines had a significantly greater leaf area than C, as early as 16 days after planting (Figure 6b). Furthermore, plants over-expressing all three transgenes (SFH) were shown to have a significantly larger leaf area when compared to all other transgenic lines including G and SF, indicating a cumulative advantage from combining these transgenes at this stage in development. This growth trend continued through to 15 days postplanting (Figure S6a). By 20 days after planting (Figure S6b), plants over-expressing the glycine decarboxylase-H protein (H) were shown to be significantly bigger than S, F and SF at the same time point, and triple over-expressing lines SFH remained significantly bigger than all other lines studied (Figure 6b).

Plants were allowed to continue growing until harvest at 38 days (Figure S7). At this stage of development, no significant difference in leaf area or dry weight could be observed between S, F, H or SF lines when compared to each other (Figure 6c). However, all lines attained a significantly larger leaf area and dry weight when compared to C. Notably, at this stage, the triple over-expressing lines SFH were significantly larger with a higher dry weight (+70%) than all other transgenic and C plants. Furthermore, lines SF and SFH both showed a significant increase in leaf number after 38 days (Figure S8).

Increased SBPase and FBPA activity and expression of the glycine decarboxylase-H protein impacts on the carbohydrate profile of selected lines

To determine how the over-expression of these key proteins impacts on downstream processes, leaf tissue was harvested and starch and sugar content were evaluated. No significant difference in starch levels were detected at the end of the day in any of the transgenic lines compared to C (Figure 7). Interestingly, slightly higher starch levels were detected 1 h before sunrise (dark) in transgenic lines F, H and SFH compared to C. All transgenic lines were shown to have consistently higher levels of sucrose, with these levels being significantly higher than C in F and SF lines. SF lines were also shown to have a significantly higher amount of glucose (Figure 7) compared to C, although other lines were consistently elevated but not significantly so.

Impact of light intensity on biomass and seed yield

A subset of plants was allowed to seed in either low or high light, and final vegetative biomass and seed yield determined per plant. In low-light grown plants, the final vegetative biomass was higher in all of the transgenic lines compared to C; however, no significant differences were observed between the different transgenic lines (Figure 8a). Furthermore, seed yield was

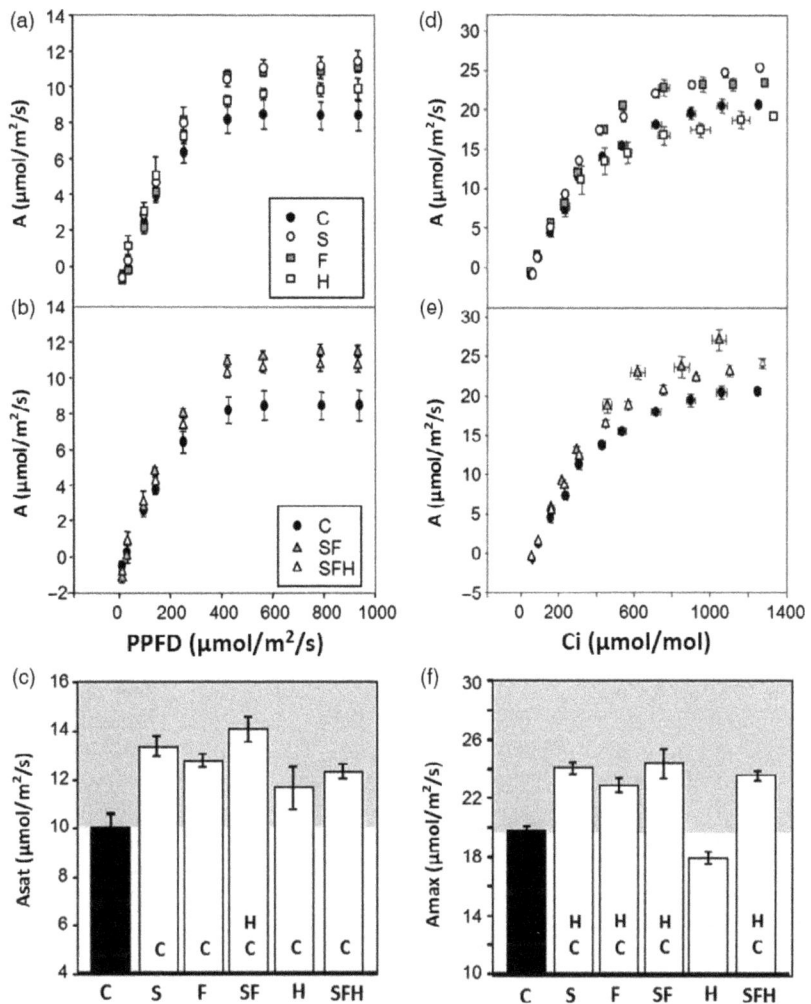

Figure 4 Photosynthetic responses of C and transgenic plants. (a, b) Photosynthesis carbon fixation rates were determined as a function of increasing light intensity. (c) A_{sat} determined from light response curves. (d, e) Photosynthetic carbon fixation rates were determined as a function of increasing CO_2 concentrations (A/C_i) at saturating-light levels (1000 $\mu mol/m^2/s$). (f) A_{max} determined from A/C_i response curves. C and transgenic plants were grown in controlled environment conditions with a light intensity 130 $\mu mol/m^2/s$, 8-h light/16-h dark cycle for 4 weeks. Lines over-expressing SBPase (S), FBPA (F), GDC-H protein (H), SBPase and FBPA (SF) and SBPase, FBPA and GDC-H (SFH) are represented. Columns represent mean values, and standard errors are displayed. Significant differences between lines ($P < 0.05$) are represented as capital letters indicating whether each specific line is significantly different from another. Results are based on 4–7 plants per line. (i.e. SBPase lines (S) are significantly different to controls (C)).

increased by 35%–53% in transgenic lines S, SF and SFH (Figure 8b). Interestingly, no increase in seed yield was observed in the H plants.

We next compared the impact of growth of plants in high light to explore further the potential positive impact of these transgenic manipulations on growth. In high-light-grown plants, an increase in vegetative biomass from 14% to 51% was observed (Figures 8c and S9). Notably, the H and SFH plant produced significantly more vegetative biomass than the S, F, SF or C plants. Furthermore, seed yield in high-light-grown plants was increased by 39%–62% in transgenic lines S, F, SF and SFH, when compared to C (Figure 8d). Although the highest increase in seed yield was observed in lines SFH in high light, no increase in seed yield was observed in the H plants in high-light-grown plants. The seed yield for individual plants can be seen in Figure S10.

Discussion

In this study, we have shown that simultaneously increasing the levels of two enzymes of the CB cycle, SBPase and FBPA, and the H protein of the glycine decarboxylase enzyme of the photorespiratory pathway in the same plant, resulted in a substantial and significant increase in both vegetative biomass and seed yield of Arabidopsis grown in controlled environment conditions. An increase in both biomass and yield was also observed in plants

overexpressing SBPase or FBPA alone or in combination. However, although overexpression of GDC-H alone resulted in an increase in vegetative biomass, no increase in seed yield was evident in these plants, grown in either low- or high-light conditions. The reasons for this differential effect on seed yield have not yet been elucidated but may be due to changes in carbon status brought about by altered source/sink allocation which is supported by changes to starch and sucrose levels at the end of the night period in some of these lines. Higher levels of sucrose (and fructose, maltose) have also been observed in GDC-L over-expressers (Timm et al., 2015), and the over-expression of GDC-L enhances the metabolic capacity of photorespiration and is believed to alter the carbon flow through the TCA cycle (Timm et al., 2015).

It was shown in earlier studies that over-expression of FBPA or SBPase alone in tobacco results in a stimulation of photosynthesis and biomass, with the greatest effect being seen in plants grown under elevated CO_2 (Lefebvre et al., 2005; Rosenthal et al., 2011; Uematsu et al., 2012). Furthermore, when FBPA was over-expressed in combination with SBPase in tobacco, this led to a cumulative increase in biomass in plants grown in ambient CO_2 under greenhouse conditions (Simkin et al., 2015). Interestingly, in this current study, we have shown that in Arabidopsis that the over-expression of FBPA alone, under current atmospheric CO_2 levels, results in a stimulation of photosynthesis and increase in

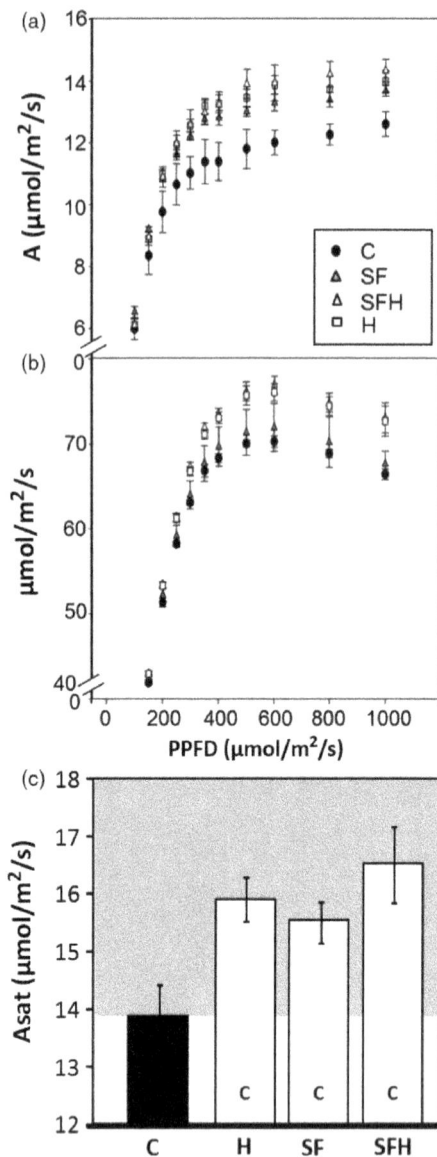

Figure 5 Photosynthetic responses of the transgenic plants at 2% [O_2] (a) and (b) chloroplast electron transport rates in transgenic plants at 2% [O_2]. (c) Mean values of A_{sat} determined from light response curves. C and transgenic plants were grown in controlled environment conditions with a light intensity 130 μmol/m^2/s, 8-h light/16-h dark cycle for 4 weeks. Lines over-expressing the GDC-H protein (H), SBPase and FBPA (SF) and SBPase, FBPA and GDC-H (SFH) are represented. Columns represent mean values, and standard errors are displayed. Significant differences between lines ($P < 0.05$) are represented as capital letters. Results are based on 5–6 plants per line compared to six controls.

biomass on a similar level to that observed by over-expression of SBPase alone. However, contrary to the results obtained in tobacco, the co-expression of SBPase and FBPA in Arabidopsis did not lead to a further significant increase in either leaf area or biomass when compared to plants independently expressing SBPase (resulting in higher endogenous FBPA activity) or FBPA. This lack of differential effect of the co-overexpression of SBPase and FBPA in this study can likely be explained by the fact that over-expression of SBPase in Arabidopsis also led to a small but significant increase in endogenous FBPA protein levels and activity

(25%–36%). Given that no increase in SBPase was present in the FBPA plants, this would suggest that in Arabidopsis, the stimulation in the SBPase, FBPA and the SF over-expression lines may be due to increased FBPA activity. This is in contrast to tobacco where over-expression of SBPase alone led to an increase in biomass and no increases in endogenous FBPA activity, highlighting the differences between species (Lefebvre et al., 2005; Rosenthal et al., 2011; Simkin et al., 2015).

Detailed analysis of a range of photosynthesis parameters revealed a similar increase in A_{sat} at low [O_2] for all of the transgenic lines studied. The most significant increase was observed in SF lines which showed a 44% increase over control plants, with the lowest increase of 19% being observed in the H plants. An evaluation of the electron transport rates at low [O_2] in a subset of these plants showed that lines over-expressing GDC-H (both H and SFH) displayed higher photosynthetic electron transport rates compared to C and plants over-expressing SBPase and FBPA (SF). These results are in keeping with the previous study by Timm et al. (2012). All of the transgenic lines analysed here showed an increase in photosynthesis under high light and ambient CO_2 conditions. However, under high light and saturated levels of CO_2 the rate of assimilation in the H plants was similar to C, and this is in contrast to all other transgenic lines. This observation is in keeping with the proposal that over-expression of the H protein stimulates the flow of carbon through the photorespiratory pathway, thereby reducing steady-state levels of inhibitory photorespiratory metabolites, which in turn stimulates flux through the CB cycle. Whilst this hypothesis is supported by metabolite data and the observation that growth of GDC-H plants is not stimulated when these plants are grown in elevated CO_2 conditions (Timm et al., 2012), the exact mechanism of such feedback from photorespiration to the CB cycle is not yet known. The effect of these manipulations on photosynthesis was also determined at the growth light intensity where small differences in A are observed even at light levels as low as 150 μmol/m^2/s. This together with the increased leaf area observed at early stages in development provides evidence that the small differences in photosynthesis lead to an increase in leaf area. The cumulative impact of this over time results in increased biomass and yield.

Conclusion

In this proof-of-concept study in Arabidopsis, we have demonstrated that the simultaneous over-expression of two CB cycle enzymes leads to an increase in photosynthesis and an increase in overall biomass and seed yield. We also show that when the transgenic SF lines were crossed with GDC-H over-expressing plants (Timm et al., 2012), the combined effects of these three transgenes (SFH) resulted in a cumulative impact on biomass (+71%) which was significantly higher than H (+50%) and SF (+41%) under low light. Importantly, the work here also allowed a parallel comparative analysis between the different manipulations under different environmental conditions.

Although it is still necessary to address the importance of these manipulations in crop species and also under field conditions, this study provides additional evidence that multigene manipulation of photosynthesis and photorespiration can form an important tool to improve crop yield. These results also provide new information indicating that it will be necessary to tailor the targets for manipulation for different crops and for either biomass or seed yield.

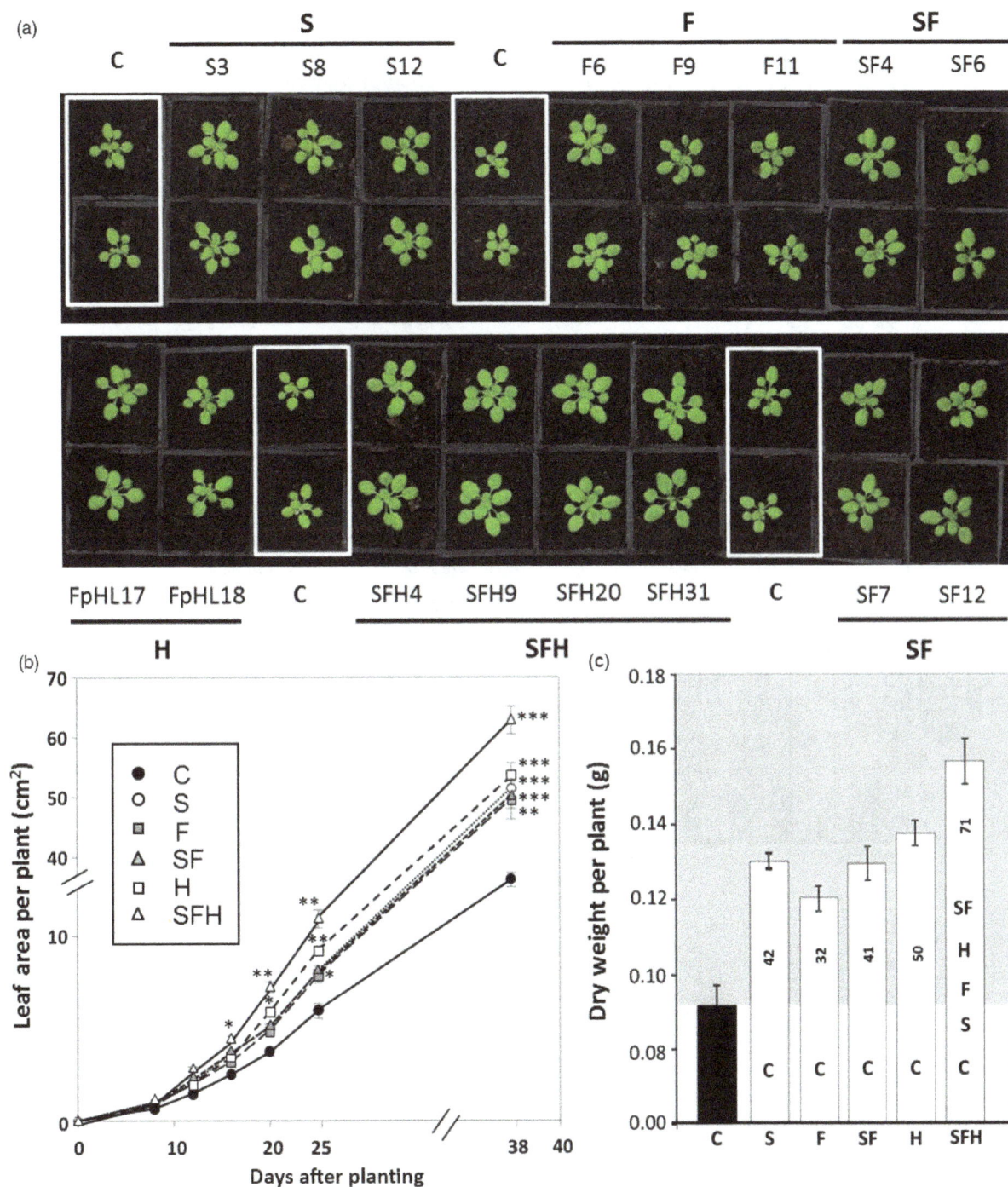

Figure 6 Growth analysis of C and transgenic lines grown in low light. (a) Plants were grown at 130 µmol/m²/s light intensity in short days (8 h/16 h days) for 15 days. (b) Plant growth rate evaluated over the first 38 days. Significant differences *(P < 0.05), **(P < 0.01), ***(P < 0.001) are indicated. (c) Final dry weight (g) after 38 days of development and statistical differences between lines. % increases over C are indicated within the columns. Lines overexpressing SBPase (S), FBPA (F), GDC-H protein (H), SBPase and FBPA (SF) and SBPase, FBPA and GDC-H (SFH) are represented. Columns represent mean values, and standard errors are displayed. Significant differences between lines (P < 0.03) are represented as capital letters indicating whether each specific line is significantly different from another. Results are representative of 9–12 plants from two (H), three (S, F, SF) or four (SFH) independent lines (C plants including wild type and azygous lines segregated from primary transformants).

Materials and methods

Generation of pGW photosynthetic tissue-specific destination vector pGWPTS1

pGWB1 (Nakagawa et al., 2007: AB289764) was cut with HindIII at 37 °C. Following purification, digested vectors were treated with alkaline phosphatase (BioLabs, UK) for 60 min at 37 °C. The rbcS2B (1150 bp; At5 g38420) promoter was amplified with primers Pr_rbcS2B_F_HindIII (5′CACCaagcttATgACATCATAgCAAgCAAggACACg′3) and Pr_rbcS2B_R_HindIII (5′CTGAGAaagcttTACTTCTTCTTgTTgTTTCTCTTCTTC′3). The amplicon was

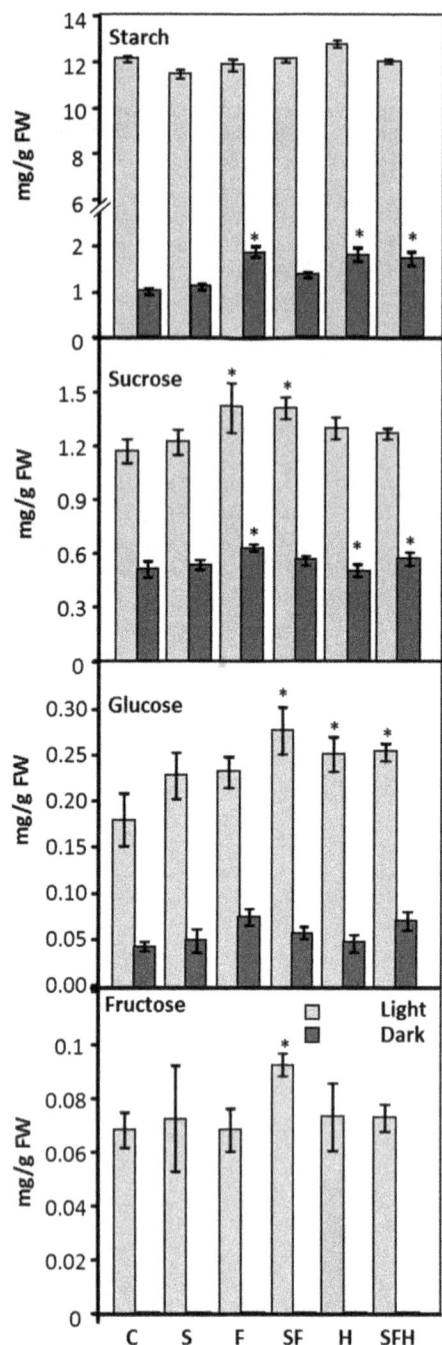

Figure 7 Leaf starch and sugar content at end of 8-h light period (light grey) and end of 16-h dark period (dark grey). Results are mean values based on 12–18 individual plants from two (H), three (S, F, SF) or four (SFH) independent transgenic lines. Columns represent mean values, and standard errors are displayed. Lines over-expressing SBPase (S), FBPA (F), GDC-H protein (H), SBPase and FBPA (SF) and SBPase, FBPA and GDC-H (SFH) are represented. Significant differences between C and over-expressing lines (*P < 0.01) are represented.

SB, PTS1-FB and PTS1-SBFB construct assembly can be seen in the supplementary data. Construct maps are shown in Figure S1b–d.

Generation of transgenic plants

The recombinant plasmids PTS1-SB, PTS1-FB and PTS1-SBFB were introduced into wild-type Arabidopsis by floral dipping (Clough and Bent, 1998) using *Agrobacterium tumefaciens* GV3101. Positive transformants were regenerated on MS medium containing kanamycin (50 mg/L) and hygromycin (20 mg/L). Kanamycin-/hygromycin-resistant primary transformants (T1 generation) with established root systems were transferred to soil and allowed to self-fertilize. Plants over-expressing SBPase, FBPA and the GDC-H protein were generated by floral inoculation of two SBPase + FBPA lines (SF6 and SF12) with the pollen from two GDC-H protein over-expressing lines (*Fp*H17 and *Fp*H18) provided by Timm et al. (2012). Lines *Fp*H17 and 18 were originally generated by floral dipping and over-expressing the *Flaveria pringlei* GDC-H protein (Kopriva and Bauwe, 1995) under the control of the leaf-specific and light-regulated *Solanum tuberosum ST-LS1* promoter (Stockhaus et al., 1989). Following initial characterization of generated lines, three lines for SBPase (S3, S8, S12), FBPA (F6, F9, F11) and SF (SF6, SF7, SF12) were selected for further study from all lines generated.

Plant growth conditions

Wild-type T2 Arabidopsis plants resulting from self-fertilization of transgenic plants were germinated in sterile agar medium containing Murashige and Skoog salts (plus kanamycin 50 mg/L for the transformants) and grown to seed in soil (Levington F2, Fisons, Ipswich, UK). Lines of interest were identified by immunoblot and qPCR. For experimental study, T3 progeny seeds from selected lines were germinated on soil in controlled environment chambers at an irradiance of 130 µmol photons/m^2/s, 22 °C, relative humidity of 60%, in an 8-h/16-h square-wave photoperiod. Plants were sown randomly, and trays rotated daily. Four leaf discs (0.6 cm diameter) from two individual leaves, for the analysis of SBPase and FBPA activities, were taken and immediately plunged into liquid N$_2$, and stored at −80 °C. Leaf areas were calculated using standard photography and ImageJ software (imagej.nih.gov/ij). Wild-type plants and null segregants (azygous) used in this study were initially evaluated independently. However, once it was determined that no significant difference were observed between these two groups (see figures and supplementary figures), wild-type plants and null segregants were combined (null segregants from the transgenic lines verified by PCR for nonintegration of the transgene) and used as a combined 'control' group (C).

Protein extraction and immunoblotting

Leaf discs sampled as described above were ground in liquid nitrogen. Total protein was extracted in extraction buffer (50 mM 4-(2-hydroxyethyl)piperazine-1-ethanesulphonic acid (HEPES) pH 8.2, 5 mM MgCl2, 1 mM ethylenediaminetetraacetic acid tetra-sodium salt (EDTA), 10% glycerol, 0.1% Triton X-100, 2 mM benzamidine, 2 mM aminocaproic acid, 0.5 mM phenylmethane-sulphonyl fluoride (PMSF) and 10 mM DTT), and the insoluble material was removed by centrifugation at 14 000 g for 10 min (4 °C) and protein quantification determined (Harrison et al., 1998). Samples were loaded on an equal protein basis, separated using 12% (w/v) SDS-PAGE, transferred to polyvinylidene difluoride membrane and probed using antibodies raised against SBPase, FBPA and the GDC-H protein (Timm et al., 2012).

digested with HindIII and cloned into the corresponding site of pGWB1 to make vector pGWPTS1 (Figure S1a).

Constructs were generated using Gateway cloning technology and vector pGWPTS1. All transgenes were under the control of the rbcS2B (1150 bp; *At5 g38420*) promoter. Full details of PTS1-

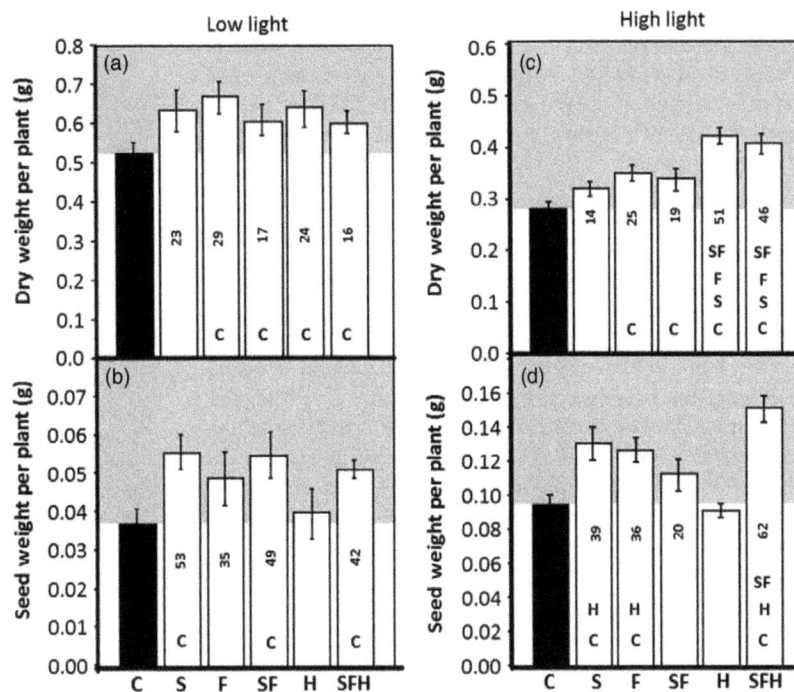

Figure 8 GDC-H and GDC-H with SBPase and FBPA overexpression in Arabidopsis differentially impact biomass and seed yield. (a, c) Dry weight and (b, d) seed weight were determined at seed harvest. C and transgenic plants were grown in controlled environment conditions at either 130 μmol/m²/s, 8-h light/16-h dark cycle (a., b.) or 390 μmol/m²/s, 8-h light/16-h dark cycle (c., d.). Lines over-expressing SBPase (S), FBPA (F), GDC-H protein (H), SBPase and FBPA (SF) and SBPase, FBPA and GDC-H (SFH) are represented. The data were obtained using 10–17 individual plants from two (H), three (S, F, SF) or four (SFH) independent transgenic lines (2 H lines. See Timm et al., 2012) compared to 12–13 C. Columns represent mean values and standard errors are displayed. Significant differences between lines ($P < 0.05$) are represented as capital letters indicating whether each specific line is significantly different from another. Numbers indicate % increases over C. Seed weights per plant and full statistical evaluation between groups can be seen in Figure S10.

Proteins were detected using horseradish peroxidase conjugated to the secondary antibody and ECL chemiluminescence detection reagent (Amersham, Buckinghamshire, UK). SBPase antibodies are previously characterized in Lefebvre et al. (2005), and FBPA antibodies were raised against a peptide from a conserved region of the protein [C]-ASIGLENTEANRQAYR-amide, Cambridge Research Biochemicals, Cleveland, UK (Simkin et al., 2015). In addition to the aforementioned antibodies, samples were probed using antibodies raised against the phosphoribulokinase (AS09464), ssAGPase (AS111739), purchased from Agrisera (via Newmarket Scientific, UK) and FBPase (see Lefebvre et al., 2005), transketolase (Henkes et al., 2001) and Rubisco (Foyer et al., 1993).

Determination of SBPase activity by phosphate release

SBPase activity was determined by phosphate release. Immediately after photosynthesis measurement, leaf discs were isolated from the same leaves and frozen in liquid nitrogen. For analysis, leaf discs were ground to a fine powder in liquid nitrogen in extraction buffer (50 mM HEPES, pH8.2; 5 mM MgCl₂; 1 mM EDTA; 1 mM EGTA; 10% glycerol; 0.1% Triton X-100; 2 mM benzamidine; 2 mM aminocaproic acid; 0.5 mM phenylmethyl-sulphonylfluoride; 10 mM dithiothreitol), and the resulting solution was centrifuged 1 min at 14 000 g, 4 °C. The resulting supernatant was desalted through an NAP-10 column (Amersham) and eluted, aliquoted and stored in liquid nitrogen. For the assay, the reaction was started by adding 20 μL of extract to 80 μL of assay buffer (50 mM Tris, pH 8.2; 15 mM MgCl₂; 1.5 mM EDTA; 10 mM dithiothreitol; 2 mM SBP) and incubated at 25 °C for 30 min as described previously (Simkin et al., 2015). The reaction was stopped by the addition of 50 μL of 1 M perchloric acid and centrifuged for 10 min at 14 000 g, 4 °C. Samples (30 μL) and standards (30 μL, 0.125–4 nmol PO_4^{3-}) in triplicate were incubated 30 min at room temperature following the addition of 300 μL of Biomol Green (Affiniti Research Products, Exeter, UK), and the A_{620} was measured using a microplate reader (VERSAmax, Molecular Devices, Sunnyvale, CA).

Determination of FBPA activity

Desalted protein extracts, as described above, were evaluated for FBPA activity as described previously (Haake et al., 1998).

Chlorophyll fluorescence imaging

Measurements were performed on 2-week-old Arabidopsis seedlings that had been grown in a controlled environment chamber providing 130 μmol/mol²/s PPFD and ambient CO_2. Chlorophyll fluorescence parameters were obtained using a chlorophyll fluorescence (CF) imaging system (Technologica, Colchester, UK; Barbagallo et al., 2003; Baker and Rosenqvist, 2004). The operating efficiency of photosystem two (PSII) photochemistry, F_q'/F_m', was calculated from the measurements of steady-state fluorescence in the light (F') and maximum fluorescence in the light (F_m') since $F_q'/F_m' = (F_m' - F')/F_m'$. Images of F' were taken when fluorescence was stable at 130 μmol/m²/s PPFD, whilst images of maximum fluorescence were obtained after a saturating 600 ms pulse of 6200 μmol/m²/s PPFD (Baker et al., 2001; Oxborough and Baker, 1997). Parallel measurements of plants grown in high light (390 μmol/mol²/s PPFD and ambient CO_2) were also performed on 2-week-old Arabidopsis (Supporting Information).

Gas exchange measurements

The response of net photosynthesis (A) to intracellular CO_2 (C_i) was measured using a portable gas exchange system (CIRAS-1, PP Systems Ltd, Ayrshire, UK). Leaves were illuminated with an integral red-blue LED light source (PP systems Ltd) attached to the gas exchange system, and light levels were maintained at saturating photosynthetic photon flux density (PPFD) of 1000 μmol/m²/s for the duration of the A/C_i response curve. Measurements of A were made at ambient CO_2 concentration (C_a) at 400 μmol/mol, before C_a was decreased to 300, 200, 150, 100 and 50 μmol/mol before returning to the initial value and increased to 500, 600, 700, 800, 900, 1000, 1100 and 1200 μmol/mol. Leaf temperature and vapour pressure deficit

(VPD) were maintained at 22 °C and 1 ± 0.2 kPa, respectively. The maximum rates of Rubisco- (Vc_{max}) and the maximum rate of electron transport for RuBP regeneration (J_{max}) were determined and standardized to a leaf temperature of 25 °C based on equations from Bernacchi et al. (2001) and McMurtrie and Wang (1993), respectively.

Photosynthetic light response curves

A/Q response curves were measured using a CIRAS-1 portable gas exchange system (PP Systems (CIRAS-1, PP Systems Ltd). Cuvette conditions were maintained at a leaf temperature of 22 °C, relative humidity of 50%–60% and ambient growth CO_2 concentration (400 mmol/mol for plants grown in ambient conditions). Leaves were initially stabilized at saturating irradiance 1000 µmol/m²/s, after which A and g_s were measured at the following PPFD levels: 0, 50, 100, 150, 200, 250, 300, 350, 400, 500, 600, 800 and 1000 µmol/m²/s. Measurements were recorded after A reached a new steady state (1–2 min) and before stomatal conductance (g_s) changed to the new light levels. A/Q analyses were performed at 21% and 2% O_2.

Determination of sucrose and starch

Carbohydrates and starch were extracted from 20 mg leaf tissue, and samples were collected at two time points, 1 h before dawn (15 h into the dark period) and 1 h before sunset (7 h into the light period). Four leaf discs collected from two different leaves were ground in liquid nitrogen, and 20 mg/FW of tissue was incubated in 80% (v/v) ethanol for 20 min at 80 °C and then repeated three times with ethanol 80% (v/v) at 80 °C. The resulting solid pellet and pooled ethanol samples were freeze-dried. Suc was measured from the extracts in ethanol using an enzyme-based protocol (Stitt et al., 1989), and the starch contents were estimated from the ethanol-insoluble pellet according to Stitt et al. (1978), with the exception that the samples were boiled for 1 h and not autoclaved.

Statistical analysis

All statistical analyses were performed by comparing ANOVA, using Sys-stat, University of Essex, UK. The differences between means were tested using the post hoc Tukey test (SPSS, Chicago, IL).

Acknowledgements

We thank Dr Lorna McAusland and Mr Jack Matthews for help with gas exchange, Dr Stuart Fisk for help with enzyme assays and James E Fox for help with pigment analysis. A.J.S, L.R.H and P.E.L were supported by the British Biological Sciences Research Council (BBSRC, grant: BB/J004138/1 awarded to C.A.R and T.L).

Author contributions

C.A.R. conceived this project, provided the funding and led the supervision of this research with input from T.L. A.J.S generated transgenic plants and performed molecular, biochemical and plant phenotypic analysis. L.R.H and P.E.L contributed to the generation and analysis of the transgenic plants. A.J.S and P.A.D carried out data analysis on their respective contributions. S.T and H.B generated and provided glycine decarboxylase over-expressing lines used for crosses. The manuscript was drafted by A.J.S and finalised by C.A.R. All authors reviewed and commented on the final manuscript.

References

Anderson, L.E. (1971) Chloroplast and cytoplasmic enzymes. 2. Pea leaf triose phosphate isomerases. Biochim. Biophys. Acta 235, 237–244.

Baker, N.R. (2008) Chlorophyll fluorescence: a probe of photosynthesis in vivo. Annu. Rev. Plant Biol. 59, 89–113.

Baker, N.R. and Rosenqvist, E. (2004) Applications of chlorophyll fluorescence can improve crop production strategies: an examination of future possibilities. J. Exp. Bot. 55, 1607–1621.

Baker, N.R., Oxborough, K., Lawson, T. and Morison, J.I. (2001) High resolution imaging of photosynthetic activities of tissues, cells and chloroplasts in leaves. J. Exp. Bot. 52, 615–621.

Barbagallo, R.P., Oxborough, K., Pallett, K.E. and Baker, N.R. (2003) Rapid, non-invasive screening for perturbations of metabolism and plant growth using chlorophyll fluorescence imaging. Plant Physiol. 132, 485–493.

Bernacchi, C.J., Singsaas, E.L., Pimentel, C., Portis, A.R. Jr and Long, S.P. (2001) Improved temperature response functions for models of Rubisco-limited photosynthesis. Plant, Cell Environ. 24, 253–260.

Betti, M., Bauwe, H., Busch, F.A., Fernie, A.R., Keech, O., Levey, M., Ort, D.R. et al. (2016) Manipulating photorespiration to increase plant productivity: recent advances and perspectives for crop improvement. J. Exp. Bot. 67, 2977–2988.

Bowes, G., Ogren, W.L. and Hageman, R.H. (1971) Phosphoglycolate production catalysed by ribulose diphosphate carboxylase. Biochem. Biophys. Res. Commun. 45, 716–722.

von Caemmerer, S. and Evans, J.R. (2010) Enhancing C_3 Photosynthesis. Plant Physiol. 154, 589–592.

von Caemmerer, S. and Farquhar, G.D. (1981) Some relationships between the biochemistry of photosynthesis and the gas exchange of leaves. Planta, 153, 376–387.

Cai, Z., Liu, G., Zhang, J. and Li, Y. (2014) Development of an activity-directed selection system enabled significant improvement of the carboxylation efficiency of Rubisco. Protein Cell, 5, 552–562.

Carmo-Silva, E., Scales, J.C., Madgwick, P.J. and Parry, M.A. (2015) Optimizing Rubisco and its regulation for greater resource use efficiency. Plant, Cell Environ. 38, 1817–1832.

Clough, S.J. and Bent, A.F. (1998) Floral dip: a simplified method for Agrobacterium-mediated transformation of Arabidopsis thaliana. Plant J. 16, 735–743.

Dalal, J., Lopez, H., Vasani, N.B., Hu, Z., Swift, J.E., Yalamanchili, R., Dvora, M. et al. (2015) A photorespiratory bypass increases plant growth and seed yield in biofuel crop Camelina sativa. Biotechnol. Biofuels 8, 175.

Ding, F., Wang, M., Zhang, S. and Ai, X. (2016) Changes in SBPase activity influence photosynthetic capacity, growth, and tolerance to chilling stress in transgenic tomato plants. Sci. Rep. 6, 32741.

Eisenhut, M., Bauwe, H. and Hagemann, M. (2007) Glycine accumulation is toxic for the cyanobacterium Synechocystis sp. strain PCC 6803, but can be compensated by supplementation with magnesium ions. FEMS Microbiol. Lett. 277, 232–237.

Engel, N., van den Daele, K., Kolukisaoglu, U., Morgenthal, K., Weckwerth, W., Pärnik, T., Keerberg, O. et al. (2007) Deletion of glycine decarboxylase in Arabidopsis is lethal under nonphotorespiratory conditions. Plant Physiol. 144, 1328–1335.

Evans, J.R. (2013) Improving photosynthesis. Plant Physiol. 162, 1780–1793.

Fischer, R.A., Rees, D., Sayre, K.D., Lu, Z.M., Condon, A.G. and Saavedra, A.L. (1998) Wheat yield progress associated with higher stomatal conductance and photosynthetic rate, and cooler canopies. Crop Sci. 38, 1467–1475.

Foyer, F.H., Nurmi, A., Dulieu, H. and Parry, M.A.J. (1993) Analysis of two rubisco-deficient tobacco mutants, H7 and Sp25; evidence for the production of rubisco large subunits in the Sp25 mutant that form clusters and are inactive. J. Exp. Bot. 44, 1445–1452.

Gifford, R.M. and Evans, L. (1981) Photosynthesis, carbon partitioning, and yield. Annu. Rev. Plant Physiol. 32, 485–509.

Haake, V., Zrenner, R., Sonnewald, U. and Stitt, M. (1998) A moderate decrease of plastid aldolase activity inhibits photosynthesis, alters the levels of sugars and starch, and inhibits growth of potato plants. Plant J. 14, 147–157.

Haake, V., Geiger, M., Walch-Liu, P., Engels, C., Zrenner, R. and Stitt, M. (1999) Changes in aldolase activity in wild-type potato plants are important for acclimation to growth irradiance and carbon dioxide concentration, because plastid aldolase exerts control over the ambient rate of photosynthesis across a range of growth conditions. Plant J. **17**, 479–489.

Harrison, E.P., Willingham, N.M., Lloyd, J.C. and Raines, C.A. (1998) Reduced sedoheptulose-1,7-bisphosphatase levels in transgenic tobacco lead to decreased photosynthetic capacity and altered carbohydrate accumulation. Planta, **204**, 27–36.

Harrison, E.P., Ölçer, H., Lloyd, J.C., Long, S.P. and Raines, C.A. (2001) Cell and molecular biology, biochemistry and molecular physiology-small decreases in SBPase cause a linear decline in the apparent RuBP regeneration rate, but do not affect Rubisco carboxylation. J. Exp. Bot. **52**, 1779–1784.

Heineke, D., Bykova, N., Gardeström, P. and Bauwe, H. (2001) Metabolic response of potato plants to an antisense reduction of the P-protein of glycine decarboxylase. Planta, **212**, 880–887.

Henkes, S., Sonnewald, U., Badur, R., Flachmann, R. and Stitt, M. (2001) A small decrease of plastid transketolase activity in antisense tobacco transformants has dramatic effects on photosynthesis and phenylpropanoid metabolism. Plant Cell, **13**, 535–551.

Kebeish, R., Niessen, M., Thiruveedhi, K., Bari, R., Hirsch, H.J., Rosenkranz, R., Stäbler, N. et al. (2007) Chloroplastic photorespiratory bypass increases photosynthesis and biomass production in Arabidopsis thaliana. Nat. Biotechnol. **25**, 593–599.

Kelly, G.J. and Latzko, E. (1976) Inhibition of spinach-leaf phosphofructokinase by 2-phosphoglycollate. FEBS Lett. **68**, 55–58.

Kopriva, S. and Bauwe, H. (1995) H-protein of glycine decarboxylase is encoded by multigene families in Flaveria pringlei and F. cronquistii (Asteraceae). Mol. Gen. Genet. **248**, 111–116.

Kossmann, J., Sonnewald, U. and Willmitzer, L. (1994) Reduction of the chloroplastic fructose-1,6-bisphosphatase in transgenic potato plants impairs photosynthesis and plant growth. Plant J. **6**, 637–650.

Lawson, T., Bryant, B., Lefebvre, S., Lloyd, J.C. and Raines, C.A. (2006) Decreased SBPase activity alters growth and development in transgenic tobacco plants. Plant, Cell Environ. **29**, 48–58.

Lefebvre, S., Lawson, T., Zakhleniuk, O.V., Lloyd, J.C. and Raines, C.A. (2005) Increased sedoheptulose-1,7-bisphosphatase activity in transgenic tobacco plants stimulates photosynthesis and growth from an early stage in development. Plant Physiol. **138**, 451–460.

Lin, M.T., Occhialini, A., Andralojc, P.J., Parry, M.A. and Hanson, M.R. (2014a) A faster Rubisco with potential to increase photosynthesis in crops. Nature, **513**, 547–550.

Lin, M.T., Occhialini, A., Andralojc, P.J., Devonshire, J., Hines, K.M., Parry, M.A. and Hanson, M.R. (2014b) β-Carboxysomal proteins assemble into highly organized structures in Nicotiana chloroplasts. Plant J. **79**, 1–12.

Lin, H., Karki, S., Coe, R.A., Bagha, S., Khoshravesh, R., Balahadia, C.P., Ver Sagun, J. et al. (2016) Targeted knockdown of GDCH in rice leads to a photorespiratory-deficient phenotype useful as a building block for C4 rice. Plant Cell Physiol. **57**, 919–932.

Long, S.P., Zhu, X.G., Naidu, S.L. and Ort, D.R.. (2006) Can improvement in photosynthesis increase crop yields? Plant, Cell Environ. **29**, 315–330.

Long, S.P., Marshall-Colon, A. and Zhu, X.G. (2015) Meeting the global food demand of the future by engineering crop photosynthesis and yield potential. Cell, **161**, 56–66.

Lu, Y., Li, Y., Yang, Q., Zhang, Z., Chen, Y., Zhang, S. and Peng, X.X. (2014) Suppression of glycolate oxidase causes glyoxylate accumulation that inhibits photosynthesis through deactivating Rubisco in rice. Plant Physiol. **150**, 463–476.

Maier, A., Fahnenstich, H., von Caemmerer, S., Engqvist, M.K., Weber, A.P., Flügge, U.I. and Maurino, V.G. (2012) Transgenic introduction of a glycolate oxidative cycle into a. thaliana chloroplasts leads to growth improvement. Front Plant Sci. **3**, 38.

McGrath, J.M. and Long, S.P. (2014) Can the cyanobacterial carbon-concentrating mechanism increase photosynthesis in crop species? A theoretical analysis Plant Physiol. **164**, 2247–2261.

McMurtrie, R.E. and Wang, Y.P. (1993) Mathematical models of the photosynthetic response of tree stands to rising CO_2 concentrations and temperature. Plant, Cell Environ. **16**, 1–13.

Meyer, M.T., McCormick, A.J. and Griffiths, H. (2016) Will an algal CO2-concentrating mechanism work in higher plants? Curr. Opin. Plant Biol. **31**, 181–188.

Miyagawa, Y., Tamoi, M. and Shigeoka, S. (2001) Over-expression of a cyanobacterial fructose-1,6/sedoheptulose-1,7-bisphosphatase in tobacco enhances photosynthesis and growth. Nat. Biotechnol. **19**, 965–969.

Montgomery, B.L., Lechno-Yossef, S. and Kerfeld, C.A. (2016) Interrelated modules in cyanobacterial photosynthesis: the carbon-concentrating mechanism, photorespiration, and light perception. J. Exp. Bot. **67**, 2931–2940.

Murchie, E.H. and Lawson, T. (2013) Chlorophyll fluorescence analysis: guide to good practice and understanding some new applications. J. Exp. Bot. **64**, 3983–3998.

Nakagawa, T., Kurose, T., Hino, T., Tanaka, K., Kawamukai, M., Niwa, Y., Toyooka, K. et al. (2007) Development of series of gateway binary vectors, pGWBs, for realizing efficient construction of fusion genes for plant transformation. J. Biosci. Bioeng. **104**, 34–41.

Nolke, G., Houdelet, M., Kreuzaler, F., Peterhänsel, C. and Schillberg, S. (2014) The expression of a recombinant glycolate dehydrogenase polyprotein in potato (Solanum tuberosum) plastids strongly enhances photosynthesis and tuber yield. Plant Biotechnol. J. **12**, 734–742.

Orr, D.J., Alcantara, A., Kapralov, M.V., Andralojc, P.J., Carmo-Silva, E. and Parry, M.A. (2016) Surveying Rubisco diversity and temperature response to improve crop photosynthetic efficiency. Plant Physiol. **172**, 707–717. doi:10.1104/pp.16.00750

Oxborough, K. and Baker, N.R. (1997) An instrument capable of imaging chlorophyll a Fluorescence from intact leaves at very low irradiance and at cellular and subcellular levels. Plant, Cell Environ. **20**, 1473–1483.

Peterhänsel, C., Krause, K., Braun, H.P., Espie, G.S., Fernie, A.R., Hanson, D.T., Keech, O. et al. (2013) Engineering photorespiration: current state and future possibilities. Plant Biol. **15**, 754–758.

Poolman, M.G., Fell, D.A. and Thomas, S. (2000) Modelling photosynthesis and its control. J. Exp. Bot. **51**, 319–328.

Raines, C.A. (2003) The Calvin cycle revisited. Photosynth. Res. **75**, 1–10.

Raines, C.A. (2006) Transgenic approaches to manipulate the environmental responses of the C3 carbon fixation cycle. Plant, Cell Environ. **29**, 331–339.

Raines, C.A. (2011) Increasing photosynthetic carbon assimilation in C3 plants to improve crop yield: current and future strategies. Plant Physiol. **155**, 36–42.

Raines, C.A. and Paul, M.J. (2006) Products of leaf primary carbon metabolism modulate the developmental programme determining plant morphology. J. Exp. Bot. **57**, 1857–1862.

Raines, C.A., Lloyd, J.C. and Dyer, T.A. (1999) New insights into the structure and function of sedoheptulose-1, 7-bisphosphatase; an important but neglected Calvin cycle enzyme. J. Exp. Bot. **50**, 1–8.

Rojas-González, J.A., Soto-Súarez, M., García-Díaz, Á., Romero-Puertas, M.C., Sandalio, L.M., Mérida, Á., Thormählen, I. et al. (2015) Disruption of both chloroplastic and cytosolic FBPase genes results in a dwarf phenotype and important starch and metabolite changes in Arabidopsis thaliana. J. Exp. Bot. **66**, 2673–2689.

Rosenthal, D.M., Locke, A.M., Khozaei, M., Raines, C.A., Long, S.P. and Ort, D.R. (2011) Over-expressing the C3 photosynthesis cycle enzyme Sedoheptulose-1-7 Bisphosphatase improves photosynthetic carbon gain and yield under fully open air CO2 fumigation (FACE). BMC Plant Biol. **11**, 123.

Sahrawy, M., Avila, C., Chueca, A., Canovas, F.M. and Lopez-Gorge, J. (2004) Increased sucrose level and altered nitrogen metabolism in Arabidopsis thaliana transgenic plants expressing antisense chloroplastic fructose-1,6-bisphosphatase. J. Exp. Bot. **55**, 2495–2503.

Sharwood, R.E., Ghannoum, O. and Whitney, S.M. (2016) Prospects for improving CO_2 fixation in C3-crops through understanding C4-Rubisco biogenesis and catalytic diversity. Curr. Opin. Plant Biol. **31**, 135–342.

Simkin, A.J., McAusland, L., Headland, L.R., Lawson, T. and Raines, C.A. (2015) Multigene manipulation of photosynthetic carbon assimilation increases CO_2 fixation and biomass yield. *J. Exp. Bot.* **66**, 4075–4090.

Stitt, M., Bulpin, P.V. and ap Rees, T.. (1978) Pathway of starch breakdown in photosynthetic tissues of *Pisum sativum*. *Biochim. Biophys. Acta* **544**, 200–214.

Stitt, M., Lilley, R.M., Gerhardt, R. and Heldt, H.W. (1989) Metabolite levels in specific cells and subcellular compartments of plant tissues. *Methods Enzymol.* **174**, 518–552.

Stockhaus, J., Schell, J. and Willmitzer, L. (1989) Correlation of the expression of the nuclear photosynthetic gene ST-LS1 with the presence of chloroplast. *EMBO J.* **8**, 2445–2451.

Timm, S., Florian, A., Arrivault, S., Stitt, M., Fernie, A.R. and Bauwe, H. (2012) Glycine decarboxylase controls photosynthesis and plant growth. *FEBS Lett.* **586**, 3692–3697.

Timm, S., Wittmiβ, M., Gamlien, S., Ewald, R., Florian, A., Frank, M., Wirtz, M. et al. (2015) Mitochondrial dihydrolipoyl dehydrogenase activity shapes photosynthesis and photorespiration of Arabidopsis thaliana. *Plant Cell* **27**, 1968–1984.

Timm, S., Florian, A., Fernie, A.R. and Bauwe, H. (2016) The regulatory interplay between photorespiration and photosynthesis. *J. Exp. Bot.* **67**, 2923–2929.

Tolbert, N.E. (1997) The C2 oxidative photosynthetic carbon cycle. *Annu. Rev. Plant Physiol. Plant Mol. Biol.* **48**, 1–25.

Uematsu, K., Suzuki, N., Iwamae, T., Inui, M. and Yukawa, H. (2012) Increased fructose 1,6-bisphosphate aldolase in plastids enhances growth and photosynthesis of tobacco plants. *J. Exp. Bot.* **63**, 3001–3009.

Walker, B.J., Van Locke, A., Bernacchi, C.J. and Ort, D.R. (2015) The costs of photorespiration to food production now and in the future. *Annu. Rev. Plant Biol.* **67**, 107–129.

Walker, B.J., South, P.F. and Ort, D.R. (2016) Physiological evidence for plasticity in glycolate/glycerate transport during photorespiration. *Photosynth. Res.* **129**, 93–103.

Whitney, S.M., Houtz, R.L. and Alonso, H. (2011) Advancing our understanding and capacity to engineer nature's CO_2-sequestering enzyme, Rubisco. *Plant Physiol.* **155**, 27–35.

Xin, C.P., Tholen, D., Devloo, V. and Zhu, X.G. (2015) The benefits of photorespiratory bypasses: how can they work? *Plant Physiol.* **167**, 574–585.

Zhu, X.G., de Sturler, E. and Long, S.P. (2007) Optimizing the distribution of resources between enzymes of carbon metabolism can dramatically increase photosynthetic rate: a numerical simulation using an evolutionary algorithm. *Plant Physiol.* **145**, 513–526.

Zhu, X.G., Long, S.P. and Ort, D.R. (2010) Improving photosynthetic efficiency for greater yield. *Annu. Rev. Plant Biol.* **61**, 235–261.

13

Elevated acetyl-CoA by amino acid recycling fuels microalgal neutral lipid accumulation in exponential growth phase for biofuel production

Lina Yao[1], Hui Shen[1], Nan Wang[2], Jaspaul Tatlay[2], Liang Li[2], Tin Wee Tan[3,4] and Yuan Kun Lee[1,*]

[1]Department of Microbiology and Immunology, Yong Loo Lin School of Medicine, National University of Singapore, Singapore, Singapore
[2]Department of Chemistry, University of Alberta, Edmonton, Alberta, Canada
[3]Department of Biochemistry, Yong Loo Lin School of Medicine, National University of Singapore, Singapore, Singapore
[4]National Supercomputing Centre (NSCC), Singapore, Singapore

*Correspondence
email yuan_kun_lee@nuhs.edu.sg

Summary

Microalgal neutral lipids [mainly in the form of triacylglycerols (TAGs)], feasible substrates for biofuel, are typically accumulated during the stationary growth phase. To make microalgal biofuels economically competitive with fossil fuels, generating strains that trigger TAG accumulation from the exponential growth phase is a promising biological approach. The regulatory mechanisms to trigger TAG accumulation from the exponential growth phase (TAEP) are important to be uncovered for advancing economic feasibility. Through the inhibition of pyruvate dehydrogenase kinase by sodium dichloroacetate, acetyl-CoA level increased, resulting in TAEP in microalga *Dunaliella tertiolecta*. We further reported refilling of acetyl-CoA pool through branched-chain amino acid catabolism contributed to an overall sixfold TAEP with marginal compromise (4%) on growth in a TAG-rich *D. tertiolecta* mutant from targeted screening. Herein, a three-step α loop-integrated metabolic model is introduced to shed lights on the neutral lipid regulatory mechanism. This article provides novel approaches to compress lipid production phase and heightens lipid productivity and photosynthetic carbon capture via enhancing acetyl-CoA level, which would optimize renewable microalgal biofuel to fulfil the demanding fuel market.

Keywords: microalga, TAG, growth phase, acetyl-CoA, BCAA, biofuel.

Introduction

To replace traditional fossil fuels and develop sustainable energy production, identifying sources of biologically derived fuels is increasingly urgent. Microalgae is recognized as a promising alternative source as they can accumulate neutral lipid, mainly in the form of triacylglycerol (TAG), which can be converted into biodiesels readily (Hossain *et al.*, 2008). In recent years, many attempts have been undertaken for the enhancement of TAG overproduction in microalgae (Radakovits *et al.*, 2010). These approaches mainly focus on biochemical and genetic engineering of lipid biosynthesis pathways and blocking of competing pathways (such as carbohydrate formation), so as to increase the pool of metabolites available for TAG biosynthesis (Courchesne *et al.*, 2009; Sharma *et al.*, 2012). However, almost all these approaches led to TAG accumulation in the stationary growth phase at the expense of biomass accumulation (Chiu *et al.*, 2009; Wang *et al.*, 2009) and overall lipid productivity. Vigorous growth and TAG accumulation appear to be mutually exclusive as TAG is a secondary (storage) metabolite and the pyruvate to acetyl-CoA (AcCoA) pathway is tightly regulated by the growth-dependent pyruvate dehydrogenase complex activity (Li *et al.*, 2014; Oliver *et al.*, 2009).

The oleaginous diatom *Fistulifera solaris* JPCC DA0580 was the first to be reported to have a temporal overlap of TAG accumulation and cell growth during the exponential growth phase (Satoh *et al.*, 2013). Such a feature that triggers TAG accumulation while maintaining high growth rate is a critical advantage in the large-scale cultivation of oleaginous microalgae for TAG production. To further exploit this potential in microalgae, fast-growing, TAG-rich, easily cultivated *Dunaliella tertiolecta* was used as the experimental organism (Rismani-Yazdi *et al.*, 2011; Shin *et al.*, 2015; Yao *et al.*, 2015).

In microalgae, AcCoA, malonyl-CoA and NADPH are the major substrates in the plastid supporting fatty acid synthesis. Malonyl-CoA is also generated from carboxylation of AcCoA. Thus, AcCoA is the primary precursor for fatty acid synthesis (Garrett and Grisham, 2013). The AcCoA balance in an algal cell could be described by the following equation:

$$[AcCoA_T] - [AcCoA_B] = [AcCoA_{NL}]$$

Total AcCoA ($[AcCoA_T]$) and reduced NADH are produced via glycolysis. In the exponential growth phase, NADH is mainly oxidized through respiration to yield ATP, and AcCoA is used predominantly for biomass growth ($[AcCoA_B]$), including that for structural lipid (glycerophospholipids) synthesis, while a minor fraction of AcCoA ($[AcCoA_{NL}]$) and reduced NAD(P)H is used for fatty acid synthesis to accumulate neutral lipids (TAG). When microalgal cells enter the stationary growth phase, carbon metabolism for biomass growth diminished, which leads to accumulation of AcCoA and reduced NADH. Thus, in stationary phases or growth hindering stress conditions, a conspicuous fraction of AcCoA and reduced NAD(P)H is channelled to fatty acids biosynthesis, resulting in TAG accumulation in cells

(Carpinelli *et al.*, 2014). Recent studies also suggested that intracellular membrane remodelling contributed to TAG accumulation during stationary phase or nitrogen starvation (Simionato *et al.*, 2013; Urzica *et al.*, 2013; Yoon *et al.*, 2012). To accelerate TAG accumulation in exponential growth phase (TAEP) while maintaining cell growth, AcCoA ([AcCoA$_T$]) level should be elevated over a certain set point that is needed for biomass growth ([AcCoA$_B$]).

There are three principal sources of AcCoA during growth phase, namely fatty acid oxidation, glycolysis pathway and amino acid degradation (Garrett and Grisham, 2013). Fatty acid oxidation is the reversal of fatty acid synthesis and does not generate *de novo* AcCoA. Instead, it is thought that AcCoA is largely derived from the glycolytic pathway via pyruvate. Pyruvate is converted to AcCoA by PDHC in mitochondria and chloroplasts, and this step has been suggested as the key rate limiting step (Garrett and Grisham, 2013; Oliver *et al.*, 2009; Tovar-Méndez *et al.*, 2003). One approach to increase AcCoA production is to relieve pyruvate dehydrogenase kinase (PDK) control of pyruvate dehydrogenase complex (PDHC) resulting in the activation of PDHC. This would facilitate the bioconversion of pyruvate to AcCoA and enhance the metabolic flux towards both cell growth via the TCA cycle, and fatty acid biosynthesis in the growth phase. The third source of AcCoA, which derived from amino acid degradation, has largely been ignored as a relevant pathway for bioengineering. Despite the fact that it bypasses pyruvate and the highly controlled PDHC/PDK regulatory process, it was considered insufficient for fatty acid biosynthesis (Garrett and Grisham, 2013).

We hypothesized that increase of AcCoA pool by multiple routes could trigger TAEP. In our study, from the activation of pyruvate to AcCoA reaction by addition of sodium dichloroacetate (DCA) to release the PDHC/PDK regulatory process, we achieved TAEP in the wild-type (WT) *D. tertiolecta*. Besides this conventional *de novo* synthetic pathway, we questioned the contribution of amino acid degradation on TAEP, although it has largely been ignored. Through performing genetic engineering, we generated mutants, which exhibited pronounced TAEP with little compromise on growth rate. By employing transcriptomics and metabolomics, key phenotypic regulatory characteristics of lipogenesis in this microalga were uncovered, implying that a secondary contributor of AcCoA derived from amino acid catabolism, in particular branched-chain amino acid catabolism, contributed to TAEP. Although no direct transport of AcCoA between subcellular compartments was reported in plant cells, a PDHC bypass pathway from activation of free acetate into AcCoA exists (Li-Beisson *et al.*, 2013; Lin and Oliver, 2008). These two major approaches were proposed in our three-step α loop model. The results highlight the complex interplay between microalgal cellular proliferation and carbon flux in lipogenesis and suggested that genetic and metabolic manipulations targeted at amino acid catabolism could be used to increase accumulation of fuel-relevant molecules in microalgae in the exponential growth phase.

Results

DCA treatment elevated AcCoA pool

After addition of DCA to the WT *D. tertiolecta*, TAG was found to be accumulated in the exponential growth phase, as shown in Figure S1a, with marginal comprise on growth (Figure S1b).

AcCoA, the primary precursor for growth and fatty acid synthesis, was found 1.8-fold that in the control (Figure S1c).

AcCoA is *de novo* converted from pyruvate, which is tightly regulated by PDHC, which catalyses the oxidative decarboxylation of pyruvate. PDHC could be deactivated by PDK through reversible ATP-dependent phosphorylation mainly in mitochondria (Kato *et al.*, 2007). To deactivate PDK, DCA was added to the algal culture medium. DCA is a by-product of chlorine disinfection process, which inhibit PDK, through formation of DCA helix bundle in the N-terminal domain of PDK (Kato *et al.*, 2007; Miller and Uden, 1983). Bound DCA promotes local conformational changes that are communicated to both nucleotide-binding and lipoyl-binding pockets of PDK, leading to the inactivation of kinase activity (Kato *et al.*, 2007). Thus, when DCA was included in the culture medium, PDHC became active as PDK was blocked resulting in an elevation of AcCoA (Figure S1c).

FACS enriched a pool of mutant strains with higher TAG content

We generated TAG-rich mutant library via genetic engineering and two rounds of fluorescence-activated cell sorting (FACS) (Terashima *et al.*, 2015). All the 27 isolated strains showed reproducible increase in Nile red signal (Figure S2e). Further observation on top six mutants showed consistent higher TAG content with a statistical significant *P* value <0.01 (*t*-test) comparing to the WT (Figure S2f). Among all the mutants, we selected the stable mutant strain G11_7, which was one of the higher TAG producers at its exponential growth phase for characterization. G11_7 mutant accumulated TAG at its exponential growth phase (Figure 1b) with marginal difference on growth (Figure 1a). On culture day 4, *P* value of TAG accumulation per biomass from the two strains is 0.0013 (**), and the rest of the time points are all less than 0.001 (***), indicating that TAG accumulation per biomass between G11_7 and WT is significantly different throughout all the culture points (Figure 1b). It consistently showed enhanced TAG production (about twofold to sevenfold) compared to WT *D. tertiolecta* (WT). In addition, the mutant had a significant better photosynthetic performance at the same light condition (Figure 1c, 33% higher than WT). It is evidence that the mutant has enhanced energy/carbon capture capacity. Lipid droplets (LPs) were visualized by fluorescent microscopy (Figure 1d) in G11_7 mutant, showing golden-yellow fluorescence. The TAG accumulation in cultures grown under high-light condition showed a similar trend as low-light condition (Figure 1e,f). The fatty acid composition of the mutant and WT is presented in Figure S3a, with a typical profile of unsaturated fatty acids 16 : 1 and 18 : 3(n-3) being the predominant fatty acids. There is a significant increase in the monounsaturated fatty acid (MUFA) with most others remained similarly (Figure S3b).

Altered expression level of genes in amino acid catabolism in G11_7 at the exponential growth phase

Differential expressed genes are presented in Supplementary Data Set 1. KEGG enrichment scores were calculated and shown in Figure 2 (Chen *et al.*, 2015; Kanehisa *et al.*, 2016). There are in total 105 KEGG enrichment scores featured in the three predicted analyses, among which 19 KEGG pathways were found significant in at least two enrichment analyses according to scores. The important genes that altered in expression levels in the mutant as

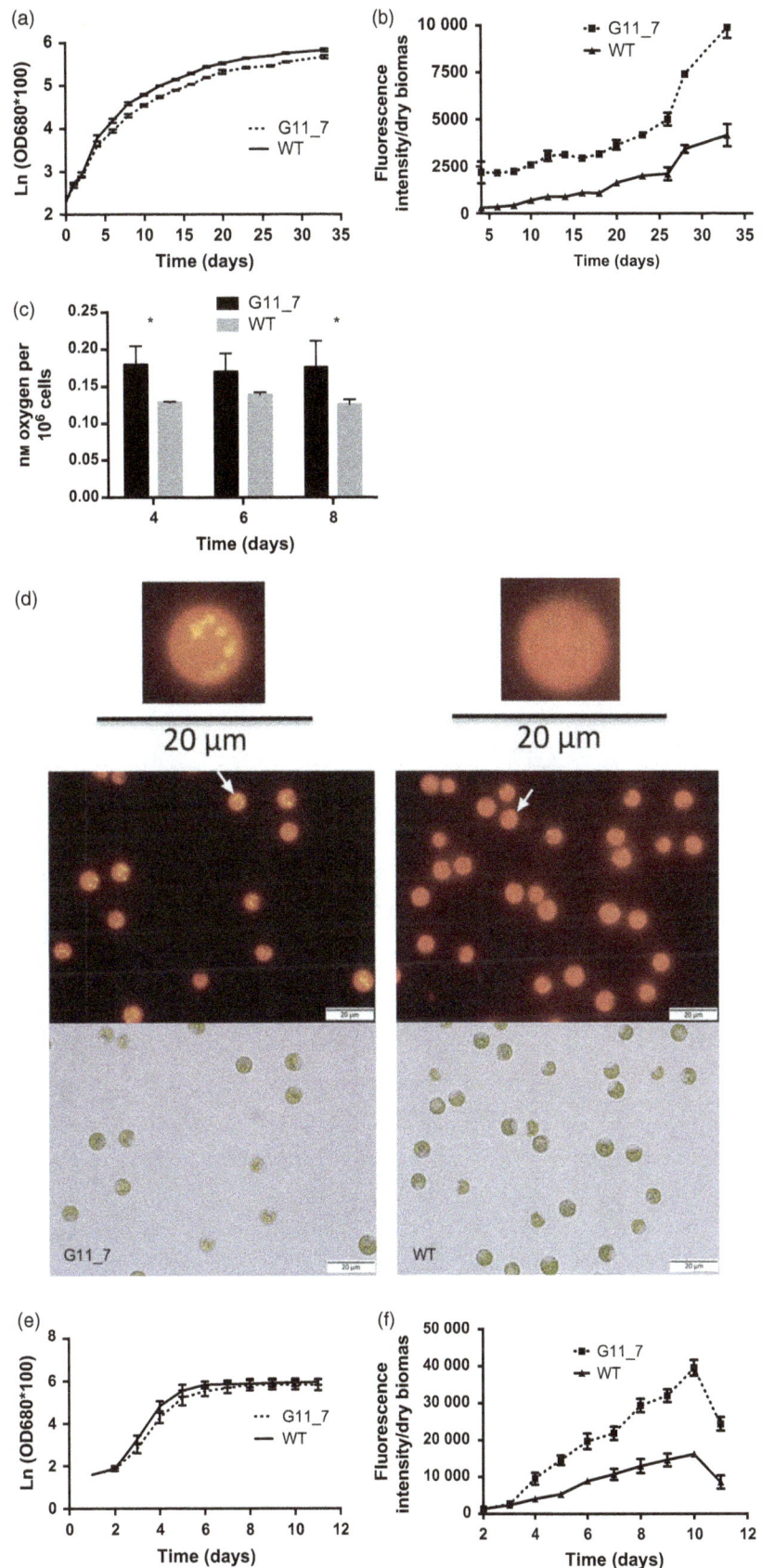

Figure 1 Physiological performance of G11_7 mutant versus WT *Dunaliella tertiolecta*. (a) Growth curve monitored by spectrophotometry under low light. (b) TAG quantitative assay by Nile red staining method under low light. (c) Photosynthetic rate of G11_7 mutant and WT *D. tertiolecta*. (d) Microscopy images (above, Nile red fluorescence; bottom, bright field) of G11_7 mutant (left) and WT (right) under low light (under 100× objective). (e) Growth curve monitored by spectrophotometry under high light. (f) TAG quantitative assay by Nile red staining method under high light. (LL: 30 μmol photons m^{-2} s^{-1}, HL: 320 μmol photons m^{-2} s^{-1}). Error bars, SEM. Statistical analyses were performed using Student's *t*-test, * $0.01 \leq P < 0.05$; ** $0.001 \leq P < 0.01$; *** $P < 0.001$.

compared to the WT are summarized in Supplementary Data Set 1. Valine, leucine and isoleucine (branched-chain amino acids, BCAA) degradation pathway is the most significantly affected

pathway detected in all three analyses. *CuAO* (or *AMX1*, K00276) (copper amine oxidase family), *IVD* (K00253) (isovaleryl-CoA-dehydrogenase) and *MCCB* (K01969) (3-methylcrotonyl-CoA carboxylase) were the top hits at expression level in both Partek analysis and in-house workflow.

The complete cDNA sequence of *D. tertiolecta* copper amine oxidase gene (1524-bp encoding for 507 amino acids), isovaleryl-CoA dehydrogenase gene (1053-bp encoding for 350 amino acids) and 3-methylcrotonyl-CoA carboxylase beta subunit gene (1725-bp encoding for 574 amino acids) was obtained using RACE PCR. The candidate *D. tertiolecta* copper amine oxidase gene contained a copper amine oxidase enzyme domain and showed the highest (60%) amino acid homology compared to copper amine oxidase of *Volvox carteri* f. nagariensis, and was designated *DtCuAO* (*AMX1*). The candidate *D. tertiolecta* iso-valeryl-CoA dehydrogenase gene contained an isovaleryl-CoA dehydrogenase domain and showed the highest (76%) homology

in amino acid sequence compared to that of *Chlamydomonas reinhardtii*, and was designated *DtIVD* (*ACAD*). The candidate *D. tertiolecta* 3-methylcrotonyl-CoA carboxylase beta subunit gene contained a 3-methylcrotonyl-CoA carboxylase beta chain domain and showed the highest (67%) homology in amino acid sequence compared to that of *C. reinhardtii*, and was designated *DtMCCB*. Amino acid sequences of *CuAO*, *IVD* and *MCCB* from other species were obtained by BLAST search in NCBI database with the putative DtCuAO, DtIVD, DtMCCB. The phylogenetic tree constructed by MEGA5 demonstrated that the putative DtCuAO, DtIVD and DtMCCB showed high homology with CuAO, IVD and MCCB, respectively, from other species (Figure S4). Using the predicted sequence of these three genes from the *D. tertiolecta* database, we designed primers and

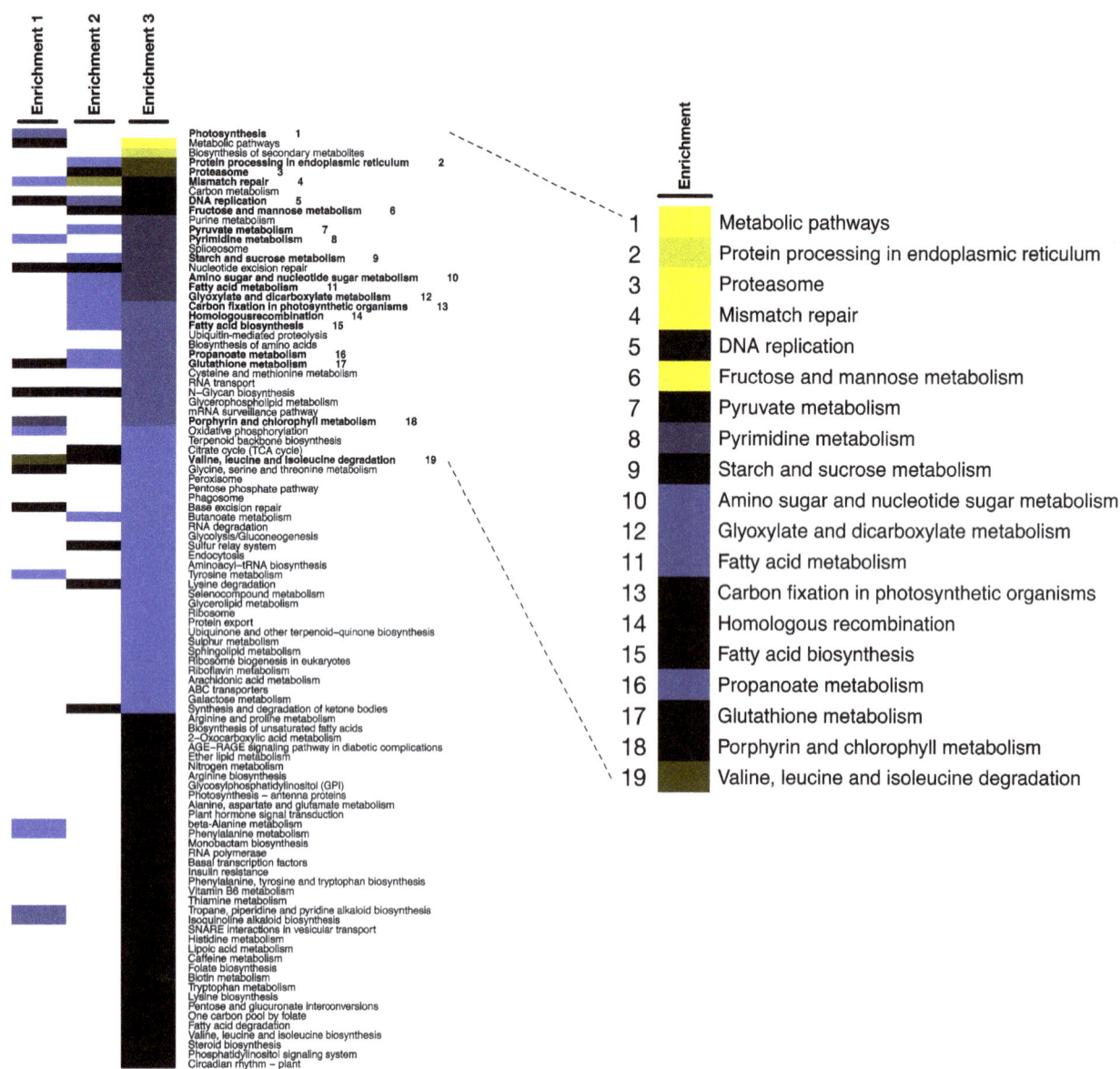

Figure 2 Heat map of the KEGG profiles. Colours represent abundance of KEGG enrichment score (blue to yellow, ascending enrichment score). Rows of the heat map represent pathways from three analyses (Enrichment 1: Partek workflow; Enrichment 2: in-house workflow), background KEGG gene that derived from *C. reinhardtii* as the reference was set at 3321; Enrichment 3: in-house workflow, background KEGG gene derived from *D. tertilecta* transcriptome nonredundant database was set at 10,579. Images were made with R (http://www.r-project.org/). Significant KEGG pathways are marked by bold font with number on the left and expansion on the right.

confirmed by high-fidelity DNA polymerases sequencing. The cDNA sequences of the three genes were also attached as Supplementary Data Set 3a-c. We arbitrarily tested two genes for their expression levels by quantitative real-time PCR at different growth phases (Figure S5a) using the *D. tertiolecta* beta-tubulin gene (DtTUB) as the internal standard for normalization (Lin et al., 2013). It showed a similar tendency with that of RNA-Seq data. Interestingly, the fold changes in the gene expression level and lipid level are correlated, suggesting that they are the key genes regulating the lipid accumulation process. The *DtPDK* gene that subsequently measured in the DCA treatment experiment was discovered by the prediction from the *D. tertiolecta* in-house database (known as Locus_5000_7Transcript_1/1_Confidence_1.000_Length_2194) and confirmed by experimental sequencing using high-fidelity DNA polymerases and shown in Supplementary Data Set 3d. Interestingly, we found mRNA expression levels of these three important genes, *DtIVD*, *DtCuAO* and *DtMCCB*, were all up-regulated in other high-TAG strains (G11_18, G11_20, G11_25) as shown in Figure S5b.

In the upstream pathways, up-regulation of *ACCA* (K00626) (acetyl-CoA C-acetyltransferase) was also detected,

which resulted in the accumulation of AcCoA. Up-regulation of the downstream *FabD* (K00645) (malonyl-CoA:acyl-carrier-protein transacylase) contributed to accumulation of fatty acids in chloroplasts. The fast fatty acid accumulation might cause a drawn-down in AcCoA, thus initiating a pull-down from upstream photosynthetic pathways. The pull-down may have caused an enhancement of photosynthetic rate (Figure 1c) to support AcCoA *de novo* synthesis, which was supported by the up-regulation of upstream photosynthesis and glycolysis genes, including *petE* (K02638) (photosynthetic electron transport), *petC* (K02636) (cytochrome b6-f complex), *LHCA1* (K08907) and *LHCB1* (K08912) (light-harvesting complex I chlorophyll a/b binding protein 1), *pfkA* (K00850) (6-phosphofructokinase 1), *PPC* (K01610) (phosphoenolpyruvate carboxykinase (ATP) and *FBP* (K03841) (fructose-1,6-bisphosphatase I).

AcCoA pool was maintained at high level in the mutant

According to the congruent transcriptome and metabolome analysis (Table S1), the tentative pathways for channelling of metabolites towards fatty acid syntheses are depicted in Figure 3. AcCoA was found 1.3-fold in G11_7 mutant.

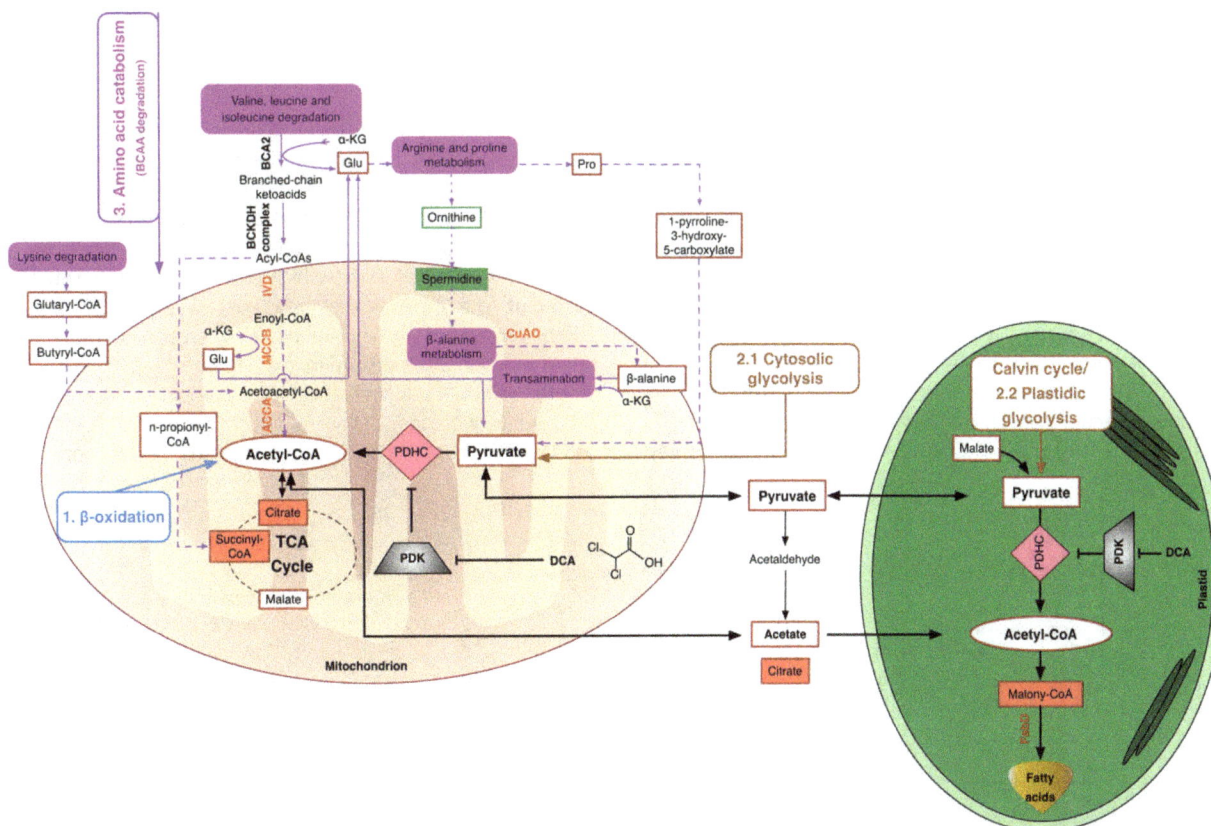

Figure 3 Hypothesized *D. tertiolecta* G11_7 mutant fatty acid metabolic pathways. Enzyme names and intermediates are abbreviated in the figure. Validated overexpressed gene-coding enzymes are shown in red colour font sitting on the pathway lines. Validated significantly increased intermediates are surrounded by red solid line rectangles, and decreased intermediates are surrounded by green solid line rectangles. The increased/decreased intermediates with a fold change ≥2 or ≤−2 are shown in a filled rectangles with red/green colour. The general three sources of acetyl-CoA are 1, β-oxidation, 2, glycolysis and 3: amino acid catabolism. The specific core pathways use solid arrows for one-step reactions and dash arrows for multiple-step reactions. The DCA treatment-related pathways have two critical enzymes, PDH and PDK in pink prismatic and grey trapezoid shapes, respectively. Note: The above showed enzyme and intermediate abbreviations are listed below: BCA2, branched-chain amino acid aminotransferase; BCKDH, branched-chain ketoacid dehydrogenase; IVD, isovaleryl-CoA dehydrogenase; MCCB, 3-methylcrotonyl-CoA carboxylase beta unit; ACCA: acetyl-CoA C-acetyltransferase; AMX1, copper amine oxidase family protein. PDHC, pyruvate dehydrogenase complex; PDK, pyruvate dehydrogenase kinase; FabD: malonyl-CoA:acyl-carrier-protein transacylase; α-KG, α-ketoglutarate; Glu, Glutamate; Pro, Proline.

Pyruvate, malate, proline and 1-pyrroline-3-hydroxy-5-carboxylate were accumulated in G11_7 mutant, suggesting the relief of PDHC/PDK regulatory process was necessary for efficient bioconversion of the pyruvate-dependent fatty acid precursors. To achieve this, DCA was further added to the mutant culture medium. TAG accumulation and growth curve in the DCA-treated G11_7 mutant under both low-light and high-light conditions are shown in Figure 4a-b. There was a significant enhancement of TAG accumulation detected in DCA-treated mutant cells from exponential growth phase, with an elevated AcCoA level (Figure 4c). Consistently, mRNA expression level of the *DtPDK* was significantly reduced in DCA-treated groups (Figure 4d). It was also reasoned that under the condition of excess energy supply (in HL), the overflow of AcCoA towards TAG would be more pronounced. Indeed, it was observed that TAG accumulation was further induced when the G11_7 cells (after DCA treatment) entered the stationary phase from day 6 in the HL culture.

BCAAs fuelled acetyl-CoA production for TCA metabolism and lipogenesis

In contrast to the WT, the mutant G11_7 has an activated BCAA catabolism via intensive up-regulation of IVD, MCCB and ACCA, resulting in an enhancement of the AcCoA pool. The increase in glutaryl-CoA (1.8-fold) and butyryl-CoA (1.3-fold) participating in the lysine degradation pathway also contributed to AcCoA via the up-regulation of ACCA. The acetyl residue of AcCoA enters the TCA cycle by reaction with oxaloacetate and subsequently incorporated into citrate (2.12-fold). According to the targeted CoA analyses, n-propionyl-CoA, a prerequisite for succinyl-CoA, had a 1.7-fold increase in the mutant. Propionyl-CoA is known to be carboxylated to generate methylmalonyl-CoA, which is racemized and then isomerized to form succinyl-CoA (5.9-fold), a member of the TCA cycle (Ge *et al.*, 2014). The flux from both citrates in the TCA cycle and propionyl-CoA accounts for the massive increase of succinyl-CoA in the mutant.

Thus, besides priming the TCA cycle for cell growth, the overproduced AcCoA was channelled into the fatty acid reservoir. The enhancement of malonyl-CoA (2.1-fold), which plays a key

role in the fatty acid biosynthesis and chain elongation, provides the evidence for this.

BCAA catabolic flux was activated by the aforementioned genes, particularly in the leucine degradation pathway. To confirm the contribution of leucine in TAG synthesis in BCAA catabolism, G11_7 mutant and WT were given a spike of different concentrations of leucine in the normal ATCC medium. In contrast to the WT, which did not metabolize leucine as the precursor for cell growth and lipid production, G11_7 mutant strain showed increase TAG production in a dose-dependent manner (Figure 5). Increase in TAG content with addition of leucine in the medium accounted for as much as 39% of TAG pools.

These data provide experimental evidence for the functional role of BCAA catabolism, reinforcing its importance in regulating lipogenesis in the exponential growth phase. There might be bottlenecks in the synthesis of structural carbohydrates, proteins and lipids for additional cell growth, as no increase in cell growth rate was observed. Instead, pronounced increase of fatty acid was detected providing a hint that fatty acid biosynthesis-related processes might not be rate limiting during the exponential cell growth phase, and AcCoA generated from the BCAA pathway was transported into chloroplasts for fatty acid accumulation in the mutant.

Discussion

The primary physiological purpose of amino acids is to serve as the building blocks for protein biosynthesis in eukaryotic cells, and as a consequence, the amount of free amino acids is trivial under most circumstances especially in cultures under high growth rate (Garrett and Grisham, 2013). Amino acids are derived from the TCA cycle, which provides carbon skeletons via 2-oxoglutarate or oxaloacetate (Kanehisa *et al.*, 2016; Lane, 2009; Wagner, 2014). New amino acids can also be formed from transamination by transferring the amino group to a ketoacid (Booth, 2000). In our case, a significant proportion of free amino acids were degraded, driven by transcriptional up-regulation of *DtIVD*, *DtMCCB* and *DtACCA* encoding for the key enzymes in

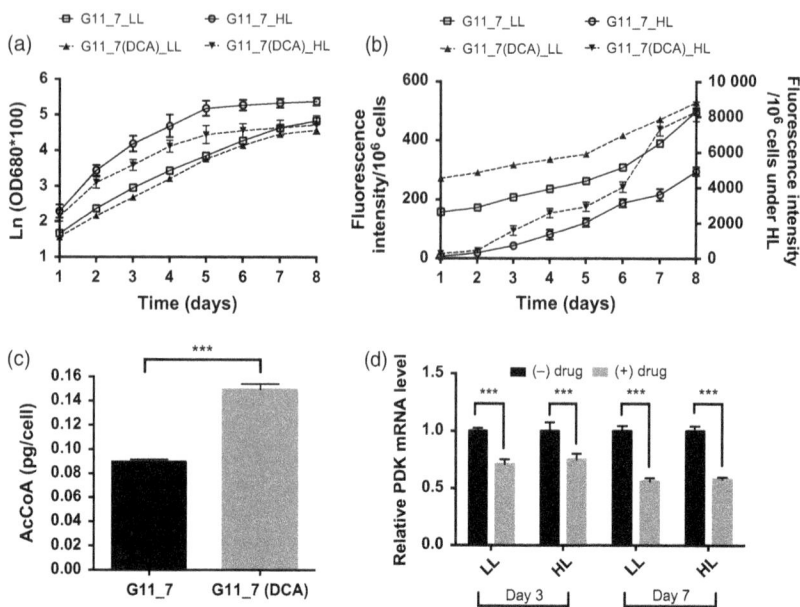

Figure 4 Physiological study and DtPDK mRNA expression levels of G11_7 mutant after DCA treatment. (a) Growth curve monitored by spectrophotometry under low-light (LL) and high-light (HL) conditions (LL: 30 μmol photons m^{-2} s^{-1}, HL: 320 μmol photons m^{-2} s^{-1}). (b) TAG quantitative assay by Nile red staining method under low-light (LL) and high-light (HL) conditions (left y axis shows the LL fluorescent performance, right y axis shows the HL fluorescent performance). (c) Acetyl-CoA level change after addition of DCA to G11_7 under the low-light condition. (d) Relative DtPDK mRNA expression levels on day 3 and day 7 under different light conditions (LL: 30 μmol photons m^{-2} s^{-1}, HL: 320 μmol photons m^{-2} s^{-1}). Error bars, SEM. Statistical analyses were performed using Student's *t*-test, * $0.01 \leq P < 0.05$; ** $0.001 \leq P < 0.01$; *** $P < 0.001$.

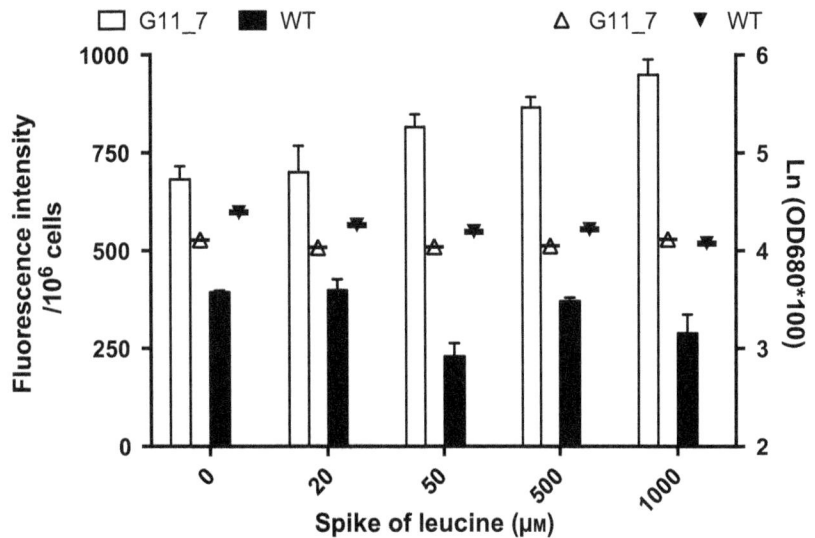

Figure 5 Effect of leucine spike to TAG accumulation and culture growth. TAG contents of G11_7 and WT are presented in bars (primary axis, white bar—G11_7, black bar—WT), and the corresponding culture growth is presented in scatter dot plots (secondary axis, white upward triangle—G11_7, black downward triangle—WT) under different leucine concentration, with all P value <0.001.

the BCAA catabolic pathway, leading to AcCoA synthesis. This strategy shows the feasibility of using the aforementioned third source for AcCoA, which was previously ignored.

The functional role of BCAA catabolic process in lipogenesis has been demonstrated in other various organisms. In the diatom, *Phaeodactylum tricornutum*, inhibition of *MCCB* expression by RNA interference disturbed the carbon flux, resulting in decreased TAG accumulation and impaired biomass growth (Ge et al., 2014). Green et al. (2016) highlighted the contribution of BCAAs to adipocyte metabolism in mouse cell line (3T3-L1 cells) and demonstrated that amino acids (BCAAs in particular) from both extracellular sources and protein catabolism were highly utilized by differentiated adipocytes. Inhibition of BCAA catabolism negatively influenced 3T3-L1 adipogenesis. In the study of Peng et al. (2015), BCAA catabolic mutants defective in enzymes both upstream and downstream of IVD displayed enhanced senescence in prolonged darkness, showing that function of BCAA catabolism in providing TCA cycle substrates in energy-limited conditions. It also demonstrated that IVD influences energy homeostasis in multiple ways, providing BCAA catabolic CoA intermediates to the mitochondrial electron transport chain, as well as catabolizing additional substrates such as phytanoyl-CoA and aromatic amino acids (Araújo et al., 2010; Ishizaki et al., 2005).

Interestingly, other amino acid catabolic pathways were found in concordant in contributing to the TCA cycle and AcCoA production. Lysine metabolism was previously demonstrated to interact with plant energy metabolism (Angelovici et al., 2011). Vorapreeda et al. (2012) also reported that leucine and lysine degradation in oleaginous fungi provided the alternative substrate for AcCoA as the precursor for lipid production, by contrast to that in nonoleaginous fungi. This is confirmed by the free amino acid (leucine) uptake study reported here. Free amino acid uptake (transport, assimilation /accumulation) and excretion had been observed in microalgae (Flynn and Butler, 1986; Huo et al., 2011). Three exogenous transamination and deamination cycles introduced by Huo et al. (2011) also reengineered carbon flux for fuel-convertible amino acids and enabled protein hydrolysates to be used for fuel production.

In conclusion, we have studied a TAG-rich mutant strain of *D. tertiolecta* under controlled laboratory conditions to advance

our understanding of lipid metabolic pathways in the growth phase. Our study revealed a 'three-step α loop' model to elucidate lipogenesis in the exponential growth phase as shown in Figure 6 and summarized below.

The vertical path: Under normal conditions, a number of key enzymes in microalgae supply carbon precursors for *de novo* fatty acid synthesis, which include those involved in PDHC, glycolysis and suites of specific transporters. They were found substantially up-regulated under nitrogen deprivation conditions (Li et al., 2014). This pyruvate-dependent glycolysis pathway is tightly regulated by cell growth via PDHC/PDK cascade (Figure 6, route 1, 2). AcCoA and NADPH produced from this pathway are readily used by the TCA cycle to produce amino acids, macromolecules and energy for biomass growth. During this process, a minor fraction of AcCoA and NADPH is used for fatty acids synthesis (Carpinelli et al., 2014).

The back loop: While in G11_7 mutant, genes involved in the amino acid catabolism in mitochondria were enhanced. The overflow free amino acids and catabolized proteins were channelled into AcCoA. This provides an additional source of AcCoA for fatty acid synthesis (Figure 6, route 5, 6). The recycling of amino acids at a moderate level led to a temporal increase in AcCoA concentration, with little compromise on the biomass growth rate (4%). This elevated AcCoA concentration (30%) exceeds the demand (set point) for biomass growth, which shunts the carbon and energy precursors to the fatty acid synthesis route (Figure 6, route 7). Genes responsible for fatty acid synthesis were constantly overexpressed at the mRNA level, and this equilibrium leads to ultimate generation of TAG in the growth phase.

The pull-down: In response to the drawn-down of [AcCoA] for lipogenesis, a pull-down of carbon flux from photosynthetic process takes effect, leading to the increase in photosynthetic rate (33%), and overexpression of genes participating in photosynthesis and glycolysis. The inclusion of DCA removed the regulation of PDHC, leading to an increase in AcCoA and accumulation in TAG (42%) (Figure 6, route 8). In consistency with our observation, bioengineering manipulation, such as antisense knock-down of PDK in the diatom *P. tricornutum*,

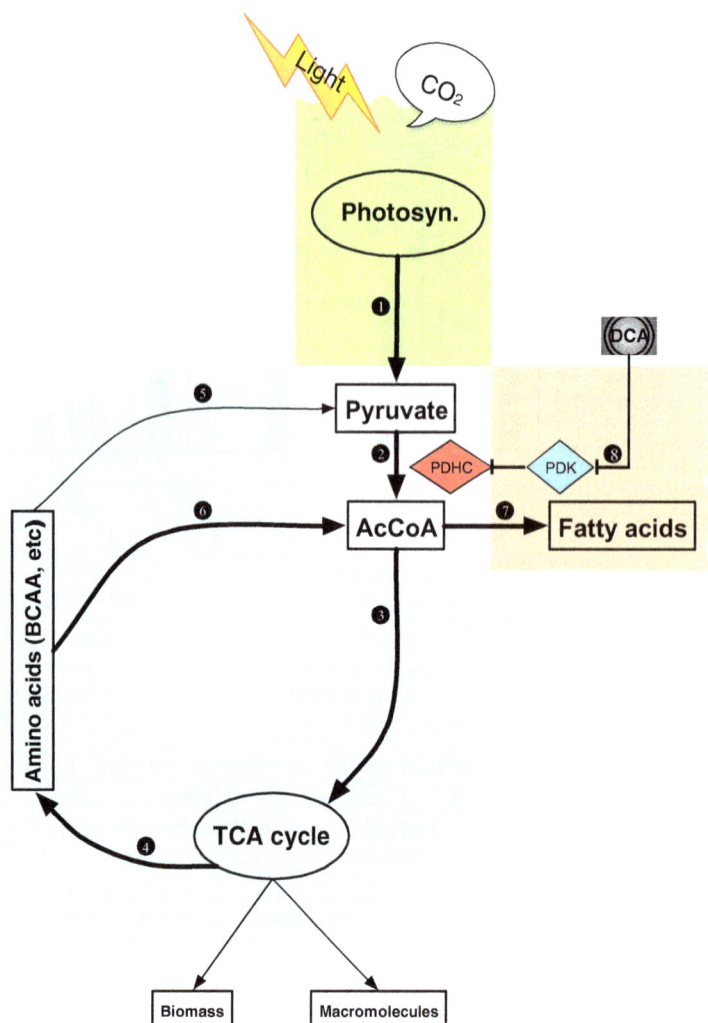

Figure 6 Regulation of metabolic pathways related to energy/carbon capture and conversion in *Dunaliella tertiolecta* mutant G11_7.

was reported to promote TAG production by 82% (Ma *et al.*, 2014).

The further increase in the TAG accumulation in the stationary growth phase in HL under the treatment of DCA (PDHC activated) could be explained by the following: in the stationary phase, as cells no longer drawn-down AcCoA for growth, the carbon and energy generated from photosynthesis was channelled to TAG production. Moreover, the cell membrane lipid also serves as the source of BCAA, which is degraded and contributes to the AcCoA reservoir in the stationary growth phase.

The proposed 'three-step α loop' model suggests that the rate of lipogenesis in the growth phase is determined by the balance between the carbon/energy supply, biomass synthesis (growth) and amino acid catabolism. When the carbon metabolism is at high gears, a rapid catabolic rate would lead to accumulation of AcCoA and fatty acid synthesis, without compromise on biomass growth. However, an overrun amino acid catabolism would result in reduced growth rate, lower biomass concentration and ultimately lower TAG productivity of the culture. Collectively, TAEP in microalgae could be stimulated by elevated AcCoA level through multiple approaches. Besides aforementioned two major approaches, a balance for fatty acid oxidation is also of great interest to be investigated. To our knowledge, no genetic manipulation has been achieved in promoting TAG production

in the growth phase. Collectively, the deliberated investigation provides targets for metabolic engineering of eukaryotic microalgae for efficient lipid production and may inspire novel biofuel production technology based on growth-phase lipid-producing oleaginous microalgae as alternative biofuel feedstocks.

Experimental procedures

Strains and culture conditions

The algal culture *Dunaliella tertiolecta* strain UTEX LB-999 was obtained from the UTEX Culture Collection of Algae (University of Texas at Austin, TX). The microalgal cells were cultivated using ATCC-1174 DA liquid medium (American Type Culture Collection at Manassas, Virginia) containing 0.5M NaCl in flask with shaking at 100 r.p.m. at 25 °C, under 12-h light/12-h dark with light intensity of 30 μmol photons m^{-2} s^{-1}. Culture was supplied with 2% CO_2 every 2 days. For the purpose of clarity, work conducted throughout the article is based on biological and technical triplicates unless otherwise stated.

TAG content was measured using quick Nile red assay (Bertozzini *et al.*, 2011; Chen *et al.*, 2009, 2011; Yao *et al.*, 2015). The growth of microalgal cells was monitored by counting cell number and measuring absorbance at OD680 nm. The specific growth rate (μ) was calculated from the equation (Wahidin *et al.*, 2013):

$$\mu = \frac{\ln{(N2 - N1)}}{t2 - t1}$$

where $N2$ and $N1$ are the cell number concentration measured at times $t2$ and $t1$, respectively. The specific growth rate of the mutant was traced and presented in Figure S6.

Mutant isolation and optimization

Dunaliella tertiolecta transformation and random mutagenesis was conducted via electroporation (Shimogawara *et al.*, 1998) (double consecutive pulses were delivered at intervals of 10–15 s in a Bio-Rad Gene Pulser Xcell™ Electroporation Systems, with capacitance = 500 µFD, resistance = 400 Ω, voltage = 400 V), and 500 µL of cells was incubated with 50 µL of linearized plasmid (1 µg/10 µL, total amount 1–5 µg) and 5 µL of carrier DNA (fish sperm DNA) in 4-mm gap cuvette. The plasmid with a cassette conferring resistance to zeocin served to generate random insertional mutagenesis (Yao *et al.*, 2015). The Nile red data showed a relatively good correlation with data from total fatty acid methyl esters analysed by gas chromatography mass spectrometry (GC–MS) (R^2 = 0.97 for G11_7, R^2 = 0.92 for WT, Figure S7), indicating such an assay could be used as a quick high-throughput screening method for TAG-overproducing mutant strains, which was also tested and suggested by Xu *et al.* (2013).

Flow cytometric analyses of *Dunaliella* strains were performed on Nile red-stained cells on Beckman Coulter CyAN. High-TAG sorting strategy was applied to enable enrichment of high-TAG mutant (0.G11) (5th day of cultivation), by two rounds of FACS on pooled mutants stained with Nile red using Beckman Coulter Mo-Flo Legacy Cell Sorter (Terashima *et al.*, 2015). The same strain treated with acetone was used as the background. The distribution of an identically treated culture of WT was used as the control to determine the TAG gates for mutant pool sorting. A high-TAG gate that captured about 1.91% of WT with maximal Nile red fluorescence signal for a given chlorophyll fluorescence signal was performed (Figure S2a). The same gate captured about 18.99% of the cells in the 0.G11 mutant cells (Figure S2b). All cells that fell into this gate were collected into vials containing 0.5 mL 0.5 M NaCl ATCC medium. The cells were spun down at 750×***g*** for 5 min, resuspended in fresh 0.5M ATCC medium, and plated onto 0.5 M ATCC 1.5% agar plates and incubated for 2 weeks. The colonies were subsequently transferred to a flask for subculturing. Mutant pool (1.G11) and WT pool on day 5 were collected and used for second round of FACS. The second cell sorting on 1.G11 mutant and WT pools was carried out using the same amount of cells at a high-TAG gate (Figure S2c) and captured 11.07% of the cells in the mutant pool (Figure S2d). A wider Nile red and chlorophyll signal distribution in the pool of mutants indicates the presence of mutants that accumulate higher and lower amounts of TAG and chlorophyll. All cells that fell into this gate were collected into 96-well plates containing 0.5 M NaCl ATCC media and incubated under the same culture condition. The cells from each well were then transferred to flasks for further TAG analysis. Kinetic studies of TAG accumulation in mutants and WT *D. tertiolecta* were carried out using Nile red quantitative assay together with GC-MS technique (Yao *et al.*, 2015). On culture day 6, G11_7 mutant and WT were harvested to visualize the LPs using fluorescent microscopy (Olympus BX63, Tokyo, Japan) and measure the photosynthetic rate using an oxygen electrode according to the operating manual (Rank Brothers, Bottisham, Cambridge, UK).

RNA extraction and cDNA preparation

Dunaliella tertiolecta was collected for total RNA extraction using an RNeasy plant mini kit (Qiagen, Valencia, CA), according to the manufacturer's instructions. DNase was added to eliminate genomic DNA contamination. For quantitative real-time PCR, total RNA was used to synthesize random hexamer-primed cDNA according to the manufacturer's instructions.

Next-generation sequencing analysis

RNA was extracted from G11_7 and WT on culture day 6 in their exponential growth phase (LL). The quality was verified using Agilent RNA 6000 Nano Kit in Agilent 2100 bioanalyzer (Agilent Technologies, Palo Alto, CA) and gel electrophoresis. The samples were linearized with 0.1N NaOH into single-stranded forms, and they were then neutralized and diluted into 200pM loading concentration with Examp master mix (EPX1 to 3) and loaded into one lane of paired-end flowcell using the Illumina cBOT machine. The DNA was attached and amplified simultaneously inside each oligo well on the flowcell surfaces, as a proprietary clustering method known as exclusion amplification. This method ensures that only a single DNA template binds and forms a cluster within a single well, reducing the occurrence of polyclonal wells thus increasing the usable reads. The sequencing primer was then attached to the reads, preparing for sequencing run. The flowcell was then loaded into the Illumina HISEQ4000 High Sequencers with the sequencing reagents and run at 2 × 151 cycles, and the second read turnaround was carried out using the sequencers after the first read was completed. The images were captured by the HiSeq Control Software (HCS), and the Real Time Analysis (RTA) software converted the images into Cycle Intensity Files (CIF) and later Basecall (bcl) files. All the bcl files were then transferred to the server for storage and primary analysis. In the primary analysis, the bcl files were converted into fastq files using the bcl2fastq pipeline. After the conversion, the fastq reads were filtered to remove all the reads that did not pass filtering, leaving only useable passed filtered (PF) reads. The usable reads were then analysed and bin to each of the barcode file known as demultiplexing, and those that did not pass the filtering are not demultiplexed. The primary analysis result was then generated as the demultiplexed report and reviewed. The paired-end raw data (150 bps in length/read) of G11_7 and WT were trimmed of adaptors and examined with pre-alignment QA/QC (trimmed from both ends based on the parameter setting: Min read length = 25; Quality encoding = Auto detect; End min quality level (Phred) = 20) (Table S2) in Partek® Flow® software (version 4.0, Partek Inc., St. Louis, MO) with Dt_v10 *D. tertiolecta* transcriptome database (available on author's website: https://github.com/SPURc-Lab) as the reference using reads per kilobase of transcript per million mapped reads (RPKM). Subsequently, the different expressed transcripts were imported into Partek® Genomics Suite® software (version 6.6, Partek Inc.) for gene annotation and KEGG pathway analysis (Yao *et al.*, 2015). We also used an in-house pipeline to analyse the RNA-Seq data. The transcriptome database was enlarged via Trinity assembler (Grabherr *et al.*, 2011) and BLASTX against reference protein sequences from all plants and bacterial from National Center for Biotechnology Information (NCBI; Shin *et al.*, 2015). Transcript level comparison was performed using RSEM (Li and Dewey, 2011) with default settings. The count data from RSEM were imported for normalization in the Ebseq pipeline (Leng *et al.*, 2013), which was used for differential expression analysis, with the design matrix formulated to fit the experimental conditions. National Supercomputing Centre (NSCC) was used for running the aforementioned software. The sequencing data were deposited into GEO with accession number of GSE82121.

DCA treatment and amino acid spike

In the DCA-treated experiment, WT *D. tertiolecta* with spike of 750 µM DCA (Sodium dichloroacetate, 98%, ACROS Organics™) and its net control were cultured under low-light condition as aforementioned. TAG content was measured using quick Nile red assay as aforementioned. *PDK* mRNA levels were measured using quantitative real-time PCR under different light conditions. AcCoA levels were quantified using Acetyl-Coenzyme A Assay Kit (Sigma, MAK039). We did the same on G11_7 mutant both under low-light (30 µmol photons m^{-2} s^{-1}) and high-light conditions (320 µmol photons m^{-2} s^{-1}), respectively, with a net G11_7 strain as the control.

In the amino acid spike experiment, different concentrations of leucine (20, 50, 500, 1000 µM) were spiked into the microalgal culture medium of WT and G11_7 mutant *D. tertiolecta*, with its medium as blank control. Their TAG amount was quantified by the quick Nile red assay.

Cloning and analysis of important genes

The 5′ and 3′ ends of the *D. tertiolecta* templates were cloned using a SMART™ RACE cDNA amplification kit (Clontech, Mountain View, CA) based on the total RNA extracted aforementioned. The full lengths of the coding region of putative *DtCuAO*, *DtIVD*, *DtMCCB* (detected in both next-generation sequencing analyses) genes were amplified from *D. tertiolecta* cDNA by RACE PCR via the predicted region from the *D. tertiolecta* database. All the primers used in this study were listed in Table S3. Phylogenetic tree of protein clusters from various species was constructed by neighbor-joining (NJ) method using software MEGA 5 [20]. Subcellular localization of the related genes was predicted by SignalP 4.1 (http://www.cbs.dtu.dk/services/SignalP/), ChloroP (http://www.cbs.dtu.dk/services/ChloroP/), MITOPROT (https://ihg.gsf.de/ihg/mitoprot.html), Hectar (http://webtools.sb-roscoff.fr) online.

Metabolome analysis

Two hundred millilitre algal cells of G11_7 and WT were collected on day 6, respectively, and freeze-dried for metabolome analysis. Metabolite extraction method was first optimized using a test sample. Four solvent mixtures were tested, and the optimal condition was chosen to ensure the maximum number of peak pairs that could be detected.

Metabolome profiles of the two genotypes of microalgae with biological triplicates were analysed using chemical isotope labelling (CIL) LC-MS method. The CIL LC-MS analysis focused on amine/phenol compounds and carboxylic acid submetabolomes extracted from the samples. Three aliquots of each sample were weighed out, and each aliquot was labelled according to protocols reported previously (Guo and Li, 2009, 2010; Peng and Li, 2013). Samples were then mixed appropriately and analysed on the Bruker Impact HD quadrupole time-of-flight (Q-TOF) mass spectrometer (Bruker, Billerica, MA) connected to a Dionex Ultimate 3000 (Dionex, Sunnyvale, CA). The samples were injected onto an Agilent reversed-phase Eclipse Plus C18 column (2.1 mm × 10 cm, 1.8 µm, 95 Å) for separation. Solvent A was 0.1% (v/v) formic acid/5% (v/v) ACN /H_2O, and solvent B was 0.1% (v/v) formic acid/ACN. The chromatographic conditions for

dansyl labelling were as follows: $t = 0$ min, 20% B; $t = 3.5$ min, 35% B; $t = 18$ min, 65% B; $t = 21$ min, 95% B; $t = 26$ min, 95% B. The gradient for DmPA labelling was $t = 0$ min, 20% B; $t = 9$ min, 50% B; $t = 22$ min, 65% B; $t = 26$ min, 80% B; $t = 29$ min, 98% B; $t = 40$ min, 98% B. Column temperature was set at 30 °C, and a flow rate of 180 µL/min was applied. All MS spectra were obtained in the positive ion mode. The mass range was set as 220–1000 m/z for danzyl labelling and 110–1000 m/z for acid labelling. The MS spectral rate was 1.0 Hz.

For targeted acyl-CoA analysis, LC-MS with standard addition method was used to quantify the CoAs in the algae samples (Friis *et al.*, 2014). A mixture of fifteen CoA standards was used. 30 µL of each biological triplicate sample was injected onto a Kinetex Coreshell HILIC column (100 mm × 2.1 mm, 1.7 µm) for separation. LC-MS analysis was carried out using a Dionex Ultimate 3000 (Dionex) connected to a Bruker maXis II Q-TOF instrument (Bruker). Column temperature was set at 40 °C, and a solvent gradient of 5 min at 500 µL/min was used with an equilibration of 8 min at 600 µL/min. Solvent A was 10 mM ammonium acetate (pH 5.6), and solvent B was 95% ACN/5% 10 mM ammonium acetate (pH 5.6). The gradient was as follows: 0–3.0 min (90% B-5% B); 3.0–5.5 min (5% B-5% B). For the Q-TOF instrument, the mass range was set as 750–1100 m/z and the spectral acquisition rate was 3 Hz. All MS spectra were obtained in the positive ion mode. Peak areas of all the CoAs were extracted using Bruker Target Analysis software for quantification.

Profiling the amine- and phenol-containing metabolites (i.e. the amine/phenol submetabolome) was carried out using differential chemical isotope labelling liquid chromatography mass spectrometry (CIL LC-MS) (Figure S8). A total of 2246 metabolites (isotope peak pairs) were detected in the 18 LC-MS runs. Among those metabolites, 276 were considered significant contributors to the differentiation of the two genotypes from Volcano Plot, and the two groups of samples can be well separated through either principal component analysis (PCA) or partial least squares discriminant analysis (PLS-DA) (Figure S9a-c). Among those 276 metabolites, 18 were positively identified against dansyl standards library based on both mass and retention time matches; 46 could be putatively identified using MyCompoundID search engine against HMDB based on 0 reaction and Metlin. The detailed amine- and phenol-containing metabolite report is shown in Supplementary Data Set 2a.

In the carboxylic acid submetabolome profiling, 2246 metabolites (isotope peak pairs) were detected in 18 LC-MS runs. Among those metabolites, 142 were considered significant contributors to the differentiation of the two genotypes (Figure S9d-f). Separation of the two groups of samples can be very well observed in PCA and PLS-DA plots. Fold changes for some of the targeted acids are listed in Supplementary Data Set 2b.

Importantly, fifteen targeted acyl-CoAs with different acyl groups were also performed using LC-MS, and their fold changes between the two groups, along with CoA concentrations, are shown in Supplementary Data Set 2c. Our emphases were mainly in the fatty acid biosynthesis pathway, protein metabolic pathway, citrate acid cycle (TCA cycle) and their precursors.

Acknowledgements

This work was funded by the National Research Foundation (NRF), Prime Minister's Office, Singapore under its Campus for Research Excellence and Technological Enterprise (CREATE) Programme (Grant No. R-182-000-205-592). We thank Dr Ong Chin Thing Jo,

Dr Chua Ming Lai Ivan, Dr Yi-Kai Ng, Mr Chin-Seng Low, Mr Guo Hui Teo, Dr Paul Edward Hutchinson and Mr Yuan Yao for their suggestions and technical help. L.L. thanks Genome Canada, Alberta Innovates, Natural Science and Engineering Research Council of Canada and Canada Research Chairs Program for supporting the analytical metabolomics research program. L.Y. was supported by the NUS Research Scholarship.

References

Angelovici, R., Fait, A., Fernie, A.R. and Galili, G. (2011) A seed high-lysine trait is negatively associated with the TCA cycle and slows down Arabidopsis seed germination. *New Phytol.* **189**, 148–159.

Araújo, W.L., Ishizaki, K., Nunes-Nesi, A., Larson, T.R., Tohge, T., Krahnert, I., Witt, S. *et al.* (2010) Identification of the 2-hydroxyglutarate and isovaleryl-CoA dehydrogenases as alternative electron donors linking lysine catabolism to the electron transport chain of Arabidopsis mitochondria. *Plant Cell*, **22**, 1549–1563.

Bertozzini, E., Galluzzi, L., Penna, A. and Magnani, M. (2011) Application of the standard addition method for the absolute quantification of neutral lipids in microalgae using Nile red. *J. Microbiol. Methods*, **87**, 17–23.

Booth, G. (2000) *Ullmann's encyclopedia of industrial chemistry.* Oldham, UK: Wiley-VCH.

Carpinelli, E.C., Telatin, A., Vitulo, N., Forcato, C., D'Angelo, M., Schiavon, R., Vezzi, A. *et al.* (2014) Chromosome scale genome assembly and transcriptome profiling of *Nannochloropsis gaditana* in nitrogen depletion. *Mol. Plant*, **7**, 323–335.

Chen, W., Zhang, C., Song, L., Sommerfeld, M. and Hu, Q. (2009) A high throughput Nile red method for quantitative measurement of neutral lipids in microalgae. *J. Microbiol. Methods*, **77**, 41–47.

Chen, M., Tang, H., Ma, H., Holland, T.C., Ng, K.S. and Salley, S.O. (2011) Effect of nutrients on growth and lipid accumulation in the green algae *Dunaliella tertiolecta. Bioresour. Technol.* **102**, 1649–1655.

Chen, L., Chu, C., Lu, J., Kong, X., Huang, T. and Cai, Y.-D. (2015) Gene ontology and KEGG pathway enrichment analysis of a drug target-based classification system. *PLoS ONE*, **10**, e0126492.

Chiu, S.-Y., Kao, C.-Y., Tsai, M.-T., Ong, S.-C., Chen, C.-H. and Lin, C.-S. (2009) Lipid accumulation and CO$_2$ utilization of *Nannochloropsis oculata* in response to CO$_2$ aeration. *Bioresour. Technol.* **100**, 833–838.

Courchesne, N.M.D., Parisien, A., Wang, B. and Lan, C.Q. (2009) Enhancement of lipid production using biochemical, genetic and transcription factor engineering approaches. *J. Biotechnol.* **141**, 31–41.

Flynn, K. and Butler, I. (1986) Nitrogen-sources for the growth of marine microalgae-role of dissolved free amino-acids. *Mar. Ecol. Prog. Ser.* **34**, 281–304.

Friis, R.M.N., Glaves, J.P., Huan, T., Li, L., Sykes, B.D. and Schultz, M.C. (2014) Rewiring AMPK and mitochondrial retrograde signaling for metabolic control of aging and histone acetylation in respiratory-defective cells. *Cell Rep.* **7**, 565–574.

Garrett, R.H. and Grisham, C.M. (2013) *Biochemistry*, 5th ed. CA, USA: Brooks/Cole, Cengage Learning.

Ge, F., Huang, W., Chen, Z., Zhang, C., Xiong, Q., Bowler, C., Yang, J. *et al.* (2014) Methylcrotonyl-CoA carboxylase regulates triacylglycerol accumulation in the model diatom *Phaeodactylum tricornutum. Plant Cell*, **26**, 1681–1697.

Grabherr, M.G., Haas, B.J., Yassour, M., Levin, J.Z., Thompson, D.A., Amit, I., Adiconis, X. *et al.* (2011) Full-length transcriptome assembly from RNA-Seq data without a reference genome. *Nat. Biotechnol.* **29**, 644–652.

Green, C.R., Wallace, M., Divakaruni, A.S., Phillips, S.A., Murphy, A.N., Ciaraldi, T.P. and Metallo, C.M. (2016) Branched-chain amino acid catabolism fuels adipocyte differentiation and lipogenesis. *Nat. Chem. Biol.* **12**, 15–21.

Guo, K. and Li, L. (2009) Differential 12C-/13C-isotope dansylation labeling and fast liquid chromatography/mass spectrometry for absolute and relative quantification of the metabolome. *Anal. Chem.* **81**, 3919–3932.

Guo, K. and Li, L. (2010) High-performance isotope labeling for profiling carboxylic acid-containing metabolites in biofluids by mass spectrometry. *Anal. Chem.* **82**, 8789–8793.

Hossain, A.S., Salleh, A., Boyce, A.N., Chowdhury, P. and Naqiuddin, M. (2008) Biodiesel fuel production from algae as renewable energy. *Am. J. Biochem. Biotechnol.* **4**, 250–254.

Huo, Y.-X., Cho, K.M., Rivera, J.G.L., Monte, E., Shen, C.R., Yan, Y. and Liao, J.C. (2011) Conversion of proteins into biofuels by engineering nitrogen flux. *Nat. Biotechnol.* **29**, 346–351.

Ishizaki, K., Larson, T.R., Schauer, N., Fernie, A.R., Graham, I.A. and Leaver, C.J. (2005) The critical role of Arabidopsis electron-transfer flavoprotein: ubiquinone oxidoreductase during dark-induced starvation. *Plant Cell*, **17**, 2587–2600.

Kanehisa, M., Sato, Y., Kawashima, M., Furumichi, M. and Tanabe, M. (2016) KEGG as a reference resource for gene and protein annotation. *Nucleic Acids Res.* **44**, D457–D462.

Kato, M., Li, J., Chuang, J.L. and Chuang, D.T. (2007) Distinct structural mechanisms for inhibition of pyruvate dehydrogenase kinase isoforms by AZD7545, dichloroacetate, and radicicol. *Structure*, **15**, 992–1004.

Lane, N. (2009) *Life Ascending: The Ten Great Inventions of Evolution.* New York: W.W. Norton & Co.

Leng, N., Dawson, J.A., Thomson, J.A., Ruotti, V., Rissman, A.I., Smits, B.M., Haag, J.D. *et al.* (2013) EBSeq: an empirical Bayes hierarchical model for inference in RNA-seq experiments. *Bioinformatics*, **29**, 1035–1043.

Li, B. and Dewey, C.N. (2011) RSEM: accurate transcript quantification from RNA-Seq data with or without a reference genome. *BMC Bioinform.* **12**, 323–338.

Li, J., Han, D., Wang, D., Ning, K., Jia, J., Wei, L., Jing, X. *et al.* (2014) Choreography of transcriptomes and lipidomes of *Nannochloropsis* reveals the mechanisms of oil synthesis in microalgae. *Plant Cell*, **26**, 1645–1665.

Li-Beisson, Y., Shorrosh, B., Beisson, F., Andersson, M.X., Arondel, V., Bates, P.D., Baud, S. *et al.* (2013) Acyl-lipid metabolism. *The Arabidopsis Book*, **11**, e0161.

Lin, M. and Oliver, D.J. (2008) The role of acetyl-coenzyme a synthetase in Arabidopsis. *Plant Physiol.* **147**, 1822–1829.

Lin, H., Fang, L., Low, C.S., Chow, Y. and Lee, Y.K. (2013) Occurrence of glycerol uptake in *Dunaliella tertiolecta* under hyperosmotic stress. *FEBS J.* **280**, 1064–1072.

Ma, Y.-H., Wang, X., Niu, Y.-F., Yang, Z.-K., Zhang, M.-H., Wang, Z.-M., Yang, W.-D. *et al.* (2014) Antisense knockdown of pyruvate dehydrogenase kinase promotes the neutral lipid accumulation in the diatom *Phaeodactylum tricornutum. Microb. Cell Fact.* **13**, 100–108.

Miller, J.W. and Uden, P.C. (1983) Characterization of nonvolatile aqueous chlorination products of humic substances. *Environ. Sci. Technol.* **17**, 150–157.

Oliver, D.J., Nikolau, B.J. and Wurtele, E.S. (2009) Acetyl-CoA—life at the metabolic nexus. *Plant Sci.* **176**, 597–601.

Peng, J. and Li, L. (2013) Liquid–liquid extraction combined with differential isotope dimethylaminophenacyl labeling for improved metabolomic profiling of organic acids. *Anal. Chim. Acta*, **803**, 97–105.

Peng, C., Uygun, S., Shiu, S.-H. and Last, R.L. (2015) The impact of the branched-chain ketoacid dehydrogenase complex on amino acid homeostasis in Arabidopsis. *Plant Physiol.* **169**, 1807–1820.

Radakovits, R., Jinkerson, R.E., Darzins, A. and Posewitz, M.C. (2010) Genetic engineering of algae for enhanced biofuel production. *Eukaryot. Cell*, **9**, 486–501.

Rismani-Yazdi, H., Haznedaroglu, B.Z., Bibby, K. and Peccia, J. (2011) Transcriptome sequencing and annotation of the microalgae *Dunaliella tertiolecta*: pathway description and gene discovery for production of next-generation biofuels. *BMC Genom.* **12**, 148.

Satoh, A., Ichii, K., Matsumoto, M., Kubota, C., Nemoto, M., Tanaka, M., Yoshino, T. *et al.* (2013) A process design and productivity evaluation for oil production by indoor mass cultivation of a marine diatom, *Fistulifera* sp. JPCC DA0580. *Bioresour. Technol.* **137**, 132–138.

Sharma, K.K., Schuhmann, H. and Schenk, P.M. (2012) High lipid induction in microalgae for biodiesel production. *Energies*, **5**, 1532–1553.

Shimogawara, K., Fujiwara, S., Grossman, A. and Usuda, H. (1998) High-efficiency transformation of *Chlamydomonas reinhardtii* by electroporation. *Genetics*, **148**, 1821–1828.

Shin, H., Hong, S.-J., Kim, H., Yoo, C., Lee, H., Choi, H.-K., Lee, C.-G. *et al.* (2015) Elucidation of the growth delimitation of *Dunaliella tertiolecta* under nitrogen stress by integrating transcriptome and peptidome analysis. *Bioresour. Technol.* **194**, 57–66.

Simionato, D., Block, M.A., La Rocca, N., Jouhet, J., Maréchal, E., Finazzi, G. and Morosinotto, T. (2013) The response of *Nannochloropsis gaditana* to nitrogen starvation includes *de novo* biosynthesis of triacylglycerols, a decrease of chloroplast galactolipids, and reorganization of the photosynthetic apparatus. *Eukaryot. Cell*, **12**, 665–676.

Terashima, M., Freeman, E.S., Jinkerson, R.E. and Jonikas, M.C. (2015) A fluorescence-activated cell sorting-based strategy for rapid isolation of high-lipid *Chlamydomonas* mutants. *Plant J.* **81**, 147–159.

Tovar-Méndez, A., Miernyk, J.A. and Randall, D.D. (2003) Regulation of pyruvate dehydrogenase complex activity in plant cells. *Eur. J. Biochem.* **270**, 1043–1049.

Urzica, E.I., Vieler, A., Hong-Hermesdorf, A., Page, M.D., Casero, D., Gallaher, S.D., Kropat, J. *et al.* (2013) Remodeling of membrane lipids in iron-starved *Chlamydomonas. J. Biol. Chem.* **288**, 30246–30258.

Vorapreeda, T., Thammarongtham, C., Cheevadhanarak, S. and Laoteng, K. (2012) Alternative routes of acetyl-CoA synthesis identified by comparative genomic analysis: involvement in the lipid production of oleaginous yeast and fungi. *Microbiology*, **158**, 217–228.

Wagner, A. (2014) *Arrival of the Fittest*, 1st ed., p. 100. New York: Penguin Group.

Wahidin, S., Idris, A. and Shaleh, S.R.M. (2013) The influence of light intensity and photoperiod on the growth and lipid content of microalgae *Nannochlopsis* sp. *Bioresour. Technol.* **129**, 7–11.

Wang, Z.T., Ullrich, N., Joo, S., Waffenschmidt, S. and Goodenough, U. (2009) Algal lipid bodies: stress induction, purification, and biochemical characterization in wild-type and starchless *Chlamydomonas reinhardtii. Eukaryot. Cell*, **8**, 1856–1868.

Xu, P., Gu, Q., Wang, W., Wong, L., Bower, A.G., Collins, C.H. and Koffas, M.A. (2013) Modular optimization of multi-gene pathways for fatty acids production in *E. coli. Nat. Commun.* **4**, 1409–1416.

Yao, L., Tan, T.W., Ng, Y.-K., Ban, K.H.K., Shen, H., Lin, H. and Lee, Y.K. (2015) RNA-Seq transcriptomic analysis with Bag2D software identifies key pathways enhancing lipid yield in a high lipid-producing mutant of the non-model green alga *Dunaliella tertiolecta. Biotechnol. Biofuels*, **8**, 191–206.

Yoon, K., Han, D., Li, Y., Sommerfeld, M. and Hu, Q. (2012) Phospholipid: diacylglycerol acyltransferase is a multifunctional enzyme involved in membrane lipid turnover and degradation while synthesizing triacylglycerol in the unicellular green microalga *Chlamydomonas reinhardtii. Plant Cell*, **24**, 3708–3724.

14

Double overexpression of DREB and PIF transcription factors improves drought stress tolerance and cell elongation in transgenic plants

Madoka Kudo[1], Satoshi Kidokoro[1], Takuya Yoshida[1,2], Junya Mizoi[1], Daisuke Todaka[1], Alisdair R. Fernie[2], Kazuo Shinozaki[3] and Kazuko Yamaguchi-Shinozaki[1,*]

[1]Graduate School of Agricultural and Life Sciences, University of Tokyo, Tokyo, Japan
[2]Max-Planck-Institut für Molekulare Pflanzenphysiologie, Golm, Germany
[3]RIKEN Center for Sustainable Resource Science, Tsurumi-ku, Yokohama, Japan

*Correspondence
email akys@mail.ecc. u-tokyo.ac.jp

Keywords: Dehydration-Responsive Element-Binding protein 1A (DREB1A), Rice Phytochrome-Interacting factor-Like 1 (OsPIL1), Drought stress tolerance, Cell elongation, Flowering, Arabidopsis.

Summary

Although a variety of transgenic plants that are tolerant to drought stress have been generated, many of these plants show growth retardation. To improve drought tolerance and plant growth, we applied a gene-stacking approach using two transcription factor genes: *DEHYDRATION-RESPONSIVE ELEMENT-BINDING 1A* (*DREB1A*) and rice *PHYTOCHROME-INTERACTING FACTOR-LIKE 1* (*OsPIL1*). The overexpression of *DREB1A* has been reported to improve drought stress tolerance in various crops, although it also causes a severe dwarf phenotype. *OsPIL1* is a rice homologue of Arabidopsis *PHYTOCHROME-INTERACTING FACTOR 4* (*PIF4*), and it enhances cell elongation by activating cell wall-related gene expression. We found that the OsPIL1 protein was more stable than PIF4 under light conditions in Arabidopsis protoplasts. Transactivation analyses revealed that DREB1A and OsPIL1 did not negatively affect each other's transcriptional activities. The transgenic plants overexpressing both *OsPIL1* and *DREB1A* showed the improved drought stress tolerance similar to that of *DREB1A* overexpressors. Furthermore, double overexpressors showed the enhanced hypocotyl elongation and floral induction compared with the *DREB1A* overexpressors. Metabolome analyses indicated that compatible solutes, such as sugars and amino acids, accumulated in the double overexpressors, which was similar to the observations of the *DREB1A* overexpressors. Transcriptome analyses showed an increased expression of abiotic stress-inducible DREB1A downstream genes and cell elongation-related OsPIL1 downstream genes in the double overexpressors, which suggests that these two transcription factors function independently in the transgenic plants despite the trade-offs required to balance plant growth and stress tolerance. Our study provides a basis for plant genetic engineering designed to overcome growth retardation in drought-tolerant transgenic plants.

Introduction

Drought is one of the most serious environmental stresses that affect global agriculture. In recent years, prolonged droughts have caused severe damage to crops in many of the most agriculturally productive areas of the world. At present, approximately 800 million people are undernourished (FAO, IFAD and WFP, 2015). In addition, the world population is expected to experience dramatic grow; thus, the world food crisis appears to be worsening. Therefore, to ensure a stable supply of foods and biomass materials, it is imperative to improve drought stress tolerance in crops.

Plant growth is repressed in response to environmental stresses, indicating that there are trade-offs between growth and stress tolerance (Claeys and Inze, 2013). Although a variety of transgenic plants that are tolerant to drought stress have been generated, many of these plants show growth retardation (Yamaguchi-Shinozaki and Shinozaki, 2006). To reduce the negative effects on plant growth, stress-inducible promoters have been used to drive the expression of transgenes in transgenic plants (Bhatnagar-Mathur *et al.*, 2007; Kasuga *et al.*, 1999; Pino *et al.*, 2007; Suo *et al.*, 2012). However, growth

retardation appears to be unavoidable under the long-term drought stress conditions because of the prolonged overexpression of transgenes, even when using stress-inducible promoters. Additionally, it is necessary to select the optimal stress-inducible promoters for each plant species. For example, an Arabidopsis *RESPONSIVE TO DEHYDRATION 29A* (*RD29A*) promoter is useful for overexpression during abiotic stress conditions in both Arabidopsis and tobacco, whereas it functions in the roots, but not the leaves, in rice (Ito *et al.*, 2006; Kasuga, 2004). Therefore, additional approaches are required to improve the growth of stress-tolerant plants.

Pyramiding (gene-stacking) breeding is essential for biotechnology applications (Halpin, 2005). For example, plants that produce several *Bacillus thuringiensis* (Bt) proteins have been shown to improve insect resistance (Carriere *et al.*, 2015). 'Golden rice' plants similarly contain three carotenoid biosynthesis genes for accumulating pro-vitamin A (Ye *et al.*, 2000). The co-expression of 9-cis-epoxycarotenoid dioxygenase (NCED) and D-arabinono-1,4-lactone oxidase (ALO) increases abscisic acid (ABA) and ascorbic acid levels and improves tolerance to drought and chilling *in planta* (Bao *et al.*, 2016). Many genetically modified (GM) crops cultivated worldwide have been generated by gene stacking (Halpin, 2005). However, few studies have focused on engineering plants to improve both drought stress tolerance and growth using gene-stacking approaches.

DREB1A is an APETALA2/ethylene-responsive element-binding factor (AP2/ERF)-type transcription factor that specifically binds dehydration-responsive elements (DREs) and up-regulates stress-inducible target gene expression. The overexpression of *DREB1A* improves the tolerance to drought, salt and freezing stress by enhancing late embryogenesis-abundant (LEA) protein levels and the compatible solute contents in Arabidopsis (Kasuga *et al.*, 1999; Maruyama *et al.*, 2004, 2009). Overexpression of the Arabidopsis *DREB1A* gene was reported to enhance abiotic stress tolerance in many crops, such as rice, soybean, peanut and wheat (Bhatnagar-Mathur *et al.*, 2014; Ito *et al.*, 2006; Pellegrineschi *et al.*, 2004; Suo *et al.*, 2012). Moreover, the mechanism of improved drought stress tolerance was revealed by transcriptome and metabolome analyses as presented above. Therefore, *DREB1A* appears to be one of the most agriculturally useful genes for improving abiotic stress tolerance in crops. However, *DREB1A* overexpression causes dwarfism and late flowering in plants, and additional investigations are required before it can be used in efficient agricultural applications.

Arabidopsis phytochrome-interacting factor (PIF) family proteins, which are basic helix-loop-helix (bHLH)-type transcription factors, were initially isolated through their interaction with phytochrome (Ni *et al.*, 1999). PIFs function to promote seedling skotomorphogenesis, shade avoidance and floral induction and regulate the expression of many downstream genes (Leivar and Quail, 2011). In rice, six PIF-like genes were identified by *in silico* analysis (Nakamura *et al.*, 2007). Of these, OsPIL1 was reported to enhance cell elongation through the activation of cell wall synthesis-related genes in rice (Todaka *et al.*, 2012). Additionally, *OsPIL1* expression levels are repressed under drought and low-temperature conditions; thus, OsPIL1 appears to act as a key regulator of plant growth in response to abiotic stress.

In this study, to overcome the trade-offs between growth and stress tolerance, we generated transgenic plants overexpressing both *OsPIL1* and *DREB1A* and characterized these double overexpressors by phenotypic analyses, metabolome analyses and genomewide transcriptome analyses. We propose that *OsPIL1* partially enhances plant growth and accelerates flowering time, even in the double overexpressors, without negative effects on DREB1A-mediated drought stress tolerance.

Results

Stability of the rice OsPIL1 protein and transactivation activity of OsPIL1 with DREB1A in Arabidopsis and rice protoplasts

Arabidopsis PIF family proteins have been reported to enhance cell elongation, and their protein levels are regulated by light stimulation (Leivar and Quail, 2011). Light triggers the interaction of PIF family proteins with phytochrome B (PhyB), which results in the degradation of the PIF proteins by the 26S proteasome. However, we previously reported that one of the four amino acid residues important for PhyB binding in Arabidopsis PIF proteins is not conserved in the rice PIF, including protein OsPIL1, which does not interact with OsPhyB in yeast two-hybrid and bimolecular fluorescence complementation (BiFC) assays (Todaka *et al.*, 2012). Thus, we analysed the OsPIL1 protein stability in Arabidopsis mesophyll protoplasts under light conditions. sGFP-fused OsPIL1 and PIF4 proteins were transiently expressed in

Arabidopsis protoplasts, and the levels of these proteins were then analysed by immunoblotting (Figure S1). In addition, 3 × Flag-tagged sGFP was co-expressed in protoplasts as an internal control. To determine whether the OsPIL1 protein is degraded under light conditions, we incubated the transfected protoplasts in the dark and then transferred the protoplasts to light conditions in the absence or presence of the 26S proteasome inhibitor MG132 (Figure S1a). In the absence of MG132, the PIF4 protein was degraded in a light- and time-dependent manner as previously reported (Nozue *et al.*, 2007), whereas the OsPIL1 protein was more stable under the same conditions (Figure S1b). In addition, PIF4 and OsPIL1 were not degraded in the presence of MG132 (Figure S1b). These results demonstrate that OsPIL1 is more stable than PIF4 and show that OsPIL1 does not appear to be degraded by the 26S proteasome in response to light irradiation. Taken together, these results suggest that OsPIL1 may enhance plant growth more effectively than PIF4 in Arabidopsis under light conditions.

OsPIL1 and DREB1A have been reported to recognize the *cis*-acting promoter elements G box (CACGTG) and DRE (A/GCCGAC), respectively (Liu *et al.*, 1998; Todaka *et al.*, 2012). To examine whether these transcription factors affect each other's transactivation activity, we performed transactivation assays using Arabidopsis and rice mesophyll protoplasts (Figures 1 and S2). *35SΩ:OsPIL1* and *35SΩ:DREB1A* were used as effector constructs, and G box:*GUS* or DRE:*GUS* was used as a reporter construct (Liu *et al.*, 1998; Todaka *et al.*, 2012). The expression of *OsPIL1* with the G box:*GUS* reporter construct showed high transactivation activity (Figures 1b and S2b). Similarly, when both *OsPIL1* and *DREB1A* were co-expressed, high transactivation activity was observed. When DRE:*GUS* was used as the reporter, the transactivation activity was enhanced by the co-expression of both OsPIL1 and DREB1A in a manner similar to the expression of DREB1A (Figures 1c and S2c). These results indicate that OsPIL1 and DREB1A act independently and do not interfere with each other's transactivation activities in Arabidopsis and rice protoplasts.

Enhanced cell elongation and improved drought stress tolerance in transgenic plants overexpressing both *OsPIL1* and *DREB1A* (double overexpressors)

Next, we generated transgenic Arabidopsis plants overexpressing both *OsPIL1* and *DREB1A* (double overexpressors) by crossing *DREB1A* overexpressors with *OsPIL1* overexpressors (Figure 2a, Table S1). *OsPIL1* or *DREB1A* was overexpressed under the control of the *CaMV 35S* promoter and the TMV Ω sequence in transgenic plants. To explore whether the double overexpressors have phenotypes related to the overexpression of each *OsPIL1* and *DREB1A*, we observed the morphology of these transgenic plants. Overexpression of rice *OsPIL1* was reported to enhance hypocotyl elongation in Arabidopsis (Nakamura *et al.*, 2007; Todaka *et al.*, 2012). Thus, we measured the hypocotyl length of 7-day-old double overexpressors compared with that of *35SΩ:OsPIL1* and *35SΩ:DREB1A* plants (Figure 2b, c). Significant differences were not observed in hypocotyl length between the two vector control plants, pGKX and pGHX (Figure 2c). The hypocotyl length of the double overexpressors was intermediate between the *35SΩ:OsPIL1* and the vector control plants. The hypocotyl length of the *35SΩ:DREB1A* plants was equivalent to that of the vector control plants. We also observed the lengths of the hypocotyl cells in these transgenic plants and found that hypocotyl length correlated with cell length (Figure S3). In

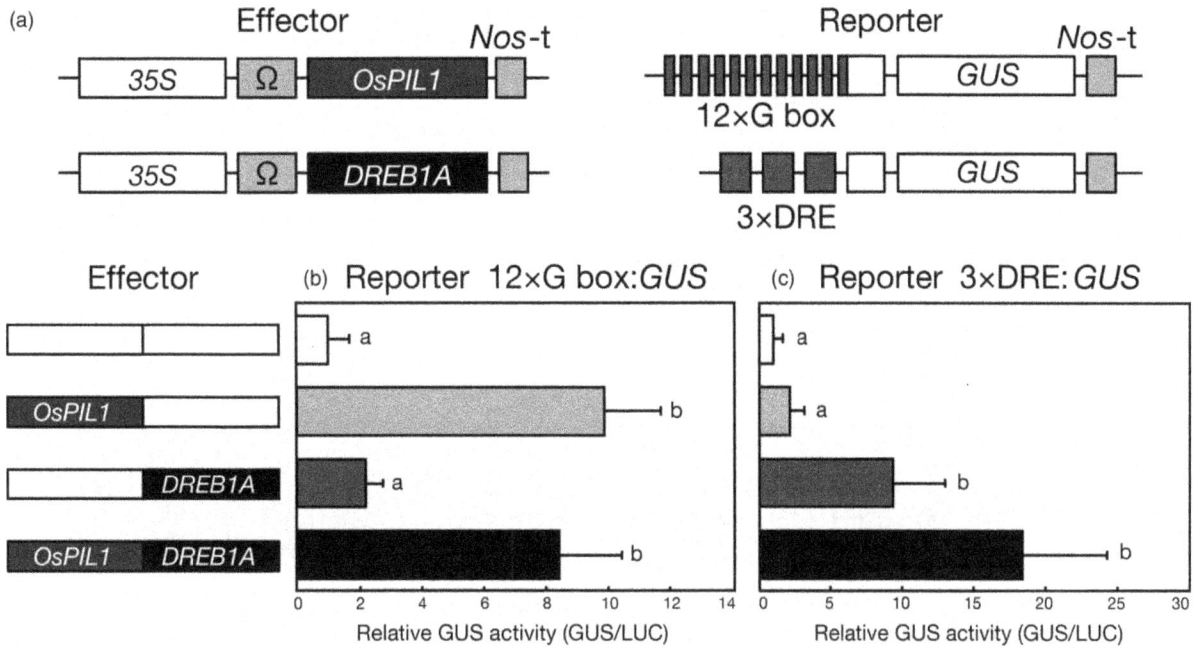

Figure 1 Effect of co-expressing both OsPIL1 and DREB1A on each transactivation activity in Arabidopsis protoplasts. (a) Schematic diagram of the effector and reporter constructs used in the transactivation analysis of the Arabidopsis protoplasts. The effector construct contains a *CaMV 35S* promoter and a TMV Ω sequence fused to the coding sequence of *OsPIL1* or *DREB1A*. *Nos*-t indicates the polyadenylation signal of the gene for nopaline synthetase. The reporter construct 12 × G box:*GUS* contains 12 tandem 28-bp G box-containing fragments. The reporter construct 3 × DRE:*GUS* contains three tandem 71-bp DRE-containing fragments. (b, c) Transactivation effects of the co-expression of OsPIL1 and DREB1A. The reporter 12 × G box:*GUS* (b) or 3 × DRE:*GUS* (c) and the effectors were co-transfected into Arabidopsis protoplasts. To normalize for transfection efficiency, an *emerald luciferase* (*ELUC*) reporter gene driven by the *CaMV 35S* promoter was co-transfected as a control in each experiment. Bars show the standard deviation (SD) of more than four replicates. The letters indicate significant differences among the assays (*P* < 0.05 according to Games–Howell's multiple range test).

addition, we found that the internode and shoot lengths of 8-week-old double overexpressors were longer than those of the *35SΩ:DREB1A* plants (Figures S4 and S5). These results indicate that overexpressing *OsPIL1* enhances cell elongation of the hypocotyl and stem, even in the double overexpressors, and show that overexpressing *DREB1A* has a negative effect on OsPIL1-mediated cell elongation. Furthermore, *DREB1A* overexpression has been reported to improve drought stress tolerance while causing a severe dwarf phenotype (Liu *et al.*, 1998). Under normal growth conditions, the double overexpressors also showed small rosette leaves and short primary roots, which is similar to the *35SΩ:DREB1A* plants (Figure S6a, b, c, d). Next, we weighed the total dry biomass of the transgenic Arabidopsis plants. The double overexpressors had a dry weight similar to the *35SΩ:DREB1A* plants (Figure S6e). In addition, we evaluated the seed yield of the plants. Significant differences were not observed in average silique number between the vector control, *35SΩ: DREB1A* and double overexpressor plants, but this number was decreased in the *35SΩ:OsPIL1* plants compared with the other

plants (Figure S6f). The average seed number per silique was similar in all the transgenic plants (Figure S6g). Therefore, overexpression of *OsPIL1* decreased the seed yield of the transgenic plants. This may be due to the early flowering and senescence phenotypes of the *35SΩ:OsPIL1* plants. Overexpression of Arabidopsis *PIF4* is also known to induce early senescence in transgenic plants (Sakuraba *et al.*, 2014). However, overexpression of *OsPIL1* did not affect the seed yield in the double overexpressors.

In addition to growth retardation, the overexpression of *DREB1A* causes late-flowering phenotypes in transgenic Arabidopsis plants (Seo *et al.*, 2009), whereas the overexpression of *PIF4*, an Arabidopsis homologue of *OsPIL1*, causes an extremely early flowering phenotype by elevating *FLOWERING LOCUS T* (*FT*) expression levels (Kumar *et al.*, 2012). Therefore, we analysed the flowering time of the transgenic plants (Figure 2d, e). Compared with the vector control plants, the *35SΩ:OsPIL1* plants showed early flowering and the *35SΩ:DREB1A* plants showed late flowering. The flowering time of the double overexpressors was

Figure 2 Generation of *OsPIL1* and *DREB1A* double-overexpressing Arabidopsis plants. (a) Expression levels of the transgenes in the single- or double-overexpressing plants were analysed by quantitative RT-PCR. The plants were grown on agar medium for 2 weeks. The error bars indicate the SD of more than four samples. ND represents not detected and VC and W-OE represent vector control and *OsPIL1 DREB1A* double overexpressor, respectively. (b) Morphology of the transgenic Arabidopsis seedlings. The plants were grown on agar medium for 7 days. Bars = 1 mm. (c) Hypocotyl length calculated from the seedlings grown as in (b). The error bars show the SD of more than 50 seedlings. The letters indicate significant differences among the seedlings (*P* < 0.01 according to Games–Howell's multiple range test). (d, e) Flowering time of the single- or double-overexpressing Arabidopsis plants. (d) Growth of 6-week-old transgenic plants. The plants were grown on agar medium for 2 weeks and then in soil pots. Bars = 10 cm. (e) Days to flowering of the plants grown as in (d). The error bars show the SD of more than four plants. The letters indicate significant differences among the plants (*P* < 0.05 according to Games–Howell's multiple range test).

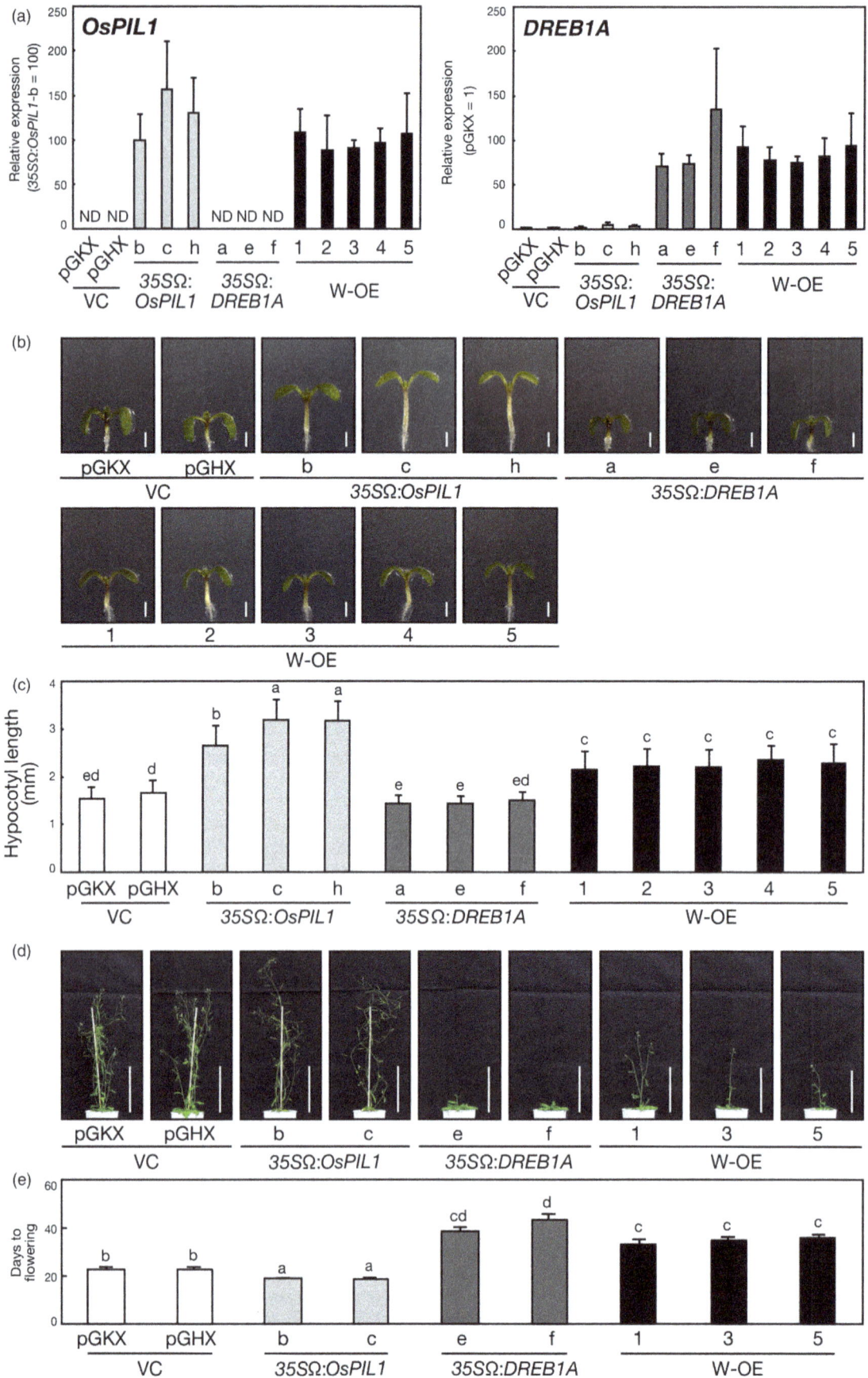

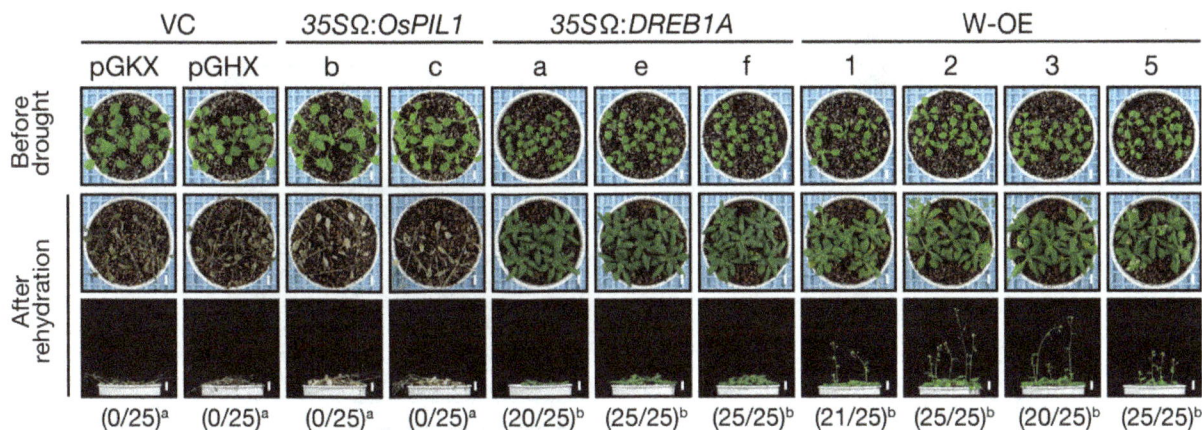

Figure 3 Drought stress tolerance of the single or double overexpressors. Number codes show the number of surviving plants out of the total number. A total of 25 plants were used in the survival test. Bars = 1 cm. The letters indicate significant differences among the plants ($P < 0.05$ according to Tukey's multiple range test of the ratio).

earlier than that of the $35S\Omega$:*DREB1A* plants, but later than that of the vector control and $35S\Omega$:*OsPIL1* plants. These results suggest that OsPIL1 and DREB1A antagonistically regulate flowering time in the double overexpressors. Thus, *OsPIL1* is expected to be useful for weakening the late-flowering phenotypes caused by *DREB1A* overexpression. Taken together, the double overexpressors present phenotypes caused by the overexpression of each *OsPIL1* and *DREB1A*.

We then evaluated the drought stress tolerance of the double overexpressors by withholding water for 2 weeks. No improvement in drought stress tolerance was observed in the $35S\Omega$:*OsPIL1* plants, whereas the double overexpressors showed the improved drought stress tolerance similar to that of the $35S\Omega$:*DREB1A* plants (Figure 3), indicating that overexpressing *OsPIL1* has no negative effects on the DREB1A-mediated improvement of drought stress tolerance in the double overexpressors.

OsPIL1 and DREB1A affect sugar and amino acid metabolism in the double overexpressors

To reveal the metabolite profiles of the double overexpressors, we performed gas chromatography–time-of-flight mass spectrometry (GC-TOF-MS) analyses of 2-week-old (vegetative stage) transgenic plants and identified 43 metabolites, including sugars, amino acids and organic acids (Figures 4, S7a). We extracted the soluble metabolites from whole plants harvested at the subjective dawn (Zeitgeber time [ZT] = 0) and the subjective sundown (ZT = 16). Because the metabolite profiles of the double overexpressors harvested at these two sampling points were nearly equivalent, the differences in the metabolite profiles among the tested transgenic plants were independent of their daily changes (Figure S7b, c). According to principal components analysis (PCA), the contribution ratios of the first (PC1) and second principal components (PC2) were 30.1% and 21.8%, respectively (Figure 4a). The $35S\Omega$:*DREB1A* and $35S\Omega$:*OsPIL1* plants were separated from the vector control plants by PC1 and PC2. Interestingly, the metabolite profiles of the double overexpressors were clearly different from those of each single overexpressor. The score plot of the double overexpressors was described by the metabolites that contributed to the positive side of the PC1 axis and the negative side of the PC2 axis (Figure 4a). We then compared the loadings of PC1 and PC2 to clarify the metabolites that contributed to the separation of the PCA score (Figure 4b).

The top five metabolites contributing to the positive side of PC1 were raffinose, glycerol-3-phosphate, β-alanine, proline and galactinol, and those contributing to the negative side were *myo*-inositol, aspartic acid, fumaric acid, maltotriose and dehydroascorbic acid dimer. However, the top five positive contributors to PC2 were glucose, arginine, fructose, ornithine and guanidine, and the top five negative contributors were erythritol, citric acid, glutamic acid, 4-hydroxyproline and 4-aminobutyric acid (GABA).

In the $35S\Omega$:*DREB1A* plants, the amounts of raffinose, galactinol, β-alanine, proline, ornithine, arginine and glycerol-3-phosphate were significantly increased relative to the vector control plants (Figure 4c). Galactinol and raffinose have been reported to act as osmoprotectants in drought stress tolerance in plants (Taji *et al.*, 2002). Proline and β-alanine are precursors of quaternary ammonium compounds (QAC), which also act as osmoprotectants (Hanson *et al.*, 1994). Following the overexpression of *OsPIL1*, the amount of citric acid was increased, whereas the amounts of glucose, fructose, ornithine, arginine and guanidine were decreased relative to the vector control plants (Figure 4c). The metabolism of ornithine and arginine appeared to be oppositely affected by the overexpression of *DREB1A* and *OsPIL1*, and the levels of these metabolites in the double overexpressors were the average of those in the single overexpressors. The amounts of GABA and glutamine were slightly increased in the $35S\Omega$:*DREB1A* plants and clearly increased in the double overexpressors (Figure 4c). These results suggest that both OsPIL1 and DREB1A independently regulate sugar and amino acid metabolism in the double overexpressors.

Downstream genes of both OsPIL1 and DREB1A were activated in the double overexpressors

To reveal the transcript profiles, we performed microarray experiments using 2-week-old $35S\Omega$:*OsPIL1* plants and double overexpressors harvested at ZT = 0. The overexpression of *OsPIL1* alone and *OsPIL1* and *DREB1A* in combination up-regulated the expression of 30 and 356 genes, respectively (fold change > 2, false discovery rate [FDR] $P < 0.05$) (Figure 5a, left). We compared these up-regulated genes with those in the $35S\Omega$:*DREB1A* plants (Kidokoro *et al.*, 2015) and found 110 common genes between the $35S\Omega$:*DREB1A* plants and the double overexpressors, and these genes included many abiotic stress-inducible

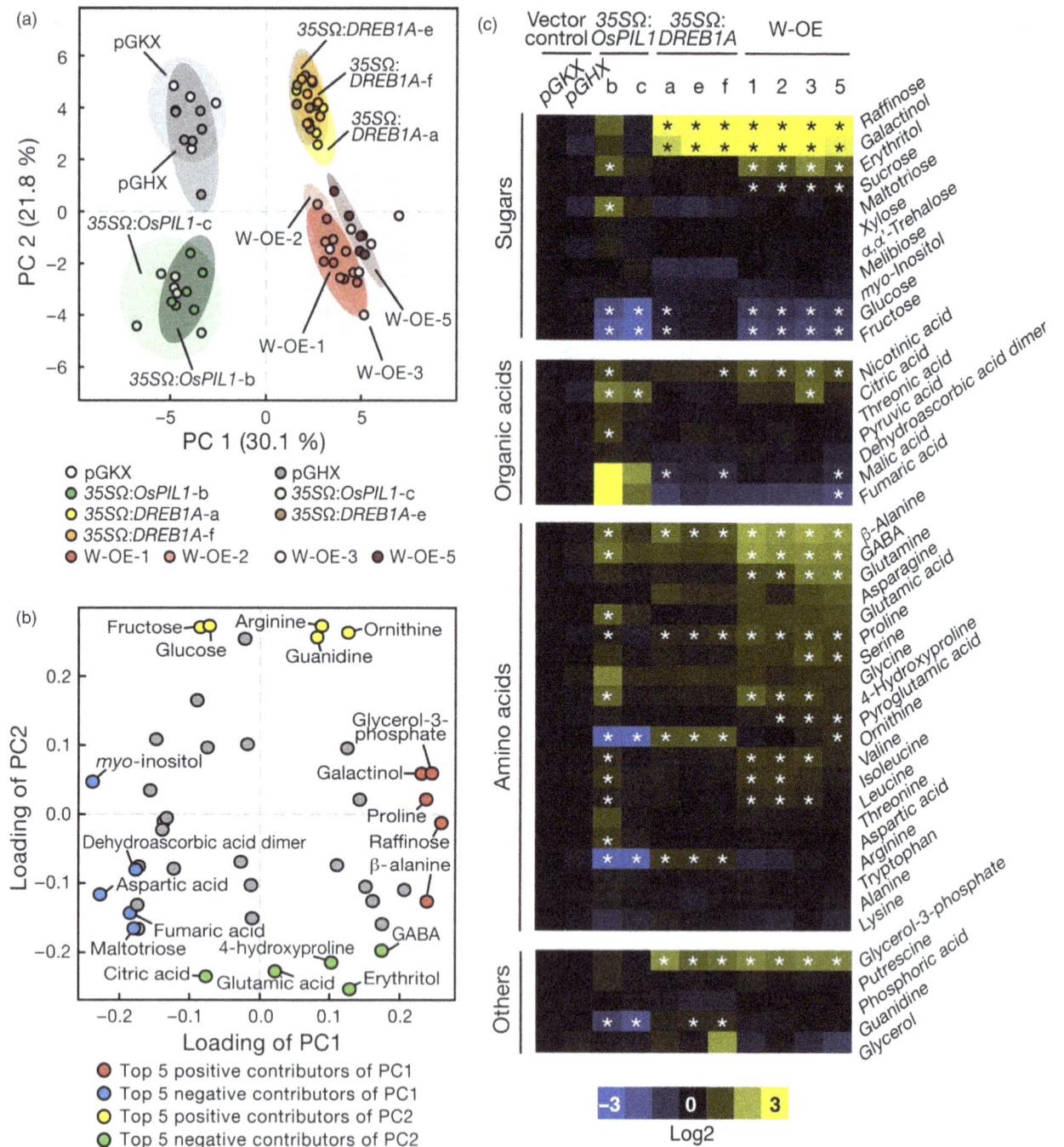

Figure 4 Metabolite profiles of the single- or double-overexpressing plants. (a, b) Principal components analysis (PCA) of metabolite profile in the transgenic plants. Score plot (a) and loading plot (b) of the metabolite profiles in the transgenic plants. The plants were grown on agar medium for 2 weeks and harvested at Zeitgeber time (ZT) = 0. GABA, 4-aminobutyric acid. (c) Heat maps of the metabolite profile in the transgenic plants. The relative metabolite amounts were normalized with vector control plants (pGKX) and transformed to log2. Yellow and blue colours show increased and decreased levels, respectively. Asterisks indicate significant differences between the vector control plants (both pGKX and pGHX) (*P* < 0.05 according to Tukey's multiple range test, n = 6).

genes (Table S2). Ten genes overlapped between the *35SΩ: OsPIL1* plants and the double overexpressors, including many cell elongation-related genes (Table S3). No genes overlapped among all three transgenic Arabidopsis plants. The number of down-regulated genes in the *35SΩ:DREB1A* plants, *35SΩ:OsPIL1* plants and double overexpressors was 235, 29 and 437, respectively (fold change < 0.5, FDR *P* < 0.05) (Figure 5a, right). Fifty-nine

genes were common between the *35SΩ:DREB1A* plants and the double overexpressors, including cell wall-related genes (Table S4). Seven genes overlapped between the *35SΩ:OsPIL1* plants and the double overexpressors, although the functions of these genes were not relevant (Table S5). We observed many up-regulated or down-regulated genes in the double overexpressors, but not in the *35SΩ:OsPIL1* or *35SΩ:DREB1A* plants alone.

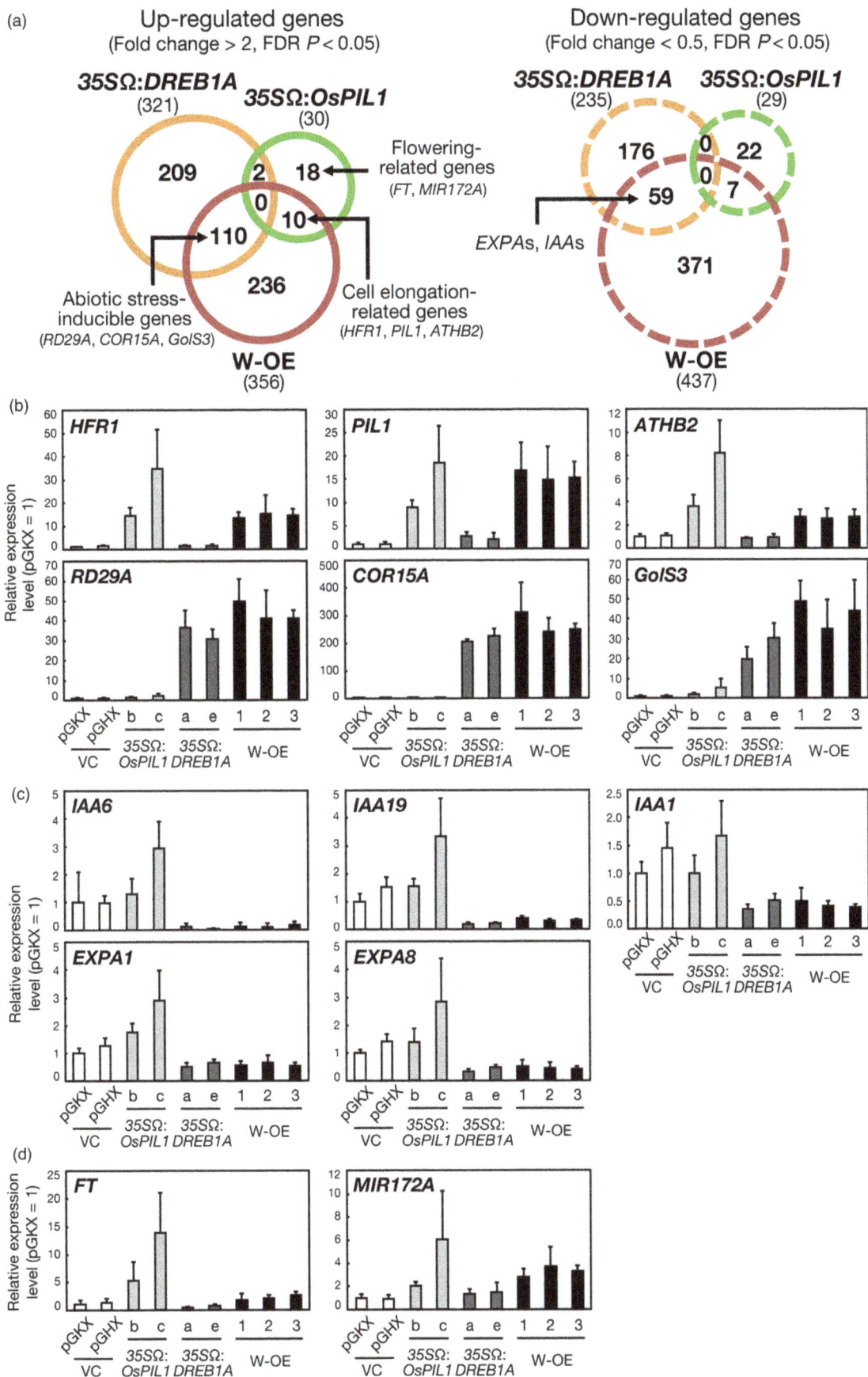

(a)

Up-regulated genes
(Fold change > 2, FDR *P* < 0.05)

35SΩ:DREB1A
(321)

35SΩ:OsPIL1
(30)

209

2

18

Flowering-related genes
(*FT, MIR172A*)

0

110

10

236

Abiotic stress-inducible genes
(*RD29A, COR15A, GolS3*)

Cell elongation-related genes
(*HFR1, PIL1, ATHB2*)

W-OE
(356)

Down-regulated genes
(Fold change < 0.5, FDR *P* < 0.05)

35SΩ:DREB1A
(235)

35SΩ:OsPIL1
(29)

176

0

22

0

59

7

EXPAs, IAAs

371

W-OE
(437)

(b)

(c)

(d)

Figure 5 Comparative analyses of the transcriptomes of single- or double-overexpressing plants. (a) Venn diagrams comparing up-regulated and down-regulated genes among the *35SΩ:DREB1A* plants (orange circles), the *35SΩ:OsPIL1* plants (green circles) and the double overexpressors (red circles). The plants were grown on agar medium for 2 weeks and harvested at ZT = 0. Up-regulated and down-regulated genes are in solid circles and dashed circles, respectively. The total numbers of up-regulated and down-regulated genes are shown in parentheses. (b, c, d) Expression levels of up-regulated genes (b), down-regulated genes (c) and flowering regulator genes (d) in the transgenic Arabidopsis plants were analysed by quantitative RT-PCR. The plants were grown on agar medium for 2 weeks and harvested at ZT = 0. The error bars indicate the SD of more than four samples.

However, the majority of these genes were either up-regulated or down-regulated at low levels. The fold change values for approximately 70% of the up-regulated or down-regulated genes were less than threefold or more than 0.4-fold (Tables S6 and S7).

Next, we analysed the expression patterns of up-regulated and down-regulated genes in each transgenic Arabidopsis plant using the microarray database at Genevestigator (https://genevestigator.com/gv/). The downstream genes in the double overexpressors were responsive to ABA, cold, drought, osmotic and salt stress, which is similar to the effects in the *35SΩ:DREB1A* plants (Figure S8). Certain up-regulated genes in the *35SΩ:OsPIL1* plants were responsive to ABA, indole-3-acetic acid (IAA) and far-red (FR) light (Figure S8). To obtain additional information on the molecular functions of the downstream genes, we categorized these genes according to overrepresentation analyses using PageMan software (http://mapman.mpimp-golm.mpg.de/). Compared with the categorization of the complete Arabidopsis genome, the up-regulated genes in the *35SΩ:DREB1A* plants showed higher ratios of the categories secondary metabolism (12.7%), stress (11.7%), hormone metabolism (6.4%) and sugar metabolism (4.5%) (Figure S9). The ratios of the genes in the categories RNA (27.1%), hormone metabolism (12.9%) and signalling (8.6%) were increased in the *35SΩ:OsPIL1* plants (Figure S9). The ratios of the genes in the categories hormone metabolism (13.6%) and cell wall (5.0%) were higher among the down-regulated genes in the *35SΩ:DREB1A* plants (Figure S10). Genes in the categories secondary metabolism (11.5%), sugar metabolism (6.4%) and cell wall (11.5%) were overrepresented among the down-regulated genes in the *35SΩ:OsPIL1* plants (Figure S10). The ratios of each category of up-regulated and down-regulated genes in the double overexpressors were similar to the average of the values in the *35SΩ:OsPIL1* and *35SΩ:DREB1A* plants (Figures S9 and S10). Furthermore, because raffinose and galactinol were accumulated in the double overexpressors, sugar metabolism in the transgenic plants was analysed using the MapMan database (http://mapman.gabipd.org/). In the sugar metabolism pathway, genes for galactinol synthases and raffinose synthase were up-regulated in *35SΩ:DREB1A* and the double overexpressors (Figure S11). However, a gene for sucrose invertase was down-regulated in *35SΩ:OsPIL1* (Figure S11).

We confirmed the expression levels of the downstream genes in the transgenic plants by using quantitative RT-PCR (Figure 5b, c, d). *LONG HYPOCOTYL IN FAR-RED* (*HFR1*), *PHYTOCHROME-INTERACTING FACTOR 3-LIKE 1* (*PIL1*) and *ARABIDOPSIS THALIANA HOMEOBOX PROTEIN 2* (*ATHB2*) are cell elongation-related target genes of PIF4 in Arabidopsis (Hornitschek et al., 2012; Kunihiro et al., 2011). The expression levels of these three genes increased in the *35SΩ:OsPIL1* plants and the double overexpressors (Figure 5b). *RD29A*, *COLD-REGULATED 15A* (*COR15A*) and *GALACTINOL SYNTHASE 3* (*GolS3*) are major abiotic stress-inducible targets of DREB1A (Maruyama et al., 2009), and the expression of these genes was elevated in the *35SΩ:DREB1A*

plants and the double overexpressors (Figure 5b). These up-regulated genes may be important for cell elongation and drought stress tolerance in the double overexpressors. However, *INDOLE-3-ACETIC ACID-INDUCIBLE* (*IAA*) genes and *EXPANSIN A* (*EXPA*) genes were down-regulated in the *35SΩ:DREB1A* plants and the doble overexpressors (Figure 5c). *IAA* genes are auxin inducible and act as repressors in auxin signalling (Reed, 2001). *EXPA*s are known to enhance cell wall enlargement by cell wall loosening (Cosgrove, 2000). In addition, we analysed the expression levels of flowering-related genes, such as *FT* and *MICRORNA172A* (*MIR172A*), in the transgenic plants (Figure 5d). *FT* encodes florigen, whose overexpression causes early flowering (Turck et al., 2008). Overexpression of *MIR172A* also causes early flowering by down-regulating AP2-like floral repressors (Aukerman and Sakai, 2003). The expression levels of *FT* and *MIR172A* were elevated in the *35SΩ:OsPIL1* plants (Figure 5d). However, the expression levels of *FT* in the *35SΩ:DREB1A* plants were approximately half of those in the vector control plants. In the double overexpressors, the expression levels of *FT* and *MIR172A* were higher than those in the control plants, which suggests that OsPIL1 enhances floral induction by regulating *FT* and *MIR172A* expression in the transgenic plants.

Furthermore, we analysed the expression levels of other PIF4 downstream genes, namely *XYLOGLUCAN ENDOTRANSGLYCO-SYLASE 7* (*XTR7*), *PACLOBUTRAZOL RESISTANCE1* (*PRE1*) and *IAA29* (Hornitschek et al., 2012; de Lucas et al., 2008; Nomoto et al., 2012; Oh et al., 2012); however, significant differences were not observed in the *35SΩ:OsPIL1* plants and double overexpressors compared with the vector control plants (Figure S12a). These results suggest that certain OsPIL1 target genes differ from those of PIF4. We then analysed the expression levels of *SMALL AUXIN-UPREGULATED RNA* (*SAUR*) genes (Figure S12b). SAURs positively regulate cell elongation to promote hypocotyl growth, and PIF4 has been shown to up-regulate the expression of *SAUR19-24* at high temperatures (Franklin et al., 2011; Ren and Gray, 2015). The expression levels of *SAUR19*, *SAUR22* and *SAUR23* were elevated in the *35SΩ:OsPIL1* plants and the double overexpressors (Figure S12b), indicating that OsPIL1 also enhances cell elongation through the up-regulation of *SAUR* genes, which is similar to the action of PIF4 in the transgenic Arabidopsis plants. These results suggest that OsPIL1 acts as a positive regulator of auxin signalling.

To characterize the promoters of the up-regulated and down-regulated genes in the double overexpressors, we assessed the enrichment of all hexamer motifs (from AAAAAA to TTTTTT, $4^6 = 4096$) in the promoters of the genes in each transgenic plant (Figure 6). We compared the frequency of each hexamer sequence in the promoters of the genes with their normalized frequencies in Arabidopsis promoters (Maruyama et al., 2012). In the *35SΩ:OsPIL1* plants, G box (CACGTG) and ABRE sequences (ACGTGG and CGTGGG) ranked among the top 10 up-regulated genes (Figure 6a, top). For the *35SΩ:DREB1A* plants, we analysed microarray data from a previous study

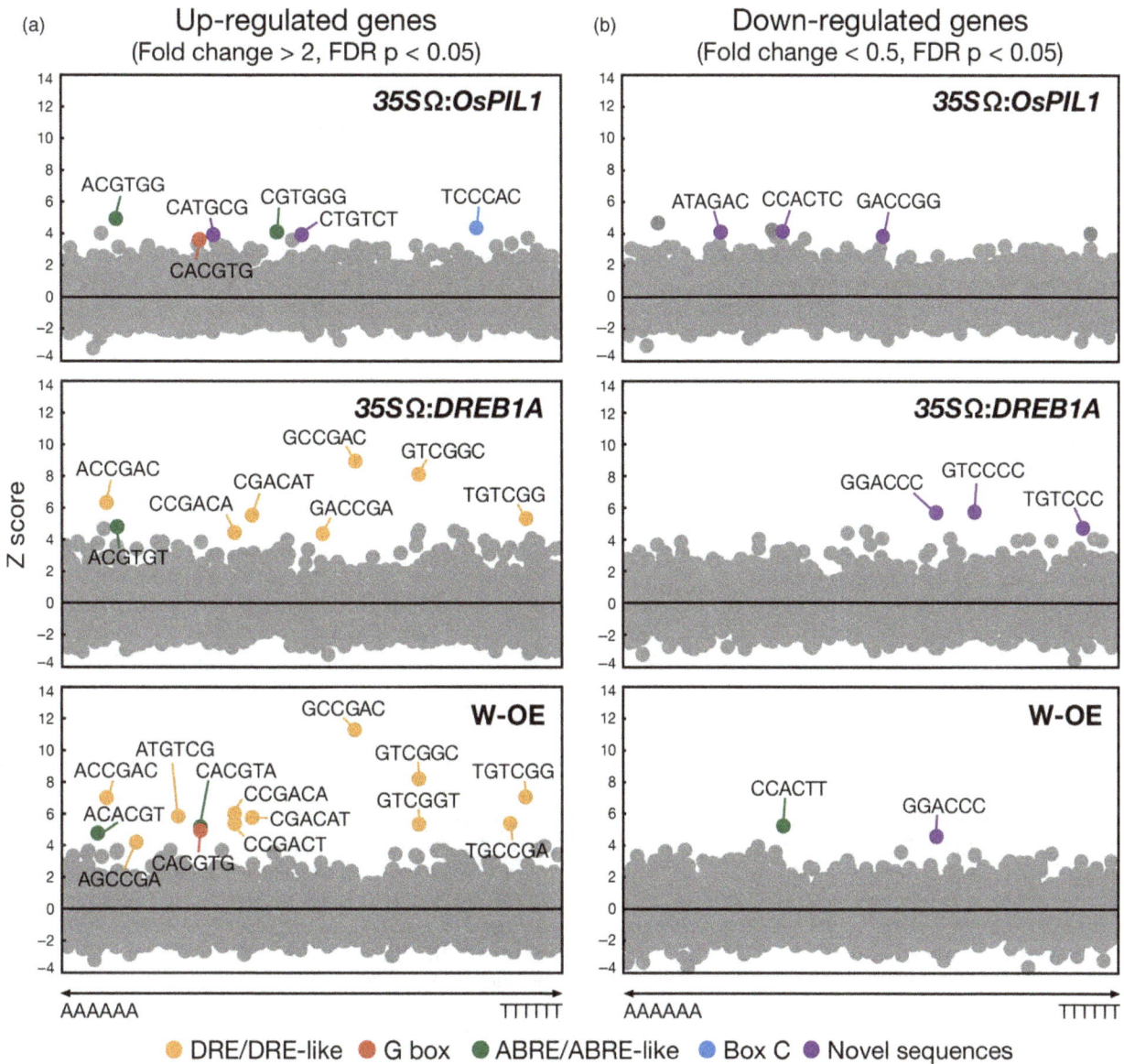

Figure 6 Overrepresentation analyses of hexamer sequences in the promoters of up-regulated and down-regulated genes in the single or double overexpressors. Scatter plots showing Z scores (y-axes) for the observed frequencies of all hexamer sequences (x-axes) in the 1-kb promoters of up-regulated (a) and down-regulated genes (b) in each transgenic plant. DRE/DRE-like (orange), G box (red), ABRE/ABRE-like (green), Box C (blue) and novel sequences (purple) are shown.

(Figure 6a, middle; Kidokoro et al., 2015). Consistent with previous work, the most overrepresented sequence in the 35SΩ: DREB1A plants was DRE (Maruyama et al., 2012). In the double overexpressors, DRE sequences occupied the top 10, and G box or ABRE sequences were ranked within the top 15 among the up-regulated genes (Figure 6a, bottom). The double overexpressor-specific elements were not found in the promoter regions of the up-regulated genes. These results indicate that OsPIL1 and DREB1A activate the expression of each downstream gene independently via each cis-acting element in the double overexpressors. However, known cis elements were not identified among the enriched hexamer motifs in the promoters of the down-regulated genes in the 35SΩ:OsPIL1 plants and the double overexpressors (Figure 6b). In the 35SΩ:DREB1A plants, known hexamer motifs were also not observed, suggesting that

IAA and EXPA gene expression is indirectly repressed by DREB1A.

Discussion

Considerable effort has been devoted to producing GM plants with improved tolerance to abiotic stress, such as drought and high salinity, by using molecular breeding strategies. The overexpression of DREB1A induces many stress-inducible target genes and enhances drought stress tolerance in various crops, although it also causes the growth retardation (Liu et al., 1998). To improve drought stress tolerance and plant growth, we generated transgenic plants overexpressing two transcription factors: DREB1A and OsPIL1. OsPIL1 is reported to be a homologue of Arabidopsis PIF4 and acts as a key regulator of internode elongation in rice by up-regulating many cell elongation-related genes, including cell wall-related genes (Todaka et al., 2012). We observed that the overexpression of OsPIL1 also up-regulated several cell elongation-related genes and increased hypocotyl and stem lengths in Arabidopsis plants (Figures 2b, c, 5b, S3, S4 and S5). The double overexpressors showed improved drought stress tolerance and enhanced hypocotyl elongation compared with the vector control plants (Figures 2b, c and 3), suggesting that plant genetic engineering using gene-stacking approaches will be useful for generating transgenic plants that exhibit improved drought stress tolerance and enhanced growth.

The results of our transcriptome and metabolome analyses of the transgenic plants supported the idea that the two transcription factors function additively in the double overexpressors (Figures 4, 5 and 6). Regarding the improvement of drought stress tolerance, we observed an accumulation of compatible solutes, such as galactinol and raffinose, and the up-regulation of the corresponding biosynthesis genes in the double overexpressors (Figures 4c and S11, Table S2). Because these data were also reported in the DREB1A overexpressors (Maruyama et al., 2009), the drought stress tolerance of the double overexpressors appears to be increased via the same mechanisms in the DREB1A overexpressors. OsPIL1 overexpression increased the amount of citric acid and decreased the amount of glucose, fructose, ornithine, arginine and guanidine, and similar alterations were observed in the amounts of these metabolites in the double overexpressors (Figure 4c). Transcriptome analyses indicated that the expression levels of many downstream genes of OsPIL1 and DREB1A were also up-regulated in the double overexpressors (Figures 5b and S12b). The major downstream genes of the two transcription factors (RD29A, COR15A, GolS3 and PIL1) were up-regulated by more than 20-fold (Tables S2 and S3), while the expression levels of the up-regulated and down-regulated genes only in the double overexpressors were low (Tables S6 and S7). Our data suggest that cross-regulation among the downstream genes may not be critical in the double overexpressors. In addition, we found that both the G box and DRE (which target the cis-acting elements of OsPIL1 and DREB1A, respectively) were enriched in the promoters of the up-regulated genes (Figure 6a). These results indicate that OsPIL1 and DREB1A act additively on the promoters of each target gene and suggest that these two transcription factors are unlikely to negatively affect each other in the transgenic plants. Thus, the co-expression of OsPIL1 does not affect the DREB1A-mediated improvement of drought stress tolerance in the double overexpressors, indicating that OsPIL1 has the potential to improve the growth of transgenic plants tolerant to abiotic stress.

A late-flowering phenotype was one of the negative effects of DREB1A overexpression in Arabidopsis (Figure 2d, e), and it has also been reported in other DREB1-overexpressing plants, such as soybean and barley (Gilmour et al., 2004; Morran et al., 2011; Seo et al., 2009; Suo et al., 2012), indicating that DREB1 overexpression results in a prolonged vegetative stage and delays the harvest time of crops. In the present study, we demonstrated that the delayed flowering time of DREB1A-overexpressing plants was improved by OsPIL1 co-expression through the up-regulation of the floral inducers FT and MIR172A (Figures 2d, e and 5d; Aukerman and Sakai, 2003; Turck et al., 2008). Thus, the appropriate expression of these genes is important for controlling flowering time in agricultural applications, and OsPIL1 overexpression can be used to improve the delayed flowering times caused by transgenic DREB1A overexpression.

Arabidopsis PIF4 is known to promote shade avoidance syndrome (SAS), which consists of hypocotyl elongation, petiole elongation, stem elongation and hyponasty (Franklin et al., 2011; Hornitschek et al., 2012; Koini et al., 2009; Kumar et al., 2012), through the up-regulation of auxin-related genes (IAAs and SAURs), shade-inducible genes (ATHB2, PIL1 and HFR1) and cell wall-related genes (XTR7) (Casal, 2013). We showed that the overexpression of rice OsPIL1 enhanced the hypocotyl and stem elongation in Arabidopsis (Figures 2b, c, S3 and S4). In the double overexpressors, HFR1, PIL1, ATHB2 and SAUR mRNA levels were increased (Figures 5b and S12b). These results indicate that cell elongation in the 35SΩ:OsPIL1 plants is mainly caused by the elevated expression of SAS-related genes similar to PIF4. Despite the similar expression levels of SAS-related genes in the double overexpressors (Figure 5b), the hypocotyl length was shorter than that of the 35SΩ:OsPIL1 plants, suggesting that DREB1A overexpression negatively affects the hypocotyl cell elongation by regulating the expression of genes other than SAS-related genes.

The function of PIF family proteins in plant metabolic pathways has not been reported. In this study, we observed higher levels of citric acid and lower levels of glucose, fructose, ornithine and arginine in the 35SΩ:OsPIL1 plants compared with the vector control plants (Figure 4c). Higher or lower levels of these primary metabolites have also been reported in the pseudo response regulator (prr) 9-11 prr7-10 prr5-10 triple mutant (d975) of Arabidopsis (Fukushima et al., 2009). PRR9, 7 and 5 are known to function as transcriptional repressors in the circadian clock and repress the expression of PIF genes (Nakamichi et al., 2012). The d975 mutant shows an accumulation of PIF transcripts and hypocotyl elongation (Nakamichi et al., 2009). Therefore, PIF family proteins may act to maintain primary plant metabolism under the regulation of circadian clock components. According to our transcriptome data, the gene expression level of CELL WALL INVERTASE 5 (CWINV5), an enzyme that degrades sucrose into glucose and fructose, was decreased in the 35SΩ:OsPIL1 plants (Figure S11). This reduction in CWINV5 expression partly accounted for the decrease in fructose and glucose. Our results suggest that PIF family proteins function in the regulation of primary metabolism.

Compared with the hypocotyl and stem, the rosette leaf size and dry weight in the double overexpressors were not increased and remained similar to those in the 35SΩ:DREB1A plants (Figure S6a, b, e). The overexpression of OsPIL1 has been shown to enhance the internode elongation, but not flag leaf length, in rice (Todaka et al., 2012). Therefore, the overexpression of OsPIL1 might not affect the phenotypes of rosette leaves in the double overexpressors. We found that the EXPA gene expression levels decreased in the 35SΩ:DREB1A plants and double overexpressors (Figure 5c). The negative regulation of EXPA genes may be a cause of the dwarfed rosette leaves in the 35SΩ:DREB1A plants and double overexpressors.

We demonstrated that OsPIL1 overexpression partially enhances the plant growth in transgenic plants that overexpress OsPIL1 and DREB1A. The co-overexpression of two transcription factors is considered an effective strategy for plant genetic engineering designed to improve drought stress tolerance and plant growth. In addition, gene-stacking approaches appear to be able to tilt the balance in favour of growth in abiotic stress-tolerant transgenic plants. We expect that the growth of stress-tolerant plants will be further enhanced by the use of stress-specific promoters that control transgene expression. Moreover, as the rosette leaf size in the double overexpressors was not increased, the utilization of genes encoding other growth regulators for leaf expansion may be an effective strategy for increasing the biomass and yield of stress-tolerant plants. Our study provides a basis for overcoming growth retardation in stress-tolerant plants through genetic engineering by gene stacking.

Experimental procedures

Plant materials and growth conditions

Arabidopsis (*Arabidopsis thaliana*) ecotype Columbia plants were grown on Murashige–Skoog (MS) agar medium at 22°C under a 16-h/8-h light/dark cycle at a photon flux density of 40 ± 10 µmol/m^2/s as previously described (Yamaguchi-Shinozaki and Shinozaki, 1994).

Protein stability analysis in protoplasts

A protein stability analysis using the protoplasts derived from Arabidopsis mesophyll cells was performed according to Mizoi *et al.* (2013) with minor modifications. OsPIL1-sGFP and PIF4-sGFP were expressed under the control of the *CaMV 35S* promoter and TMV Ω sequence. Approximately 1.6×10^5 protoplasts were transformed with 33.25 µg of OsPIL1-sGFP or PIF4-sGFP plasmid and 1.75 µg of 3 × FLAG-sGFP. A 14-h incubation at 22°C in the dark was followed by incubation in a solution containing 50 µM MG132 or 0.5% (v/v) dimethyl sulphoxide (DMSO, as a solvent control) under white-light conditions (50 ± 5 µmol/m^2/s). After 2 and 4 h, the protoplasts were precipitated by centrifugation and dissolved in 40 µL of sample buffer. The extracts were denatured at 95°C for 3 min and then analysed by SDS–PAGE and immunoblotting. The sGFP-fused proteins were detected with a polyclonal antibody against GFP (Tanaka *et al.*, 2012). Horseradish peroxidase-conjugated anti-rabbit IgG antibodies (Pierce, Rockford, IL), ECL Plus (GE Healthcare, Chalfont, UK) and an ImageQuant LAS 4000 system (GE Healthcare) were used to visualize the signals.

Transient reporter assays with Arabidopsis and rice protoplasts

Transient expression assays using the protoplasts derived from Arabidopsis and rice mesophyll cells were performed as previously described (Kidokoro *et al.*, 2009; Matsukura *et al.*, 2010; Mizoi *et al.*, 2013). The effector plasmids were constructed as described in Figures 1 and S2. Plasmids containing 12 tandem repeats of a 28-bp G box-containing fragment from the rice *1-AMINOCY-CLOPROPANE-1-CARBOXYLATE OXIDASE1* (*OsACO1*) promoter (G box:*GUS*) or three tandem repeats of a 71-bp DRE-containing fragment from the *RD29A* promoter (DRE:*GUS*) were used as reporters (Liu *et al.*, 1998; Todaka *et al.*, 2012). An *emerald luciferase* (*ELUC*) reporter gene driven by the *CaMV 35S* promoter (*35SΩ:ELUC*) and a *luciferase* (*LUC*) reporter gene driven by the *maize ubiquitin* promoter (*Ubi:LUC*) were used as internal controls in the Arabidopsis and rice protoplasts, respectively (Matsukura *et al.*, 2010; Mizoi *et al.*, 2013).

Construction of plasmids and generation of *OsPIL1* and *DREB1A* double-overexpressing plants

We used pGKX and pGHX vectors as binary vectors (Fujita *et al.*, 2012; Qin *et al.*, 2008). The pGKX and pGHX vectors include genes that confer antibiotic resistance to kanamycin and hygromycin, respectively. To generate *OsPIL1* or *DREB1A* single overexpressors, the coding sequences were inserted into the backbone binary vectors to construct pGKX-*35SΩ:OsPIL1* and pGHX-*35SΩ:DREB1A* plasmids. Arabidopsis plants were transformed using the floral dip method (Clough and Bent, 1998). *OsPIL1* and *DREB1A* double-overexpressing Arabidopsis plants were generated by crossing *35SΩ:DREB1A* plants with *35SΩ:OsPIL1* plants (Table S1).

Drought stress tolerance test

A dehydration treatment was performed as previously described (Liu *et al.*, 1998) with minor modifications. The plants were grown on agar medium for 2 weeks and then in pots with soil for a week. Watering was withheld from 3-week-old plants for 2 weeks, and the survival rates were determined after 4 days of recovery following rehydration.

Metabolome analysis

Metabolites were extracted from the whole plants grown on agar medium for 2 weeks. Extraction, derivatization and GC-TOF-MS analyses were performed as previously described (Lisec *et al.*, 2006). The data were processed using TagFinder (Luedemann *et al.*, 2008) and Xcalibur software (Thermo Scientific, Waltham, MA), and the relative amounts were normalized using ribitol as an internal standard and the fresh weight of the transgenic plants.

Expression analysis

Total RNA was isolated from 2-week-old plants using RNAiso plus (TaKaRa Bio, Otsu, Japan) according to the manufacturer's instructions. cDNA synthesis and quantitative RT-PCR were performed as previously described (Sato *et al.*, 2014). The relative expression levels were calculated using the delta Ct method, and the obtained values were normalized to the *18S rRNA*. The primer sequences are shown in Table S8.

Microarray experiment and data processing

Genome-wide expression analyses using the Arabidopsis 3 Oligo Microarray were performed as previously described (Mizoi *et al.*, 2013) with minor modifications. Normalization and statistical analyses of the data were performed using Subio Platform software (https://www.subio.jp/). The signal intensities were normalized by the Lowess method, and the significance of the expression changes was evaluated by one-sample *t*-tests. The p values were corrected by the Benjamini–Hochberg false discovery rate (FDR) method, and probes with an FDR of less than 0.05 were used for further analyses. All microarray data are available at Array Express under accession numbers E-MTAB-4073 and E-MTAB-4739.

Acknowledgements

We thank Y. Tanaka and S. Alseekh for providing technical assistance, E. Toma for providing skilful editorial assistance, K. Kodaira for providing material support and T. Tohge for providing fruitful discussions. This work was supported by a Grant-in-Aid for JSPS Fellows (No. 16J01053 to M.K.), for Scientific Research on Innovative Areas (No. 15H05960 to K.Y.-S.) and for Scientific Research (A) (No. 25251031 to K.Y.-S.), JSPS 'Strategic Young Researcher Overseas Visits Program for Accelerating Brain Circulation' (No. S2503) from the Japan Society for the Promotion of Science, and Technology of Japan and by the Program for the Promotion of Basic Research Activities for Innovative Biosciences (BRAIN) of Japan (to K.Y.-S.). The authors declare no conflict of interest.

References

Aukerman, M.J. and Sakai, H. (2003) Regulation of flowering time and floral organ identity by a microRNA and its APETALA2-like target genes. *Plant Cell*, **15**, 2730–2741.

Bao, G., Zhuo, C., Qian, C., Xiao, T., Guo, Z. and Lu, S. (2016) Co-expression of NCED and ALO improves vitamin C level and tolerance to drought and chilling in transgenic tobacco and stylo plants. *Plant Biotechnol. J.* **14**, 206–214.

Bhatnagar-Mathur, P., Devi, M.J., Reddy, D.S., Lavanya, M., Vadez, V., Serraj, R., Yamaguchi-Shinozaki, K. *et al.* (2007) Stress-inducible expression of At DREB1A in transgenic peanut (Arachis hypogaea L.) increases transpiration efficiency under water-limiting conditions. *Plant Cell Rep.* **26**, 2071–2082.

Bhatnagar-Mathur, P., Rao, J.S., Vadez, V., Dumbala, S.R., Rathore, A., Yamaguchi-Shinozaki, K. and Sharma, K.K. (2014) Transgenic peanut overexpressing the DREB1A transcription factor has higher yields under drought stress. *Mol. Breed.* **33**, 327–340.

Carriere, Y., Crickmore, N. and Tabashnik, B.E. (2015) Optimizing pyramided transgenic Bt crops for sustainable pest management. *Nat. Biotechnol.* **33**, 161–168.

Casal, J.J. (2013) Photoreceptor signaling networks in plant responses to shade. *Annu. Rev. Plant Biol.* **64**, 403–427.

Claeys, H. and Inze, D. (2013) The agony of choice: how plants balance growth and survival under water-limiting conditions. *Plant Physiol.* **162**, 1768–1779.

Clough, S.J. and Bent, A.F. (1998) Floral dip: a simplified method for Agrobacterium-mediated transformation of Arabidopsis thaliana. *Plant J.* **16**, 735–743.

Cosgrove, D.J. (2000) Loosening of plant cell walls by expansins. *Nature*, **407**, 321–326.

FAO, IFAD and WFP (2015) *The State of Food Insecurity in the World 2015. Meeting the 2015 International Hunger Targets: Taking Stock of Uneven Progress*. Rome: Food and Agriculture Organization of the United Nations.

Franklin, K.A., Lee, S.H., Patel, D., Kumar, S.V., Spartz, A.K., Gu, C., Ye, S. *et al.* (2011) Phytochrome-interacting factor 4 (PIF4) regulates auxin biosynthesis at high temperature. *Proc. Natl Acad. Sci. USA*, **108**, 20231–20235.

Fujita, M., Fujita, Y., Iuchi, S., Yamada, K., Kobayashi, Y., Urano, K., Kobayashi, M. *et al.* (2012) Natural variation in a polyamine transporter determines paraquat tolerance in Arabidopsis. *Proc. Natl Acad. Sci. USA*, **109**, 6343–6347.

Fukushima, A., Kusano, M., Nakamichi, N., Kobayashi, M., Hayashi, N., Sakakibara, H., Mizuno, T. *et al.* (2009) Impact of clock-associated Arabidopsis pseudo-response regulators in metabolic coordination. *Proc. Natl Acad. Sci. USA*, **106**, 7251–7256.

Gilmour, S.J., Fowler, S.G. and Thomashow, M.F. (2004) Arabidopsis transcriptional activators CBF1, CBF2, and CBF3 have matching functional activities. *Plant Mol. Biol.* **54**, 767–781.

Halpin, C. (2005) Gene stacking in transgenic plants – the challenge for 21st century plant biotechnology. *Plant Biotechnol. J.* **3**, 141–155.

Hanson, A.D., Rathinasabapathi, B., Rivoal, J., Burnet, M., Dillon, M.O. and Gage, D.A. (1994) Osmoprotective compounds in the Plumbaginaceae: a natural experiment in metabolic engineering of stress tolerance. *Proc. Natl Acad. Sci. USA*, **91**, 306–310.

Hornitschek, P., Kohnen, M.V., Lorrain, S., Rougemont, J., Ljung, K., Lopez-Vidriero, I., Franco-Zorrilla, J.M. *et al.* (2012) Phytochrome interacting factors 4 and 5 control seedling growth in changing light conditions by directly controlling auxin signaling. *Plant J.* **71**, 699–711.

Ito, Y., Katsura, K., Maruyama, K., Taji, T., Kobayashi, M., Seki, M., Shinozaki, K. *et al.* (2006) Functional analysis of rice DREB1/CBF-type transcription factors involved in cold-responsive gene expression in transgenic rice. *Plant Cell Physiol.* **47**, 141–153.

Kasuga, M. (2004) A combination of the Arabidopsis DREB1A gene and stress-inducible rd29A promoter improved drought- and low-temperature stress tolerance in tobacco by gene transfer. *Plant Cell Physiol.* **45**, 346–350.

Kasuga, M., Liu, Q., Miura, S., Yamaguchi-Shinozaki, K. and Shinozaki, K. (1999) Improving plant drought, salt, and freezing tolerance by gene transfer of a single stress-inducible transcription factor. *Nat. Biotechnol.* **17**, 287–291.

Kidokoro, S., Maruyama, K., Nakashima, K., Imura, Y., Narusaka, Y., Shinwari, Z.K., Osakabe, Y. *et al.* (2009) The phytochrome-interacting factor PIF7 negatively regulates DREB1 expression under circadian control in Arabidopsis. *Plant Physiol.* **151**, 2046–2057.

Kidokoro, S., Watanabe, K., Ohori, T., Moriwaki, T., Maruyama, K., Mizoi, J., Myint Phyu Sin Htwe, N. *et al.* (2015) Soybean DREB1/CBF-type transcription factors function in heat and drought as well as cold stress-responsive gene expression. *Plant J.*, **81**, 505–518.

Koini, M.A., Alvey, L., Allen, T., Tilley, C.A., Harberd, N.P., Whitelam, G.C. and Franklin, K.A. (2009) High temperature-mediated adaptations in plant architecture require the bHLH transcription factor PIF4. *Curr. Biol.* **19**, 408–413.

Kumar, S.V., Lucyshyn, D., Jaeger, K.E., Alos, E., Alvey, E., Harberd, N.P. and Wigge, P.A. (2012) Transcription factor PIF4 controls the thermosensory activation of flowering. *Nature*, **484**, 242–U127.

Kunihiro, A., Yamashino, T., Nakamichi, N., Niwa, Y., Nakanishi, H. and Mizuno, T. (2011) Phytochrome-interacting factor 4 and 5 (PIF4 and PIF5) activate the homeobox ATHB2 and auxin-inducible IAA29 genes in the coincidence mechanism underlying photoperiodic control of plant growth of Arabidopsis thaliana. *Plant Cell Physiol.* **52**, 1315–1329.

Leivar, P. and Quail, P.H. (2011) PIFs: pivotal components in a cellular signaling hub. *Trends Plant Sci.* **16**, 19–28.

Lisec, J., Schauer, N., Kopka, J., Willmitzer, L. and Fernie, A.R. (2006) Gas chromatography mass spectrometry-based metabolite profiling in plants. *Nat. Protoc.* **1**, 387–396.

Liu, Q., Kasuga, M., Sakuma, Y., Abe, H., Miura, S., Yamaguchi-Shinozaki, K. and Shinozaki, K. (1998) Two transcription factors, DREB1 and DREB2, with an EREBP/AP2 DNA binding domain separate two cellular signal transduction pathways in drought- and low-temperature-responsive gene expression, respectively, in Arabidopsis. *Plant Cell*, **10**, 1391–1406.

de Lucas, M., Daviere, J.-M., Rodriguez-Falcon, M., Pontin, M., Iglesias-Pedraz, J.M., Lorrain, S., Fankhauser, C. *et al.* (2008) A molecular framework for light and gibberellin control of cell elongation. *Nature*, **451**, 480–U411.

Luedemann, A., Strassburg, K., Erban, A. and Kopka, J. (2008) TagFinder for the quantitative analysis of gas chromatography - mass spectrometry (GC-MS)-based metabolite profiling experiments. *Bioinformatics*, **24**, 732–737.

Maruyama, K., Sakuma, Y., Kasuga, M., Ito, Y., Seki, M., Goda, H., Shimada, Y. *et al.* (2004) Identification of cold-inducible downstream genes of the Arabidopsis DREB1A/CBF3 transcriptional factor using two microarray systems. *Plant J.* **38**, 982–993.

Maruyama, K., Takeda, M., Kidokoro, S., Yamada, K., Sakuma, Y., Urano, K., Fujita, M. *et al.* (2009) Metabolic pathways involved in cold acclimation identified by integrated analysis of metabolites and transcripts regulated by DREB1A and DREB2A. *Plant Physiol.* **150**, 1972–1980.

Maruyama, K., Todaka, D., Mizoi, J., Yoshida, T., Kidokoro, S., Matsukura, S., Takasaki, H. *et al.* (2012) Identification of cis-acting promoter elements in cold- and dehydration-induced transcriptional pathways in Arabidopsis, rice, and soybean. *DNA Res.* **19**, 37–49.

Matsukura, S., Mizoi, J., Yoshida, T., Todaka, D., Ito, Y., Maruyama, K., Shinozaki, K. *et al.* (2010) Comprehensive analysis of rice DREB2-type genes that encode transcription factors involved in the expression of abiotic stress-responsive genes. *Mol. Genet. Genomics*, **283**, 185–196.

Mizoi, J., Ohori, T., Moriwaki, T., Kidokoro, S., Todaka, D., Maruyama, K., Kusakabe, K. *et al.* (2013) GmDREB2A;2, a canonical DEHYDRATION-RESPONSIVE ELEMENT-BINDING PROTEIN2-type transcription factor in soybean, is posttranslationally regulated and mediates dehydration-responsive element-dependent gene expression. *Plant Physiol.* **161**, 346–361.

Morran, S., Eini, O., Pyvovarenko, T., Parent, B., Singh, R., Ismagul, A., Eliby, S. *et al.* (2011) Improvement of stress tolerance of wheat and barley by modulation of expression of DREB/CBF factors. *Plant Biotechnol. J.* **9**, 230–249.

Nakamichi, N., Kusano, M., Fukushima, A., Kita, M., Ito, S., Yamashino, T., Saito, K. *et al.* (2009) Transcript profiling of an Arabidopsis PSEUDO RESPONSE REGULATOR arrhythmic triple mutant reveals a role for the circadian clock in cold stress response. *Plant Cell Physiol.* **50**, 447–462.

Nakamichi, N., Kiba, T., Kamioka, M., Suzuki, T., Yamashino, T., Higashiyama, T., Sakakibara, H. *et al.* (2012) Transcriptional repressor PRR5 directly regulates clock-output pathways. *Proc. Natl Acad. Sci. USA*, **109**, 17123–17128.

Nakamura, Y., Kato, T., Yamashino, T., Murakami, M. and Mizuno, T. (2007) Characterization of a set of phytochrome-interacting factor-like bHLH proteins in Oryza sativa. *Biosci. Biotechnol. Biochem.* **71**, 1183–1191.

Ni, M., Tepperman, J.M. and Quail, P.H. (1999) Binding of phytochrome B to its nuclear signalling partner PIF3 is reversibly induced by light. *Nature*, **400**, 781–784.

Nomoto, Y., Kubozono, S., Yamashino, T., Nakamichi, N. and Mizuno, T. (2012) Circadian clock- and PIF4-controlled plant growth: a coincidence mechanism directly integrates a hormone signaling network into the photoperiodic control of plant architectures in Arabidopsis thaliana. *Plant Cell Physiol.* **53**, 1950–1964.

Nozue, K., Covington, M.F., Duek, P.D., Lorrain, S., Fankhauser, C., Harmer, S.L. and Maloof, J.N. (2007) Rhythmic growth explained by coincidence between internal and external cues. *Nature*, **448**, 358–U311.

Oh, E., Zhu, J.-Y. and Wang, Z.-Y. (2012) Interaction between BZR1 and PIF4 integrates brassinosteroid and environmental responses. *Nat. Cell Biol.* **14**, 802–U864.

Pellegrineschi, A., Reynolds, M., Pacheco, M., Brito, R.M., Almeraya, R., Yamaguchi-Shinozaki, K. and Hoisington, D. (2004) Stress-induced expression in wheat of the Arabidopsis thaliana DREB1A gene delays water stress symptoms under greenhouse conditions. *Genome*, **47**, 493–500.

Pino, M.-T., Skinner, J.S., Park, E.-J., Jeknic, Z., Hayes, P.M., Thornashow, M.F. and Chen, T.H.H. (2007) Use of a stress inducible promoter to drive ectopic AtCBF expression improves potato freezing tolerance while minimizing negative effects on tuber yield. *Plant Biotechnol. J.* **5**, 591–604.

Qin, F., Sakuma, Y., Tran, L.-S.P., Maruyama, K., Kidokoro, S., Fujita, Y., Fujita, M. *et al.* (2008) Arabidopsis DREB2A-interacting proteins function as RING E3 ligases and negatively regulate plant drought stress-responsive gene expression. *Plant Cell*, **20**, 1693–1707.

Reed, J.W. (2001) Roles and activities of Aux/IAA proteins in Arabidopsis. *Trends Plant Sci.* **6**, 420–425.

Ren, H. and Gray, W.M. (2015) SAUR proteins as effectors of hormonal and environmental signals in plant growth. *Mol. Plant*, **8**, 1153–1164.

Sakuraba, Y., Jeong, J., Kang, M.-Y., Kim, J., Paek, N.-C. and Choi, G. (2014) Phytochrome-interacting transcription factors PIF4 and PIF5 induce leaf senescence in Arabidopsis. *Nat. Commun.* **5**, 4636.

Sato, H., Mizoi, J., Tanaka, H., Maruyama, K., Qin, F., Osakabe, Y., Morimoto, K. *et al.* (2014) Arabidopsis DPB3-1, a DREB2A interactor, specifically enhances heat stress-induced gene expression by forming a heat stress-specific transcriptional complex with NF-Y subunits. *Plant Cell*, **26**, 4954–4973.

Seo, E., Lee, H., Jeon, J., Park, H., Kim, J., Noh, Y.-S. and Lee, I. (2009) Crosstalk between cold response and flowering in Arabidopsis is mediated through the flowering-time gene SOC1 and its upstream negative regulator FLC. *Plant Cell*, **21**, 3185–3197.

Suo, H., Ma, Q., Ye, K., Yang, C., Tang, Y., Hao, J., Zhang, Z.J. *et al.* (2012) Overexpression of AtDREB1A causes a severe dwarf phenotype by decreasing endogenous gibberellin levels in soybean [Glycine max (L.) Merr]. *PLoS ONE*, **7**, e45568.

Taji, T., Ohsumi, C., Iuchi, S., Seki, M., Kasuga, M., Kobayashi, M., Yamaguchi-Shinozaki, K. *et al.* (2002) Important roles of drought- and cold-inducible genes for galactinol synthase in stress tolerance in Arabidopsis thaliana. *Plant J.* **29**, 417–426.

Tanaka, H., Osakabe, Y., Katsura, S., Mizuno, S., Maruyama, K., Kusakabe, K., Mizoi, J. *et al.* (2012) Abiotic stress-inducible receptor-like kinases negatively control ABA signaling in Arabidopsis. *Plant J.* **70**, 599–613.

Todaka, D., Nakashima, K., Maruyama, K., Kidokoro, S., Osakabe, Y., Ito, Y., Matsukura, S. *et al.* (2012) Rice phytochrome-interacting factor-like protein OsPIL1 functions as a key regulator of internode elongation and induces a morphological response to drought stress. *Proc. Natl Acad. Sci. USA*, **109**, 15947–15952.

Turck, F., Fornara, F. and Coupland, G. (2008) Regulation and identity of florigen: FLOWERING LOCUS T moves center stage. *Annu. Rev. Plant Biol.* **59**, 573–594.

Yamaguchi-Shinozaki, K. and Shinozaki, K. (1994) A novel cis-acting element in an Arabidopsis gene is involved in responsiveness to drought, low-temperature, or high-salt stress. *Plant Cell*, **6**, 251–264.

Yamaguchi-Shinozaki, K. and Shinozaki, K. (2006) Transcriptional regulatory networks in cellular responses and tolerance to dehydration and cold stresses. *Annu. Rev. Plant Biol.* **57**, 781–803.

Ye, X.D., Al-Babili, S., Kloti, A., Zhang, J., Lucca, P., Beyer, P. and Potrykus, I. (2000) Engineering the provitamin A (beta-carotene) biosynthetic pathway into (carotenoid-free) rice endosperm. *Science*, **287**, 303–305.

15

An advanced reference genome of *Trifolium subterraneum* L. reveals genes related to agronomic performance

Parwinder Kaur[1],* (iD), Philipp E. Bayer[2], Zbyněk Milec[3], Jan Vrána[3], Yuxuan Yuan[2] (iD), Rudi Appels[4], David Edwards[2], Jacqueline Batley[2], Phillip Nichols[2,5], William Erskine[1,†] and Jaroslav Doležel[3,†]

[1]*Centre for Plant Genetics and Breeding and Institute of Agriculture, The University of Western Australia, Crawley, WA, Australia*
[2]*School of Plant Biology and Institute of Agriculture, The University of Western Australia, Crawley, WA, Australia*
[3]*Institute of Experimental Botany, Centre of the Region Haná for Biotechnological and Agricultural Research, Olomouc, Czech Republic*
[4]*Murdoch University, Murdoch, WA, Australia*
[5]*Department of Agriculture and Food Western Australia, South Perth, WA, Australia*

Correspondence
email Parwinder.kaur@uwa.edu.au
†Contributed equally as senior authors.

Summary

Subterranean clover is an important annual forage legume, whose diploidy and inbreeding nature make it an ideal model for genomic analysis in *Trifolium*. We reported a draft genome assembly of the subterranean clover TSUd_r1.1. Here we evaluate genome mapping on nanochannel arrays and generation of a transcriptome atlas across tissues to advance the assembly and gene annotation. Using a BioNano-based assembly spanning 512 Mb (93% genome coverage), we validated the draft assembly, anchored unplaced contigs and resolved misassemblies. Multiple contigs (264) from the draft assembly coalesced into 97 super-scaffolds (43% of genome). Sequences longer than >1 Mb increased from 40 to 189 Mb giving 1.4-fold increase in N50 with total genome in pseudomolecules improved from 73 to 80%. The advanced assembly was re-annotated using transcriptome atlas data to contain 31 272 protein-coding genes capturing >96% of the gene content. Functional characterization and GO enrichment confirmed gene expression for response to water deprivation, flavonoid biosynthesis and embryo development ending in seed dormancy, reflecting adaptation to the harsh Mediterranean environment. Comparative analyses across Papilionoideae identified 24 893 *Trifolium*-specific and 6325 subterranean-clover-specific genes that could be mined further for traits such as geocarpy and grazing tolerance. Eight key traits, including persistence, improved livestock health by isoflavonoid production in addition to important agro-morphological traits, were fine-mapped on the high-density SNP linkage map anchored to the assembly. This new genomic information is crucial to identify loci governing traits allowing marker-assisted breeding, comparative mapping and identification of tissue-specific gene promoters for biotechnological improvement of forage legumes.

Keywords: forage legumes, advanced reference assembly, BioNano, transcriptome, gene expression, Legume comparative genomics.

Introduction

Forage legumes are highly valued feed for extensive livestock production. There is an increasing interest worldwide in using annual forage legumes as cover crops to supply soil nitrogen (Sulas, 2005; Piano *et al.*, 2010). Symbiotic nitrogen fixation in legumes leads to high protein fodder content and rejuvenated soils for a sustainable feed system. The clovers in particular are among the most effective to break the 'infernal circle of the fallow' a technique known to the Germans as 'Besömmerung' (Blanning, 2008). Subterranean clover (*Trifolium subterraneum* L.) makes the greatest contribution to livestock feed production and soil improvement in terms of total worldwide usage among annual clovers (McGuire, 1985), particularly in Australia, where it is sown over 29 mill. ha. The self-reseeding ability and grazing tolerance of subterranean clover, even under suboptimal and variable environmental conditions (Nichols *et al.*, 2013), contribute to its widespread distribution.

Subterranean clover is a diploid (2n = 2x = 16), predominantly inbreeding, annual species with a relatively small genome size of 540 Mbp (1C = 0.55 pg DNA; Vižintin *et al.*, 2006) that can be readily hybridized, and exhibits wide diversity for both qualitative and quantitative agronomic and morphological characters (Ghamkhar *et al.*, 2011). Within the genus *Trifolium*, it is established as a reference species for genetic and genomic studies. We reported a *de novo* draft genome assembly of the subterranean clover TSUd_r1.1, generated using a combination of long- and short-read sequencing platforms (Figure S1) (Hirakawa *et al.*, 2016). Genetic and genomic analyses of the internationally commercially important perennial legumes white clover and red clover are difficult, as they are outcrossing and have self-incompatible fertilization, with white clover also being an allotetraploid (2n = 4x = 32) (Abberton and Marshall, 2005; Ghamkhar *et al.*, 2011). As molecular markers developed for white and red clovers were readily transferable to subterranean clover (Ghamkhar *et al.*, 2011), it is likely that subterranean clover QTLs and genes are applicable to these other species. Close synteny with the model legume, *Medicago truncatula*, also provides opportunities for genomic comparisons and the identification of candidate genes.

A variety of approaches are required for the *de novo* assemblies to improve draft genomes, and include a method developed by Dovetail Genomics for genome scaffolding using long-range genomic information obtained by Chicago method (Putnam *et al.*, 2016). In this study, we evaluated genome mapping on nanochannel arrays employing BioNano Irys® system (www.bionanogenomics.com) to validate the initial draft assembly, anchor additional unplaced contigs and to resolve misassemblies, with the objective of improving the overall assembly coverage. BioNano genome mapping is a technique of optical mapping in which specific sequence motifs in single DNA molecules are fluorescently labelled. The labelled DNA molecules are loaded onto the IrysChip where they are electrophoretically linearized in thousands of silicon channels. Fluorescence imaging allows the construction of maps of the physical distances between occurrences of the sequence motifs (Lam *et al.*, 2012). The aim was to advance the pseudomolecule assembly of the eight chromosomes of subterranean clover, based on the direct visualizations of sequence motifs on long single DNA molecules. Gene annotations for the *de novo* draft genome assembly of the subterranean clover TSUd_r1.1 were conducted using *in silico* automated Augustus and Maker pipelines only. This was compared with direct transcriptome analysis using whole genome RNA sequencing technology across five different tissues of subterranean clover to generate a valuable resource for the identification and characterization of genes and pathways underlying plant growth and development. This also provided a basis to investigate specific processes, biological functions and gene interactions for key agronomic traits.

Results and discussion

The advanced assembly

The first draft genome assembly of the Australian subterranean clover variety, *cv.* Daliak, covers 85.4% of the estimated genome in eight pseudomolecules of 401.1 Mb length and was constructed using a linkage map consisting of 35 341 SNPs (Hirakawa *et al.*, 2016). In this study, we evaluated genome mapping *via* scaffolding using *de novo* physical maps generated using the BioNano Genomics (BNG) Irys platform to improve the genome assembly. A total of 221.7 Gb (401× genome coverage) of filtered data (molecules >150 kb) was generated on the Irys instrument. After filtering out low-quality single molecules, a total of 188.5 Gb (341× genome coverage) of data was included in the final BioNano-based *de novo* physical map assembly. This physical map assembly consisted of 309 075 individual molecules and 468 consensus maps that spanned 512 Mb (93% genome coverage) with N50 of 1.4 Mb. Multiple contigs (264) from the first draft assembly coalesced into 97 super-scaffolds (containing 43% of the total genome captured) (Figure 1). In the advanced assembly (Tsub_Refv2.0), the size of sequences longer than >1 Mb increased dramatically from 40 to 189 Mb. This resulted in a 1.4-fold increase in the N50 with the total percentage of genome captured in pseudomolecules improving from 73 to 80% with a substantial reduction of sequence gaps (Table 1).

In this study, the BNG Irys platform provided affordable, high-throughput physical maps of improved contiguity to validate the draft assembly (Hirakawa *et al.*, 2016) generated across a combination of long-read and short-read platforms and extended scaffolds. In contrast to alternative, sequencing-based approaches (e.g. Chicago method of Dovetail Genomics,

GemCode Technology of 10× Genomics, or Hi-C), it enables a highly accurate sizing of gaps in sequence assemblies and provides a real picture of genomic regions intractable to current sequencing technologies, such as long arrays of tandem repeats (Staňková *et al.*, 2016). Thus, combinations of platforms are recommended and no one platform is a perfect technology to use in answering every research question (Chaney *et al.*, 2016).

The transcriptome ATLAS and gene features

The identification and annotation of expressed genes within the *T. subterraneum* genome assembly used high-throughput whole genome RNA sequencing analyses to predict a total of 32 333 transcripts for 31 272 protein-coding genes, with evidence for their expression across five different tissue types. The process-involved annotations that combined evidence from transcriptome alignments obtained from protein homology and *in silico* gene prediction derived from different tissue types (roots, stem and peduncles, leaf and petioles, flowers and developing seeds) of *cv.* Daliak (Table S1). Phytozome and TrEMBL were the most informative databases for assigning functional annotations to subterranean clover proteins, with 29 157 (90.2%) and 29 278 (89.9%) proteins annotated, respectively (Table S2). In the draft TSUd_r1.1, the presence of 42 706 genes was predicted from *in silico* evidence using homology studies, domain searches and *ab initio* gene predictions. The reduction in genes annotated in the present study was achieved because the latter computational methods are unable to provide definitive evidence about which genes are actually expressed (Guigó *et al.*, 2000; Wheelan and Boguski, 1998).

Repetitive element analysis predicted that 64% of the genome comprises repeat sequences, of which 15.9% and 8.4% are within introns and exons, respectively. About a quarter (23.9%) of repetitive elements were classified in the unknown zone. Like most eukaryotes, transposable elements (TEs) were most commonly long terminal repeats (LTR) retrotransposons (9.3%), with *Gypsy*-like elements as the most frequently classified retrotransposons. Other TEs such as DNA, rolling circle (RC) and non-LTR long retrotransposons such as long interspersed nuclear elements (LINEs) and short interspersed nuclear elements (SINEs) comprised a relatively small proportion (2.2% and 1.9%, respectively) (Figure 2; Table S3). Non-coding RNA was estimated to comprise 0.12% of the genome, the majority being ribosomal RNA (0.02%) and transfer RNA (0.01%), with predicted snRNA and miRNA representing 0.01% and 0.07%, respectively (Table S4).

To test the quality of the advanced genome assembly and re-annotation, we conducted a CEGMA analysis (Parra *et al.*, 2007) to identify the presence of core eukaryotic genes. From the core set of 248 eukaryotic genes, there were 240 complete and 247 partial genes present in the advanced assembly, representing 96.8% and 99.6% [in comparison with the 95.6% and 98.0% for the draft assembly TSUd_r1.1] genes of the core set, respectively (Table S5).

Functional characterization and GO enrichment analyses of the advanced genome

The advanced assembly was functionally characterized by assigning gene ontology (GO) terms to proteins by manually transferring the GO terms for Swiss-Prot IDs using the UniProt-GOA database (Huntley *et al.*, 2015). All these protein-coding genes/transcripts were grouped into the three main GO categories:

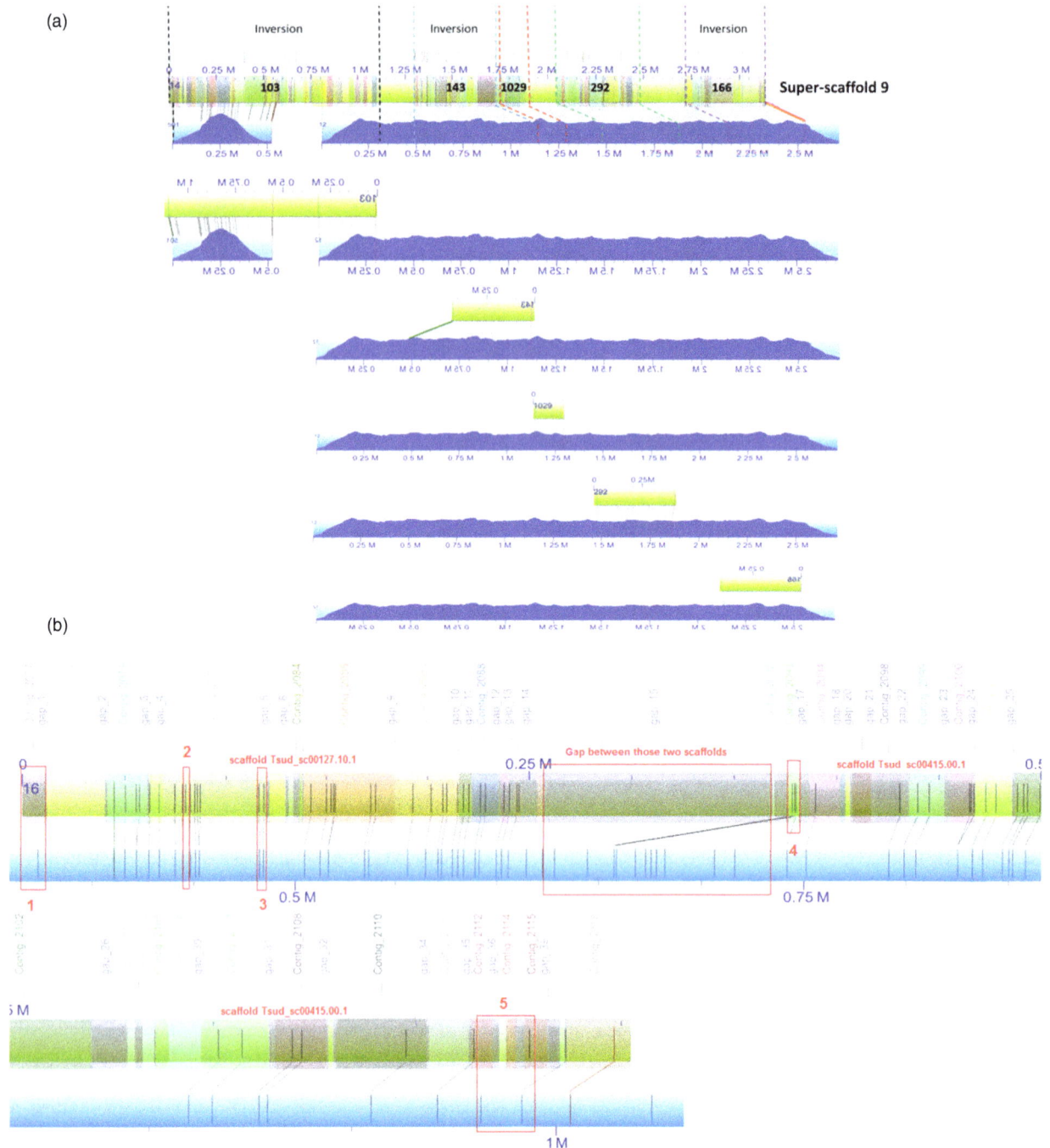

Figure 1 (a) *In silico* map of super-scaffold 9 aligned to the BioNano optical maps (401× coverage). XMAP alignments for *in silico* map of sequence scaffolds 103, 143, 1029, 292 and 166 are shown. Consensus genome maps (blue with molecule map coverage shown in dark blue) align to the *in silico* maps of scaffolds (green with contigs overlaid as translucent coloured squares). An illustration of using BioNano optical maps to assist contig placement, scaffolding and inversion correction. Sequence scaffolds 103, 143, 1029, 292 and 166 were placed within super-scaffold 9 by the optical maps and among these scaffolds; scaffolds 103, 143 and 166 were reversed in the super-scaffold. (b) Illustration of misassemblies in the genome examined using BioNano optical maps. The super-scaffold (colourful bar) contains scaffold Tsud_sc00127.10.1, scaffold Tsud_sc00415.00.1 and a gap between them. Compared with the optical map (blue bar), there are more *Bsp*QI restriction cut sites (displayed as straight lines) in the super-scaffold at positions 2; there are some *Bsp*QI restriction cut sites missing at positions 1. At positions 3, 4 and 5, the super-scaffold does not consist of BioNano consensus map. All those discordances can be examined in detail in the corresponding contigs or gaps (indicated above the super-scaffold).

biological processes, molecular function and cellular components. A total of 21 210 (65.6%) of the subterranean clover protein-coding genes/transcripts were assigned GO terms (Table S1). Of the 131 324 GO terms identified, 5,648 appear only once. There are 1013 GO terms appearing between 50 and 200 times with a total sum of 35 383 (Figure S2).

Among the 5648 GO terms appearing once, the top five most highly represented groups in the biological processes category

Table 1 BioNano genome mapping statistics (*cv.* Daliak)

N Genome Maps	468
Total Genome Map Len (Mb)	512.439
Avg. Genome Map Len (Mb)	1.095
Genome Map N50 (Mb)	1.408
Molecule Stats	
Contig Coverage (x)	367.81
Molecules Aligned to the first draft genome assembly TSUd_r1.1	
N mol align	30 9075
Mol fraction align	0.37
Tot align len (Mb)	59930.8
Avg align len (kb)	193.9

were genes/transcripts associated with signal transduction, response to abscisic acid, cellular response to DNA damage stimulus, methylation and response to water deprivation (Table S6, for enriched cellular components and molecular functions see Tables S7 and S8, respectively). More detailed classification of the biological process GO category also showed enrichment in comparison with whole UniProt for response to cold and oxidative stress, flower development, flavonoid biosynthesis and embryo development ending in seed dormancy as highly represented groups. Overall, this functional characterization and GO enrichment analysis confirmed gene expression for response to water deprivation, cold and oxidative stress, flavonoid biosynthetic process and embryo development ending in seed

dormancy reflecting the adaptation of subterranean clover to the harsh Mediterranean environment.

Global gene expression trends

To investigate the representation of genes among the five tissue types, hierarchical clustering of tissues, based on global gene expression and GO terms, was performed on 100 genes with the highest sum of all transcripts per million (TPM). The biological identity of the tissues was clearly reflected in the analysis by hierarchical clustering of the RNA-Seq-based transcriptome (Figure 3; Table S9). Predicted genes showed substantial variation in their expression over tissues, reflecting tissue-specific biological activities. For instance, in leaf and petiole tissue GO-based clusters G1 and G2 were enriched and genes-encoding proteins involved in photosynthesis, binding and chloroplast thylakoid membranes were over-represented consistent with the specialized biological function of these tissues. Likewise, the root transcriptome profile showed enriched expression for stress and defence responses with genes encoding for cysteine-type endopeptidase inhibitor activity. However, the G3 cluster indicates commonalities in the transcriptome of all above ground vegetative tissue types and showed enrichments for translational and structural ribosomal pathways.

For tissues with predominance of a specific biological activity, a large proportion of reads may have represented abundantly expressed genes and, therefore, deeper sequencing may be needed to detect genes with relatively low expression levels. However, the high correlation between biological replicates and

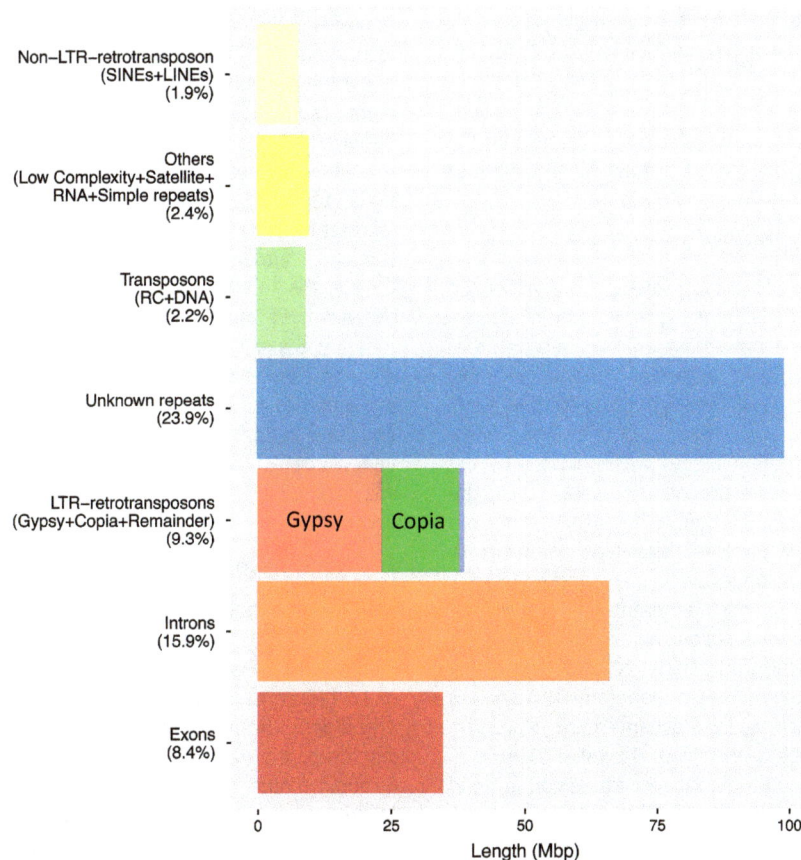

Figure 2 Functional composition of the assembled reference sequence *Trifolium subterraneum L.* genome.

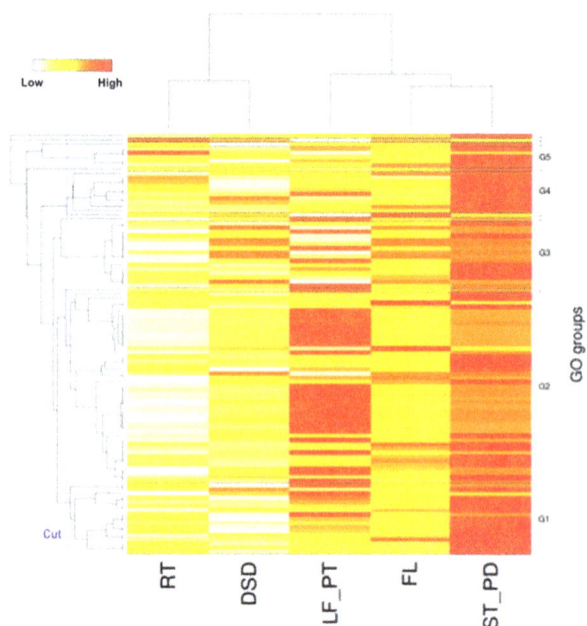

Figure 3 Heat map showing hierarchical clustering of tissues based on global gene expression and GO terms. Clustering was based on 100 genes with highest sum of all TPM (transcripts per million) of the five tissue types. The dendrogram of the selected genes [vertical axis] is visualized along with the expression patterns that are used to cluster the five tissue samples, i.e. root (RT), developing seed (DSD), leaf and petioles (LF_PT), stem and peduncles (ST_PD) and flower (FL) [horizontal axis]. There are five GO-based clusters named G1...G5 that contain more than one gene. The GO clusters are separated by a horizontal bar in the heat map. Genes without annotations are omitted from the heat map. Highly expressed genes are shown with red colour and lower expressed genes with white or yellow colour, respectively.

the clustering of tissues based on their biology indicates that the sampling depth in this study is sufficient to draw inferences about the transcriptome.

Dynamic spatial gene expression analysis

The tissues sampled in the RNA library preparations allow the determination of dynamic spatial gene expression and characterization. A comparative gene expression analysis between five tissue types (Table S1), using the advanced genome assembly (Tsub_Refv2.0) as the reference, revealed tissue-specific functional diversification of paralogous genes. Specific gene expression in each of the five different tissues was indicated by the comparison of global gene expression across tissues. To explore gene expression by tissue-type further, differentially expressed genes were identified in all pairwise tissue-type comparisons. These differentially expressed genes were then used to identify clusters of co-expressed genes, which represent spatially enriched expression patterns. Such co-expression clusters were identified by the Mfuzz package (Futschik and Carlisle, 2005) using the soft clustering approach with fuzzy c-means algorithm. This analysis generated five clusters, which revealed dynamic gene expression patterns across the five tissue types (Figure 4a; Table S10).

Cluster 1 included genes with the highest expression in floral tissue (Figure 4b). Among genes in this cluster several are of interest. For example, this cluster includes many genes defining floral organ identity such as floral homeotic protein AGAMOUS, APETALA 2, PMADS 2 and DEFICIENS. In addition, Cluster 1 also

included chromatin structure-remodelling complex protein SYD, MADS-box protein CMB1 and many chloroplast and carotenoid genes. Leaf-enriched Cluster 2 included transcripts-encoding enzymes involved in photosynthesis, Rubisco and oxygenase activity in addition to many chloroplast and plastid-regulating genes. Cluster 3 contained genes with high expression in stem and peduncle tissue, and included genes/transcripts-encoding enzymes involved in nutrient transport, structural growth with a high activity of various receptor kinases, expansins and transferases. Developing seed-enriched Cluster 4 showed genes genes/transcripts involved with embryogenesis and growth. Cluster 5 enriched with root tissue gene expression profiles included different nodulins, binding proteins, MYB-transcription factors and both stress- and defence-related genes.

Genes within different clusters are potential sources of tissue-specific promoters. For example, to improve forage nutritional quality by isoflavonoid biosynthesis or digestibility through processes such as cell wall loosening and lignin biosynthesis, promoters specific to leaf tissue are required. Options for transformation have recently broadened to include genome-editing tools such as CRISPR-cas9 (Rani et al., 2016).

Functional classification of the tissue-enriched gene expression clusters

Genes representing tissue-enriched clusters were subjected to GO enrichment analysis to further characterize and identify over-represented functional groups in subterranean clover using REVIGO web server (Supek et al., 2011) to graphically represent the results. The scatterplot depicting leaf-enriched cluster (Cluster 2) showed enrichment for genes associated with photosynthesis, transport and response to stimulus (Figure 4b; Table S11). The flower-enriched cluster (1) showed a concentration of genes associated with the processes of nitrogen compound cellular biosynthesis, multicellular organismal organization and development (Table S12). The stem- and peduncle-enriched cluster (3) displayed a high representation of genes associated with response to stimulus, stress, oxidation-reduction process, transport and signal transduction. The root-enriched cluster (5) had a high representation of genes associated with response to stress, stimulus, nutrient levels, transport and nitrate metabolism. Clearly, the GO enrichment analyses of tissue-enriched clusters showed over-representation of predicted classes of genes in different tissue types and thereby validated our tissue-enriched gene co-expression clusters. Additionally, putative genes-encoding enzymes in subterranean clover were assigned to various pathways in the KEGG database using BLASTKOALA (Kanehisa et al., 2016). This analysis distributed 4,094 genes-encoding enzymes (12.7%) into 133 different KEGG pathways (Figure S3; Table S13).

Papilionoideae whole genome phylogenetic analyses

Comparative analyses across the Papilionoideae using the published whole genome sequences for 13 legumes and the Tsub_Refv2.0 (Figure 5a and b; Table S14) revealed an overlap of 12 170 orthologous gene clusters between Papilionoideae and *Arabidopsis thaliana* as the outgroup species. Within the Papilionoideae, the number of orthologous gene clusters specific to galegoid (*Lotus japonicus, Trifolium pratense, T. subterraneum, M. truncatula* and *Cicer arietinum*), millettioid (*Glycine max, Cajanus cajan, Vigna radiata, V. angularis* and *Phaseolus vulgaris*) and dalbergoid (*Arachis ipaensis* and *A. duranensis*) was 10 873, 7976 and 4004, respectively. There were 3171 orthologous gene

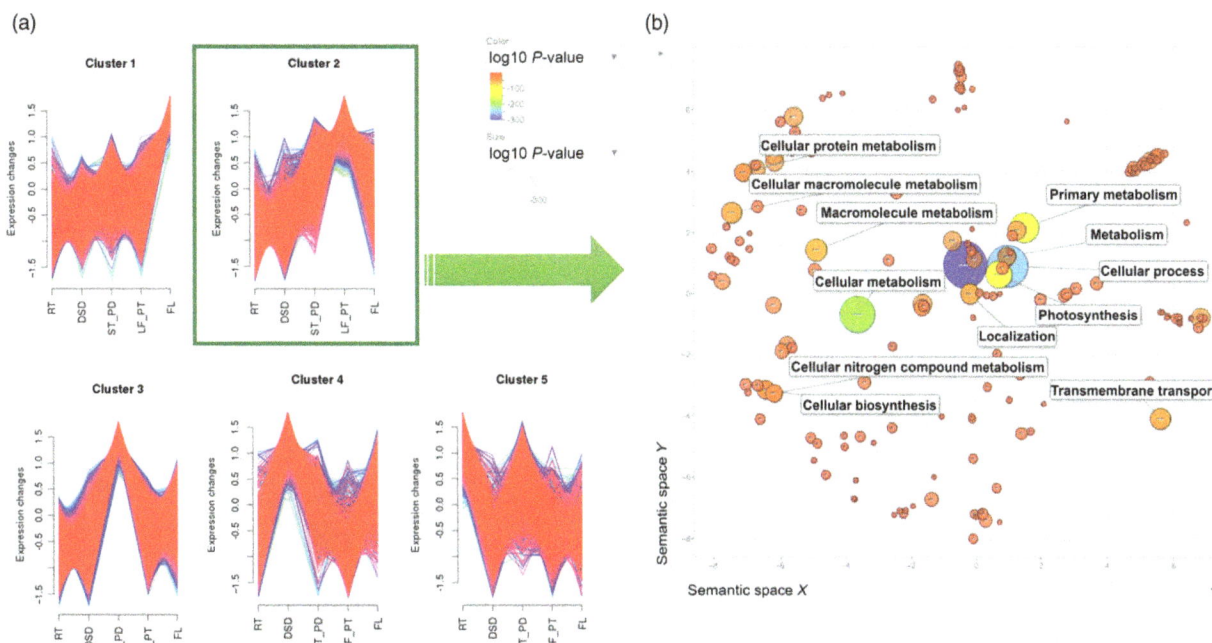

Figure 4 Dynamic tissue-specific gene co-expression clusters. (a) Five clusters were generated using Fuzzy C-means soft clustering algorithm implemented in Mfuzz. Data points on the X-axis represent root (RT), developing seed (DSD), stem and peduncles (ST_PD), leaf and petioles (LF_PT), flowers (FL) tissues, respectively. The Y-axis represents gene expression values, where gene expression values were standardized to have zero mean and one standard deviation. (b) The scatterplot generated using REVIGO web browser shows the cluster representatives (i.e. terms remaining after the redundancy reduction) for leaf and petiole tissue-enriched Cluster 2 biological processes in a two-dimensional space derived by applying multidimensional scaling to a matrix of the GO terms' semantic similarities. Bubble colour indicates the P-value (legend in upper left-hand corner); size indicates the frequency of the GO term in the wholeUniProt GOA database (bubbles of more general terms are larger).

clusters in common among galegoid, millettioid and dalbergoid species.

To find *Trifolium*-specific genes among the galegoids, we identified 18 131 genes representing 24.3% of the total 74 556 genes using BLAST searches (Table S15). This proportion is higher than reported in chickpea (10%; Garg *et al.*, 2011; Jain *et al.*, 2013) Arabidopsis (4.9%; Lin *et al.*, 2010) and rice (17.4%; Campbell *et al.*, 2007). *Trifolium*-specific gene clusters are a source of unique genes controlling important traits as drought tolerance, and disease resistance. Within the genus *Trifolium*, similar BLAST searches identified 6,325 (19.6%) and 11 806 (28.0%) genes as subterranean-clover-specific and red-clover-specific genes, respectively (Tables S16 and S17). These candidate subterranean-clover-specific genes could be mined further for such key traits as geocarpy and other factors related to the grazing tolerance of subterranean clover. Examining orthologous gene clusters provides an important foundation for comparative biology and functional inference in subterranean clover, because genes with simple orthologous relationships often have conserved functions, whereas genes duplicated more recently relative to speciation often underlie functional diversification. Comparative genomic analysis of the advanced subterranean assembly with the model legume, *Medicago truncatula*, showed close synteny and extensive collinearity of large sequence blocks (Figure 6). This provides opportunities for genomic comparisons and translation to identify candidate genes for traits of interest in the two species.

Marker-trait association studies

Phenotypic information for eight important traits described by Ghamkhar *et al.* (2011) (flowering time, hardseededness, leaf marks, calyx tube pigmentation, stem hairiness and the

isoflavonoids, formononetin, genistein, and biochanin A) was associated with specific regions in the revised high-density SNP linkage map (Tables S18 and S19) previously described by Hirakawa *et al.*, 2016;. Significant associations were then mapped onto the advanced assembly (Tsub_Refv2.0) to identify possible candidate genes (Figure 6; Table 2; Table 3). The effective anchoring of sequence scaffolds onto the high-density SNP linkage map in a high-quality chromosome-level genome assembly was a major factor in the identification of loci governing key traits. For example, QTLs for leaf marks (LM), calyx pigmentation (CP) and the isoflavonoids [formononetin (FO), genistein (GT) and biochanin A (BCA)] were found to map adjacently to a region on Chr 5. The linkage between these traits was demonstrated by Francis and Millington (1965), who used mutagenesis on *cv.* Geraldton, which has a high formononetin content, a prominent leaf mark and anthocyanin pigmentation of leaves, calyx tubes, stipules and stems to produce the low formononetin *cv.* Uniwager, which concomitantly lost its leaf mark and anthocyanin pigmentation. Co-localization of the QTLs identified for the traits in the present study explains the observation of breeders that isoflavone content is linked to pigmentation traits (leaf and calyx) (Figure 6; Table 3) (Francis and Millington, 1965). This information illustrates links between the new assembly, the high-density linkage map and key quantitative traits that can assist future marker-assisted selection and comparative mapping with other species to improve forage legumes and increase livestock productivity.

Conclusion

To improve the draft assembly of subterranean clover, the BNG Irys® system provided affordable, high-throughput physical maps

(a)

(b)

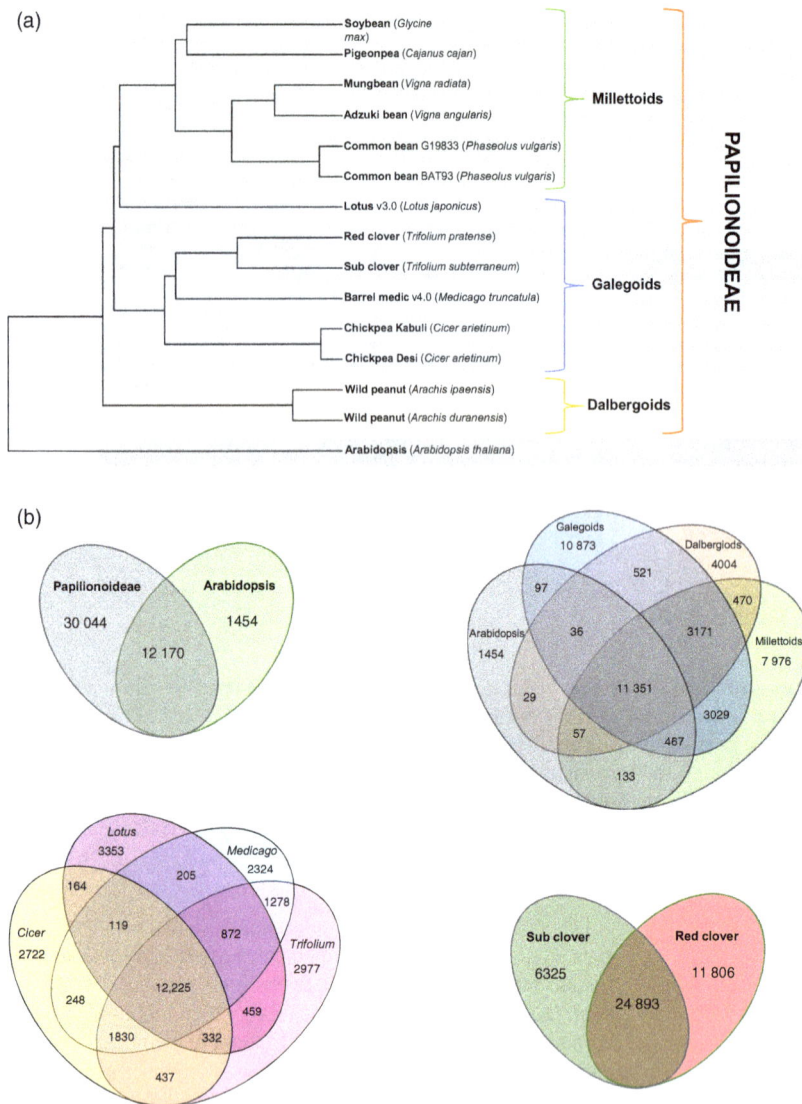

Figure 5 (a) Phylogenetic tree using the published predicted protein sequences for 13 legume genomes including the newly annotated *Trifolium subterraneum* L. and *Arabidopsis thaliana*. Mash v1.1 (preprint: http://biorxiv.org/content/early/2016/04/19/029827) was used to calculate a distance matrix and pairwise mutation distance estimation using the published whole genome sequences for all the legumes except the transcriptome assembly for *Cajanus cajan* as a whole genome assembly was not available. The phylogeny was constructed using UPGMA as implemented in the R-package 'phangorn' v2.0.3. (b) Shared and unique gene clusters in A) Papilionoideae species and *A. thaliana*; B) Millettioid, Galegoid and Dalbergioid clade or and *A. thaliana*; C) Lotus, Trifoliums, Medicago and Cicers; D) Shared and unique genes in sub-clover and red clover.

of high contiguity to validate the draft and extend existing scaffolds. Effective anchoring of sequence scaffolds to genetic linkage groups, coupled with the use of the BioNano system, resulted in a high-confidence chromosome-level genome assembly. Tissue-specific transcript profiling with RNA-Seq technology delivered gene expression data of high value for gene annotation of the assembly (Tsub_Refv2.0) and transcriptional dynamics to understand tissue-specific pathways. This new genomic information is the key to identifying loci governing traits that allow marker-assisted breeding in subterranean clover for comparative mapping with other species and the identification of tissue-specific gene promoters for biotechnological improvement of forage legumes.

Experimental procedures

Plant materials

Suspensions of intact cell nuclei were prepared according to Vrána *et al.* (2016). Briefly, mature dry seeds of subterranean clover (*Trifolium subterraneum* L.) *cv.* Daliak (approx. 20 g) were germinated on moistened paper towels in the dark at 25° ± 0.5 °C until roots achieved 2–3 cm in length. Roots were

cut to 1 cm from the apex and transferred into 25 mL formaldehyde fixative (2% v/v) for 20 min at 5 °C, followed by three 5-min washes in Tris buffer. Finally, the root tips (approx. 40) were cut, transferred into 1 mL IB buffer (Šimková *et al.*, 2003) and homogenized using a mechanical homogenizer (blender) at 13 000 RPM for 18 s. The crude homogenate was filtered through 50-μm (pore size) nylon mesh and stained with DAPI (2 μg/mL final concentration). A total of fifteen samples were prepared.

For transcriptome work, subterranean clover *cv.* Daliak plants were sown on 1 May 2015 in the field at Shenton Park, Western Australia (31°57′S, 115°50′E). Five different tissue types (roots, stem and peduncles, leaf and petioles, flowers and developing seeds) were harvested from a single Daliak plant on 24 September, when it was flowering and setting seeds. All samples were taken between 10.00 am and noon to eliminate any diurnal variations.

Preparation of high molecular weight (HMW) DNA for BioNano mapping

Cell nuclei were purified by a FACSAria II SORP flow cytometer and sorter (BD Biosciences, San Jose, CA) equipped with UV laser

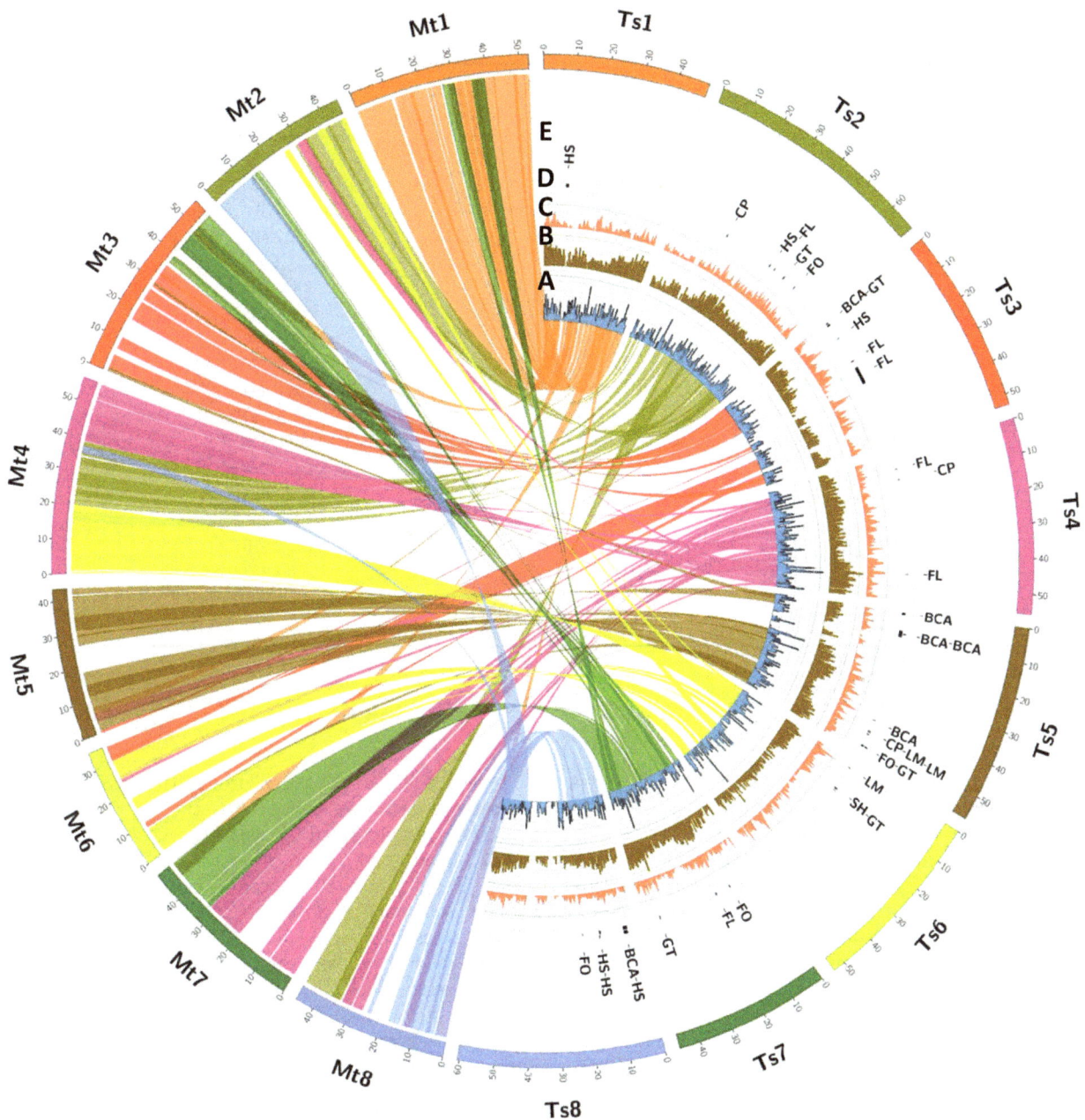

Figure 6 Graphical view of the genome structure of *Trifolium subterraneum L.* A) Syntenic relationship between *Medicago truncatula* (left) and *T. subterraneum* (right) with synteny links and synteny density histograms; B) *T. subterraneum* gene densities C) *T. subterraneum* SNP densities; D) *T. subterraneum* QTLs mapped for important traits using the high-density SNP linkage map E) Names for the QTLs mapped for important traits using the high-density SNP linkage map [FL Flowering time, LM Leaf marks, CP Calyx pigmentation, SH Stem Hairiness, FO Formononetin, GT Genistein, BCA Biochanin A and HS Hardseededness].

for DAPI excitation. Populations of G_1 nuclei were sorted in batches of 700 000 into 1.5-mL polystyrene tubes containing 660 μL IB buffer. In total, four batches of nuclei were sorted and each batch was used for preparation of one 20-μL agarose miniplug. DNA embedded in miniplugs was purified by proteinase K (Roche, Basel, Switzerland) treatment according to Šimková *et al.* (2003). The miniplugs were washed four times in wash buffer (10 mM Tris, 50 mM EDTA, pH 8.0) and five times in TE buffer (10 mM Tris, 1 mM EDTA, pH 8.0) and then melted for 5 min at 70 °C and solubilized with GELase (Epicentre, Madison, WI) for 45 min. A drop dialysis against TE buffer (Merck Millipore,

Billerica, MA) was performed for 90 min to purify DNA from any residues and subsequently quantified using a Quant-iT™ PicoGreen® dsDNA assay (Thermo Fisher Scientific, Waltham, MA).

RNA extractions and library preparation

Total RNA from all the tissue samples (Figure S4) was extracted using the Spectrum™ Plant Total RNA Kit (Sigma-Aldrich, USA) following the manufacturer's instructions. Aliquots of purified RNA were stored at −80 °C. The concentration of RNA was confirmed using Qubit fluorometer with Qubit RNA assay kit (Life

Table 2 Assembly and predicted gene statistics for advanced Tsub_Refv2.0 assembly using RNAseq supported data in comparison to the first draft genome assembly TSUd_r1.1

	TSUd_r1.1 contigs	TSUd_r1.1_pseudomolecules	Tsub_Refv2.0 scaffolds	Tsub_Refv2.0_pseudomolecules
Number of Sequences	27 424	1702	27 257	1545
Maximum length (bases)	2 878 652	63 731 624	7 089 608	67 952 282
Minimum length (bases)	300	42 658 284	300	46 363 611
Average length (bases)	17 205	50 143 517	19 481	55 397 738
Total % of genome captured in pseudomolecules (bp)*	–	72.62	–	80.23
N50 length (bases)	287 605	–	410 493	–
Number sequences >100 Kbp	1 302	–	1 135	–
Size of sequences >100 Kbp	386 665 122	–	445 838 546	–
Number sequences >1 Mbp	31	–	86	–
Size of sequences >1 Mbp	40 489 272	–	189 281 499	–
Number of sequences contained in super–scaffolds	–	–	97[†]	–
Length of sequences contained in super–scaffolds	–	–	193 110 501	–
GC%	33.3	–	33.3	–
Number of predicted genes	42 706	–	32 333	–
Total length of predicted genes (bp)	47 965 017	–	34 758 167	–
Average length of predicted genes (bp)	1123	–	1075	–
Max length of predicted genes (bp)	15 417	–	15 309	–
Min length of predicted genes (bp)	150	–	201	–
N50 length of predicted genes (bp)	1548	–	1437	–

*Estimated genome size 552.4 Mb.

[†]Consists of 264 contigs from TSUd_r1.1.

Technologies, Australia). The integrity of total RNA was determined by electrophoretic separation on 1.2% (w/v) denaturing agarose gels. Sequencing libraries were constructed using 500 ng of total RNA with a TruSeq® Stranded Total RNA Sample Prep Kits with Ribo-Zero (Illumina, San-Diego, USA) following the manufacturer's instructions. Library concentrations were measured using a Qubit fluorometer with Qubit dsDNA BR assay kit (Life Technologies, USA) and Agilent high-sensitivity DNA chips (Agilent Technologies, USA). The amplified libraries were pooled in equimolar amounts, and quality was assessed with Agilent high-sensitivity DNA chips (Agilent Technologies, USA). Paired-end 100-bp x 2 sequencing was performed with HiSeq2000 (Illumina, San-Diego, USA).

BioNano mapping

The genome sequence of subterranean clover (TSUd_r1.1) (Hirakawa et al., 2016) was analysed with Nickers software to assess the frequency of recognition sites for four different nicking enzymes (Nt.BbvC1, Nt.BsPQ1, Nb.bsm1 and Nb.BsrD1). The optimal labelling frequency was calculated for endonuclease Nt.BsPQ1 (7.6 sites/100 kb). DNA was then processed using NLRS protocol using the IrysPrep® Reagent Kit (BioNano Genomics, San Diego, CA) following manufacturer's instructions. DNA was nicked using 8U of Nt.BspQ1 (New England BioLabs, Beverly, MA) for two hours at 37 °C in NEBuffer 3. The nicked DNA was labelled with a fluorescent-dUTP nucleotide analogue using Taq polymerase (New England BioLabs) for one hour at 72 °C. After labelling, the nicks were ligated with Taq ligase (New England BioLabs) in the presence of dNTPs for 30 min at 37 °C. The backbone of the labelled DNA was stained with IrysPrep® DNA Stain (BioNano Genomics). The NLRS DNA concentration was measured again with the Quant-iT™ PicoGreen® dsDNA assay.

Labelled and stained DNA was loaded on the Irys chip, and four consecutive runs were performed (each run consisting of 30 cycles). A total of 489.8 Gb data were generated, of which 221.7 Gb exceeded 150 kb, the threshold for map assembly. These filtered data (>150 kb), corresponding to 401× coverage of the subterranean clover genome, were compiled from 994 895 molecules with N50 of 215.4 kb. De novo assembly of the BioNano map was performed by a pairwise comparison of all single molecules and graph building (Cao et al., 2014). A P-value threshold of $1e^{-9}$ was used during the pairwise assembly, $1e^{-10}$ for extension and refinement steps and $1e^{-15}$ for a final refinement.

Stitch: super scaffolding and correcting potential misassemblies

The complete pipeline of the Stitch algorithm by Shelton et al. (2015) (https://github.com/i5K-KINBRE-script-share/Irys-scaffolding/blob/master/KSU_bioinfo_lab/assemble_XeonPhi/assemble_XeonPhi_LAB.md) was run on the online cluster zythos provided by the Pawsey Supercomputing Center, Western Australia. The first draft genome assembly of T. subterraneum (TSUd_r1.1) (Hirakawa et al., 2016) was used as the reference genome. Super-scaffolds were generated, and BioNano IrysView was used to examine the new assemblies and the alignments between BioNano consensus maps and the in silico scaffolds. Based on these alignments, misassemblies were identified in TSUd_r1.1 (Figure 1b).

Reference genome-guided transcriptome sequence assembly

All RNASeq libraries were Illumina, San-Diego, USA TruSeq adapter trimmed and quality trimmed (sliding window, minimum

Table 3 Major QTLs and candidate genes mapped for the important traits using the high-density SNP linkage map anchoring to the advanced Tsub_Refv2.0 assembly

Trait Name	Trait ID	Chr	Left Marker	Right Marker	LOD	PVE (%)	Candidate genes
Flowering time	FL	3	Tsud_sc00634.00_113902	Tsud_sc01407.00_29664	36.36	61.94	87
		3	Tsud_sc00557.00_23548	Tsud_sc01682.00_2842	4.78	4.93	6
		2	Tsud_sc01286.00_17178	Tsud_sc01608.00_15349	3.25	3.14	5
		4	Tsud_sc00001.20_668469	Tsud_sc00001.20_681817	3.10	2.99	2
		4	Tsud_sc00277.00_152530	Tsud_sc00277.00_243760	2.76	2.43	8
		7	Tsud_sc00124.00_567396	Tsud_sc00124.00_311320	2.98	2.85	28
Leaf marks	LM	5	Tsud_sc01769.00_33298	Tsud_sc01806.00_29808	91.74	52.58	3
		5	Tsud_sc00041.00_165499	Tsud_sc00041.00_347862	25.67	8.90	16
		6	Tsud_sc00011.20_420838	Tsud_sc00011.20_536648	2.76	0.61	15
Calyx pigmentation	CP	5	Tsud_sc00695.00_95603	Tsud_sc00094.10_42823	111.79	96.88	13
		4	Tsud_sc00164.00_12669	Tsud_sc00164.00_104886	2.96	0.50	6
		2	Tsud_sc00213.10_37103	Tsud_sc00274.00_337396	2.70	0.48	3
Stem hairiness	SH	6	Tsud_sc00057.00_9596	Tsud_sc00057.00_401981	136.43	98.71	30
Formononetin	FO	5	Tsud_sc00724.00_157584	Tsud_sc00955.00_89974	17.25	25.53	30
		2	Tsud_sc00651.20_25763	Tsud_sc01310.00_44071	12.43	15.53	5
		7	Tsud_sc00459.00_51828	Tsud_sc00459.00_195702	7.42	9.03	13
		8	Tsud_sc00629.00_42778	Tsud_sc00629.00_69362	4.12	4.74	3
Genistein	GT	2	Tsud_sc01915.00_18283	Tsud_sc00906.00_202766	5.80	12.07	19
		3	Tsud_sc00026.00_688983	Tsud_sc00178.00_122069	5.57	11.50	34
		7	Tsud_sc00024.00_176871	Tsud_sc00024.00_366227	5.05	10.00	25
		6	Tsud_sc01326.00_38538	Tsud_sc01503.00_12748	3.12	6.17	0
		5	Tsud_sc00724.00_157584	Tsud_sc00955.00_89974	2.75	5.38	30
Biochanin_A	BCA	3	Tsud_sc00211.20_34823	Tsud_sc00026.00_63478	9.60	23.45	19
		8	Tsud_sc00654.20_18111	Tsud_sc01006.00_41002	4.53	10.40	30
		5	Tsud_sc00222.00_346664	Tsud_sc00242.00_222297	2.92	6.22	37
		5	Tsud_sc00285.00_86451	Tsud_sc00285.00_325955	2.76	5.84	10
		5	Tsud_sc02041.00_8358	Tsud_sc00674.20_21693	2.70	5.76	1
		5	Tsud_sc00073.20_492966	Tsud_sc00076.10_146128	2.64	5.60	87
Hardseededness	HS	1	Tsud_sc00068.20_1257208	Tsud_sc00855.00_29720	9.29	36.24	80
		3	Tsud_sc00009.30_108575	Tsud_sc00009.30_115638	2.76	6.62	2
		8	Tsud_sc00654.20_18111	Tsud_sc01006.00_41002	2.70	6.50	30
		8	Tsud_sc00309.00_532001	Tsud_sc00204.10_27738	2.58	6.46	10
		8	Tsud_sc00204.10_27738	Tsud_sc00431.00_91208	2.61	6.29	12
		2	Tsud_sc00682.00_120236	Tsud_sc01560.00_10224	2.55	6.16	6

quality score: 20) using Trimmomatic v 0.36 (Bolger et al., 2014). Trimmed libraries were aligned to the advanced reference sequence using HISAT2 v 2.0.1 (insert size 0 to 1000) (Kim et al., 2015). The resulting SAM files were converted to BAM format using samtools v1.3 (Li et al., 2009).

BRAKER1 v 1.9 (Hoff et al., 2016) was used to predict genes using GeneMark-ET v 4.32 (Lomsadze et al., 2014) and AUGUSTUS v 3.2.1 (Stanke et al., 2008) based on the RNASeq alignments.

Functional annotation and classification of the transcriptome

The resulting predicted proteins were aligned to several databases using blastp v 2.2.31+ (Altschul et al., 1990) (minimum e-value $1e^{-10}$). The database used were Swiss-Prot and TrEMBL downloaded on March 13 2016 (Boeckmann et al., 2003) and all 2 542 385 predicted proteins downloaded from Phytozome v11 on 14 March 2016 (Goodstein et al., 2012). For each predicted protein, the hit with the highest score and lowest e-value was chosen as annotation.

GO terms were assigned to proteins by manually transferring the GO terms for Swiss-Prot IDs using the UniProt-GOA database downloaded on 13 March 2016 (Huntley et al., 2015). KEGG K numbers were assigned to all predicted proteins using BLASTKOALA (taxonomy group: Plants, KEGG GENES database: family_eukaroytes) (Kanehisa et al., 2016).

Tissue-specific expression analysis

Tissue expression was estimated using kallisto v 0.42.4 (Bray et al., 2015). All genes with total transcripts per million (TPM) count below 1 or where more than two tissue types had a TPM of 0 were removed. To make log-normalization possible, 0.25 was added to the remaining TPM values. Expression was log-normalized and clustered into 12 clusters using Mfuzz v. 2.30.0 (Futschik and Carlisle, 2005) (m = 1.25).

Hierarchical clustering of tissues based on global gene expression

Clustering was based on 100 genes with highest sum of all TPM (transcripts per million) of the five tissue types. The dendrogram of the selected genes [vertical axis] is visualized along with the expression patterns used to cluster the five tissue samples, i.e. root (RT), developing seed (DSD), leaf and petioles (LF_PT), stem and peduncles (ST_PD) and flower (FL) [horizontal axis]. There are five GO-based clusters named G1, G2, G3, G4 and G5 that contain more than one gene. The GO clusters are separated by a horizontal bar in the heat map. Genes without annotations are omitted from the heat map. Over-expressed genes are shown with red colour and under expressed genes with white or yellow colour, respectively (Figure 3).

Genome structure and synteny analysis

Syntenic relationships between *M. truncatula* and *T. subterraneum* were calculated using SyMAP v4.2 (Soderlund *et al.*, 2006, 2011). The *T. subterraneum* genome was mined for repeats using RepeatModeler and RepeatMasker (http://www.repeatmasker.org).

Linkage map construction and quantitative trait locus (QTL) analyses

The linkage map was constructed using MultiPoint 3.3 (http://www.multiqtl.com/) as described in (Hirakawa *et al.*, 2016). The complete set of 188 F2 lines of a biparental population 92S80 (cv. Woogenellup x cv. Daliak) phenotypic data as reported by Ghamkhar *et al.*, 2011 was used for QTL screens for the following morphological and agronomic traits: levels of the oestrogenic isoflavones, formononetin (FO), genistein (GT) and biochanin A (BCA); days to first flowering (FT); leaf marks (LM); pigmentation of calyces (CP); hairiness of stems (SH); and hardseededness (HS). Levels of the isoflavones FO, GT and BCA were measured using the technique of Francis and Millington (1965). FT was measured as the number of days from sowing to appearance of the first flower. The morphological traits, LM, CP and SH, were scored 120 days after sowing, using the rating systems given in Nichols *et al.*, (1996). Hardseededness was measured in the laboratory using the method of Quinlivan (1961).

QTL screens for the traits reported (Table 3) were conducted using an inclusive composite interval mapping (ICIM) approach implemented in QTL IciMapping v4.0 (Wang *et al.*, 2014). Missing phenotypes were deleted using the 'Deletion' command in the software. The walking speed was set at 1 cm. A suitable probability for entering marker variables in stepwise regression was chosen so that the variation explained by the model approximated the trait heritability. The regression model was then used for background genetic variation control in the ICIM QTL mapping. The LOD was calculated using 1000 permutations, with a Type 1 error being 0.05, and significant QTLs were defined accordingly.

Papilionoideae whole genome phylogenetic analyses

A phylogenetic tree was constructed, based on published predicted protein sequences for 13 legume genomes, including those of the newly annotated *Trifolium subterraneum* L. and *A. thaliana*. Mash v1.1 (Ondov *et al.*, 2016) was used to calculate a distance matrix using k-mer counting and pairwise mutation distance estimation, using the predicted amino acids from the published whole genome sequences for all the legumes, except *Cajanus cajan*, for which only the transcriptome assembly was used (as the whole genome assembly is not available). The tree was then constructed using UPGMA as implemented in the R-package 'phangorn' v2.0.3. Shared and unique gene families were called using BLAST and OrthoMCL (Li *et al.*, 2003). OrthomclToVenn (https://github.com/philippbayer/orthomcltovenn) was used to calculate the number of unique and shared genes between: (i) the Papilionoideae species and *A. thaliana*; (ii) the Millettioid, Galegoid and Dalbergioid clades; (iii) *A. thaliana*

and the *Lotus*, *Trifolium*, *Medicago* and *Cicer* genera; and (iv) *T. subterranean* and *T. pratense*.

Accession code

Advanced genome sequence assembly and annotation data have been made available (under embargoed) at https://zenodo.org/record/161479, DOI: http://doi.org/10.5281/zenodo.161479

Acknowledgements

Meat & Livestock Australia provided financial support to the study. Z.M., J.V. and J.D. were supported by grant award LO1204 from the National Program of Sustainability I and by the Czech Science Foundation (award no. P501/12/G090). We thank Zdeňka Dubská for assistance with nuclei flow sorting, Helena Staňková for help with BioNano mapping and Hana Šimková for advice on BioNano mapping. We acknowledge the supercomputing resources provided by the Pawsey Supercomputing Centre with funding from the Australian Government and the Government of Western Australia. The authors declare no conflicts of interest.

Author contributions

P.K. conceived and performed the research and wrote the manuscript with contributions from P.E.B, Z.M, J.V, Y.Y, R.A, D.E, J.B, P.N, J.D and W.E. The BioNano Irys®;System genome mapping experiments were led and designed by J.D and performed and analysed by Z.M, J.V and Y.Y. P.E.B and Y.Y did the bioinformatics analysis and helped with the figure preparations. All authors read the manuscript and approved the content.

References

Abberton, M.T. and Marshall, A.H. (2005) Progress in breeding perennial clovers for temperate agriculture. *J. Agric. Sci.* **143**, 117–135.

Altschul, S.F., Gish, W., Miller, W., Myers, E.W. and Lipman, D.J. (1990) Basic local alignment search tool. *J. Mol. Biol.* **215**, 403–410.

Blanning, T. (2008) *The Pursuit of Glory: The Five Revolutions that Made Modern Europe: 1648–1815*. New York: Penguin Books.

Boeckmann, B., Bairoch, A., Apweiler, R., Blatter, M.-C., Estreicher, A., Gasteiger, E., Martin, M.J. *et al.* (2003) The SWISS-PROT protein knowledgebase and its supplement TrEMBL in 2003. *Nucleic Acids Res.* **31**, 365–370.

Bolger, A.M., Lohse, M. and Usadel, B. (2014) Trimmomatic: a flexible trimmer for Illumina sequence data. *Bioinformatics*, **30**, 2114–2120.

Bray, N., Pimentel, H., Melsted, P. and Pachter, L. (2015) *Near-optimal RNA-Seq quantification*. arXiv:1505.02710 [cs, q-bio].

Campbell, M.A., Zhu, W., Jiang, N., Lin, H., Ouyang, S., Childs, K.L., Haas, B.J. *et al.* (2007) Identification and characterization of lineage-specific genes within the Poaceae. *Plant Physiol.* **145**, 1311–1322.

Cao, H., Hastie, A.R., Cao, D., Lam, E.T., Sun, Y., Huang, H., Liu, X. *et al.* (2014) Rapid detection of structural variation in a human genome using nanochannel-based genome mapping technology. *GigaScience* **3**, 34.

Chaney, L., Sharp, A.R., Evans, C.R. and Udall, J.A. (2016) Genome Mapping in Plant Comparative Genomics. *Trends Plant Sci.* **21**, 770–780.

Francis, C.M. and Millington, A.J. (1965) Isoflavone mutations in subterranean clover. 1. Their production, characterisation and inheritance. *Aust. J. Agric. Res.* **16**, 567–573.

Futschik, M.E. and Carlisle, B. (2005) Noise-robust soft clustering of gene expression time-course data. *J. Bioinform. Comput. Biol.* **03**, 965–988.

Garg, R., Patel, R.K., Tyagi, A.K. and Jain, M. (2011) De novo assembly of chickpea transcriptome using short reads for gene discovery and marker identification. *DNA Res.* **18**, 53–63.

Ghamkhar, K., Isobe, S., Nichols, P.G.H., Faithfull, T., Ryan, M.H., Snowball, R., Sato, S. et al. (2011) The first genetic maps for subterranean clover (Trifolium subterraneum L.) and comparative genomics with T. pratense L. and Medicago truncatula Gaertn. to identify new molecular markers for breeding. Mol. Breed. 30, 213–226.

Goodstein, D.M., Shu, S., Howson, R., Neupane, R., Hayes, R.D., Fazo, J., Mitros, T. et al. (2012) Phytozome: a comparative platform for green plant genomics. Nucleic Acids Res. 40, D1178–D1186.

Guigó, R., Agarwal, P., Abril, J.F., Burset, M. and Fickett, J.W. (2000) An assessment of gene prediction accuracy in large DNA Sequences. Genome Res. 10, 1631–1642.

Hirakawa, H., Kaur, P., Shirasawa, K., Nichols, P., Nagano, S., Appels, R., Erskine, W. et al. (2016) Draft genome sequence of subterranean clover, a reference for genus Trifolium. Sci. Rep. 6, 30358.

Hoff, K.J., Lange, S., Lomsadze, A., Borodovsky, M. and Stanke, M. (2016) BRAKER1: unsupervised RNA-Seq-based genome annotation with GeneMark-ET and AUGUSTUS. Bioinformatics, 32, 767–769.

Huntley, R.P., Sawford, T., Mutowo-Meullenet, P., Shypitsyna, A., Bonilla, C., Martin, M.J. and O'Donovan, C. (2015) The GOA database: Gene Ontology annotation updates for 2015. Nucleic Acids Res. 43, D1057–D1063.

Jain, M., Misra, G., Patel, R.K., Priya, P., Jhanwar, S., Khan, A.W., Shah, N. et al. (2013) A draft genome sequence of the pulse crop chickpea (Cicer arietinum L.). Plant J. 74, 715–729.

Kanehisa, M., Sato, Y. and Morishima, K. (2016) BlastKOALA and GhostKOALA: KEGG Tools for functional characterization of genome and metagenome sequences. J. Mol. Biol. 428, 726–731.

Kim, D., Langmead, B. and Salzberg, S.L. (2015) HISAT: a fast spliced aligner with low memory requirements. Nat. Meth. 12, 357–360.

Lam, E.T., Hastie, A., Lin, C., Ehrlich, D., Das, S.K., Austin, M.D., Deshpande, P. et al. (2012) Genome mapping on nanochannel arrays for structural variation analysis and sequence assembly. Nat. Biotechnol. 30, 771–776.

Li, H., Handsaker, B., Wysoker, A., Fennell, T., Ruan, J., Homer, N., Marth, G. et al. (2009) The sequence alignment/map format and SAMtools. Bioinformatics, 25, 2078–2079.

Lin, H., Moghe, G., Ouyang, S., Iezzoni, A., Shiu, S.-H., Gu, X. and Buell, C.R. (2010) Comparative analyses reveal distinct sets of lineage-specific genes within Arabidopsis thaliana. BMC Evol. Biol. 10, 41.

Lomsadze, A., Burns, P.D. and Borodovsky, M. (2014) Integration of mapped RNA-Seq reads into automatic training of eukaryotic gene finding algorithm. Nucleic Acids Res. 42, e119. http://doi.org/10.1093/nar/gku557.

McGuire, W.S. (1985) Subterranean clover: Clover science and technology. American Society of Agronomy, Crop Science Society of America and Soil Science Society of America 1985, 515–534.

Nichols, P.G.H., Collins, W.J. and Barbetti, M.J. (1996) Registered cultivars of subterranean clover - their characteristics, origin and identification. Agriculture Western Australia Bulletin No. 4327, 61.

Nichols, P.G.H., Foster, K.J., Piano, E., Pecetti, L., Kaur, P., Ghamkhar, K. and Collins, W.J. (2013) Genetic improvement of subterranean clover (Trifolium subterraneum L.). 1. Germplasm, traits and future prospects. Crop Pasture Sci. 64, 312–346.

Ondov, B.D., Treangen, T.J., Melsted, P., Mallonee, A.B., Bergman, N.H., Koren, S. and Phillippy, A.M. (2016) Mash: fast genome and metagenome distance estimation using MinHash. Genome Biology 17, 132. doi 10.1186/s13059-016-0997-x

Parra, G., Bradnam, K. and Korf, I. (2007) CEGMA: a pipeline to accurately annotate core genes in eukaryotic genomes. Bioinformatics, 23, 1061–1067.

Piano, E., Pecetti, L., Boller, B., Posselt, U.K. and Veronesi, F. (2010) Minor Legume Species. In Fodder Crops and Amenity Grasses (Boller, F.,Posselt, U.K., and Veronesi, F., eds), pp. 477–500. New York: Springer .

Putnam, N.H., O'Connell, B.L., Stites, J.C., Rice, B.J., Blanchette, M., Calef, R. et al. (2016) Chromosome-scale shotgun assembly using an in vitro method for long-range linkage. Genome Res. 26, 342–350.

Quinlivan, B.J. (1961) The effect of constant and fluctuating temperatures on the permeability of the hard seeds of some legume species. Aust. J. Agric. Res. 12, 1009–1022.

Rani, R., Yadav, P., Barbadikar, K.M., Baliyan, N., Malhotra, E.V., Singh, B.K., Kumar, A. et al. (2016) CRISPR/Cas9: a promising way to exploit genetic variation in plants. Biotechnol. Lett. 38, 1991–2006.

Shelton, J.M., Coleman, M.C., Herndon, N., Lu, N., Lam, E.T., Anantharaman, T., Sheth, P. et al. (2015) Tools and pipelines for BioNano data: molecule assembly pipeline and FASTA super scaffolding tool. BMC Genom. 16, 734.

Šimková, H., Číhalíková, J., Vrána, J., Lysák, M.A. and Doležel, J. (2003) Preparation of HMW DNA from plant nuclei and chromosomes isolated from root tips. Biol. Plant. 46, 369–373.

Soderlund, C., Nelson, W., Shoemaker, A. and Paterson, A. (2006) SyMAP: a system for discovering and viewing syntenic regions of FPC maps. Genome Res. 16, 1159–1168.

Soderlund, C., Bomhoff, M. and Nelson, W.M. (2011) SyMAP v3.4: a turnkey synteny system with application to plant genomes. Nucleic Acids Res. 39, e68.

Stanke, M., Diekhans, M., Baertsch, R. and Haussler, D. (2008) Using native and syntenically mapped cDNA alignments to improve de novo gene finding. Bioinformatics, 24, 637–644.

Staňková, H., Hastie, A.R., Chan, S., Vrána, J., Tulpová, Z., Kubaláková, M. et al. (2016) BioNano genome mapping of individual chromosomes supports physical mapping and sequence assembly in complex plant genomes. Plant Biotech. J. 14, 1523–1531.

Sulas, L., (2005) The future role of forage legumes in Mediterranean-climate areas. In Grasslands: developments opportunities perspectives. (Reynolds, S.and Frame, J., eds), pp. 29–54. Rome: FAO and Plymouth UK: Science Publishers, Inc.

Supek, F., Bošnjak, M., Škunca, N. and Šmuc, T. (2011) REVIGO Summarizes and visualizes long lists of gene ontology terms. PLoS ONE 6, e21800.

Vižintin, L., Javornik, B. and Bohanec, B. (2006) Genetic characterization of selected Trifolium species as revealed by nuclear DNA content and ITS rDNA region analysis. Plant Sci. 170, 859–866.

Vrána, J., Cápal, P., Číhalíková, J., Kubaláková, M. and Doležel, J. (2016) Flow sorting plant chromosomes. Methods Mol. Biol. 1429, 119–134.

Wang, J.K., Li, H.H., Zhang, Y. and Meng, L. (2014) Users' Manual of QTL IciMapping v4.0. Beijing: Institute of Crop Science, CAAS.

Wheelan, S.J. and Boguski, M.S. (1998) Late-night thoughts on the sequence annotation problem. Genome Res. 8, 168–169.

16

Establishment of a tobacco BY2 cell line devoid of plant-specific xylose and fucose as a platform for the production of biotherapeutic proteins

Uri Hanania, Tami Ariel, Yoram Tekoah*, Liat Fux, Maor Sheva, Yehuda Gubbay, Mara Weiss, Dina Oz, Yaniv Azulay, Albina Turbovski, Yehava Forster and Yoseph Shaaltiel

Protalix Biotherapeutics, Carmiel, Israel

Correspondence
email yoram.tekoah@protalix.com

Keywords: glyco-engineering, bio-pharming, DNaseI, plant glycans, CRISPR/Cas9, gene editing.

Summary

Plant-produced glycoproteins contain N-linked glycans with plant-specific residues of β(1,2)-xylose and core α(1,3)-fucose, which do not exist in mammalian-derived proteins. Although our experience with two enzymes that are used for enzyme replacement therapy does not indicate that the plant sugar residues have deleterious effects, we made a conscious decision to eliminate these moieties from plant-expressed proteins. We knocked out the β(1,2)-xylosyltranferase (*XylT*) and the α(1,3)-*fucosyltransferase* (*FucT*) genes, using CRISPR/Cas9 genome editing, in *Nicotiana tabacum* L. cv Bright Yellow 2 (BY2) cell suspension. In total, we knocked out 14 loci. The knocked-out lines were stable, viable and exhibited a typical BY2 growing rate. Glycan analysis of the endogenous proteins of these lines exhibited N-linked glycans lacking β(1,2)-xylose and/or α(1,3)-fucose. The knocked-out lines were further transformed successfully with recombinant DNaseI. The expression level and the activity of the recombinant protein were similar to that of the protein produced in the wild-type BY2 cells. The recombinant DNaseI was shown to be totally free from any xylose and/or fucose residues. The glyco-engineered BY2 lines provide a valuable platform for producing potent biopharmaceutical products. Furthermore, these results demonstrate the power of the CRISPR/Cas9 technology for multiplex gene editing in BY2 cells.

Introduction

The use of plant cell suspension culture as a host for the production of recombinant proteins is gaining more and more popularity (Santos *et al.*, 2016) and the *Nicotiana tabacum* (*N. tabacum*) cv. Bright yellow 2 (BY2) cell line is among the most commonly used cells for this purpose (Mercx *et al.*, 2016). Plant-produced proteins contain an N-linked glycan core structure similar to what is found in mammalian cells, but they may also include additional core α(1,3)-fucose and β(1,2)-xylose, not found in mammalian-produced proteins (Tekoah *et al.*, 2015).

While there is no clinical evidence indicating differences in immunogenic response to plant-specific glycans versus mammalian-derived glycans (Santos *et al.*, 2016; Tekoah and Shaaltiel, 2016), extensive research efforts have been undertaken in recent years to modulate the plant-specific N-glycosylation machinery aiming at the production of recombinant proteins with mammalian-like modifications.

The transfer and the attachment of plant-specific sugar moieties to the developing glycan structure are carried out by the two Golgi resident enzymes α(1,3)-*fucosyltransferase* (*FucT*) and β(1,2)-*xylosyltransferase* (*XylT*) (Strasser *et al.*, 2004). Various strategies such as RNA interference (RNAi) technology and random mutagenesis methods have been applied, in various plant species, to interfere with the function of the *FucT* and *XylT* genes (Castilho *et al.*, 2011; Cox *et al.*, 2006; Strasser *et al.*, 2008; Weterings and Gerben, 2013; Yin *et al.*, 2011). Other genes, involved in the plant-specific N-glycosylation pathway, were also addressed. RNAi methodology was used to modulate the *GMD* gene, encoding the Guanosine 5′-diphosphate (GDP)-D-mannose 4,6-dehydratase enzyme, which is associated with GDP-L-fucose biosynthesis in *Nicotiana benthamiana* (*N. benthamiana*) plants. This resulted in N-glycans with decreased α(1,3)-fucose residues (Matsumura, 2010).

Stable expression of *GnTI* (*N*-acetylglucosaminyltransferase I) antisense in potato (*Solanum tuberosum*) and in tobacco (*N. tabacum*) showed no significant changes in the total N-glycan profiling versus the wild-type plants (Wenderoth and von Schaewen, 2000), while the down-regulation of the *GnTI* in *N. benthamiana*, using RNAi, resulted in incomplete elimination of the plant-specific glycans (Limkul *et al.*, 2016). Apparently none of the above approaches were able to completely deactivate the functions of the targeted genes, especially when more than one gene or a gene family was involved.

An alternative approach for the production of recombinant glycoproteins lacking xylose and fucose residues is to knock out the β(1,2)-*xylosyltransferase* (*XylT*) and α(1,3)-*fucosyltransferase* (*FucT*) genes using targeted genome editing. Targeted genome editing using zinc finger nucleases (ZFNs; Kumar *et al.*, 2006), transcription activator-like effector nucleases (TALENs; Christian *et al.*, 2010), or the recently developed CRISPR/Cas9 (Clustered Regularly Interspaced Short Palindromic Repeats) technology (Cong *et al.*, 2013; Jinek *et al.*, 2012) has been achieved by their ability to cleave DNA and create double-strand breaks (DSBs) at specific sites in the genome (Gaj *et al.*, 2013). DSBs at a gene target site are most often repaired by the nonhomologous end joining (NHEJ) DNA repair mechanism. This repair is often accompanied by small insertions or deletions (in-del) of nucleotides at the site of repair which can cause a mutation and knock out the targeted gene (Gorbunova and Levy, 1999).

Recently, the TALEN-mediated targeted gene editing technology was employed to manipulate N-glycosylation pathways in *N. benthamiana* (Li et al., 2016). The endogenous proteins of the knocked-out line had N-glycans that lacked β(1,2)-xylose but demonstrated only a 40% reduction in core α(1,3)-fucose levels when compared to the proteins from the wild-type cells. The authors suggested that the residual activity of the *FucT* enzyme in the mutated line might be explained by the presence of multiple copies of the *FucT* genes in the *N. benthamiana* genome.

The CRISPR/Cas9 technology comprises small guide RNAs (gRNA), which identify and locate the targeted DNA sequence, and an associate DNA endonuclease (Cas9), which execute the sequence-specific cleavage (Jinek et al., 2012). Two types of gRNAs—designated as CRISPR RNA (crRNA or the 'protospacer') and *trans*-acting RNA (tracrRNA)—are known to be involved in the guiding process. The crRNA guides the Cas9 endonuclease to the specific DNA target, while the actual binding of the enzyme to the DNA is mediated by the tracrRNA (Jinek et al., 2012). The DNA sequence encoding the crRNA is about 20 nucleotides long and is terminated, at its 3′ end, with a 3-base-pair proto spacer adjacent motif (PAM) of the sequence NGG (N can be any nucleotide) required for the correct recognition of the target. To facilitate the practical application of the system, the two separate gRNAs were assembled into a single chimeric molecule—designated single guide RNA (sgRNA; Jinek et al., 2012). Because of their small size, multiple sgRNAs can be co-constructed with the Cas9 on the same vector to achieve 'multiplex gene editing' (Belhaj et al., 2013).

The *N. tabacum* genome is known to contain two *XylT* genes (Ntab-*XylT*) and five *FucT* genes (Ntab-*FucT*; Table 1). Thus, to achieve *N. tabacum* BY2 cells devoid of any activity of these enzymes, simultaneous editing of seven genes and two alleles per gene is needed. Based on the results recently achieved with the CRISPR/Cas9 technology in accomplishing targeted DNA modifications in a wide variety of organisms (Cong et al., 2013; Mali et al., 2013) including several plant species (Brooks et al., 2014; Jacobs et al., 2015; Jia and Wang, 2014; Jiang et al., 2013, 2014; Mercx et al., 2016; Schiml and Puchta, 2016), this technology was chosen to knock out the *XylT* and the *FucT* genes.

The goal of this project was to completely eliminate the β(1,2)-xylose and α(1,3)-fucose plant-specific glycans in the Protalix's BY2 cell system, enabling the production of recombinant glycoproteins lacking these sugar moieties.

Results

Isolation of the targeted genes in the BY2 cells

Based on the published sequences of the Ntab-*XylT*-A and Ntab-*XylT*-B (Table 1), two primers were designed (Table S1, items 1, 2) and were used in a polymerase chain reaction (PCR) to isolate a partial sequence of *XylT* genes from the BY2 cells. The reaction revealed two PCR products. The first 2161-bp product was identical to a fragment of the published Ntab-*XylT*-A sequence and was labelled BY2-*XylT*-A (Figure S1). The second 2145-bp product was identical to a fragment of the published Ntab-*XylT*-B sequence and was labelled BY2-*XylT*-B (Figure S2).

Five Ntab-*FucT* genes are publicly known (Table 1). Based on an alignment analysis of these five variants and for the purpose of this work, we clustered the five genes into two groups. The first group contained the Ntab-*FucT*-A, Ntab-*FucT*-B and Ntab-*FucT*-C genes that share 96% of identity between the sequences of their first three exons (Table 1; Figure 1). The second group included the Ntab-*FucT*-D and Ntab-*FucT*-E genes that share 98% identity between their third exons (Table 1; Figure 1). Accordingly, two different sets of primers were designed (Table S1, items 3–6) and were used in a PCR to isolate parts of the sequence of each of the five *FucT* genes from the BY2 cells genome.

Using the first pair of primers (Table S1, items 3,4) resulted in two PCR products: a 3089-bp PCR product, showing 99.9% identity with exons 1,2,3 and introns 1 and 2 of both the Ntab-*FucT*-A and Ntab-*FucT*-B genes and a 5343-bp PCR product sharing 99.9% identity with exons 1,2,3 and introns 1 and 2 of the Ntab-*FucT*-C gene. Based on the identity between the first three exons of the Ntab-*FucT*-A and Ntab-*FucT*-B genes, the first PCR product can correspond to either gene and therefore was referred to as BY2-*FucT*-A and BY2-*FucT*-B (Figure S3). The second product was referred as BY2-*FucT*-C (Figure S4).

Using the second pair of primers (Table S1, items 5, 6) resulted in another two DNA fragments: a 834-bp product identical to the sequence of the final part of intron 2, exon 3 and the initial part of intron 3 of the Ntab-*FucT*-D gene and a 832-bp product identical to the final part of intron 2, exon 3 and the initial part of

Table 1 Details of the *N. tabacum XylT* and *FucT* genes

Gene family	Name	Accession numbers*	Remarks
XylT	Ntab-*XylT*-A	Ntab-BX_AWOK-SS596	Alignment analysis of the open reading frame showed 94% identity between the two genes
	Ntab-*XylT*-B	Ntab-BX_AWOK-SS12784	
FucT	Ntab-*FucT*-A	Ntab-K326_AWOJ-SS19752	These three *FucT* genes were grouped into 'Group 1' based on the high percentage of identity between the nucleotides sequence of their first and second exons
	Ntab-*FucT*-B	Ntab-BX_AWOK-SS16887	
	Ntab-*FucT*-C	Ntab-K326_AWOJ-SS16744	
	Ntab-*FucT*-D	Ntab-K326_AWOJ-SS19661	These two *FucT* genes were grouped into 'Group 2' based on the 98% identity between their third exons
	Ntab-*FucT*-E	Ntab-K326_AWOJ-SS19849	

*The Sol Genomic Network (www.solgenomic.net).

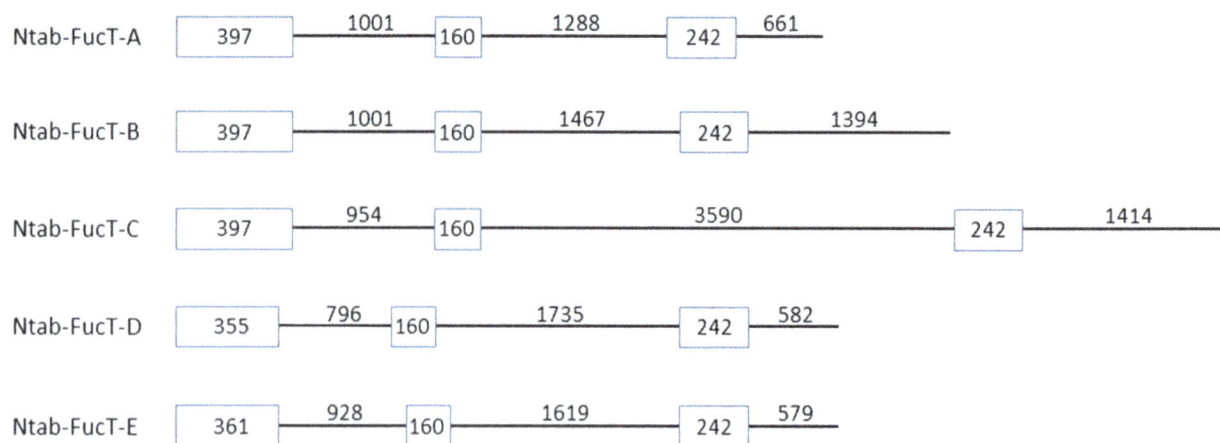

Figure 1 Schematic illustration of the first three exons and introns of the five *FucT* genes of *N. Tabaccum*. Exons 1, 2, 3 are shown as boxes and introns 1, 2, 3 as lines. The size of each exon and intron is indicated in base pairs (bp).

intron 3 of the N.tab-*FucT*-E gene. These fragments were labelled BY2-*FucT*-D (Figure S5) and BY2-*FucT*-E (Figure S6), respectively.

Construction of the Cas9/sgRNA constructs to knock out the XylT and the FucT genes

Following the identification of the BY2-*XylT* and the BY2-*FucT* genes in the BY2 cells, various DNA sequences, all starting with nucleotide G and tailed with the required PAM at their 3' ends, were selected as the Cas9 targets. Accordingly, the following five crRNAs were defined (Table S2): **crRNA1**—a 20-bp DNA sequence shared between the BY2-*XylT*-A and the BY2-*XylT*-B genes (Table S2, item 1); **crRNA2** and **crRNA3**, 18 and 21 bp long, respectively, were chosen based on DNA sequences shared between the BY2-*FucT*-A, B and C genes (Table S2, items 2,3); **crRNA4** and **crRNA5**, each 20 bp long, were chosen based on two DNA sequences shared between the BY2-*FucT*-D and the BY2-*FucT*-E (Table S2, items 4,5).

For the application of the CRISPR/Cas9 technology, the five crRNAs were each fused to the tracrRNA backbone sequence (Figure S7) resulting in the construction of five sgRNAs (designated sgRNA1—sgRNA5).

Three binary vectors namely **phCas9-XylT** (Figure 2a), **phCas9-FucT** (Figure 2b) and **phCas9-XylT-FucT** (Figure 2c) were then constructed and used in three separate cell transformations aiming at the knockout of either the BY2-*XylT* genes, the BY2-*FucT* genes or both groups of genes within the same cell.

The **phCas9-XylT** vector (Figure 2a) comprised three cassettes: a *Hygromycin phophotransferase (hpt)* selectable marker under an internal ribosome entry site (IRES) sequence, followed by a human-optimized Cas9 (Nekrasov *et al.*, 2013) attached to a nuclear localization signal (NLS) at its 3' terminal end, driven by 35S cauliflower mosaic virus promoter (35S) and the sgRNA1 under the Arabidopsis U6 promoter.

The **phCas9-FucT** (Figure 2b) contained the following six cassettes: *hpt* selectable marker under the IRES; a human-optimized synthetic Cas9 (Nekrasov *et al.*, 2013) attached to a nuclear localization signal (NLS) at its 3' terminal end, driven by 35S promoter; and four sgRNA cassettes, each under the Arabidopsis U6 promoter, and each containing one of the gRNA2 through gRNA5 fused to a tracrRNA.

The **phCas9-XylT-FucT** (Figure 2c) contained the following seven cassettes: a *hpt* selectable marker under IRES; a human-optimized synthetic Cas9 (Nekrasov *et al.*, 2013) attached to a

nuclear localization signal (NLS) at its 3' terminal end, driven by 35S promoter; and all the five sgRNAs that were previously described and used in this project, each under the Arabidopsis U6 promoter.

Knocking out of the XylT and the FucT genes in the BY2 cells

The three constructed vectors were used in three separate stable transformations of BY2 cells. In the first transformation, aiming at the knockout of the BY2-*XylT* genes and using the phCas9-XylT (Figure 2a), a total of 110 individual cell lines were isolated and screened. Total soluble protein was extracted from the cells and separated on SDS-PAGE, followed by transfer to a nitrocellulose membrane and detection by Western blot analysis using anti-xylose antibodies. About 30% of the screened lines were found negative to the anti-xylose antibody.

In the second transformation, aiming at the knockout of the BY2-*FucT* genes and using the phCas9-FucT (Figure 2b) about 100 individual cell lines were isolated and screened. Total soluble protein was extracted from the cells and separated on SDS-PAGE, followed by a Western blot analysis using anti-fucose antibodies. About 60% of the screened lines were found negative to the anti-fucose antibody.

In the third transformation, using the phCas9-XylT-FucT vector (Figure 2c) and aiming at knocking out both groups of genes (the *XylT* and the *FucT*), a total of 250 individual lines were isolated and screened sequentially for the absence of fucose and xylose. Screening for fucose-free lines was conducted by using an ELISA test. Total soluble protein from the putative fucose-free lines (about 30% of the screened lines) was separated by SDS-PAGE followed by a Western blot analysis using anti-xylose. Twenty five (10% of the total 250 lines) lines were assumed to be knocked out for both *XylT* and *FucT*.

Western blots of the total soluble protein extracted from three selected glyco-engineered cell lines and the nontransgenic BY2 cells, using anti-xylose, anti-fucose or anti-HRP antibodies, are presented in Figures 3 and S8.

Identification and characterization of the mutations generated by the Cas9 multiplexed targeting of the XylT and the FucT genes

The mutations that were generated by the Cas9 were characterized in three different mutated cell lines, that is in a *XylT*

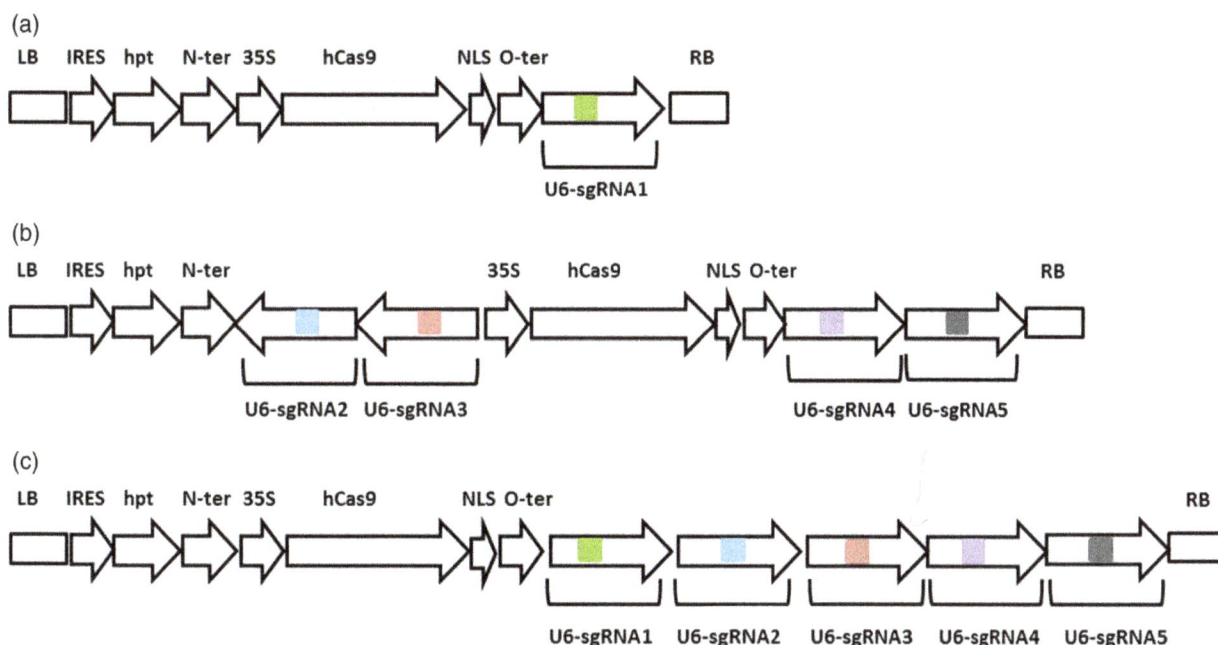

Figure 2 Schematic description of the three binary vectors used for the transformation of the BY2 cells. (a) The phCas9-XylT; (b) the phCas9-FucT; (c) the phCas9-XylT/FucT. LB, left border; IRES, internal ribosome entry site; hpt, hygromycin phosphotransferase; N-ter, nopaline synthase terminator; 35S, 35S cauliflower mosaic virus promoter with omega enhancer; hCas9, human-optimized Cas9; NLS, SV40 nuclear localization signal; O-ter, octopine synthase terminator; U6, Arabidopsis U6 promoter; sgRNA, chimeras of the various crRNAs with tracrRNA, the colour boxes represent the five different crRNAs that were used; RB, right border.

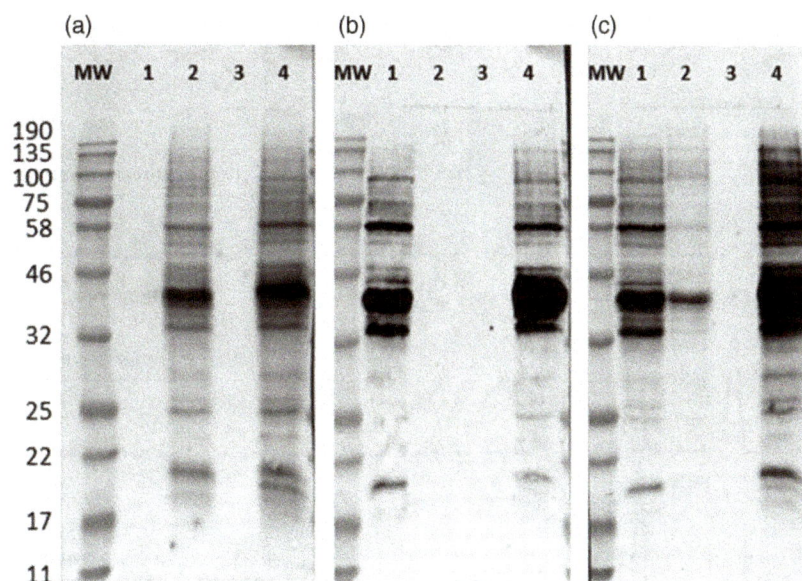

Figure 3 Western blots using anti-xylose, anti-fucose or anti-HRP antibodies. The total soluble protein was extracted from the glyco-engineered and the nontransgenic BY2 cells and 10 µg of protein from each sample were loaded on SDS-PAGE followed by Western blot using (a) anti-xylose, (b) anti-fucose and (c) anti-HRP antibodies. 1, ΔXT cell line; 2, ΔFT cell line; 3, ΔXFT cell line; 4, the wild-type nontransgenic BY2 cells. MW, molecular weight marker in kDa.

knocked-out cell line (ΔXT), in a *FucT* knocked-out cell line (ΔFT) and in a double knocked-out cell line (ΔXFT).

For each of the three cell lines, a PCR was performed using primers flanking the Cas9 target sites (Table S3). The obtained PCR products were cloned into a pGEMT vector and 36–60 clones for each sample were sequenced revealing the presence of assorted insertions and deletions (in-dels). No wild-type products were detected among any of the tested clones.

Three types of mutations were demonstrated for the ΔXT cell line: an identical 36-bp deletion in both alleles of *XylT*-A and a 1-bp insertion and a 13-bp deletion in *XylT*-B (Figure S9).

Seven assorted in-dels mutations were identified in the ΔFT cell line: an identical 1213-bp deletion in the *FucT*-A; an identical 2-bp insertion in the *FucT*-B; a 7-bp and a 15-bp deletions in the *FucT*-C; a 3-bp and a 7-bp deletions in the *FucT*-D and an identical 5-bp deletion in the *FucT*-E (Figure S10).

In the ΔXFT cell line, three mutations were identified for the *XylT* genes and ten for the *FucT* genes (Figure S11). An identical 7-bp deletion was found in both alleles of the *XylT*-A; a 1-bp and a 13-bp deletions were identified in the *XylT*-B. A 7-bp deletion and a 2-bp insertion were found in *FucT*-A; a 22-bp and a 21-bp deletions in *FucT*-B; a 7-bp and a 47-bp deletions were demonstrated in *FucT*-C; a 72-bp and a 37-bp deletions in *FucT*-D and a 56-bp insertion and a 37-bp deletion were found in *FucT*-E.

Glycan analysis of proteins derived from the XylT, the FucT and the XylT/FucT knocked-out lines

Glycan analysis of the total soluble proteins extracted from the three various knocked-out cell lines (ΔXT—Δ*XylT* cell line, ΔFT—Δ*FucT* cell line and ΔXFT—Δ*XylT*/Δ*FucT* cell line) was compared with the glycans found in the wild-type BY2 control cells. The glycan pools, separated on a normal phase UPLC system coupled with a fluorescence detector, showed a clear shift to shorter retention times [smaller glucose unit (GU) values] in the three mutated cell lines (Figure 4) compared to the control sample (Figure 4, top panel). The main glycan peaks were identified and are annotated on the top of each glycan profile. The samples of the released glycan pools were then subjected to digestion by various exoglycosidases and analysis by matrix-assisted laser desorption ionization (MALDI) time-of-flight (ToF) mass spectrometry (MS), confirming the glycan pool assignments of the total soluble proteins from the various cell lines. Results from exoglycosidase digestion and MS from the double-mutated BY2 cell line (ΔXFT) are presented in Figures S12 and S13.

The prevalence of xylose, fucose and xylose-/fucose-containing glycans among the total glycans derived from each of the tested samples is summarized in Table 2. Whereas 80% and 77% of the various glycans that were released from the proteins derived from the wild-type BY2 cells contained either xylose or fucose residues, respectively, the glycans that were released from the proteins derived from the ΔXT cell lines contained no xylose; those released from the ΔFT cell lines contained no fucose; and neither xylose nor fucose were identified in the glycans that were released from the proteins derived from the ΔXFT cell lines. Additional glycans, identified as high mannose structures, did not

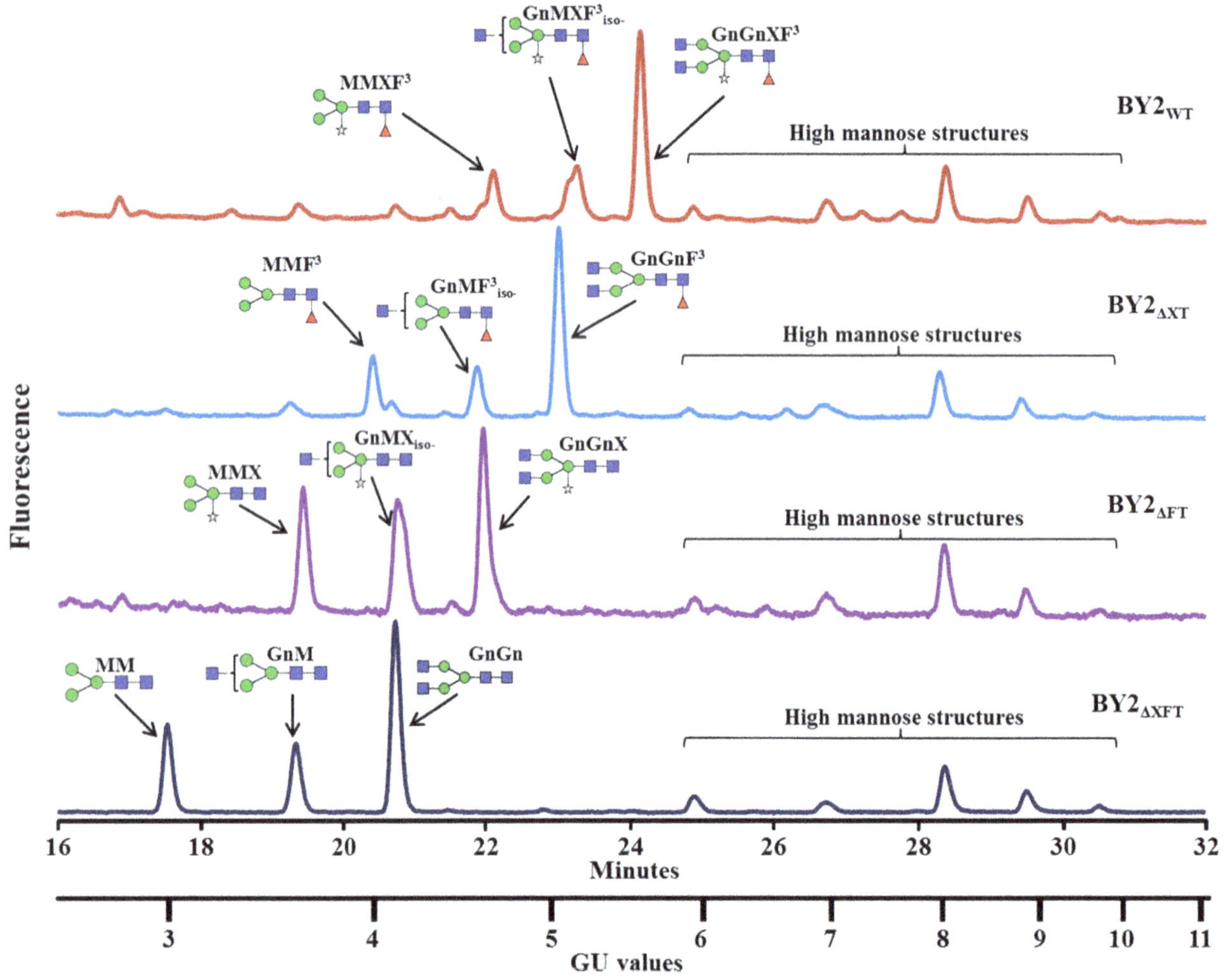

Figure 4 UPLC chromatograms presenting the peaks of the main glycans derived from the glycoproteins derived from (top to bottom): BY2 control cells (WT); a *XylT* knocked-out line (ΔXT); a *FucT* knocked-out line (ΔFT); and a *XylT*/*FucT* knocked-out cell line (ΔXFT). Glycans are annotated with cartoons and acronyms above main peaks. Symbols are as follows: ●-Mannose; ☆-Xylose; ■-*N*-acetylglucosamine; ▲-Fucose. Glycan acronyms are based on the nomenclature indicated at the following website http://www.proglycan.com/sites/default/public/pdf/nomen_2007.pdf.

Table 2 Distribution of the glycans in the various BY2 cell lines

Glucose unit value	Glycan acronym*	WT	ΔXT	ΔFT	ΔXFT
4.4	MM	–	–	–	27[†]
4.9	GnM	–	4	–	13
4.9	MMX	3	–	21	–
5.3	MMF[3]	–	14	–	–
5.3	GnMX	–	–	25	–
5.4	Man4	4	5	–	–
5.5	GnGn	–	–	–	34
5.8	GnGnX	–	–	32	–
5.9	GnMF[3]	–	13	–	–
5.9	MMXF[3]	17	–	–	–
6.3	GnGnF[3]	–	43	–	–
6.3	GnMXF[3]	22	–	–	–
6.8	GnGnXF[3]	38	–	–	–
7.1	Man6	3	–	2	5
8.0	Man7	4	4	4	4
8.9	Man8	7	12	11	11
9.6	Man9	3	5	4	6

*Based on nomenclature from http://www.proglycan.com/sites/default/public/pdf/nomen_2007.pdf.

[†]Results presented are of glycans above 2% of total pool.

change between the various cell lines, as expected, because they did not include any fucose or xylose sugars. Exoglycosidase digestion and MALDI-ToF mass spectrometry data of the ΔXFT cell lines verify that the predominant glycans are of the paucimannose type (MM) or paucimannose, substituted with either one or two N-acetylglycosamine sugars on the terminal mannose (GnM and GnGn, respectively). In addition, a variety of high mannose glycans (Man6-Man9) can be found. All glycans did not contain xylose or fucose.

Glycan analysis of recombinant DNaseI expressed in the XylT, FucT and XylT/FucT knocked-out cell lines

To demonstrate that the knocked-out cell lines can be used to produce recombinant proteins that do not contain plant-specific sugar moieties, the glyco-engineered cells were transformed with human recombinant DNaseI (Figure S14) and compared with the control BY2 cells that underwent the same transformation. The various cell lines underwent transformation at typical rates and expressed comparable amounts of total and active DNaseI (0.92 ± 0.11; 0.97 ± 0.03; 0.91 ± 0.07; 0.89 ± 0.08 µg active DNAseI per µg of total DNaseI for the WT BY2 cells, the ΔXT, the ΔFT and the ΔXFT cells, respectively), measured in crude extract. The recombinant DNaseI was purified from the culture medium, and the specific activity was evaluated using a methyl green-based activity assay at 37 °C. Results show that all variants were equally active, with specific activity values of 4.1 ± 0.2, 3.9 ± 0.1, 4.4 ± 0.1 and 4.1 ± 0.3 µg DNA/min/µg DNase for the WT BY2 cells, the ΔXT, the ΔFT and the ΔXFT cells, respectively. The various purified DNaseI were run on SDS-PAGE, and the glycans were separated from the isolated protein band on the gel. The glycans derived from the recombinant DNaseI produced in the glyco-engineered cell lines and in the control BY2 cells were then separated on UPLC (Figure 5). The results showed a significant difference between the glycan profiles of the various samples. While the glycans derived from the DNaseI produced in the control BY2 WT line (DNaseI_WT) all contained xylose and

fucose (Figure 5, top panel), the glycans derived from the DNaseI produced by the ΔXT cells, by the ΔFT cells and by the ΔXFT cells (DNaseI_ΔXT, DNaseI_ΔFT, DNaseI_ΔXFT, respectively) contained either no xylose, no fucose or neither xylose nor fucose (Figure 5, three lower panels), respectively. A summary of the various glycans is presented in Table 3. Exoglycosidase digestion and MALDI-ToF mass spectrometry data (S15 and S16, respectively) verify that the predominant glycans of the DNaseI_ΔXFT protein are of the paucimannose type (MM) or paucimannose, substituted with either one or two N-acetylglycosamine sugars on the terminal mannose (GnM and GnGn, respectively). The DNaseI glycan did not contain any high mannose glycans (Man6-Man9) as expected from a secreted protein.

The DNaseI_BY2 had two major peaks (Figure 5 top)—a peak identified as a bianntenary core-fucosylated and xylosylated glycan (GnGnXF[3], 51%) and a monoanntenary core-fucosylated and xylosylated peak (GnMXF[3], 45%). A third peak (4%) was identified as a paucimannose structure substituted with xylose and fucose (MMXF[3]). In the DNaseI_ΔXT, the main peak (92%) was assigned as a bianntenary core-fucosylated glycan (GnGnF[3]) and the minor peak (8%) as a monoanntenary core-fucosylated glycan (GnMF[3]; Figure 5, DNaseI_ΔXT). In the DNaseI_ΔFT, the main peak (56%) was assigned as a bianntenary glycan with xylose (GnGnX) and the minor peaks (40%, 3%) as a monoanntenary glycan containing xylose (GnMX) or a paucimannose structure substituted with a bisecting xylose (MMX), respectively (Figure 5, DNaseI_ΔFT). The DNaseI_ΔXFT main peak (67%) was assigned as a nonfucosylated/nonxylosylated bianntenary glycan (GnGn), and its minor peaks (30% and 3%) were identified as a nonfucosylated/nonxylosylated monoanntenary glycan (GnM) or a paucimannose structure (MM; Figure 5, DNaseI_ΔXFT).

Discussion

Plants and plant cell suspensions in particular are considered as a promising platform for the production of biopharmaceuticals proteins (Santos et al., 2016). Furthermore, plant cell suspension was the first plant-based system that produced an ERT product approved by the FDA and other authorities around the world (Fox, 2012).

Aiming at eliminating the plant-specific glycans β(1,2)-xylose and α(1,3)-fucose from recombinant glycoproteins that are produced in N. tabacum BY2 cells, we utilized the CRISPR/Cas9 technology to produce various cell lines devoid of these specific plant glycans. Three BY2 glyco-engineered cell lines were established: XylT, FucT and both XylT and FucT knocked-out lines. All three lines demonstrated complete eradication of the targeted glycans, respectively. The knocked-out cell lines were stable, viable and exhibited a typical BY2 growing rate. As the presence of glycans β(1,2)-xylose and α(1,3)-fucose in glycoprotein is conserved in plants, the absence of an apparent difference between the WT and the knockout cell lines should be further studied. It might be that fucosylation and xylosylation of plant glycoproteins play a role under certain stress conditions, or during development in planta. In this respect, we show here, for the first time, that the presence of xylose or fucose sugars in glycoproteins is not vital for the growth of BY2 cells in culture.

We were then able to utilize the new cell lines to produce a specific biotherapeutic protein. A subsequent transformation of the mutated lines with recombinant DNaseI showed that the biotherapeutic did not contain the β(1,2)-xylose and/or the α(1,3)-fucose in its glycans. This was used as a model to show that any

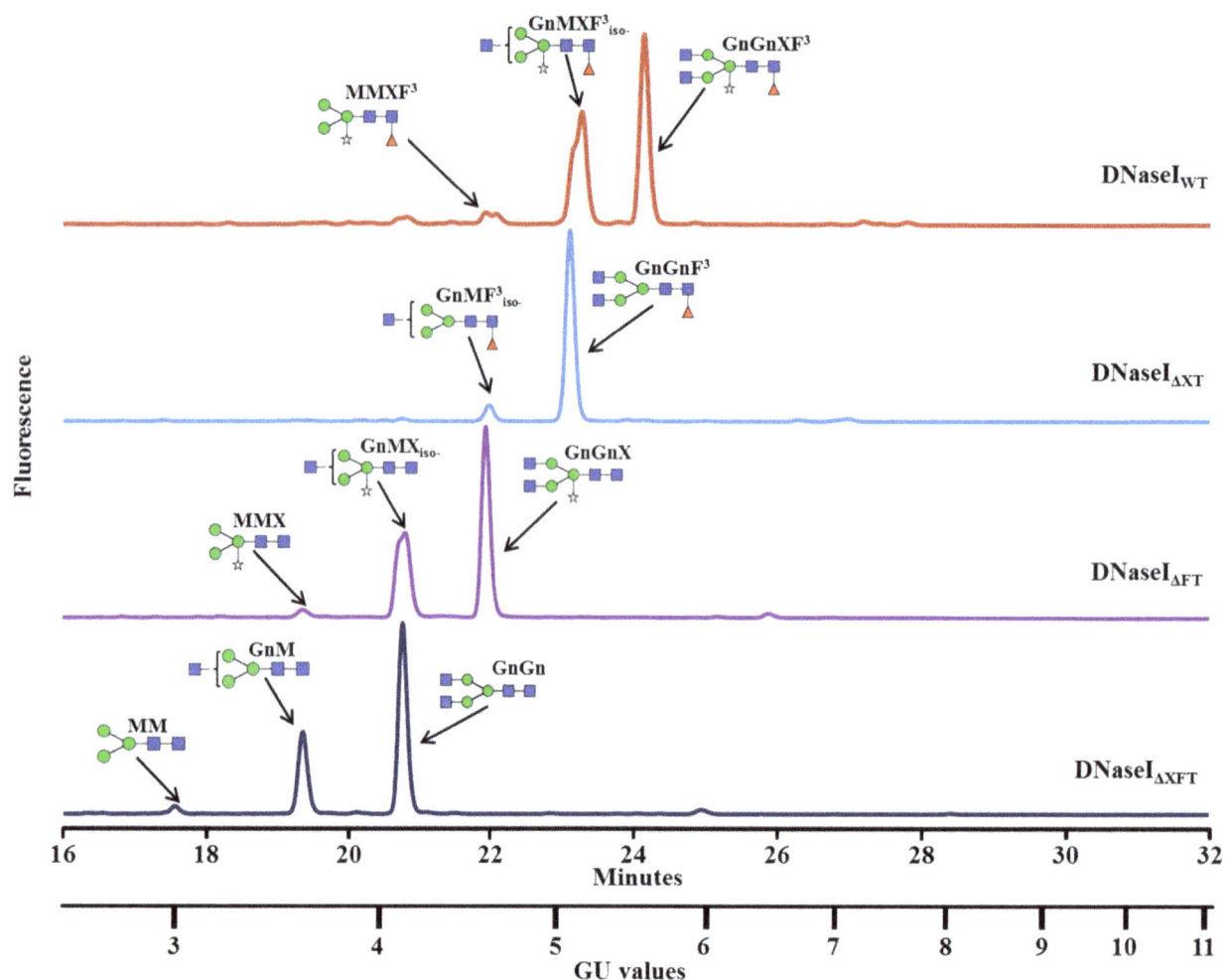

Figure 5 UPLC chromatograms presenting the peaks of the main glycans released from a DNaseI protein (top to bottom): Glycan pool derived from DNaseI produced by control cells (WT); glycan pool from a *XylT* knocked-out line (ΔXT); glycan pool released from DNaseI produced from a *FucT* knocked-out line (ΔFT); and a glycan pool released from DNaseI from a *XylT/FucT* knocked-out cell line (ΔXFT). Glycans are annotated with cartoons and acronyms above main peaks. Symbols are as follows: ●-Mannose; ★-Xylose; ■-N-acetylglucosamine; ▲-Fucose. Glycan acronyms are based on the nomenclature indicated at the following website http://www.proglycan.com/sites/default/public/pdf/nomen_2007.pdf.

recombinant glycoprotein can be produced in Protalix's glyco-engineered BY2 cells.

This study has added the *N. tabacum* cv. BY2 cells to the list of plant systems that were shown to be amenable to targeted DNA modifications by applying the CRISPR/Cas9 technology. Furthermore, by simultaneously addressing and mutating seven genes, which are involved in the fucosylation and xylosylation of plant glycoproteins, and using one vector, it was demonstrated that the CRISPR/Cas9 technology can be efficiently used for multiplex genome editing in the BY2 cells. Note that in order to achieve loss of function, we had to produce bi-allelic mutations in seven genes; therefore, a total of 14 genetic loci were knocked out. Earlier studies reported on the knockout of multiple loci in polyploid plants (Wang *et al.*, 2014). To our knowledge, this is the highest number reported so far in plants.

The glyco-engineered BY2 cell lines lacking fucose and xylose provide a valuable platform for producing potent biopharmaceutical products that can be similar to the mammalian proteins. The CRISPR/Cas9 technology can be further utilized for knocking out other unwanted genes.

Experimental procedures

Plant cell suspensions

Nicotiana tabacum cv. BY2 cells (Nagata, 2004) were cultured as a suspension culture in liquid MS-BY2 medium (Nagata and Kumagai, 1999) at 25 °C with constant agitation on an orbital shaker (85 r.p.m.). The suspensions were grown at 50 mL of volume in 250 mL erlenmeyers and were subcultured weekly at 2.5% (v/v) concentration.

Isolation and verification of the targeted genes

To isolate fragments of the *XylT* and *FucT* genes from the BY2 cells genome, genomic DNA was extracted from the cells and the appropriate pair of primers (Table S1) were used in a PCR performed according to the PWO DNA polymerase protocol (Roche).

Construction of the Cas9/sgRNA plasmid and transformation of BY2 cells

Three binary vectors, that is phCas9-XylT, phCas9-FucT and phCas9-XylT-FucT (Figure 2), were constructed using the pBIN19

Table 3 Distribution of the glycans in the various produced DNase I products

Glucose unit value	Glycan acronym*	WT	ΔXT	ΔFT	ΔXFT
4.4	MM	–	–	–	3[†]
4.9	GnM	–	–	–	30
4.9	MMX	–	–	3	–
5.3	MMF[3]	–	–	–	–
5.3	GnMX	–	–	40	–
5.4	Man4	–	–	–	–
5.5	GnGn	–	–	–	67
5.8	GnGnX	–	–	56	–
5.9	GnMF[3]	–	8	–	–
5.9	MMXF[3]	4	–	–	–
6.3	GnGnF[3]	–	92	–	–
6.3	GnMXF[3]	45	–	–	–
6.8	GnGnXF[3]	51	–	–	–
7.1	Man6	–	–	–	–
8.0	Man7	–	–	–	–
8.9	Man8	–	–	–	–
9.6	Man9	–	–	–	–

*Based on nomenclature from http://www.proglycan.com/sites/default/public/pdf/nomen_2007.pdf.
[†]Results presented are of glycans above 2% of total pool.

backbone vector and were used to transform the tobacco BY2 cells via the Agrobacterium plant transformation procedure (An, 1985). Once a stable transgenic cell suspension was established, it was used for isolating and screening individual cell lines (clones). Establishing of individual cell lines was conducted by using highly diluted aliquots of the transgenic cell suspension and spreading them on solid medium. The cells were allowed to grow until small calli (plant cell mass) developed. Each callus, representing a single clone, was then resuspended in liquid medium and sampled.

Screening for the knocked-out lines

Individual lines were screened for the absence of fucose and xylose by extracting their total soluble proteins and loading 10 µg of protein on SDS-PAGE followed by transfer to a nitrocellulose membrane and a Western blot analysis using anti-fucose or anti-xylose antibodies (Agrisera AS07-268 and Agrisera AS07-267). Using the fucose antibodies, a direct ELISA was developed to enable a high-throughput screening. Lines that were assumed to be knocked out for the FucT or XylT at the screening stage were then sampled for glycan analysis.

Detection of the Cas9-induced mutations in the knocked-out cell lines

Genomic DNA was extracted using the DNeasy plant mini kit (Qiagen). The DNA was amplified by PCR using specific primers for XylT and FucT genes (Table S3). The PCR products were subcloned into pGEMT vector. For each sample, 36–60 colonies were sequenced and were aligned with the wild-type target sequences to determine the mutations.

Transformation of the mutated lines with DNaseI and purification of DNaseI

Using a binary plasmid and the agrobacterium cell transformation procedure (An, 1985), the three mutated cell lines (i.e. the

BY2$_{ΔXT}$, BY2$_{ΔFT}$, BY2$_{ΔXFT}$) and nontransformed BY2 cells were transformed with DNAseI (Figure S14). The content and activity of DNaseI in the harvested medium was assessed using indirect sandwich ELISA and methyl green-based activity (Sinicropi et al., 1994) assays, respectively. The DNaseI was purified from the filtered medium using 2 column purification steps and was then used for measuring specific activity, using the methyl green-based activity assay, and for glycan analyses.

Glycan analyses

Glycan analysis of the knocked-out lines versus the wild-type BY2 cells was done based on the procedure described by Tekoah et al. (2004) with slight modifications. Briefly, sample preparation and analysis were as follows: total soluble proteins were extracted from the cells using sample buffer followed by 10-min boiling of the samples in a water bath. The extracts were centrifuged, and the supernatant was transferred to a clean tube. The total protein concentration of each sample was estimated by dot blot analysis and 200 µg of protein from each sample was then further reduced, alkylated and centrifuged to remove residual precipitates and the supernatant was run on a 12.5% SDS-PAGE until all loaded samples entered the separation part of the gel, without further separation (a run of about 40 min). For DNaseI samples, 200 µg protein was reduced, alkylated and run on a similar SDS-PAGE system. After staining with coomassie, the protein bands were excised, cut into small pieces and digested by trypsin. The resulting peptides were extracted from the gel pieces, and the glycans were released by digestion of peptides with PNGase A (an endoglycosidase that releases all types of glycans including α1-3 fucose). The released glycans were cleaned and fluorescently labelled using 2-anthranilamide (2AB), followed by removal of the excess 2AB. The labelled glycans were then analysed by separation on a UPLC system using a Waters BEH amide glycan column coupled with a fluorescence detector (Waters, Milford, MA). Dextran hydrolysate was used as a glucose unit (GU) ladder for assignment of the glycans. Further exoglycosidase digestion using a fucosidase and hexosaminidase (Prozyme, CA), followed by HILIC separation, combined with MALDI-ToF MS analysis was used to confirm the various glycans in the original glycan pool of each sample.

Author Contribution

Uri Hanania, Tami Ariel and Yoram Tekoah contributed equally to this study, designed the research and were involved in writing the original manuscript and all subsequent revisions.

Acknowledgements

The authors thank Sivan Gelley, Roey Mizrachi, Denisa Rozitsky, Roy Shadmon and Shay Zamin for the various biochemical analyses.

References

An, G. (1985) High efficiency transformation of cultured tobacco cells. *Plant Physiol.* **79**, 568–570.

Belhaj, K., Chaparro-Garcia, A., Kamoun, S. and Nekrasov, V. (2013) Plant genome editing made easy: targeted mutagenesis in model and crop plants using the CRISPR/Cas system. *Plant Methods*, **9**, 39 DOI: 10.1186/1746-4811-9-39.

Brooks, C., Nekrasov, V., Lippman, Z.B. and Van Eck, J. (2014) Efficient gene editing in tomato in the first generation using the clustered regularly interspaced short palindromic repeats/CRISPR-associated9 system. *Plant Physiol.* **166**, 1292–1297.

Castilho, A., Gattinger, P., Grass, J., Jez, J., Pabst, M., Altmann, F., Gorfer, M. *et al.* (2011) N-Glycosylation engineering of plants for the biosynthesis of glycoproteins with bisected and branched complex N-glycans. *Glycobiology*, **21**, 813–823.

Christian, M., Cermak, T., Doyle, E.L., Schmidt, C., Zhang, F., Hummel, A., Bogdanove, A.J. *et al.* (2010) Targeting DNA double-strand breaks with TAL effector nucleases. *Genetics*, **186**, 757–761.

Cong, L., Ran, F.A., Cox, D., Lin, S., Barretto, R., Habib, N., Hsu, P.D. *et al.* (2013) Multiplex genome engineering using CRISPR/Cas systems. *Science*, **339**, 819–823.

Cox, K.M., Sterling, J.D., Regan, J.T., Gasdaska, J.R., Frantz, K.K., Peele, C.G., Black, A. *et al.* (2006) Glycan optimization of a human monoclonal antibody in the aquatic plant Lemna minor. *Nat. Biotechnol.* **24**, 1591–1597.

Fox, J.L. (2012) First plant-made biologic approved. *Nat. Biotechnol.* **30**, 472.

Gaj, T., Gersbach, C.A. and Barbas, C.F. (2013) ZFN, TALEN, and CRISPR/Cas-based methods for genome engineering. *Trends Biotechnol.* **31**, 397–405.

Gorbunova, V.V. and Levy, A.A. (1999) How plants make ends meet: DNA double-strand break repair. *Trends Plant Sci.* **4**, 263–269.

Jacobs, T.B., LaFayette, P.R., Schmitz, R.J. and Parrott, W.A. (2015) Targeted genome modifications in soybean with CRISPR/Cas9. *BMC Biotechnol.* **15**, 16. doi:10.1186/s12896-2015-10131-12892.

Jia, H. and Wang, N. (2014) Targeted genome editing of sweet orange using Cas9/sgRNA. *PLoS ONE*, **9**, 4:e93806.

Jiang, W., Zhou, H., Bi, H., Fromm, M., Yang, B. and Weeks, D.P. (2013) Demonstration of CRISPR/Cas9/sgRNA-mediated targeted gene modification in Arabidopsis, tobacco, sorghum and rice. *Nucleic Acids Res.* **41**, 20, e188. doi:10.1093/nar/gkt780

Jiang, W., Yang, B. and Weeks, D.P. (2014) Efficient CRISPR/Cas9-mediated gene editing in *Arabidopsis thaliana* and inheritance of modified genes in the T2 and T3 generations. *PLoS ONE*, **9** (6), e99225 doi:10.1371/journal.pone.0099225.

Jinek, M., Chylinski, K., Fonfara, I., Hauer, M., Doudna, J.A. and Charpentier, E. (2012) A programmable dual-RNA-guided DNA endonuclease in adaptive bacterial immunity. *Science*, **337**, 816–821.

Kumar, S., Allen, G.C. and Thompson, W.F. (2006) Gene targeting in plants: fingers on the move. *Trends Plant Sci.* **11**, 159–161.

Li, J., Stoddard, T.J., Demorest, Z.L., Lavoie, P.O., Luo, S., Clasen, B.M., Cedrone, F. *et al.* (2016) Multiplexed, targeted gene editing in *N. benthamiana* for glyco-engineering and monoclonal antibody production. *Plant Biotechnol. J.* **14**, 533–542.

Limkul, J., Iizuka, S., Sato, Y., Misaki, R., Ohashi, T. and Fujiyama, K. (2016) The production of human glucocerebrosidase in glyco-engineered N. benthamiana plants. *Plant Biotechnol. J.* **14**, 1682–1694. doi:10.1111/pbi.12529.

Mali, P., Yang, L., Esvelt, K.M., Aach, J., Guell, M., DiCarlo, J.E., Norville, J.E. *et al.* (2013) RNA-guided human genome engineering via Cas9. *Science*, **339**, 823–826.

Matsumura, K.M.A.T. (2010) Deletion of fucose residues in plant N-glycans by repression of the GDP-mannose 4,6-dehydratase gene using virus-induced gene silencing and RNA interference. *Plant Biotechnol. J.* **9**, 264–281.

Mercx, S., Tollet, J., Magy, B., Navarre, C. and Boutry, M. (2016) Gene inactivation by CRISPR-Cas9 in *N. tabacum* BY-2 suspension cells. *Front. Plant Sci.* **7**, 40. doi:10.3389/fpls.2016.00040.

Nagata, T. (2004) When I encountered Tobacco BY-2 cells. In *Tobacco BY-2 Cells* (Nagata, T., Hasezawa, S. and Inze, D., eds), pp. 1–6. Berlin: Springer.

Nagata, T. and Kumagai, F. (1999) Plant cell biology through the window of the highly synchronized tobacco BY-2 cell line. *Methods Cell Sci.* **21**, 123–127.

Nekrasov, V., Staskawicz, B., Weigel, D., Jones, J.D. and Kamoun, S. (2013) Targeted mutagenesis in the model plant N. benthamiana using Cas9 RNA-guided endonuclease. *Nat. Biotechnol.* **31**, 691–693.

Santos, R.B., Abranches, R., Fischer, R., Sack, M. and Holland, T. (2016) Putting the spotlight back on plant suspension cultures. *Front. Plant Sci.* **7**, 297. doi:10.3389/fpls.2016.00297.

Schiml, S. and Puchta, H. (2016) Revolutionizing plant biology: multiple ways of genome engineering by CRISPR/Cas. *Plant Methods*, **12**, doi:10.1186/s13007-016-0103-0.

Sinicropi, D., Baker, D.L., Prince, W.S., Shiffer, K. and Shak, S. (1994) Colorimetric determination of DNase I activity with a DNA-methyl green substrate. *Anal. Biochem.* **222**, 351–358.

Strasser, R., Altmann, F., Mach, L., Glossl, J. and Steinkellner, H. (2004) Generation of *Arabidopsis thaliana* plants with complex N-glycans lacking beta1,2-linked xylose and core alpha1,3-linked fucose. *FEBS Lett.* **561**, 132–136.

Strasser, R., Stadlmann, J., Schahs, M., Stiegler, G., Quendler, H., Mach, L., Glossl, J. *et al.* (2008) Generation of glyco-engineered N. benthamiana for the production of monoclonal antibodies with a homogeneous human-like N-glycan structure. *Plant Biotechnol. J.* **6**, 392–402.

Tekoah, Y. and Shaaltiel, Y. (2016) Plant specific N-glycan do not have proven adverse effects in humans. *Nat. Biotechnol.* **34**, 17–18.

Tekoah, Y., Ko, K., Koprowski, H., Harvey, D.J., Wormald, M.R., Dwek, R.A. and Rudd, P.M. (2004) Controlled glycosylation of therapeutic antibodies in plants. *Arch. Biochem. Biophys.* **426**, 266–278.

Tekoah, Y., Shulman, A., Kizhner, T., Ruderfer, I., Fux, L., Nataf, Y., Bartfeld, D. *et al.* (2015) Large-scale production of pharmaceutical proteins in plant cell culture-the protalix experience. *Plant Biotechnol. J.* **13**, 1199–1208.

Wang, Y., Cheng, X., Shan, Q., Zhang, Y., Liu, J., Gao, C. and Qiu, J.L. (2014) Simultaneous editing of three homoeoalleles in hexaploid bread wheat confers heritable resistance to powdery mildew. *Nat. Biotechnol.* **32**, 947–951.

Wenderoth, I. and von Schaewen, A. (2000) Isolation and characterization of plant N-acetyl glucosaminyltransferase I (GntI) cDNA sequences. Functional analyses in the Arabidopsis cgl mutant and in antisense plants. *Plant Physiol.* **123**, 1097–1108.

Weterings, K. and Gerben, V.E. (2013) *N. benthamiana* plants deficient in fucosyltransferase activity. Icon Genetic Gmbh. Patent No. WO 2013050155 A1.

Yin, B.J., Gao, T., Zheng, N.Y., Li, Y., Tang, S.Y., Liang, L.M. and Xie, Q. (2011) Generation of glyco-engineered BY2 cell lines with decreased expression of plant-specific glycoepitopes. *Protein Cell*, **2**, 41–47.

17

Feruloylation and structure of arabinoxylan in wheat endosperm cell walls from RNAi lines with suppression of genes responsible for backbone synthesis and decoration

Jackie Freeman, Jane L. Ward, Ondrej Kosik, Alison Lovegrove, Mark D. Wilkinson, Peter R. Shewry and Rowan A.C. Mitchell* (iD)

Plant Biology and Crop Science, Rothamsted Research, Harpenden, Hertfordshire, UK

*Correspondence
email rowan.mitchell@rothamsted.ac.uk

Summary

Arabinoxylan (AX) is the major component of the cell walls of wheat grain (70% in starchy endosperm), is an important determinant of end-use qualities affecting food processing, use for animal feed and distilling and is a major source of dietary fibre in the human diet. AX is a heterogeneous polysaccharide composed of fractions which can be sequentially extracted by water (WE-AX), then xylanase action (XE-AX) leaving an unextractable (XU-AX) fraction. We determined arabinosylation and feruloylation of AX in these fractions in both wild-type wheat and RNAi lines with decreased AX content (TaGT43_2 RNAi, TaGT47_2 RNAi) or decreased arabinose 3-linked to mono-substituted xylose (TaXAT1 RNAi). We show that these fractions are characterized by the degree of feruloylation of AX, <5, 5–7 and 13–19 mg bound ferulate (g^{-1} AX), and their content of diferulates (diFA), <0.3, 1–1.7 and 4–5 mg (g^{-1} AX), for the WE, XE and XU fractions, respectively, in all RNAi lines and their control lines. The amount of AX and its degree of arabinosylation and feruloylation were less affected by RNAi transgenes in the XE-AX fraction than in the WE-AX fraction and largely unaffected in the XU-AX fraction. As the majority of diFA is associated with the XU-AX fraction, there was only a small effect (TaGT43_2 RNAi, TaGT47_2 RNAi) or no effect (TaXAT1 RNAi) on total diFA content. Our results are compatible with a model where, to maintain cell wall function, diFA is maintained at stable levels when other AX properties are altered.

Keywords: hydroxycinnamic acids, xylan acylation, grass xylan, wheat endosperm cell wall.

Introduction

The composition of the cell walls of wheat grain is unusual in that they are largely comprised of arabinoxylan (AX), with no lignin and low amounts of cellulose in starchy endosperm and aleurone tissues. Starchy endosperm cell walls are primary cell walls containing 70% AX, 20% β-(1,3;1,4) glucan, 2%–7% gluco-mannan and 2%–4% cellulose. The composition of aleurone cell walls is similar to that of starchy endosperm tissue comprising 65% AX, 30% β-(1,3;1,4) glucan and 2% glucomannan and cellulose, whereas, whilst AX is still the major component of pericarp cell walls (60% AX), they also contain 30% cellulose and 12% lignin (Shewry et al., 2014). AX consists of a linear backbone of (1→4)-linked xylopyranose (Xylp) residues which may be mono-substituted with α-(1,3) linked arabinofuranose (Araf) or di-substituted with both α-(1,2) and α-(1,3) linked Araf. The hydroxycinnamic acids ferulic acid (FA) and p-coumaric acid (pCA) can occur ester-linked to the mono-substituted α-(1,3) linked Araf. Ester-linked pCA is not detected in pure starchy endosperm tissue dissected from wheat grain but is concentrated in the aleurone layer of wheat grain (Barron et al., 2007). Ester-linked ferulic acid can undergo oxidative dimerization forming ether or C–C bonds linking chains of AX or glucuronoarabinoxy-lan (GAX) (Ishii, 1991), or of (G)AX to lignin in lignified tissues (Ralph et al., 1995).

The structure and solubility of wheat cell walls have important effects on a number of end users including consumption by humans and animals, and alcohol and biofuel production. The cell walls also provide mechanical strength and protection against pathogen attack. As the major component of wheat grain cell walls, the structure and properties of AX are of great importance. The solubility of AX molecules is affected by the degree of Araf substitution and the distribution pattern of Araf residues on the xylan backbone (Hoije et al., 2008): Araf substitution hinders hydrogen bonding between xylan chains and favours solubility of the polymer. The degree of feruloylation also affects the solubility of AX. The formation of the covalent diferulate cross-links between AX molecules occurs via oxidative coupling using free radicals, probably generated by peroxidases (Ralph, 2010). The greater the degree of feruloylation the more cross-linking is likely to occur, decreasing the solubility of AX. In starchy endosperm cell walls, the degree of feruloylation is much lower than in other tissues of the grain being fivefold less per unit Xylp than in outer tissues (Barron et al., 2007; Saulnier et al., 2007). However, as starchy endosperm cell walls have little cellulose and no lignin, it is possible that, despite the low amount, feruloylation of AX is critical in maintaining the structural integrity of starchy endosperm cell walls; heterologous expression of a feruloyl esterase in starchy endosperm resulted in endosperm collapse in some transgenic wheat lines (Harholt et al., 2010). There is also some variation in AX structure between different parts of the endosperm (Saulnier et al., 2009; Toole et al., 2010). These differences may reflect a requirement for different cell wall properties in different parts of the starchy endosperm.

Arabidopsis genes encoding IRX9 and IRX14, in the glycosyl transferase (GT) 43 family, and IRX10, in the GT47 family, were identified in genetic screens as being essential for normal extension of the xylan backbone (Brown et al., 2007, 2009; Pena et al., 2007). Subsequent studies demonstrated that the IRX10 proteins can extend xylan chains in vitro without IRX9 or IRX14; therefore, it is likely that IRX9 and IRX14 are accessory proteins required for xylan extension in planta and that IRX10 is the catalytic unit directly responsible for extension of the xylan chain (Jensen et al., 2014; Urbanowicz et al., 2014). Recent evidence supports the concept that IRX9, IRX14 and IRX10 participate in a xylan synthase complex (Zeng et al., 2016).

Based on greater expression in grasses than in dicots, we previously identified candidate genes for AX biosynthesis in the GT43, GT47, GT61 gene families and for feruloylation of AX in a grass-specific clade of the BAHD acyl-coA transferase superfamily (Mitchell et al., 2007). Close homologues of IRX9, IRX14 and IRX10 are highly expressed throughout wheat starchy endosperm development (Pellny et al., 2012). We have shown that in transgenic wheat lines with RNAi constructs targeting suppression of the most highly expressed homologues of IRX9 (TaGT43_2) and IRX10 (TaGT47_2) in starchy endosperm, AX amount is decreased by up to 50% (Lovegrove et al., 2013). Amongst GT gene families, the greatest expression bias in favour of grasses compared to dicots was in the GT61 gene family (Mitchell et al., 2007). We also showed that wheat and rice genes (XAT1, 2 and 3) in the GT61 family are responsible for addition of α-(1,3) linked Araf to xylan; RNAi suppression in wheat starchy endosperm of the most highly expressed GT61 gene (TaXAT1) results in a 70%–80% reduction in α-(1,3) Araf substitution of mono-substituted Xylp in AX (Anders et al., 2012). Three independent transgenic wheat lines per target, in which expression of the TaGT43_2, TaGT47_2 or TaXAT1 genes was suppressed in starchy endosperm, were used to study the effects on chain length of water-soluble AX (WE-AX) and AX solubilized by alkaline extraction (AE-AX) and on extract viscosity (Freeman et al., 2016). Suppression of TaGT43_2 and TaGT47_2 genes resulted in decreased AX chain length in both fractions and a decrease in extract viscosity of up to sixfold whereas suppression of TaXAT1 resulted in a population shift towards shorter chain length in WE-AX with little effect on AE-AX and a more modest decrease in extract viscosity by twofold.

Feruloylation of AX is a key property of grass cell walls, allowing cross-linking which confers structural strength and determines digestibility (de Oliveira et al., 2015), and ferulate content of wheat grain is important with respect to diet because it may confer health benefits (Shewry and Hey, 2015). We have therefore determined whether AX feruloylation and ferulate dimerization were affected in the wheat RNAi lines with radically altered AX structure. To this end, we examined feruloylation and arabinosylation in different fractions extracted from wheat white flour (which is essentially equivalent to starchy endosperm tissue) in these RNAi lines. Using enzymatic digestion to solubilize water-unextractable AX without removing ester-linked FA, we determined the distribution of AX between different fractions (water-extractable, xylanase-extractable and xylanase-unextractable), the nature of arabinosylation

and the degree of feruloylation and dimerization of ferulate in these fractions in the TaGT43_2, TaGT47_2 and TaXAT1 RNAi lines.

Results

Effect of suppression of AX biosynthetic genes on feruloylation and structure of AX in white flour

We have previously demonstrated that transgenic RNAi lines suppressing TaGT43_2 and TaGT47_2 genes encoding components of xylan synthase have 40%–50% decreases in total AX, whereas RNAi lines suppressing TaXAT1 which encodes an arabinosyl transferase have only 0%–15% decreases (Anders et al., 2012). The abundance of AX oligosaccharides (AXOS) containing xylose mono- and di-substituted by Araf was measured in solubilized fractions and mono- and di-substituted AXOS were shown to be similarly decreased in TaGT43_2 and TaGT47_2 RNAi lines (Lovegrove et al., 2013), whereas in TaXAT1 RNAi lines, there was no effect on di-substituted but a 70-80% decrease in mono-substituted AXOS (Anders et al., 2012). However, the differences in FA which occurs ester-linked to a proportion of these 3-linked Araf have not been previously reported. The bound FA and FA dimer contents of multiple transgenic lines are summarized in Table 1. It is assumed that all bound ferulate in wheat endosperm is ester-linked to AX (Saulnier et al., 2007) so an overall value of FA and diFA per unit AX where AX is estimated from monosaccharide is also given. There is a decrease in amount of FA monomer in all TaGT43_2 and TaGT47_2 lines and in two of the three TaXAT1 RNAi lines per unit dry weight (dwt), but per unit AX, there was no consistent effect of the transgenes. FA dimer was less affected than monomer by transgenes per unit dwt and was consistently increased per unit AX. We decided to look at these changes in AX structure in selected representative lines in more detail by examining different fractions of AX to provide insight into the properties of endosperm AX.

Composition of fractions sequentially extracted from wild-type wheat white flour

White flour was separated into three fractions: a water-extractable fraction (WE); a fraction released by digestion with endoxylanse GH11 and lichenase (XE) and the remaining fraction which was insoluble in water and not solubilized by digestion (XU) (Figure 1). The composition of total white flour and of the different fractions isolated from white flour of wheat cv. Cadenza was analysed (Table 2). All galactose present in the WE fraction is assumed to be from arabinogalactan (AGP). Glucose and mannose in the XU fraction are most likely from glucomannan. WE-AX content is consistent with amounts reported by (Gebruers et al., 2010) for Cadenza grown at multiple sites and years (4.3–6.9 mg/g dry weight), but TOT-AX content is less than reported by these authors (23.5–27.3 mg/g dry weight). The sum of AX content measured in the three fractions (18.4 mg/g dry weight) is similar to that measured in the unfractionated white flour (18.7 mg/g dry weight) indicating good recovery of AX during extraction. As expected, the majority of the AX is in the XE fraction (64%) and more than 90% of the AX is extracted with only 8% in the XU fraction (28% in the WE fraction). The A/X ratio is lowest in the WE fraction, slightly but significantly greater in the XE fraction ($P = 0.012$), but in the XU fraction, there is approximately three times the amount of arabinosylation as in the WE and XE fractions. The A/X ratios of the WE and XE fractions

Table 1 Bound ferulic acid monomer (FA) and dehydrodimer (diFA) content of white flour from transgenic wheat lines homozygous (H) for RNAi constructs suppressing AX biosynthetic genes, and azygous (A) control null segregant lines. Experiment number (Expt.) indicates which lines were grown together in glasshouse experiments. Expt. 1 was a randomized block design (*n* = 4), Expt. 2 and Expt. 4 were randomized design (*n* = 4) and Expt. 3 did not have biological replication. Contents are expressed per unit dry weight flour and per unit arabinoxylan (AX). Ratio of H/A is shown as mean % ± SE (calculated using variance of H and A) with *P*-value for H-A comparison in parentheses

Expt.	Line	Trans-gene	FA µg/g d.wt.		diFA µg/g d.wt.		FA µg/g AX		diFA µg/g AX	
			Mean ± SE	H/A % (*P*-value)	Mean ± SE	H/A % (*P*-value)	Mean ± SE	H/A % (*P*-value)	Mean ± SE	H/A % (*P*-value)
1	TaGT43_2-3	A	94 ± 2	69 ± 3 (<0.001)	19.5 ± 0.4	82 ± 2 (0.008)	4464 ± 75	126 ± 5 (0.007)	927 ± 19	149 ± 3 (<0.001)
		H	65 ± 2		16.0 ± 0.3		5637 ± 179		1385 ± 11	
1	TaGT43_2-6	A	111 ± 5	71 ± 4 (<0.001)	20.5 ± 0.9	90 ± 5 (0.085)	5663 ± 380	88 ± 6 (0.085)	1043 ± 65	110 ± 8 (<0.001)
		H	79 ± 2		18.4 ± 0.8		4958 ± 142		1147 ± 34	
2	TaGT43_2-5	A	161 ± 2	45 ± 5 (<0.001)	29.0 ± 5.7	82 ± 18 (0.436)	6855 ± 403	85 ± 9 (0.169)	1190 ± 199	162 ± 31 (0.031)
		H	72 ± 8		23.8 ± 2.4		5794 ± 548		1921 ± 169	
1	TaGT47_2-1	A	81 ± 4	78 ± 5 (0.003)	17.1 ± 1.1	86 ± 7 (0.061)	3812 ± 381	145 ± 16 (<0.001)	807 ± 94	159 ± 20 (0.21)
		H	64 ± 2		14.8 ± 0.6		5523 ± 257		1282 ± 50	
1	TaGT47_2-4	A	117 ± 6	59 ± 4 (<0.001)	25.2 ± 1.2	74 ± 5 (<0.001)	5968 ± 325	92 ± 6 (0.218)	1290 ± 63	116 ± 7 (0.014)
		H	68 ± 2		18.6 ± 0.8		5514 ± 135		1501 ± 54	
3	TaGT47_2-7	A	133	67	30.5	80	5537	99	1272	120
		H	89		24.3		5503		1520	
4	TaXAT1-1*	A	123 ± 8	86 ± 8 (0.071)	18.1 ± 1.0	87 ± 6 (0.006)	5606 ± 353	99 ± 9 (0.906)	821 ± 43	100 ± 6 (0.994)
		H	106 ± 8		15.8 ± 0.6		5540 ± 398		820 ± 30	
4	TaXAT1-2	A	99 ± 6	101 ± 10 (0.922)	14.9 ± 0.4	107 ± 5 (0.179)	4798 ± 422	139 ± 15 (0.004)	721 ± 40	148 ± 9 (<0.001)
		H	100 ± 8		16.0 ± 0.6		6652 ± 456		1066 ± 29	
4	TaXAT1-3	A	117 ± 1	73 ± 4 (0.002)	15.5 ± 0.2	93 ± 3 (0.167)	5574 ± 140	102 ± 9 (0.885)	743 ± 15	130 ± 8 (<0.001)
		H	85 ± 5		14.5 ± 0.5		5656 ± 506		964 ± 59	

*No biological replication of AX content. AX content is taken from Freeman *et al.* (2016).

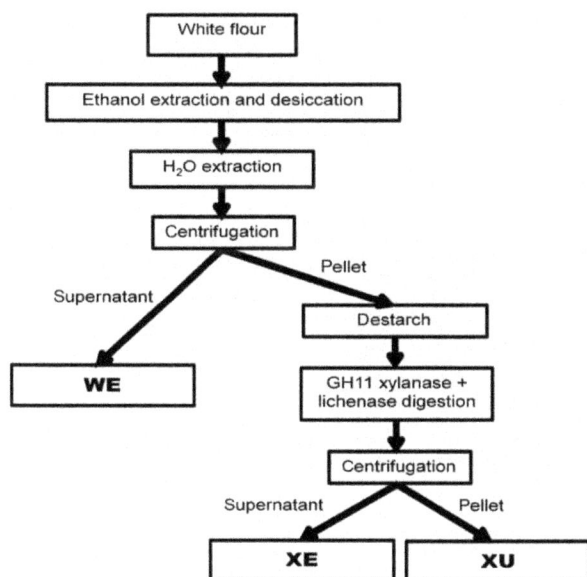

Figure 1 Scheme representing the method for sequential extraction of fractions containing arabinoxylan with different properties from alcohol insoluble residue from wheat white flour. WE = water-extractable fraction; XE = xylanase-extractable fraction solubilized by digestion with GH11 endoxylanse and lichenase; XU = xylanase-unextractable fraction.

are consistent with those reported for twenty wheat cultivars by (Ordaz-Ortiz and Saulnier, 2005).

The degree of feruloylation of AX varied widely between the different AX fractions. Despite only accounting for 8% of the total AX content, the fraction which was not solubilized by enzymatic digestion (XU) contained 36% of the bound ferulate monomer and 49% of the dimers meaning that the AX in this fraction had 10 times the amount of bound FA monomer and 30 times the amount of dimer than in the WE-AX fraction (Table 2). The per cent dimerization of bound FA was 7.6%, 12.9% and 19.2% in the WE, XE and XU fractions, respectively. The values for FA and FA dimers are similar to those previously reported for WE and water-unextractable fractions (equivalent to our XE + XU fractions combined) of wheat flour AX (Dervilly-Pinel *et al.*, 2001).

Effect of suppression of AX biosynthetic genes on the distribution of AX between the fractions and the structure of AX in the fractions

Consistent with our previously published results (Anders *et al.*, 2012; Lovegrove *et al.*, 2013), TOT- and WE-AX are decreased by approximately 50% in the TaGT43_2 and TaGT47_2 transgenic lines and by approximately 10% in the TaXAT1 line compared to control lines (Figure 2; Table S1). In the XE-AX fraction, there is a similar decrease in AX content in the TaGT43_2 and TaGT47_2 lines to that in TOT-AX and WE-AX, but in the TaXAT1 line, AX content of this fraction is unaffected by the transgene. The amount of AX in the XU-AX fraction is, however, largely unaffected by suppression of any of the three target biosynthetic genes (Figure 2). For the TaGT43_2 and TaGT47_2 transgenic lines, bound FA monomer content is decreased in the WE-AX and XE-AX fractions by similar amounts to AX content but bound FA dimer content is less affected (Figures 3 and 4). Therefore, the AX in these fractions of these transgenic lines appears to be more highly cross-linked than in control lines. In contrast, the WE-AX is much less feruloylated in the TaXAT1 transgenic lines than in the control line but largely unaffected in the XE-AX fraction. The

	Fraction			
	WE	XE	XU	TOT
Arabinose	2.9 ± 0.02	4.3 ± 0.08	1.0 ± 0.06	10.0 ± 0.18
Xylose	3.4 ± 0.03	7.5 ± 0.15	0.6 ± 0.05	11.6 ± 0.28
Galactose	1.6 ± 0.01	Trace	0.3 ± 0.03	4.2 ± 0.07
Glucose	6.4*	8.6 ± 0.28	3.0 ± 0.05	728.1 ± 11.04
Mannose	Trace	Trace	0.6 ± 0.04	1.2 ± 0.14
AX[‡]	5.2 ± 0.02	11.7 ± 0.22	1.5 ± 0.10	18.7 ± 0.40
A/X	0.52 ± 0.009	0.57 ± 0.009	1.65 ± 0.091	0.86 ± 0.011
Ferulic acid (µg/g d.wt.)	12.0 ± 0.44	54.2 ± 2.81	36.6 ± 0.58	103.4 ± 1.19
Diferulic acid[†] (µg/g d.wt.)	1.0 ± 0.07	8.1 ± 0.45	8.71 ± 0.17	21.0 ± 0.59
Ferulic acid (µg/g AX)	2313 ± 92	4634 ± 281	24022 ± 1373	5540 ± 64
Diferulic acid[†] (µg/g AX)	191 ± 12	688 ± 41	5705 ± 308	1131 ± 31
Percentage of bound ferulic acid as dimer	8%	13%	19%	17%

Table 2 Composition of fractions sequentially extracted from wheat white flour cv Cadenza. Sugar contents are in mg/g dry weight flour, bound ferulic acid monomer and total dehydrodimers contents are in µg/g dry weight flour (d.wt.) or µg/g arabinoxylan (AX) from the same fraction. WE = water-extractable fraction; XE = xylanase-extractable fraction solubilized by digestion with GH11 endoxylanse and lichenase; XU = xylanase-unextractable fraction; TOT = total without fractionation. Values are average ± SEM, n = 4 replicate extractions except * where n = 2

[†]sum of amounts of 8-5′ benzofuran, 8-O-4′, 5-5′ and 8-5′ diferulates. Three further diferulate standards (8-8′, 8-8′ aryltetralin and 8-8′ tetrahydrofuran) were not detected in these samples.
[‡]Sum of arabinose and xylose corrected for arabinogalactan content.

degree of feruloylation in the XU-AX fraction is, like AX content, largely unaffected by suppression of any of the three target genes (Figure 3; Table S2), particularly expressed per unit AX (Figure 4).

Feruloylation of endosperm AX mostly occurs via esterified Araf 3-linked to mono-substituted Xylp, (Saulnier et al., 2007) although a smaller proportion may occur as feruloylated Araf 3-linked to di-substituted Xylp (Veličković et al., 2014). We therefore also examined the effect of the RNAi transgenes on the arabinosylation pattern in the WE-AX and XE-AX fractions by quantification of AX oligosaccharides (AXOS) by HPAEC and by proton NMR analyses. AXOS abundances present in the XE-AX fraction or generated by subsequent endoxylanase digestion for

the WE-AX fraction are shown in Figure S1. For TaGT43_2 and TaGT47_2 RNAi lines, the AXOS are decreased more than the AX amount (Figure S1; compare with Figure 2a, b). This is because the substitution pattern is changed in these lines such that larger AXOS are released by GH11 digestion which are not identified by the HPAEC method (Lovegrove et al., 2013). In the TaXAT1 WE-AX and XE-AX fractions AXOS containing di-substituted Xylp show smaller decreases than other AXOS, with AXOS containing di-substituted Xylp actually increased relative to controls in XE-AX (Fig. S1).

Interpretation of the HPAEC data, which is limited to analyses of AXOS with low DP, is complicated by the changed structure of

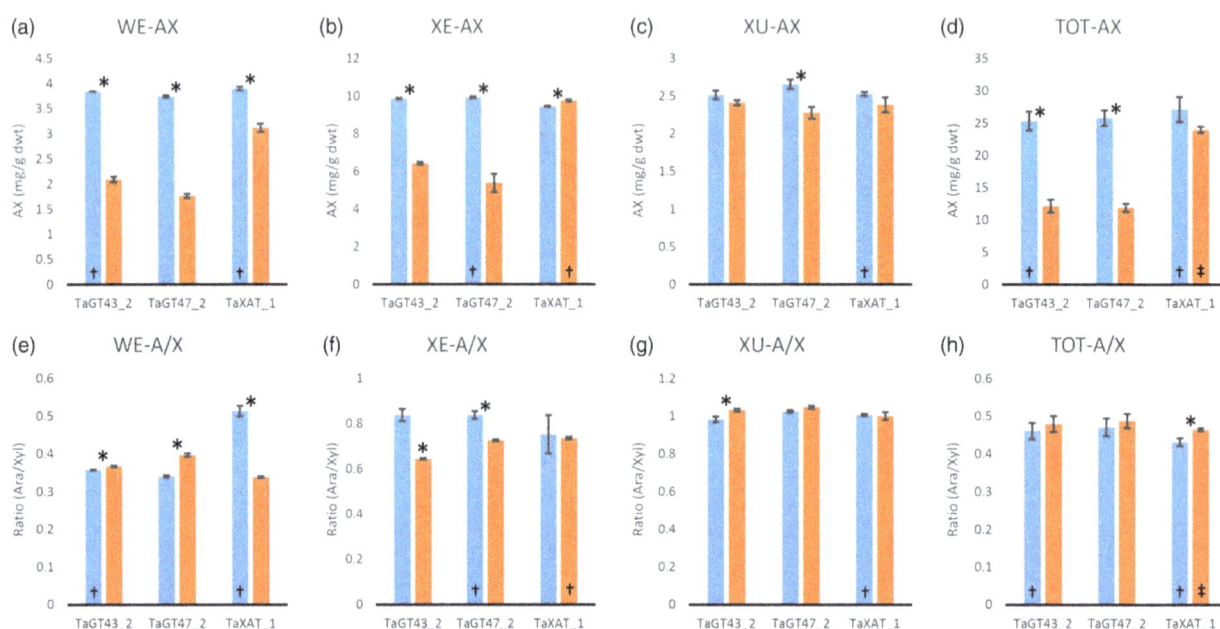

Figure 2 a to d: arabinoxylan content of white flour (TOT-AX) and fractions sequentially extracted from white flour from transgenic wheat lines homozygous for RNAi constructs suppressing AX biosynthetic genes (orange bars), and azygous control lines (blue bars). e to h: ratio of arabinose to xylose in AX in the same samples as in a to d. Values are average ± SEM, n = 4 replicate extractions except bars marked † where n = 3 and ‡ where n = 2; asterisks denote significant differences at P < 0.05 from Student's t-test.

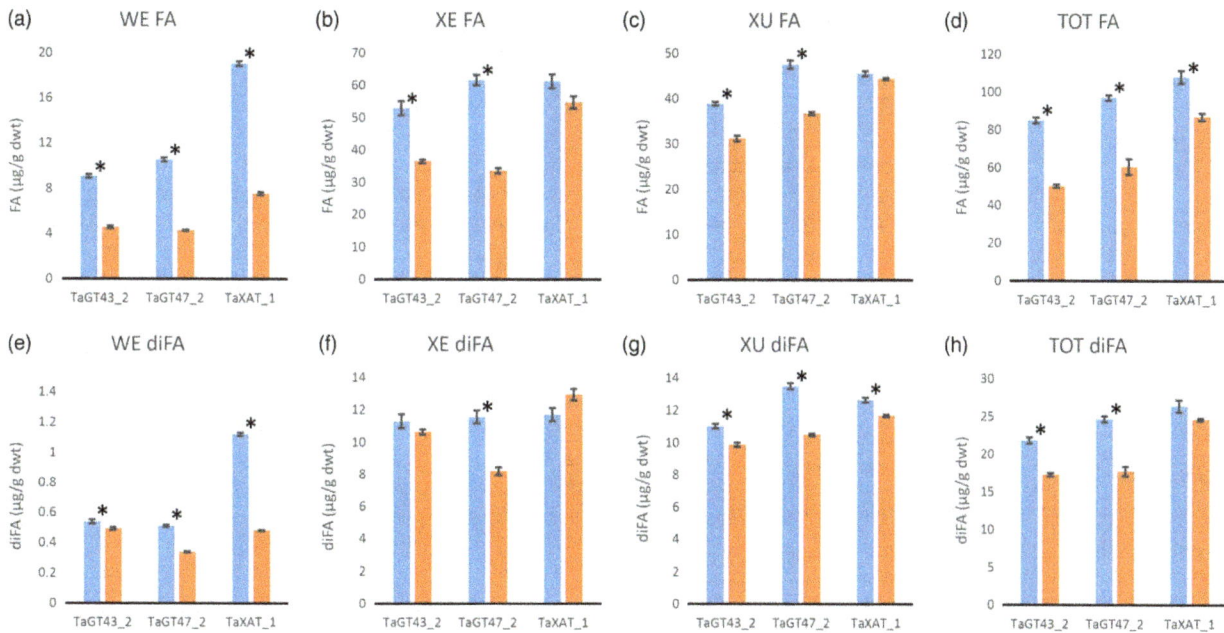

Figure 3 a to d: Bound ferulic acid monomer (FA) content of white flour (TOT-AX) and fractions sequentially extracted from white flour from transgenic wheat lines homozygous for RNAi constructs suppressing AX biosynthetic genes (orange bars), and azygous control lines (blue bars). e to h: Bound ferulic acid dehydrodimer (diFA) content in the same samples as in a to d. Dimers are the sum of amounts of 8-5′ benzofuran, 8-O-4′, 5-5′ and 8-5′ diferulates (separate values given in Table S2). Values are average ± SEM; for WE, XE and XU $n = 4$ replicate sequential extractions for TOT $n = 3$; asterisks denote significant differences at $P < 0.05$ from Student's t-test.

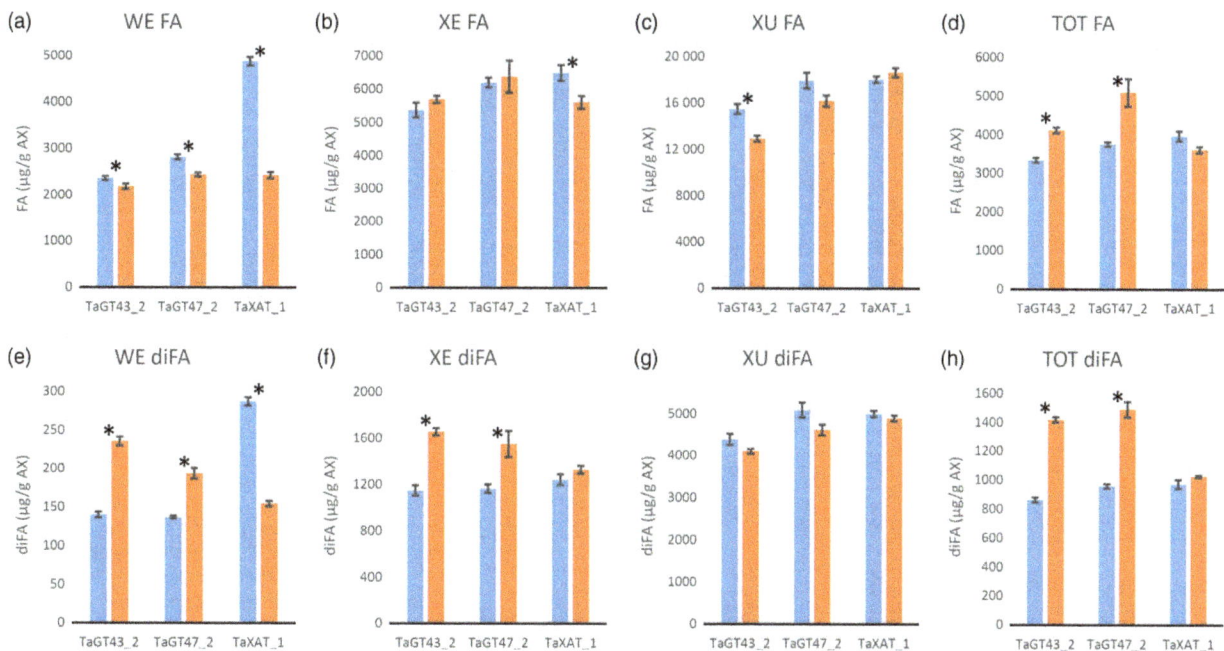

Figure 4 Bound ferulic acid monomer (FA) (a-d) and dimer content (diFA) (e-h) per unit of arabinoxylan (AX) in white flour (TOT-AX) and fractions sequentially extracted from white flour from transgenic wheat lines homozygous for RNAi constructs suppressing AX biosynthetic genes (orange bars), and azygous control lines (blue bars). Values are average ± SEM; for WE, XE and XU $n = 4$ replicate sequential extractions for TOT $n = 3$; asterisks denote significant differences at $P < 0.05$ from Student's t-test.

the AX with fewer GH11 xylanase cleavage sites in the transgenic lines. NMR spectra of the region corresponding to the anomeric H1 resonances of the Ara$f$ linked to the xylan backbone give an overview of all the Ara$f$ 3-linked to mono-substituted Xyl$p$ (A^3-Xmono), 3-linked to di-substituted Xyl$p$ (A^3-Xdi) and 2-linked to di-substituted Xyl$p$ (A^2-Xdi). Chemical shifts occur at around 5.40, 5.28 and 5.23 ppm, respectively, for A^3-Xmono, A^3-Xdi and A^2-Xdi in all AXOS arising from GH11 digestion, although the exact position varies according to substitutions of neighbouring Xyl$p$ (Hoffmann et al., 1991; Petersen et al., 2014). Intact AX molecules give rise to broad peaks at these positions due to the multiplicity of contexts for the Ara*f* present (Petersen *et al.*, 2014), whereas mixtures of specific AXOS from GH11 digestion give sets of overlapping but distinct peaks and this is seen here for the WE-AX and XE-AX fractions (Figure 5). For all RNAi lines, the NMR shows bigger decreases in both mono- and di-substitution in WE-AX than in XE-AX; the effects are quantified by integration of peak areas for all spectra in Table 3. The TaXAT1 line shows far more decreased mono-substitution than decreased di-substitution; for the XE-AX fraction, di-substitution is actually increased by 30%–40% in TaXAT1 RNAi whilst mono-substitution is decreased by 27% (Table 3). The increase in di-substitution in XE-AX explains why there is no overall decrease in Ara*f* content in TaXAT1 RNAi lines in this fraction (Figure 2).

Discussion

Our results show that in wheat RNAi lines with substantially decreased amount and chain length of endosperm AX (TaGT43_2 and TaGT47_2 RNAi lines) and altered AX structure (all RNAi lines), the magnitude of decreases in amounts of FA and particularly of diFA are much smaller or absent (Table 1). We assume, as is common, that all the bound FA and diFA in endosperm are ester-linked to Ara*f* on AX which is supported by the lack of reports of ferulate ester-linked to residues other than Ara*f* using approaches where the ester bonds are preserved on cell wall fractions from cereal grain (Saulnier *et al.*, 1995) and grasses in general (Ishii, 1997; Mueller-Harvey and Hartley, 1986).

By separation of the endosperm AX into the WE-AX, XE-AX and XU-AX fractions, we show that the smaller effects on FA and diFA than on AX are because the effects of the transgenes on amount of AX are greatest for the lowly feruloylated WE-AX fraction, less for the moderately feruloylated XE-AX and absent for the highly feruloylated XU-AX fraction (Figures 2 and 3). However, the compositions of these fractions are not completely unaffected by the transgenes, at least for WE-AX and XE-AX. For TaGT43_2 and TaGT47_2 RNAi lines, diFA content per unit AX (Figure 4e, f) is increased although FA monomer (Figure 4a, b) is not. Thus, FA dimerization [diFA/(diFA + FA)] in WE-AX is increased from 6% in azygous controls to 9% in TaGT43_2 and TaGT47_2 RNAi lines and in XE-AX from 16% to 22%. For the TaXAT1 line, FA monomer and diFA content per unit AX are decreased to a similar degree in the WE fraction; hence, the degree of dimerization is unaffected by the transgene (6%) whereas in the XE fraction, there is a small but significant decrease in FA monomer per unit AX and no effect on diFA content resulting in an increase in dimerization from 16% in azygous controls to 19% in the homozygous line. Dimerization in the XU-AX fraction is however unchanged at around 22% for all lines. Ultimately, the amount of dimerization that is possible depends on the fraction of FA monomer that are sufficiently close to oxidatively couple and it may be these are at a maximum for dimerization in the XU-AX fractions.

The solubility of wheat grain AX is an important characteristic for end use as it is the WE-AX that determines extract viscosity; a negative trait for nonfood uses of wheat, but desirable in food as

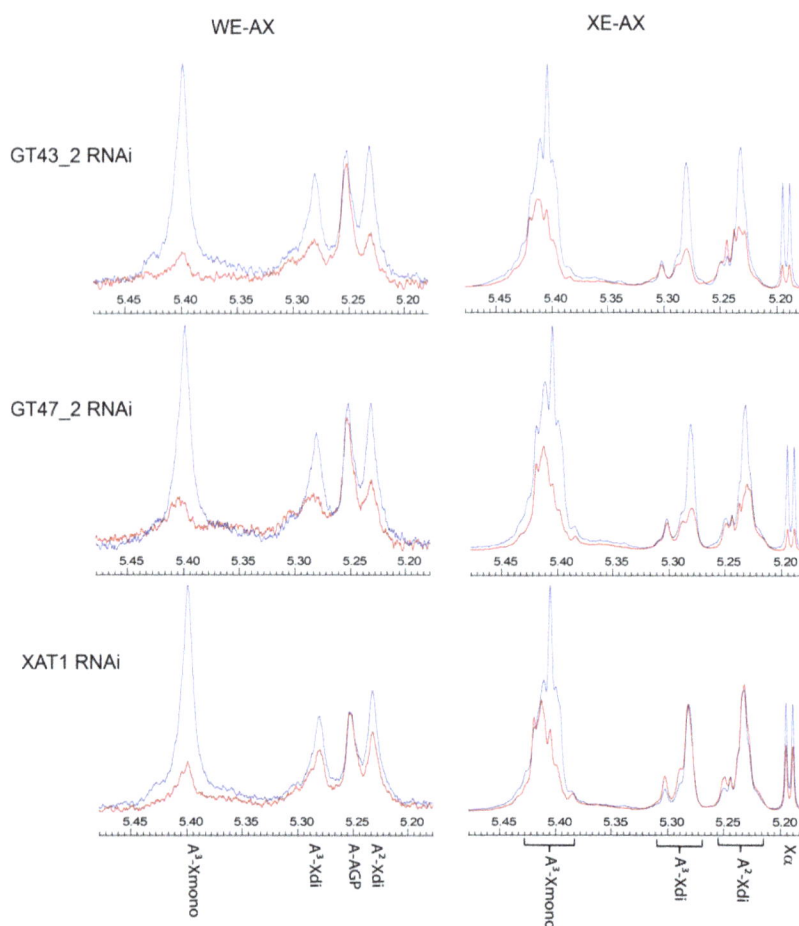

Figure 5 Representative 1H-NMR spectra for transgenic (red) and azygous control (blue) samples; signal size is normalized to internal standard added to extract from a fixed dwt of endosperm. H1 resonances for Ara*f* in AX are indicated: α-(1,3)-linked to mono-substituted Xyl*p* (A3-Xmono), α-(1,3)-linked (A3-Xdi), and α-(1,2)-linked (A2-Xdi) to di-substituted Xyl*p* and for Ara*f* in arabinogalactan peptide (A-AGP; WE-AX samples only). For XE-AX samples, ranges are shown within which peaks for A3-Xmono, A3-Xdi and A2-Xdi are known to occur from distinct oligosaccharides generated by GH11 digestion (Hoffman *et al.*, 1991: Petersen *et al.*, 2014); also a doublet from H1 α resonance for Xyl*p* β-(1,4)-linked (Xα) is shown.

Table 3 Estimates of Araf content with different linkages in mg/g dry weight flour from integration of peaks in H1-NMR spectra, assuming peak area per ^{1}H is the same as for the internal standard. Integrations were performed in regions A^3-X mono δ5.435–5.375, A^3-Xdi δ5.310–5.265, A^2-Xdi δ 5.255–5.210. For the region of A^2-Xdi peak in WE-AX which overlaps with A-AGP at around δ5.255 (Figure 6), area was estimated using values from the other side of the peak as the peak is symmetrical. Mean values are shown ± SEM, n = 3. H/A is ratio of transgenic mean to azygous control mean as %, with P-value from Student's t-test for H-A comparison shown in parentheses

Fraction	RNAi target	Trans-gene	A^3-Xmono		A^3-Xdi		A^2-Xdi	
			Mean	H/A	Mean	H/A	Mean	H/A
WE	TaGT43_2	A	0.37 ± 0.03	30% (0.002)	0.21 ± 0.01	62% (0.002)	0.23 ± 0.01	51% (0.000)
		H	0.11 ± 0.01		0.13 ± 0.01		0.12 ± 0.00	
	TaGT47_2	A	0.32 ± 0.03	30% (0.004)	0.18 ± 0.02	63% (0.014)	0.19 ± 0.00	57% (0.000)
		H	0.10 ± 0.01		0.12 ± 0.00		0.11 ± 0.01	
	TaXAT1	A	0.47 ± 0.02	29% (0.000)	0.24 ± 0.00	69% (0.000)	0.28 ± 0.00	69% (0.000)
		H	0.14 ± 0.01		0.17 ± 0.00		0.19 ± 0.00	
XE	TaGT43_2	A	4.21 ± 0.11	51% (0.000)	1.52 ± 0.04	53% (0.000)	1.84 ± 0.04	67% (0.000)
		H	2.15 ± 0.05		0.80 ± 0.01		1.23 ± 0.01	
	TaGT47_2	A	4.27 ± 0.10	58% (0.000)	1.47 ± 0.02	64% (0.000)	1.85 ± 0.03	72% (0.002)
		H	2.46 ± 0.12		0.94 ± 0.03		1.34 ± 0.07	
	TaXAT1	A	3.84 ± 0.12	73% (0.002)	1.25 ± 0.04	141% (0.000)	1.60 ± 0.05	130% (0.001)
		H	2.80 ± 0.05		1.76 ± 0.01		2.07 ± 0.04	

soluble fibre. It has previously been suggested that the dominant characteristic determining AX solubility is amount of diFA (Saulnier et al., 2007). This is supported by the evidence here; AX from TaGT43_2 and TaGT47_2 RNAi lines has substantially shorter chain length (Freeman et al., 2016) and not much change in overall Araf/Xylp ratio (Figure 2h) so might be expected to have a higher proportion of AX in the WE-AX fraction. In fact, this proportion is not much changed and this could be due to the overall greater amount of diFA per unit AX (Figure 4h).

We were also interested in the pattern of arabinosylation in the WE-AX and XE-AX fractions, especially since ferulate is believed to be exclusively linked to Araf 3-linked to Xylp [since this is the only linkage found in feruloylated AX (Ishii, 1991; Lequart et al., 1999; Rhodes et al., 2002)]. HPAEC quantification of AXOS suggested that mono-substitution and di-substitution were equally decreased in TaGT43_2 and TaGT47_2 RNAi lines, whereas mono-substitution was much more decreased than di-substitution in TaXAT1 RNAi lines; the magnitude of decreases was greater in WE-AX than XE-AX fractions for all three RNAi lines (Figure S1). However, the AX structural changes in TaGT43_2 and TaGT47_2 RNAi lines also led to a shift to larger AXOS not detected by the HPAEC (Lovegrove et al., 2013). This problem is effectively overcome by NMR analysis, where A^3-Xmono, A^3-Xdi and A^2-Xdi signals are seen from all AXOS and whole AX chains at similar positions (Figure 5) with the exact position depending mostly on the substitution of immediately neighbouring Xylp (Petersen et al., 2014). The NMR spectra confirm that in the TaGT43_2 and TaGT47_2 RNAi lines, there are decreases in the A^3-Xmono, A^3-Xdi and A^2-Xdi signals whereas in the TaXAT1 RNAi line, the decrease in A^3-Xmono signal is greater than for the A^3-Xdi and A^2-Xdi signals which are slightly decreased in WE-AX and increased in XE-AX (Table 3). In all NMR spectra for XE-AX fractions, it is noticeable that some regions within the sets of peaks ascribed to A^3-Xmono, A^3-Xdi and A^2-Xdi are relatively unaffected by the transgenes even when the others are substantially decreased (Figure 5). AXOS where the neighbouring Xylp is di-substituted with Araf are known to give rise to peaks in these regions [namely 5.42–5.44 for A^3-Xmono,

5.31 for A^3-Xdi and 5.25 for A^2-Xdi (Petersen et al., 2014)]. As A-Xdi peak areas in Table 3 include these regions, they tend to be somewhat less affected than A^3-Xmono even for TaGT43_2 and TaGT47_2 RNAi lines. We know that AX deposited earlier in grain development is richer in di-substitution (Pellny et al., 2012; Toole et al., 2010) and that RNAi transgenes driven by the HMW promoter tend to have more effect on cell wall polysaccharides later in development (Lovegrove et al., 2013; Nemeth et al., 2010). Therefore, these relatively unaffected regions may reflect AX rich in di-substitution that is deposited too early in development to be affected by the RNAi suppression.

By contrast, feruloylation of AX appears to be maintained at a fairly constant level during development (Pellny et al., 2012) so a lack of effect on the degree of feruloylation of AX cannot be attributed to early deposited AX. In the XE-AX fraction, there was no significant decrease in FA or diFA in the TaXAT1 RNAi line (Figure 3b, f; Table S2) despite the fact that there were substantial decreases in Xylp mono-substituted with 3-linked Araf to which this FA is linked, (around 40%) as indicated by NMR (Figure 5). In one model of feruloylation of AX, feruloylated Araf is transferred onto the AX in the Golgi by the action of a glycosyl transferase (Buanafina, 2009; de Oliveira et al., 2015); the lack of effect of suppressing TaXAT1 on feruloylation here indicates that TaXAT1 protein may only transfer nonferuloylated Araf. This is compatible with the expression pattern of TaXAT1 which is highest in starchy endosperm, a tissue with uniquely low levels of AX feruloylation (Pellny et al., 2012).

Our data show that FA and diFA content of starchy endosperm cell walls is substantially independent of the amount of AX when decreased by TaGT43_2 and TaGT47_2 suppression, and of the amount of Araf 3-linked to mono-substituted Xylp when decreased by TaXAT1 suppression. The particular maintenance of diFA levels suggests that diFA content may be vital to the integrity of the endosperm cell walls. This is plausible because FA dimerization probably represents the dominant mode of covalent cross-linking between chains in the cell walls of this tissue where 70% of the polysaccharide is AX and there is no lignin. When AX ferulate in wheat endosperm was decreased by heterologous expression of a fungal feruloyl esterase in another study, grain morphology was severely affected with an apparent collapse of endosperm tissue (Harholt et al., 2010). Cell wall integrity is sensed in higher plants, and compensatory changes occur in response to weakening, for example, by production of reactive oxygen species (ROS) to promote cell wall stiffening by increased peroxidative coupling (Voxeur and Höfte, 2016). Similar mechanisms could operate here and when the wall is weakened by the low AX content and shorter chains in the TaGT43_2 and TaGT47_2 lines, FA dimerization could be increased (as seen in WE-AX and XE-AX fractions) by ROS production and increased peroxidase activity, but greater feruloylation per unit AX may also be required as is observed overall (Figure 4d). This hypothesis could explain how the endosperm cell walls that are half as thick as normal in the TaGT43_2 and TaGT47_2 RNAi lines (Lovegrove et al., 2013) are able to maintain their form.

The wheat endosperm cell wall is a tractable system, and our results here may provide some general insight into AX feruloylation; however, it is a special case with simple, lowly feruloylated AX. Studies on the highly feruloylated GAX typical of the majority of grass tissue (e.g. in maize pericarp; Chateigner-Boutin et al., 2016) are needed to establish the control of this key process in determining cell wall properties and grass biomass digestibility.

Experimental procedures

Plant material and growth conditions

Wild-type wheat cv. Cadenza was grown in the field at Rothamsted Research farm. All other wheat (Triticum aestivum L. var. Cadenza) plants were grown in temperature-controlled glasshouse rooms as previously described (Nemeth et al., 2010) in 25-cm pots with five plants per pot. Within experiments, homozygous and azygous segregants descended from the same original transformant were grown at the same time in the same glasshouse room. All grains were harvested, and biological replicates consisted of grain from single pots and 5-g grains were milled. Generation of transgenic lines is described in (Anders et al., 2012) for TaXAT1 RNAi lines and (Lovegrove et al., 2013) for TaGT43_2 and TaGT47_2 RNAi lines.

Sequential extraction of fractions from white flour

White flour from wild-type wheat cv. Cadenza was prepared using a Chopin CD1 mill (Calibre Control International Limited). All other white flour samples were prepared as previously described (Anders et al., 2012). For quantification of monosaccharide and phenolic acid content, fractions were prepared as represented in Figure 1. Alcohol insoluble residue (AIR) was prepared from 200 mg of white flour by extraction in 80% ethanol as previously described (Pellny et al., 2012) and dried by centrifugation under vacuum. The water-extractable fraction (WE) was prepared by suspending the AIR in 1.5 mL water, mixing by rotation at 30 rpm for 30 min and centrifugation for 10 min at 10 000 g. Samples of supernatant (WE) were taken and dried as above. After washing with 1.5 mL water, the pellet was destarched by digestion at 80 °C for 30 min in 1 mL 0.1 M sodium acetate, pH5.2, 4 mM calcium chloride containing 250 units of α-amylase (Sigma-Aldrich, A3403), cooled to 50 °C before the addition of 10 μL pullulanase (Sigma-Aldrich, P2986) and incubation at 50 °C for 30 min. Released glucose was removed by centrifugation, and the pellet washed twice with 1 mL water and dried as above. The xylanase-extractable fraction (XE) was prepared by digestion with 4 μL xylanase GH11

(Prozomix, PRO-E0062) and 1.9 μL lichenase (Prozomix, PRO-E0017) in 1 mL water following the procedure described by Nemeth et al. (2010). The residue (xylanase-unextractable fraction, XU) was washed with 1 mL water, dispersed in 1 mL water and dried as above. Samples for NMR analysis were prepared essentially as described above except that the WE fraction was digested with 0.5 units of α-amylase in 4 mM calcium chloride for 15 min at 100 °C and arabinoxylan (AX) recovered by precipitation as described by Ordaz-Ortiz and Saulnier (2005); the XE fraction was not destarched. For analysis of oligosaccharides (AXOS) released by digestion with xylanase (GH11) and lichenase in WE and XE fractions, AIR was prepared from 100 mg white flour as described by Saulnier et al. (2009), WE and XE fractions as for NMR analysis without drying the XE fraction, and then the WE fraction was digested with xylanase and lichenase as for the XE fraction except the total volume was 0.5 mL.

Cell wall analyses of white flour and fractions sequentially extracted from white flour

Quantification of ferulic acid monomer (FA) and dehydrodimer (diFA) content

Cell wall bound phenolics were extracted from AIR prepared from white flour, or from WE, XE and XU fractions as previously described (Pellny et al., 2012).Quantification was by HPLC using a binary gradient with acetonitrile (solvent A) and 2% acetic acid (solvent B) either as described by Pellny et al. (2012) or essentially as described by (Pellny et al., 2012) except separating 20 μL of sample on a UPLC Kinetex Phenyl-Hexyl (150 × 4.6 mm, 5 μm) column with the following gradient: linear 100% to 30% B, 0–12 min; isocratic 30% B, 12–14 min; Linear 30% to 100% B, 14 to 14.1 min; isocratic 100% B, 14.1 to 18 min and a flow rate of 2 mL/min. Quantification of FA was by integration of peak areas at 280 nm with reference to calibrations made using known amounts of pure compounds. Peaks of the major diFAs were identified by comparison with retention times and absorption spectra with pure standards kindly supplied by Professor John Ralph (Lu et al., 2012). Values reported for diFAs are the sum of amounts of 8-5′ benzofuran (BF), 8-O-4′, 5-5′ and 8-5′ diFAs. Three further diFA standards (8-8′, 8-8′ aryltetralin and 8-8′ tetrahydrofuran) were not detected in these samples. DiFA content of samples, relative to FA monomer, was calculated using areas of these peaks and response factors (relative response factors for 8-5′BF, 8-O-4′, 5-5′ and 8-5′ diFAs were 0.501, 0.692, 0.417 and 0.456 respectively).

Monosaccharide analysis by high-performance anion-exchange chromatography (HPAEC)

For total neutral sugar content of white flour, 10 mg of flour was finely ground using 3 × 3 mm diameter stainless steel ball bearings (Atlas Ball and Bearing Company Limited) and agitation for 1 min at 15 000 rpm in a Genogrinder (Spex Sample Prep®), dilution to 1 mg/mL and 200 μL aliquots (equivalent to 200 μg flour) dried by centrifugation under vacuum. For WE, XE and XU fractions, 37.5 μL, 12.5 μL and 50 μL aliquots (equivalent to extract from 5 mg, 2.5 mg and 10 mg white flour), respectively, were dried as above. Sugars were released by acid hydrolysis by dissolution in 400 μL 2 M trifluoroacetic acid and heating to 120 °C for 60 min and then dried as above and washed by dissolution in 500 μL water and drying. Calibration curves were generated by subjecting a range of amounts of commercially available sugars (Sigma-Aldrich UK) to acid hydrolysis following the same protocol as for samples. Dried samples and standards were dissolved in 400 µL water and filtered through 0.45 µm PVDF disposable filters (Whatman) before separation by HPAEC (Dionex ICS-5000+ HPIC, Thermo Scientific) on a CarboPac PA20 analytical column (3 × 150 mm) with a CarboPac PA20 guard column (3 × 30 mm) at 30 °C and equipped with an eluent generator with an EGC 500 KOH cartridge. The flow rate was 0.5 mL/min, and the gradient was isocratic 4.5 mM KOH, 0–13 min; linear 4.5–10 mM KOH, 13–14 min; linear 10–13 mM KOH, 14–15 min; linear 13–20 mM, 15–16 min; isocratic 20 mM 16–17 min; linear 20–4.5 mM KOH, 17–18 min followed by isocratic 4.5 mM KOH 18–23 min to equilibrate the column for the next injection. Detection was by electrochemical detector, and chromatograms were analysed and data calculated using Chromeleon 7.2 software (Thermo Scientific). AX content was calculated as the sum of arabinose and xylose except in WE-AX fractions and flour which were corrected for arabinogalactan content assuming an arabinose to galactose ratio of 0.7 (Ordaz-Ortiz and Saulnier, 2005).

Analysis of arabinoxylan (AX) and (1-3):(1-4)-β-D-glucan content by HPAEC

AXOS and glucans were separated and chromatograms analysed as described by (Kosik et al., 2017). Integrated peak areas were recorded for each AXOS and for G3 and G4 peaks derived from (1-3):(1-4)-β-D-glucan and the ratio of peak area in transgenic lines to that in corresponding azygous control lines calculated.

¹H-NMR analysis

Samples were suspended in D_2O (1 mL) containing 0.05% (wt/vol) d_4-trimethylsilyl propionic acid. ¹H-NMR spectra were recorded at 300 K on a Bruker Avance NMR spectrometer (Bruker Biospin), operating at 600.05 MHz, equipped with a selective inverse probe. Spectra were collected using 128 scans using a zgpr pulse sequence with a 90° pulse angle. The residual water signal was suppressed by presaturation during a 5-s delay. Spectra consisted of 64-K data points over a sweep width of 12 ppm. Free induction decays were Fourier transformed using an exponential window with a line broadening of 0.5 Hz. Phasing and baseline correction were carried out automatically within the instrument's TOPSPIN v. 2.1 software. AX peak assignments were based on (Anders et al., 2012), and the H1 Araf peak assignment of arabinogalactan peptide was made by comparison with an authentic standard.

Acknowledgements

We thank Delia Corol for the preparation of NMR samples, Steve Powers for statistical advice, John Ralph for supplying standards of the ferulate dehydrodimers and Luc Saulnier for help with interpretation of NMR. This work was funded out of the 'Designing Seeds' strategic award to Rothamsted Research by the UK Biotechnology and Biosciences Research Council. The authors declare no conflict of interest.

References

Anders, N., Wilkinson, M.D., Lovegrove, A., Freeman, J., Tryfona, T., Pellny, T.K., Weimar, T. et al. (2012) Glycosyl transferases in family 61 mediate arabinofuranosyl transfer onto xylan in grasses. Proc. Natl. Acad. Sci. USA, 109, 989–993.

Barron, C., Surget, A. and Rouau, X. (2007) Relative amounts of tissues in mature wheat (Triticum aestivum L.) grain and their carbohydrate and phenolic acid composition. J. Cereal Sci. 45, 88–96.

Brown, D.M., Goubet, F., Vicky, W.W.A., Goodacre, R., Stephens, E., Dupree, P. and Turner, S.R. (2007) Comparison of five xylan synthesis mutants reveals new insight into the mechanisms of xylan synthesis. Plant J. 52, 1154–1168.

Brown, D.M., Zhang, Z.N., Stephens, E., Dupree, P. and Turner, S.R. (2009) Characterization of IRX10 and IRX10-like reveals an essential role in glucuronoxylan biosynthesis in Arabidopsis. Plant J. 57, 732–746.

Buanafina, M.M.D. (2009) Feruloylation in grasses: Current and future perspectives. Molecular Plant, 2, 861–872.

Chateigner-Boutin, A.-L., Ordaz-Ortiz, J.J., Alvarado, C., Bouchet, B., Durand, S., Verhertbruggen, Y., Barrière, Y. et al. (2016) Developing pericarp of maize: a model to study arabinoxylan synthesis and feruloylation. Front. Plant Sci. 7, 1476.

Dervilly-Pinel, G., Rimsten, L., Saulnier, L., Andersson, R. and Aman, P. (2001) Water-extractable arabinoxylan from pearled flours of wheat, barley, rye and triticale. Evidence for the presence of ferulic acid dimers and their involvement in gel formation. J. Cereal Sci. 34, 207–214.

Freeman, J., Lovegrove, A., Wilkinson, M.D., Saulnier, L., Shewry, P.R. and Mitchell, R.A.C. (2016) Effect of suppression of arabinoxylan synthetic genes in wheat endosperm on chain length of arabinoxylan and extract viscosity. Plant Biotechnol. J. 14, 109–116.

Gebruers, K., Dornez, E., Bedo, Z., Rakszegi, M., Fras, A., Boros, D., Courtin, C.M. et al. (2010) Environment and genotype effects on the content of dietary fiber and its components in wheat in the HEALTHGRAIN diversity screen. J. Agric. Food Chem. 58, 9353–9361.

Harholt, J., Bach, I.C., Lind-Bouquin, S., Nunan, K.J., Madrid, S.M., Brinch-Pedersen, H., Holm, P.B. et al. (2010) Generation of transgenic wheat (Triticum aestivum L.) accumulating heterologous endo-xylanase or ferulic acid esterase in the endosperm. Plant Biotechnol. J. 8, 351–362.

Hoffmann, R.A., Leeflang, B.R., de Barse, M.M.J., Kamerling, J.P. and Vliegenthart, J.F.G. (1991) Characterisation by 1H-n.m.r. spectroscopy of oligosaccharides, derived from arabinoxylans of white endosperm of wheat, that contain the elements –>4)[[alpha]-l-Araf-(1-ar3)]-[beta]-d-Xylp-(1–> or – >4)[[alpha]-l-Araf-(1->2)][[alpha]-lAraf-(1->3)]-[beta]-d-Xylp-(1–>. Carbohyd. Res. 221, 63–81.

Hoije, A., Sternemalm, E., Heikkinen, S., Tenkanen, M. and Gatenholm, P. (2008) Material properties of films from enzymatically tailored arabinoxylans. Biomacromol, 9, 2042–2047.

Ishii, T. (1991) Isolation and characterization of a diferuloyl arabinoxylan hexasaccharide from bamboo shoot cell-walls. Carbohyd. Res. 219, 15–22.

Ishii, T. (1997) Structure and functions of feruloylated polysaccharides. Plant Sci. 127, 111–127.

Jensen, J.K., Johnson, N.R. and Wilkerson, C.G. (2014) Arabidopsis thaliana IRX10 and two related proteins from psyllium and Physcomitrella patens are xylan xylosyltransferases. Plant J. 80, 207–215.

Kosik, O., Powers, S.J., Chatzifragkou, A., Prabhakumari, P.C., Charalampopoulos, D., Hess, L., Brosnan, J. et al. (2017) Changes in the arabinoxylan fraction of wheat grain during alcohol production. Food Chem. 221, 1754–1762.

Lequart, C., Nuzillard, J.M., Kurek, B. and Debeire, P. (1999) Hydrolysis of wheat bran and straw by an endoxylanase: production and structural characterization of cinnamoyl-oligosaccharides. Carbohyd. Res. 319, 102–111.

Lovegrove, A., Wilkinson, M.D., Freeman, J., Pellny, T.K., Tosi, P., Saulnier, L., Shewry, P.R. et al. (2013) RNA interference suppression of genes in glycosyl transferase families 43 and 47 in wheat starchy endosperm causes large decreases in arabinoxylan content. Plant Physiol. 163, 95–107.

Lu, F.C., Wei, L.P., Azapira, A. and Ralph, J. (2012) Rapid syntheses of dehydrodiferulates via biomimetic radical coupling reactions of ethyl ferulate. J. Agric. Food Chem. 60, 8272–8277.

Mitchell, R.A.C., Dupree, P. and Shewry, P.R. (2007) A novel bioinformatics approach identifies candidate genes for the synthesis and feruloylation of arabinoxylan. Plant Physiol. 144, 43–53.

Mueller-Harvey, I. and Hartley, R.D. (1986) Linkage of p-coumaroyl and feruloyl groups to cell wall polysaccharides of barley straw. Carbohyd. Res. 148, 71–85.

Nemeth, C., Freeman, J., Jones, H.D., Sparks, C., Pellny, T.K., Wilkinson, M.D., Dunwell, J. *et al.* (2010) Down-regulation of the CSLF6 gene results in decreased (1,3;1,4)-beta-D-Glucan in endosperm of wheat. *Plant Physiol.* **152**, 1209–1218.

de Oliveira, D.M., Finger-Teixeira, A., Mota, T.R., Salvador, V.H., Moreira-Vilar, F.C., Molinari, H.B.C., Mitchell, R.A.C. *et al.* (2015) Ferulic acid: a key component in grass lignocellulose recalcitrance to hydrolysis. *Plant Biotechnol. J.* **13**, 1224–1232.

Ordaz-Ortiz, J.J. and Saulnier, L. (2005) Structural variability of arabinoxylans from wheat flour. Comparison of water-extractable and xylanase-extractable arabinoxylans. *J. Cereal Sci.* **42**, 119–125.

Pellny, T.K., Lovegrove, A., Freeman, J., Tosi, P., Love, C.G., Knox, J.P., Shewry, P.R. *et al.* (2012) Cell walls of developing wheat starchy endosperm: comparison of composition and RNA-seq transcriptome. *Plant Physiol.* **158**, 612–627.

Pena, M.J., Zhong, R.Q., Zhou, G.K., Richardson, E.A., O'Neill, M.A., Darvill, A.G., York, W.S. *et al.* (2007) *Arabidopsis irregular xylem8* and *irregular xylem9*: Implications for the complexity of glucuronoxylan biosynthesis. *Plant Cell*, **19**, 549–563.

Petersen, B.O., Lok, F. and Meier, S. (2014) Probing the structural details of xylan degradation by real-time NMR spectroscopy. *Carbohyd. Polym.* **112**, 587–594.

Ralph, J. (2010) Hydroxycinnamates in lignification. *Phytochem. Rev.* **9**, 65–83.

Ralph, J., Grabber, J.H. and Hatfield, R.D. (1995) Lignin-ferulate cross-links in grasses - active incorporation of ferulate polysaccharide esters into ryegrass lignins. *Carbohyd. Res.* **275**, 167–178.

Rhodes, D.I., Sadek, M. and Stone, B.A. (2002) Hydroxycinnamic acids in walls of wheat aleurone cells. *J. Cereal Sci.* **36**, 67–81.

Saulnier, L., Vigouroux, J. and Thibault, J.F. (1995) Isolation and partial characterization of feruloylated oligosaccharides from maize bran. *Carbohyd. Res.* **272**, 241–253.

Saulnier, L., Sado, P.E., Branlard, G., Charmet, G. and Guillon, F. (2007) Wheat arabinoxylans: exploiting variation in amount and composition to develop enhanced varieties. *J. Cereal Sci.* **46**, 261–281.

Saulnier, L., Robert, P., Grintchenko, M., Jamme, F., Bouchet, B. and Guillon, F. (2009) Wheat endosperm cell walls: spatial heterogeneity of polysaccharide structure and composition using micro-scale enzymatic fingerprinting and FT-IR microspectroscopy. *J. Cereal Sci.* **50**, 312–317.

Shewry, P.R. and Hey, S.J. (2015) The contribution of wheat to human diet and health. *Food Energy Security*, **4**, 178–202.

Shewry, P.R., Saulnier, L., Gebruers, K., Mitchell, R.A.C., Freeman, J., Nemeth, C. and Ward, J.L. (2014) Optimising the content and composition of dietary fibre in wheat grain for end-use quality. In: *Genomics of Plant Genetic Resource* (Tuberosa, R. ed), pp. 455–465. Dordecht: Springer Science

Toole, G.A., Le Gall, G., Colquhoun, I.J., Nemeth, C., Saulnier, L., Lovegrove, A., Pellny, T. *et al.* (2010) Temporal and spatial changes in cell wall composition in developing grains of wheat cv. Hereward. *Planta*, **232**, 677–689.

Urbanowicz, B.R., Pena, M.J., Moniz, H.A., Moremen, K.W. and York, W.S. (2014) Two Arabidopsis proteins synthesize acetylated xylan *in vitro*. *Plant J.* **80**, 197–206.

Veličković, D., Ropartz, D., Guillon, F., Saulnier, L. and Rogniaux, H. (2014) New insights into the structural and spatial variability of cell-wall polysaccharides during wheat grain development, as revealed through MALDI mass spectrometry imaging. *J. Exp. Bot.* **65**, 2079–2091.

Voxeur, A. and Höfte, H. (2016) Cell wall integrity signaling in plants: "To grow or not to grow that's the question". *Glycobiology*, **26**, 950–960.

Zeng, W., Lampugnani, E.R., Picard, K.L., Song, L.L., Wu, A.M., Farion, I.M., Zhao, J. *et al.* (2016) Asparagus IRX9, IRX10, and IRX14A are components of an active xylan backbone synthase complex that forms in the Golgi apparatus. *Plant Physiol.* **171**, 93–109.

Permissions

List of Contributors

Yingqi Cai, Elizabeth McClinchie, Ann Price, Drew Sturtevant and Kent D. Chapman
Center for Plant Lipid Research, University of North Texas, Denton, TX, USA

Thuy N. Nguyen, Satinder K. Gidda, Samantha C. Watt and Robert T. Mullen
Department of Molecular and Cellular Biology, University of Guelph, Guelph, ON, Canada

Olga Yurchenko, Sunjung Park, John M. Dyer
US Arid-Land Agricultural Research Center, USDA-ARS, Maricopa, AZ, USA

Kongqing Li
Department of Rural Development, Nanjing Agricultural University, Nanjing, China

Caihua Xing, Zhenghong Yao and Xiaosan Huang
College of Horticulture, State Key Laboratory of Crop Genetics and Germplasm Enhancement, Nanjing Agricultural University, Nanjing, China

Da Lu, Xianghe Yuan, Sung-Jin Kim, Joaquim V. Marques, P. Pawan Chakravarthy, Syed G. A. Moinuddin, Laurence B. Davin and Norman G. Lewis
Institute of Biological Chemistry, Washington State University, Pullman, WA, USA

Randi Luchterhand
Puyallup Research and Extension Center, Washington State University, Puyallup, WA, USA

Barri Herman
Institute of Biological Chemistry, Washington State University, Pullman, WA, USA
Puyallup Research and Extension Center, Washington State University, Puyallup, WA, USA

Ewelina Mnich, Nanna Bjarnholt and Mohammed S. Motawie
Plant Biochemistry Laboratory, Department of Plant Biology and Environmental Sciences, University of Copenhagen, Frederiksberg C, Denmark

Ruben Vanholme and Paula Oyarce
Department of Plant Biotechnology and Bioinformatics, Ghent University, Ghent, Belgium
Department of Plant Systems Biology, VIB, Ghent, Belgium

Geert Goeminne
Department of Plant Systems Biology, VIB, Ghent, Belgium

John Ralph, Sarah Liu and Fachuang Lu
Department of Biochemistry and DOE Great Lakes Bioenergy Research Center, Wisconsin Energy Institute, Madison, WI, USA

Bodil Jørgensen and Peter Ulvskov
Section for Plant Glycobiology, Department of Plant Biology and Environmental Sciences, University of Copenhagen, Frederiksberg C, Denmark

Jesper Harholt
Section for Plant Glycobiology, Department of Plant Biology and Environmental Sciences, University of Copenhagen, Frederiksberg C, Denmark
Carlsberg Research Laboratory, Copenhagen, Denmark

Birger L. Møller
Plant Biochemistry Laboratory, Department of Plant Biology and Environmental Sciences, University of Copenhagen, Frederiksberg C, Denmark
Carlsberg Research Laboratory, Copenhagen, Denmark

Craig J. van Dolleweerd, Leonard Both and Julian K-C. Ma
Institute for Infection and Immunity, St. George's Hospital Medical School, University of London, London, UK

Waranyoo Phoolcharoen
Institute for Infection and Immunity, St. George's Hospital Medical School, University of London, London, UK
Pharmacognosy and Pharmaceutical Botany, Faculty of Pharmaceutical Sciences, Chulalongkorn University, Bangkok, Thailand

Christophe Prehaud, Anaelle da Costa and Monique Lafon
Unité de Neuroimmunologie Virale, Département de Virologie, Institut Pasteur, Paris, France

Daniel Ponndorf, Sven Ehmke, Benjamin Walliser, Kerstin Thoss, Inge Broer, Christoph Unger and Henrik Nausch
Faculty of Agricultural and Environmental Sciences, Department of Agrobiotechnology and Risk Assessment for Bio- and Gene Technology, University of Rostock, Rostock, Germany

Solvig Görs, Gürbüz Das and Cornelia C. Metges
Leibniz Institute for Farm Animal Biology (FBN), Institute of Nutritional Physiology 'Oskar Kellner', Dummerstorf, Germany

David W. Ow
Plant Gene Engineering Center, South China Botanical Garden, Chinese Academy of Sciences, Guangzhou, China

Maryam Rajaee
Plant Gene Engineering Center, South China Botanical Garden, Chinese Academy of Sciences, Guangzhou, China
University of Chinese Academy of Sciences, Beijing, China

Erick Miguel Ramos-Martinez, Lorenzo Fimognari and Yumiko Sakuragi
Department of Plant and Environmental Sciences, Copenhagen Plant Science Centre, University of Copenhagen, Frederiksberg C, Copenhagen, Denmark

Qian Wang, Niha Dhar, Nadimuthu Kumar, Prasanna Nori Venkatesh, Chakravarthy Rajan, Deepa Panicker, Vishweshwaran Sridhar, Hui-Zhu Mao and Rajani Sarojam
Temasek Life Sciences Laboratory, National University of Singapore, Singapore, Singapore

Vaishnavi Amarr Reddy
Temasek Life Sciences Laboratory, National University of Singapore, Singapore, Singapore
Department of Biological Sciences, National University of Singapore, Singapore, Singapore

Noemi Ruiz-Lopez
IHSM-UMA-CSIC, Universidad de Malaga, Malaga, Spain
Department of Biological Chemistry, Rothamsted Research, Harpenden, Herts, UK

Richard Broughton, Sarah Usher, Richard P. Haslam, Johnathan A. Napier and Frédéric Beaudoin
Department of Biological Chemistry, Rothamsted Research, Harpenden, Herts, UK

Joaquin J. Salas
Instituto de la Grasa, Universitario Pablo de Olavide, Seville, Spain

Wendelin Schnippenkoetter, Katherine Dibley, Ricky Milne, Guoquan Liu, Wai Lung and Eunjung Kwong
CSIRO Agriculture and Food, Canberra, ACT, Australia

Evans Lagudah
CSIRO Agriculture and Food, Canberra, ACT, Australia
School of Life and Environmental Sciences, University of Sydney, Sydney, NSW, Australia

Chan and Clive Lo
School of Biological Sciences, The University of Hong Kong, Hong Kong, China

Ian Godwin
School of Agriculture and Food Sciences, The University of Queensland, St Lucia, QLD, Australia

Jodie White
Centre for Crop Health, University of Southern Queensland, Toowoomba, QLD, Australia

Alexander Zwart
CSIRO Data61, Canberra, ACT, Australia

Beat Keller, Simon G. Krattinger
Department of Plant and Microbial Biology, University of Zurich, Zurich, Switzerland

Andrew J. Simkin, Patricia E. Lopez-Calcagno, Philip A. Davey, Lauren R. Headland, Tracy Lawson and Christine A. Raines
School of Biological Sciences, University of Essex, Colchester, UK

Stefan Timm and Hermann Bauwe
Plant Physiology Department, University of Rostock, Rostock, Germany

Lina Yao and Hui Shen
Department of Microbiology and Immunology, Yong Loo Lin School of Medicine, National University of Singapore, Singapore, Singapore

Nan Wang, Jaspaul Tatlay and Liang Li
Department of Chemistry, University of Alberta, Edmonton, Alberta, Canada

Tin Wee Tan
Department of Biochemistry, Yong Loo Lin School of Medicine, National University of Singapore, Singapore, Singapore
National Supercomputing Centre (NSCC), Singapore, Singapore

Madoka Kudo, Satoshi Kidokoro, Junya Mizoi, Daisuke Todaka and Kazuko Yamaguchi-Shinozaki
Graduate School of Agricultural and Life Sciences, University of Tokyo, Tokyo, Japan

Takuya Yoshida
Graduate School of Agricultural and Life Sciences, University of Tokyo, Tokyo, Japan

Alisdair R. Fernie
Max-Planck-Institut für Molekulare Pflanzenphysiologie, Golm, Germany

Kazuo Shinozaki
RIKEN Center for Sustainable Resource Science, Tsurumi-ku, Yokohama, Japan

Parwinder Kaur and William Erskine
Centre for Plant Genetics and Breeding and Institute of Agriculture, The University of Western Australia, Crawley, WA, Australia

Philipp E. Bayer, David Edwards, Jacqueline Batley and Yuxuan Yuan
School of Plant Biology and Institute of Agriculture, The University of Western Australia, Crawley, WA, Australia

Phillip Nichols
School of Plant Biology and Institute of Agriculture, The University of Western Australia, Crawley, WA, Australia

Department of Agriculture and Food Western Australia, South Perth, WA, Australia

Zbyněk Milec, Jan Vrána and Jaroslav Doležel
Institute of Experimental Botany, Centre of the Region Hana for Biotechnological and Agricultural Research, Olomouc, Czech Republic

Rudi Appels
Murdoch University, Murdoch, WA, Australia

Uri Hanania, Tami Ariel, Yoram Tekoah, Liat Fux, Maor Sheva, Yehuda Gubbay, Mara Weiss, Dina Oz, Yaniv Azulay, Albina Turbovski, Yehava Forster and Yoseph Shaaltiel
Protalix Biotherapeutics, Carmiel, Israel

Jackie Freeman, Jane L. Ward, Ondrej Kosik, Alison Lovegrove, Mark D. Wilkinson, Peter R. Shewry and Rowan A.C. Mitchell
Plant Biology and Crop Science, Rothamsted Research, Harpenden, Hertfordshire, UK

Index

www.ingramcontent.com/pod-product-compliance
Lightning Source LLC
Chambersburg PA
CBHW050446200326
41458CB00014B/5079